Advances in

ATOMIC, MOLECULAR, AND OPTICAL PHYSICS

VOLUME 52

ADVANCES IN

ATOMIC, MOLECULAR, AND OPTICAL PHYSICS

Edited by

P.R. Berman
PHYSICS DEPARTMENT
UNIVERSITY OF MICHIGAN
ANN ARBOR, MI, USA

C.C. Lin
DEPARTMENT OF PHYSICS
UNIVERSITY OF WISCONSIN
MADISON, WI, USA

Volume 52

ELSEVIER
ACADEMIC
PRESS

Amsterdam · Boston · Heidelberg · London · New York · Oxford
Paris · San Diego · San Francisco · Singapore · Sydney · Tokyo

ELSEVIER B.V.
Sara Burgerhartstraat 25
P.O. Box 211, 1000 AE Amsterdam
The Netherlands

ELSEVIER Inc.
525 B Street, Suite 1900
San Diego, CA 92101-4495
USA

ELSEVIER Ltd
The Boulevard, Langford Lane
Kidlington, Oxford OX5 1GB
UK

ELSEVIER Ltd
84 Theobalds Road
London WC1X 8RR
UK

Notice
No responsibility is assumed by the Publisher for any injury and/or damage to persons or property as a matter of products liability, negligence or otherwise, or from any use or operation of any methods, products, instructions or ideas contained in the material herein. Because of rapid advances in the medical sciences, in particular, independent verification of diagnoses and drug dosages should be made.

First edition 2005

ISBN: 0 12 003852 8
ISSN: 1049-250X

⊗ The paper used in this publication meets the requirements of ANSI/NISO Z39.48-1992 (Permanence of Paper).

Printed in USA.

Contents

Fine Structure in High-L Rydberg States: A Path to Properties of Positive Ions

Stephen R. Lundeen

A Storage Ring for Neutral Molecules

Floris M.H. Crompvoets, Hendrick L. Bethlem and Gerard Meijer

Nonadiabatic Alignment by Intense Pulses. Concepts, Theory, and Directions

Tamar Seideman and Edward Hamilton

Relativistic Nonlinear Optics

Donald Umstadter, Scott Sepke and Shouyuan Chen

Coupled-State Treatment of Charge Transfer

Thomas G. Winter

CONTRIBUTORS

Numbers in parentheses indicate the pages on which the author's contributions begin.

IMMANUEL BLOCH (1), Johannes Gutenberg-University, Institut für Physik, 55099 Mainz, Germany

MARKUS GREINER (1), JILA, University of Colorado, Boulder, CO 80309-0440, USA

F.B. DUNNING (49), Department of Physics and Astronomy, and the Rice Quantum Institute, Rice University, MS 61, 6100 Main Street, Houston, TX 77005, USA

J.C. LANCASTER (49), Department of Physics and Astronomy, and the Rice Quantum Institute, Rice University, MS 61, 6100 Main Street, Houston, TX 77005, USA

C.O. REINHOLD (49), Physics Division, Oak Ridge National Laboratory, Oak Ridge, TN 37831-6372, USA; Department of Physics, University of Tennessee, Knoxville, TN 37996-1200, USA

S. YOSHIDA (49), Institute for Theoretical Physics, Vienna University of Technology, A-1040 Vienna, Austria

J. BURGDÖRFER (49), Department of Physics, University of Tennessee, Knoxville, TN 37996-1200, USA; Institute for Theoretical Physics, Vienna University of Technology, A-1040 Vienna, Austria

J.B. ALTEPETER (105), Department of Physics, University of Illinois at Urbana-Champaign, Urbana, IL 61801, USA

E.R. JEFFREY (105), Department of Physics, University of Illinois at Urbana-Champaign, Urbana, IL 61801, USA

P.G. KWIAT (105), Department of Physics, University of Illinois at Urbana-Champaign, Urbana, IL 61801, USA

STEPHEN R.S.R. LUNDEEN (161), Dept. of Physics, Colorado State University, USA

FLORIS M.H. CROMPVOETS (209), FOM—Institute for Plasma Physics Rijnhuizen, P.O. Box 1207, NL-3430 BE Nieuwegein, The Netherlands

HENDRICK L. BETHLEM (209), FOM—Institute for Plasma Physics Rijnhuizen, P.O. Box 1207, NL-3430 BE Nieuwegein, The Netherlands; Fritz-Haber-Institut der Max-Planck-Gesellschaft, Faradayweg 4-6, D-14195 Berlin, Germany

GERARD MEIJER (209), Fritz-Haber-Institut der Max-Planck-Gesellschaft, Faradayweg 4-6, D-14195 Berlin, Germany

TAMAR SEIDEMAN (289), Department of Chemistry, Northwestern University, 2145 Sheridan Road, Evanston, IL 60208-3113, USA

EDWARD HAMILTON (289), Department of Chemistry, Northwestern University, 2145 Sheridan Road, Evanston, IL 60208-3113, USA

DONALD UMSTADTER (331), Physics and Astronomy Department, University of Nebraska, Lincoln, NE 68588-0111, USA

SCOTT SEPKE (331), Physics and Astronomy Department, University of Nebraska, Lincoln, NE 68588-0111, USA

SHOUYUAN CHEN (331), Physics and Astronomy Department, University of Nebraska, Lincoln, NE 68588-0111, USA

THOMAS G. WINTER (391), Department of Physics, Pennsylvania State University, Wilkes-Barre Campus, Lehman, PA 18627, USA

EXPLORING QUANTUM MATTER WITH ULTRACOLD ATOMS IN OPTICAL LATTICES

IMMANUEL BLOCH[1]* *and MARKUS GREINER*[2]**

[1] *Johannes Gutenberg-University, Institut für Physik, 55099 Mainz, Germany*

[2] *JILA, University of Colorado, Boulder, CO 80309-0440, USA*

* e-mail: bloch@uni-mainz.de

** e-mail: markus.greiner@colorado.edu

1

© 2005 Elsevier Inc. All rights reserved
ISSN 1049-250X
DOI 10.1016/S1049-250X(05)52001-9

1. Introduction

The advent of ultracold bosonic (Anderson et al., 1995; Davis et al., 1995; Bradley et al., 1995) and fermionic quantum gases (deMarco and Jin, 1999; Truscott et al., 2001; Schreck et al., 2001) as well as the observation of fermionic pair condensates (Regal et al., 2004; Zwierlein et al., 2004; Chin et al., 2004) has opened the path for the exploration of fundamental questions of many-body physics. By adding standing wave optical light fields to such quantum gases, it has now also become possible to make the analogy with condensed matter physics even closer. Millions of atoms can be trapped in a perfectly periodic crystal of light, whose lattice geometry and lattice depth are under the full control of the experimentalist. Together with the tunability of the interactions between ultracold atoms via Feshbach resonances (Inouye et al., 1998; Courteille et al., 1998), almost all experimental parameters of such a system can be manipulated. In many ways this comes close to what Richard P. Feynman conceived a quantum simulator (Feynman, 1985, 1986) to be—namely a controllable quantum system, with which one is able to investigate the quantum dynamics and quantum phases of another quantum system.

Ultracold atoms in optical lattices now form an almost independent research field for ultracold quantum gases, with many stimulating inputs from condensed matter physics. During the last years the field has rapidly grown with several hundred only recent publications and it is impossible to review the field in its entirety, as this would by now require writing a whole book on the subject. Therefore we have chosen to follow in this review the path how we entered into this field. We hope that this will give the reader a first overview of the different aspects that can be covered with ultracold quantum gases in optical lattices—ranging from condensed matter physics over quantum optics to quantum information. In the review, we have chosen to focus mainly on bosonic atoms in optical lattices and to concentrate on the regime where strong correlations between the atoms become important. A recent review (Morsch and Oberthaler, 2005) covers in more detail the aspects of weakly interacting BECs in 1D and 2D optical lattices, to which the reader is advised to turn to for that subject. There, one should especially highlight the transport studies in coupled Josephson junction arrays by (Cataliotti et al., 2001) and the recent observation of Josephson oscillations and self trapping in a double well system realized in an optical lattice (Albiez et al., 2004).

The strongly correlated regime of bosons in optical lattices, which will be at the focus of the discussion here, was in fact one of the first milestone proposals (Jaksch et al., 1998) in the community to create quantum gases far beyond the description of a weakly interacting BEC via the Gross–Pitaevskii equation, which had been at the focus of research in the beginning years of investigations of ultracold quantum gases. The transition to a Mott insulator (Fisher et al., 1989; Jaksch et al., 1998; Greiner et al., 2002; Porto et al., 2003; Stöferle et al., 2004) is therefore at the center of this review and a natural starting point for the further experiments that are discussed subsequently.

2. Optical Lattices

2.1. OPTICAL DIPOLE FORCE

In the interaction of atoms with coherent light fields, two fundamental forces arise (Cohen-Tannoudji et al., 1992; Metcalf and van der Straten, 1999). The so called Doppler force is dissipative in nature and can be used to efficiently laser cool a gas of atoms and relies on the radiation pressure together with spontaneous emission. The so called dipole force on the other hand creates a purely conservative potential in which the atoms can move. No cooling can be realized with this dipole force, however if the atoms are cold enough initially, they may be trapped in such a purely optical potential (Chu et al., 1986; Grimm et al., 2000).

How does this dipole force arise? We may grasp the essential points through a simple classical model, in which we view the electron as harmonically bound to the nucleus with oscillation frequency ω_0. An external oscillating electric field of a laser $\mathbf{E}$ with frequency ω_L can now induce an oscillation of the electron resulting in an oscillating dipole moment $\mathbf{d}$ of the atom. Such an oscillating dipole moment will be in phase with the driving oscillating electric field, for frequencies much lower than an atomic resonance frequency and 180° out of phase, for frequencies much larger than the atomic resonance frequency. The induced dipole moment again interacts with the external oscillating electric field, resulting in a dipole potential V_{dip} experienced by the atom (Askar'yan, 1962; Kazantsev, 1973; Cook, 1979; Gordon and Ashkin, 1980; Grimm et al., 2000):

$$V_{\text{dip}} = -\frac{1}{2}\langle \mathbf{dE}\rangle, \tag{1}$$

where $\langle \cdot \rangle$ denotes a time average over fast oscillating terms at optical frequencies. From Eq. (1) it becomes immediately clear that for a red detuning ($\omega_L < \omega_0$), where $\mathbf{d}$ is in phase with $\mathbf{E}$, the potential is attractive, whereas for a blue detuning ($\omega_L > \omega_0$), where $\mathbf{d}$ is in 180° out of phase with $\mathbf{E}$, the potential is repulsive. By relating the dipole moment to the polarizability $\alpha(\omega_L)$ of an atom and expressing the electric field amplitude E_0 via the intensity of the laser field I, one obtains for the dipole potential:

$$V_{\text{dip}}(\mathbf{r}) = -\frac{1}{2\epsilon_0 c}\text{Re}(\alpha)I(\mathbf{r}). \tag{2}$$

A spatially dependent intensity profile $I(\mathbf{r})$ can therefore create a trapping potential for neutral atoms.

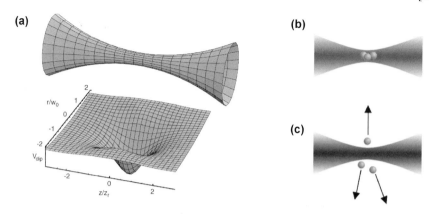

FIG. 1. (a) Gaussian laser beam together with corresponding trapping potential for a red detuned laser beam. (b) A red detuned laser beams lead to an attractive dipole potential, whereas a blue detuned laser beam leads to a repulsive potential (c).

For a two level atom a more useful form of the dipole potential may be derived within the rotating wave approximation, which is a reasonable approximation provided that the detuning $\Delta = \omega_L - \omega_0$ of the laser field ω_L from an atomic transition frequency ω_0 is small compared to the transitions frequency itself $|\Delta| \ll \omega_0$. Here one obtains (Grimm et al., 2000):

$$V_{\text{dip}}(\mathbf{r}) = \frac{3\pi c^2}{2\omega_0^3} \frac{\Gamma}{\Delta} I(\mathbf{r}), \tag{3}$$

with $\Gamma \ll |\Delta|$ being the decay rate of the excited state. Here a red detuned laser beam ($\omega_L < \omega_0$) leads to an attractive dipole potential and a blue detuned laser beam ($\omega_L > \omega_0$) leads to a repulsive dipole potential. By simply focussing a Gaussian laser beam, this can be used to attract or repel atoms from an intensity maximum in space (see Fig. 1).

For such a focussed Gaussian laser beam the intensity profile $I(r, z)$ is given by

$$I(r, z) = \frac{2P}{\pi w^2(z)} e^{-2r^2/w^2(z)}, \tag{4}$$

where $w(z) = w_0(1 + z^2/z_R^2)$ is the $1/e^2$ radius depending on the z-coordinate, $z_R = \pi w^2/\lambda$ is the Rayleigh length and P is the total power of the laser beam (Saleh and Teich, 1991). Around the intensity maximum a potential depth minimum occurs for a red detuned laser beam, leading to an approximately harmonic potential of the form:

$$V_{\text{dip}}(r, z) \approx -V_0 \left\{ 1 - 2\left(\frac{r}{w_0}\right)^2 - \left(\frac{z}{z_R}\right)^2 \right\}. \tag{5}$$

This harmonic confinement is characterized by radial ω_r and axial ω_{ax} trapping frequencies $\omega_r = (4V_0/mw_0^2)^{1/2}$ and $\omega_z = (2V_0/mz_R^2)$.

Great care has to be taken to minimize spontaneous scattering events, as they lead to heating and decoherence of the trapped ultracold atom samples. For a two-level atom, the scattering rate $\Gamma_{sc}(\mathbf{r})$ can be estimated (Grimm et al., 2000) through:

$$\Gamma_{sc}(\mathbf{r}) = \frac{3\pi c^2}{2\hbar\omega_0^3}\left(\frac{\Gamma}{\Delta}\right)^2 I(\mathbf{r}). \tag{6}$$

From Eqs. (3) and (6) it can be seen that the ratio of scattering rate to optical potential depth can always be minimized by increasing the detuning of the laser field. In practice however, such an approach is limited by the maximum available laser power. For experiments with ultracold quantum gases of alkali atoms, the detuning is typically chosen to be large compared to the excited state hyperfine structure splitting and in most cases even large compared to the fine structure splitting in order to sufficiently suppress spontaneous scattering events. Typical detunings range from several tens of nm to optical trapping in CO_2 laser fields. A laser trap formed by a CO_2 laser fields can be considered as a quasi-electrostatic trap, where the detuning is much larger than the optical resonance frequency of an atom (Takekoshi et al., 1995; Friebel et al., 1998; Barrett et al., 2001).

One final comment should be made about state dependent optical potentials. For a typical multilevel alkali atom, the dipole potential will depend both on the internal magnetic substate m_F of a hyperfine ground state with angular momentum F, as well as on the polarization of the light field $P = +1, -1, 0$ (circular $\sigma^{\pm}$ and linear polarization). One can then express the lattice potential depth through (Jessen and Deutsch, 1996; Grimm et al., 2000):

$$V_{dip}(\mathbf{r}) = \frac{\pi c^2 \Gamma}{2\omega_0^3}\left(\frac{2 + Pg_F m_F}{\Delta_{2,F}} + \frac{1 - Pg_F m_F}{\Delta_{1,F}}\right)I(\mathbf{r}). \tag{7}$$

Here g_F is the Landé factor and $\Delta_{2,F}$, $\Delta_{1,F}$ refer to the detuning relative to the transition between the ground state with hyperfine angular momentum F and the center of the excited state hyperfine manifold on the D_2 and D_1 transition, respectively. For large detunings relative to the fine structure splitting Δ_{FS}, the optical potentials become almost spin independent again. For detunings of the laser frequency in between the fine structure splitting, very special spin-dependent optical potentials can be created that will be discussed below.

2.2. OPTICAL LATTICE POTENTIALS

A periodic potential can simply be formed by overlapping two counterpropagating laser beams. Due to the interference between the two laser beams an optical

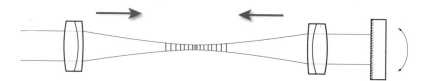

FIG. 2. One-dimensional optical lattice potential. By interfering two counterpropagating Gaussian laser beams, a periodic intensity profile is created due to the interference of the two laser fields.

standing wave with period $\lambda/2$ is formed, in which the atoms can be trapped. By interfering more laser beams, one can obtain one-, two- and three-dimensional periodic potentials (Petsas et al., 1994), which in their simplest form will be discussed below. Note that by choosing to let two laser beams interfere under an angle less than $180°$, one can also realize periodic potentials with a larger period.

2.2.1. *1D Lattice Potentials*

The simplest possible periodic optical potential is formed by overlapping two counterpropagating focussed Gaussian laser beams, which results in a trapping potential of the form:

$$V(r, z) = -V_{\text{lat}} \cdot e^{-2r^2/w^2(z)} \cdot \sin^2(kx)$$
$$\approx -V_{\text{lat}} \cdot \left(1 - 2\frac{r^2}{w^2(z)}\right) \cdot \sin^2(kz), \qquad (8)$$

where w_0 denotes the beam waist, $k = 2\pi/\lambda$ is the wave vector of the laser light and V_{lat} is the maximum depth of the lattice potential. Note that due to the interference of the two laser beams V_{lat} is four times larger than V_0 if the laser power and beam parameters of the two interfering lasers are equal (see Fig. 2).

2.2.2. *2D Lattice Potentials*

Periodic potentials in two dimensions can be formed by overlapping two optical standing waves along different directions. In the simplest form one chooses two orthogonal directions and obtains at the center of the trap an optical potential of the form (neglecting the harmonic confinement due to the Gaussian beam profile of the laser beams):

$$V(y, z) = -V_{\text{lat}}\big(\cos^2(kx) + \cos^2(ky) + 2\mathbf{e}_1 \cdot \mathbf{e}_2 \cos\phi \cos(kx)\cos(ky)\big). \qquad (9)$$

Here $\mathbf{e}_1$ and $\mathbf{e}_2$ denote the polarization vectors of the laser fields, each forming one standing wave and ϕ is the temporal phase between them. If the polarization vectors are chosen not to be orthogonal to each other, then the resulting potential will not only be the sum of the potentials created by each standing wave, but

(a)   **(b)**

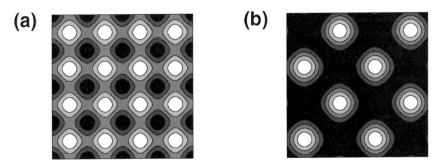

FIG. 3. 2D optical lattice potentials for a lattice with (a) orthogonal polarization vectors and with (b) parallel polarization vectors and a time phase of $\phi = 0$.

will be modified according to the temporal phase ϕ used (see Fig. 3). In such a case it is absolutely essential to stabilize the temporal phase between the two standing waves (Hemmerich et al., 1992), as small vibrations will usually lead to fluctuations of the time phase, resulting in severe heating and decoherence effects of the ultracold atom samples.

In such a two-dimensional optical lattice potential, the atoms are confined to arrays of tightly confining one-dimensional tubes (see Fig. 4(a)). For typical experimental parameters the harmonic trapping frequencies along the tube are very weak and on the order of 10–200 Hz, while in the radial direction the trapping frequencies can become as high as up to 100 kHz, thus allowing the atoms to effectively move only along the tube for deep lattice depths (Greiner et al., 2001; Moritz et al., 2003; Laburthe-Tolra et al., 2004; Paredes et al., 2004; Kinoshita et al., 2004).

2.2.3. 3D Lattice Potentials

For the creation of a three-dimensional lattice potential, three orthogonal optical standing waves have to be overlapped. Here we only consider the case of independent standing waves, with no cross interference between laser beams of different standing waves. This can, for example, be realized by choosing orthogonal polarization vectors between different standing wave light fields and also by using different wavelengths for the three standing waves. In this case the resulting optical potential is simply given by the sum of three standing waves:

$$V(\mathbf{r}) = -V_x e^{-2(y^2+z^2)/w_x^2} \sin^2(kx) - V_y e^{-2(x^2+z^2)/w_y^2} \sin^2(ky)$$
$$- V_z e^{-2(x^2+y^2)/w_z^2} \sin^2(kz). \tag{10}$$

Here $V_{x,y,z}$ are the potential depths of the individual standing waves along the different directions. In the center of the trap, for distances much smaller than the beam waist, the trapping potential can be approximated as the sum of a homogeneous periodic lattice potential and an additional external harmonic confinement due to the Gaussian laser beam profiles:

$$V(\mathbf{r}) \approx V_x \cdot \sin^2(kx) + V_y \cdot \sin^2(ky) + V_z \cdot \sin^2(kz)$$
$$+ \frac{m}{2}\left(\omega_x^2 x^2 + \omega_y^2 y^2 + \omega_z^2 z^2\right), \tag{11}$$

where $\omega_{x,y,z}$ are the effective trapping frequencies of the external harmonic confinement. They can again be approximated by:

$$\omega_x^2 = \frac{4}{m}\left(\frac{V_y}{w_y^2} + \frac{V_z}{w_z^2}\right); \qquad \omega_{y,z}^2 = \text{(cycl. perm.)}. \tag{12}$$

In addition to this harmonic confinement due to the Gaussian laser beam profiles, a confinement due to a magnetic trapping typically exists, which has to be taken into account as well for the total harmonic confinement of the atom cloud.

For sufficiently deep optical lattice potentials, the confinement on a single lattice site is also approximately harmonic. Here the atoms are very tightly confined with typical trapping frequencies ω_{lat} of up to 100 kHz. One can estimate the trapping frequencies at a single lattice site through a Taylor expansion of the sinusoidally varying lattice potential at a lattice site and obtains:

$$\omega_{\text{lat}} \approx \sqrt{\frac{V_{\text{lat}}}{E_r}\frac{\hbar^2 k^4}{m^2}}. \tag{13}$$

Here $E_r = \hbar^2 k^2 / 2m$ is the so called recoil energy, which is a natural measure of energy scales in optical lattice potentials.

2.3. SPIN-DEPENDENT OPTICAL LATTICE POTENTIALS

In order to realize a spin-dependent lattice potential, a standing wave configuration formed by two counterpropagating laser beams with linear polarization vectors enclosing an angle θ has been proposed (Finkelstein et al., 1992; Jessen and Deutsch, 1996; Brennen et al., 1999; Jaksch et al., 1999). Such a standing wave light field can be decomposed into a superposition of a σ^+ and σ^- polarized standing wave laser field, giving rise to lattice potentials $V_+(x, \theta) = V_0 \cos^2(kx + \theta/2)$ and $V_-(x, \theta) = V_0 \cos^2(kx - \theta/2)$. Here k is the wave vector of the laser light used for the standing wave and V_0 is the potential depth of the lattice. By changing the polarization angle θ one can thereby control the separation between the two potentials $\Delta x = \theta/180° \cdot \lambda_x/2$. When increasing θ, both potentials shift in opposite directions and overlap again when $\theta = n \cdot 180°$, with n being an integer. For a spin-dependent transport, two internal spin states of the atom should be used, where one spin state dominantly experiences the $V_+(x, \theta)$

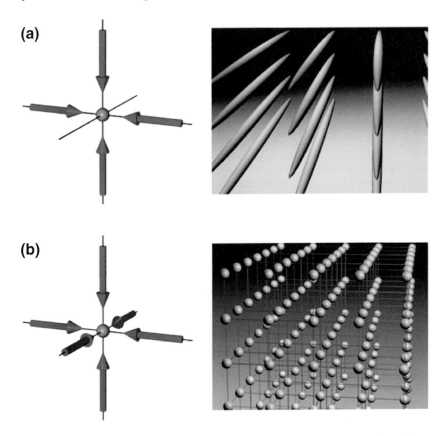

FIG. 4. Two-dimensional (a) and three-dimensional (b) optical lattice potentials formed by superimposing two or three orthogonal standing waves. For a two-dimensional optical lattice, the atoms are confined to an array of tightly confining one-dimensional potential tubes, whereas in the three-dimensional case the optical lattice can be approximated by a three-dimensional simple cubic array of tightly confining harmonic oscillator potentials at each lattice site.

potential and the other spin state mainly experiences the $V_-(x, \theta)$ dipole force potential. Such a situation can be realized in rubidium by tuning the wavelength of the optical lattice laser to a value of $\lambda_x \approx 785$ nm between the fine structure splitting of the rubidium D1 and D2 transition. Then the dipole potential experienced by an atom in, e.g., the $|1\rangle \equiv |F = 2, m_F = -2\rangle$ state is given by $V_1(x, \theta) = V_-(x, \theta)$ and that for an atom in the $|0\rangle \equiv |F = 1, m_F = -1\rangle$ state is given by $V_0(x, \theta) = 3/4V_+(x, \theta) + 1/4V_-(x, \theta)$. If an atom is now first placed in a coherent superposition of both internal states $1/\sqrt{2}(|0\rangle + i|1\rangle)$ and the polarization angle θ is continuously increased, the spatial wave packet of the atom

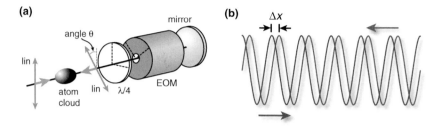

FIG. 5. Schematic experimental setup. A one-dimensional optical standing wave laser field is formed by two counterpropagating laser beams with linear polarizations. The polarization angle of the returning laser beam can be adjusted through an electro-optical modulator. The dashed lines indicate the principal axes of the wave plate and the EOM.

is split with both components moving in opposite directions. This can be used to coherently move atoms across lattices and realize quantum gates between them (see Section 2.3).

The polarization angle θ—and therefore the relative position of the two standing wave potentials (see Section 2.3)—can be dynamically controlled through the use of a quarter wave plate and an electro-optical modulator (EOM) that enable one to dynamically rotate the polarization vector of the retro-reflected laser beam through an angle θ by applying an appropriate voltage to the EOM (see Fig. 5) (Mandel et al., 2003).

3. Bose–Einstein Condensates in Optical Lattices

3.1. BLOCH BANDS

Characteristic for the movement of a particle in a periodic potential is the emergence of a band structure with energy bands and band gaps in between (Ashcroft and Mermin, 1976). In this section the wave function of a single particle in a periodic lattice potential is calculated and the band structure is investigated.

A particle in a periodic potential $V(x)$ is described by the Schrödinger equation

$$H\phi_q^{(n)}(x) = E_q^{(n)}\phi_q^{(n)}(x) \quad \text{with } H = \frac{1}{2m}\hat{p}^2 + V(x). \tag{14}$$

Solutions of this equation are called Bloch wave functions (see, e.g., Ashcroft and Mermin (1976)) and can be written as a product of a plane wave $\exp(iqx/\bar{h})$ and a function $u_q^{(n)}(x)$ with the same periodicity as the periodic potential:

$$\phi_q^{(n)}(x) = e^{iqx/\bar{h}} \cdot u_q^{(n)}(x). \tag{15}$$

Inserting this ansatz into Eq. (14) leads to a Schrödinger equation for $u_q^{(n)}(x)$:

$$H_B u_q^{(n)}(x) = E_q^{(n)} u_q^{(n)}(x) \quad \text{with } H_B = \frac{1}{2m}(\hat{p} + q)^2 + V_{lat}(x). \tag{16}$$

Since both the potential $V_{lat}(x)$ and the functions $u_q^{(n)}(x)$ are periodic with the same periodicity, they can be written as a discrete Fourier sum:

$$V(x) = \sum_r V_r e^{i2rkx} \quad \text{and} \quad u_q^{(n)}(x) = \sum_l c_l^{(n,q)} e^{i2lkx}, \tag{17}$$

with l and r integers. With these sums the potential energy term of Eq. (16) becomes

$$V(x) u_q^{(n)}(x) = \sum_l \sum_r V_r e^{i2(r+l)kx} c_l^{(n,q)} \tag{18}$$

and the kinetic term becomes

$$\frac{(\hat{p} + q)^2}{2m} u_q^{(n)}(x) = \sum_l \frac{(2\hbar k l + q)^2}{2m} c_l^{(n,q)} e^{i2lkx}. \tag{19}$$

In the experiment, a sinusoidal lattice potential is created, such that:

$$V(x) = -V_{lat} \cos^2(kx) = -\frac{1}{4V_{lat}}\left(e^{2ikx} + e^{-2ikx} + 2\right). \tag{20}$$

Thus only two terms of the Fourier sum in Eq. (17) are nonzero: $V_{-1} = V_1 = -1/4V_{lat}$ and V_0 can be set to zero. Using these results we can write the Schrödinger Eq. (16) in matrix form as:

$$\sum_l H_{l,l'} \cdot c_l^{(n,q)} = E_q^{(n)} c_l^{(n,q)}$$

$$\text{with } H_{l,l'} = \begin{cases} \left(2l + \dfrac{q}{\hbar k}\right)^2 E_r & \text{if } l = l', \\ -\dfrac{1}{4} \cdot V_0 & \text{if } |l - l'| = 1, \\ 0 & \text{else.} \end{cases} \tag{21}$$

Here q is the quasi-momentum, within the first Brillouin zone ranging from $q = -\hbar k$ to $\hbar k$. For a certain quasi-momentum q the eigenvalues $E_q^{(n)}$ of H represent the eigenenergies in the nth energy band. The corresponding eigenvector $c^{(n,q)}$ defines the appropriate Bloch wave function through Eqs. (17) and (15). These eigenstates and eigenvectors can be simply calculated if the Hamiltonian is truncated for large positive and negative l. The corresponding coefficients $c^{(n,q)}$ become very small for large enough l, e.g., a restriction to $-5 \geq l \geq 5$ is a good choice if only the lowest energy bands are considered.

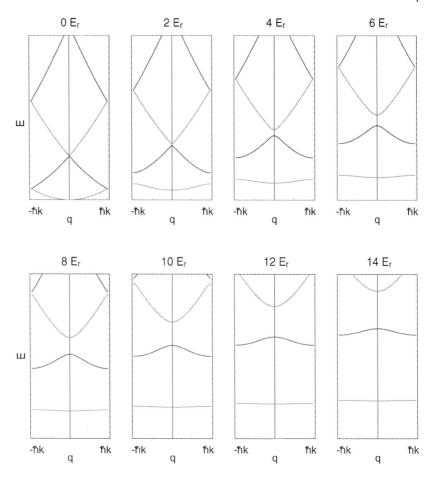

FIG. 6. Band structure of an optical lattice: Energy of the Bloch state versus quasi-momentum q in the first Brillouin zone, plotted for different lattice depths between 0 and $14E_r$. For deep lattices the lowest band becomes flat and the width of the first band gap corresponds to the level spacing $\hbar\omega$ on each lattice site.

Figure 6 shows the band structure for a one-dimensional sinusoidal lattice for different potential depths. For a vanishing lattice depth, there are no band gaps and the "bands" equal the free particle energy–momentum parabola reduced to the first Brillouin zone. When the lattice depth is increased, the band gaps become larger and the width of the energy bands becomes exponentially smaller.

Two and three-dimensional sinusoidal simple cubic lattices are in our case fully separable. Therefore the wave functions can be calculated separately for each axis and the total energy is given by the sum of the eigenenergies of all axes.

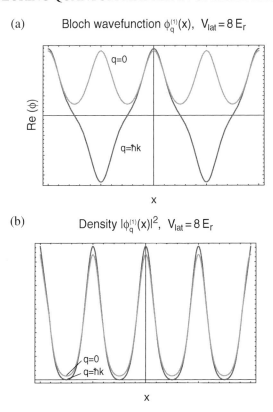

(a)   Bloch wavefunction $\phi_q^{(1)}(x)$, $V_{lat} = 8\,E_r$

q=0

Re (ϕ)

q=ℏk

x

(b)   Density $|\phi_q^{(1)}(x)|^2$, $V_{lat} = 8\,E_r$

q=0
q=ℏk

x

FIG. 7. Real part (a) and probability density (b) of the Bloch wave functions $\phi_q^{(1)}(x)$ in the lowest band, corresponding to a quasi-momentum of $q = 0$ and $q = \hbar k$. The lattice potential is a 1D sinusoidal potential with a lattice depth of $8E_r$.

In Fig. 7(a) the real part of a Bloch wave function for an $8E_r$ deep lattice is plotted. The function with $\mathrm{Re}(\phi) > 0$ is the Bloch function $\phi_{q=0}^{(n=1)}(x)$ in the lowest band with a quasi-momentum $q = 0$, and the function with alternating sign is the Bloch function with $q = \hbar k$ at the border of the Brillouin zone. Figure 7(b) shows the corresponding probability densities of the same wave functions.

3.2. WANNIER FUNCTIONS

Bloch states are completely delocalized energy eigenstates of the Schrödinger equation for a given quasi-momentum q and energy band n. In contrast to this, Wannier functions constitute an orthogonal and normalized set of wave functions that are maximally localized to individual lattice sites. The Wannier function for a localized particle in the nth energy band of the optical lattice potential is given by

$$w_n(x - x_i) = \mathcal{N}^{-1/2} \sum_q e^{-iqx_i/\hbar} \phi_q^{(n)}(x). \tag{22}$$

Here x_i is the position of the ith lattice site and $\mathcal{N}$ is a normalization constant. If a particle is in a mode corresponding to this Wannier wave function, it is well localized to the ith lattice site. Figure 8 shows the Wannier function and the density square of this function for a $3E_r$ and $10E_r$ deep lattice. Two side lobes of the Wannier function are visible which can be attributed to a nonvanishing probability to find the atom in the neighboring lattice site due to tunnelling. The tunnelling matrix element J, which describes the tunnelling between neighboring lattice sites,

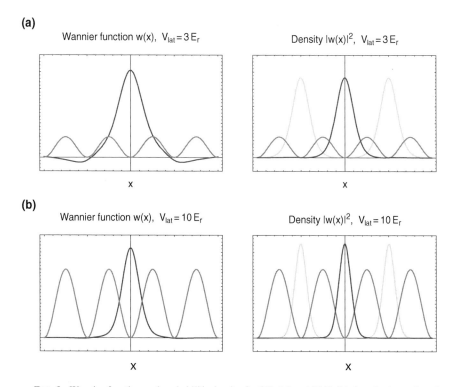

FIG. 8. Wannier function and probability density for $3E_r$ (a) and $10E_r$ (b) deep lattices, plotted together with a schematic lattice potential. Wannier functions constitute an orthogonal set of maximally localized wave functions. For $3E_r$ sidelobes are visible, for $10E_r$ the sidelobes become very small corresponding to a decreased tunnelling probability.

can be calculated by considering two Wannier functions of neighboring lattice sites i and j:

$$J = \int w_n(x - x_i) \left(-\frac{\hbar^2}{2m} \frac{\partial^2}{\partial x^2} + V(x) \right) w_n(x - x_j) \, dx. \qquad (23)$$

For deep lattices the localized Wannier wave function can be approximated by a Gaussian ground state wave function. However, due to the much weaker side lobes of a Gaussian, the tunnel matrix element J may then be underestimated by almost an order of magnitude.

The Wannier description becomes particularly important when the interaction between particles is taken into account. Local interactions of particles on a lattice site can best be described in a localized Wannier basis.

3.3. GROUND STATE WAVE FUNCTION OF A BEC IN AN OPTICAL LATTICE

3.3.1. Discretization

In the weakly interacting regime, when the tunnel coupling between neighboring lattice sites is large compared to the interaction energy between two atoms, a Bose–Einstein condensate trapped in a periodic lattice potential can be described by a macroscopic wave function, which in turn can be determined through the Gross–Pitaevskii equation. If the chemical potential is small compared to the trap depth, a tight binding picture can be used to describe the system. In this regime the extension of the ground state wave function is much smaller than the lattice spacing, and the condensate effectively consists of tiny BECs with phases that are coupled due to tunnelling between the lattice sites. Such an arrangement behaves similarly to a Josephson junction (Anderson and Kasevich, 1998; Cataliotti et al., 2001).

In the tight binding picture the atoms on a lattice site j can be described by a localized macroscopic wave function $\varphi_j(x)$. If the chemical potential on a lattice site is much smaller than the vibrational level spacing, the ground state wave function can be well approximated by the noninteracting ground state wave function of the lowest vibrational level. To a first approximation this is the Gaussian ground state wave function of a harmonic oscillator.

If the chemical potential is slightly larger, the broadening of the ground state wave function due to the repulsive interaction between the atoms has to be taken into account. In this case a better approximation for the ground state wave function can be found by determining an effective width of the wave function using a variational ansatz.

For a three-dimensional simple cubic lattice with a chemical potential much smaller than the level spacing in each direction the localized wave function can be described as the product of three Wannier functions $w(x)$ of the lowest Bloch band for each direction:

$$\varphi_j(\mathbf{x}) = w_x(x - x_j) \cdot w_y(y - y_j) \cdot w_z(z - z_j). \tag{24}$$

The situation is different for a one- or two-dimensional optical lattice, where the confinement is usually very weak in two or one direction respectively. Along the axes of weak confinement the ground state wave function Θ will generally be broadened due to the repulsive interactions between the atoms. The resulting wave function can then be written as:

$$\begin{aligned}
\text{1D lattice:} \quad & \varphi_j(\mathbf{x}) = w_x(x - x_j) \cdot \Theta^{(y)}(y - y_j) \cdot \Theta^{(z)}(z - z_j), \\
\text{2D lattice:} \quad & \varphi_j(\mathbf{x}) = w_x(x - x_j) \cdot w_y(y - y_j) \cdot \Theta^{(z)}(z - z_j).
\end{aligned} \tag{25}$$

Using these localized wave functions we can define a macroscopic wave function which describes the total system. This wave function is the sum of localized wave functions at each lattice site j

$$\Psi(\mathbf{x}) = \sum_j \psi_j \cdot \varphi_j(\mathbf{x}); \quad \psi_j = \sqrt{\bar{n}_j} \cdot e^{i\phi_j}, \tag{26}$$

each having a well defined phase ϕ_j and an amplitude $\sqrt{\bar{n}_j}$, where $\bar{n}_j$ corresponds to the average atom number on the jth lattice site. The total atom number is given by $\sum_j |\psi_j|^2 = \sum_j \bar{n}_j = N$. For this array of weakly coupled condensates, each described by ψ_j, the Hamiltonian is

$$\mathcal{H} = -J \sum_{\langle i,j \rangle} \psi_i^* \psi_j + \sum_j \epsilon_j |\psi_j|^2 + \sum_j \frac{1}{2} U |\psi_j|^4, \tag{27}$$

where the first summation is carried out over neighboring lattice sites only. This term characterizes the Josephson energy of the system. The second term describes the inhomogeneity of the trapping potential, where ϵ_j is the energy offset of the jth lattice site, given by $\epsilon_j = m/2 \cdot \omega_{\text{ext}}^2 r_j^2$. The third term describes the on-site interaction energy, where U is the on-site interaction matrix element defined by

$$U = \frac{4\pi \hbar^2 a}{m} \int d^3\mathbf{x} |\varphi_j(\mathbf{x})|^4, \tag{28}$$

with a being the scattering length of the atoms.

The dynamics of this Josephson junction array can be described by a discrete nonlinear Schrödinger equation (DNLSE), also called the discrete Gross–Pitaevskii equation (Cataliotti et al., 2001).

3.3.2. Ground State

In the inhomogeneous system it is important to find the new ground state of the combined lattice and external confining potential. For a purely periodic potential the energy is minimized if the phases of the wave functions on different lattice sites are equal, since then the Josephson energy is minimal. In the Bloch picture this corresponds to the Bloch state with zero quasi-momentum ($q = 0$).

But how will the atoms distribute over the lattice in the inhomogeneous system? In order to be in a stationary state, the phases of the wave functions on different lattice sites have to evolve with the same rate over time. This requires that their local chemical potentials are equal. In a Thomas–Fermi approximation we can neglect the contribution of the kinetic energy term (which is the Josephson term in Eq. (27)), so that the total energy on a single lattice site is given by

$$E_j \simeq \epsilon_j |\psi_j|^2 + \frac{U}{2} |\psi_j|^4 = \epsilon_j \bar{n}_j + \frac{U}{2} \bar{n}_j^2. \tag{29}$$

The local chemical potential can then be calculated as

$$\mu_j = \frac{\partial E_j}{\partial n_j} = \epsilon_j + U \cdot \bar{n}_j = \text{const}. \tag{30}$$

We ignore the discreteness of the lattice for the following calculation. This is justifiable if the lattice spacing is much larger than the extension of the atom cloud in the lattice. Furthermore we assume a spherically symmetric situation. Thus the density becomes continues $\bar{n}_i \rightarrow n(r)$ and the external confinement is given by $\epsilon_i \rightarrow \epsilon(r) = m/2 \, \omega_{\text{ext}}^2 r^2$. Note that the density is normalized to the volume of one lattice site, thus $n(r)$ is the atom number per lattice site in the distance r from the trap center. The total atom number is then given by

$$N = \int n(\mathbf{x}) \frac{1}{a_{\text{lat}}^3} d^3\mathbf{x} = \frac{4\pi}{a_{\text{lat}}^3} \int_0^\infty r^2 n(r) \, dr \tag{31}$$

where a_{lat} is the lattice spacing. The condition for a constant chemical potential in Eq. (30) directly leads to a density distribution over the lattice, which is the usual Thomas–Fermi parabola (see Fig. 9(c)):

$$n(r) = \frac{1}{U} \left(\mu - \frac{1}{2} m\omega_{\text{ext}}^2 r^2 \right). \tag{32}$$

The corresponding Thomas–Fermi radius and the chemical potential depending on atom number and trap parameters are given by

$$r_{TF} = \sqrt{\frac{2\mu}{m\omega_{\text{ext}}^2}} \quad \text{and} \quad \mu = \left(\frac{15}{16} \frac{(\lambda/2)^3 m^{3/2} N U \omega_{\text{ext}}^3}{\sqrt{2\pi}} \right)^{2/5}. \tag{33}$$

(a)

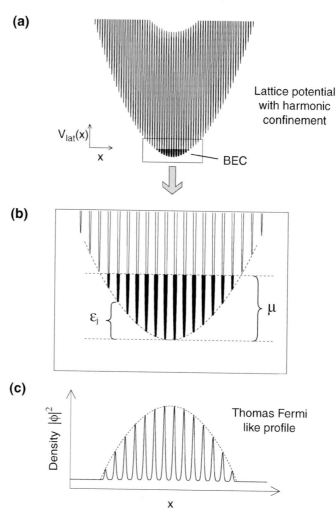

Lattice potential
with harmonic
confinement

$V_{lat}(x)$

x

BEC

(b)

ϵ_i

μ

(c)

Density $|\phi|^2$

Thomas Fermi
like profile

x

FIG. 9. (a) The atoms are trapped in a periodic lattice potential with an additional external confinement due to the Gaussian laser beam profile and the magnetic trapping potential. (b) In the ground state the BEC is distributed over the lattice in a way that the chemical potential μ is constant over the lattice. The local chemical potential on a lattice site can be approximated as the sum of the energy offset ϵ_i and the interaction energy per atom, which is proportional to the density. Therefore the BEC is distributed over the lattice with a density profile of a Thomas–Fermi parabola (c).

3.3.3. Adiabatic Mapping of Crystal Momentum to Free Particle Momentum

One of the big advantages of using optical lattice potentials is that the lattice depth can be completely dynamically controlled by simply tuning the laser power. This

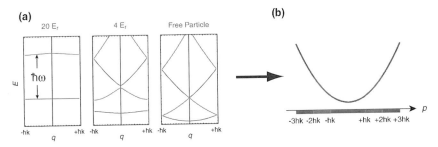

FIG. 10. (a) Bloch bands for different potential depths. During an adiabatic ramp down the quasi-momentum is conserved and (b) Bloch wave with quasi-momentum q in the nth energy band is mapped onto a free particle with momentum p in the nth Brillouin zone of the lattice.

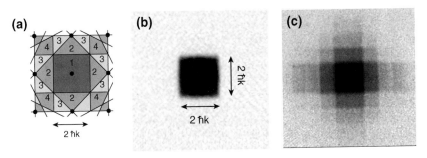

FIG. 11. (a) Brillouin zones of a 2D simple cubic optical lattice. For a homogeneously filled lowest Bloch band, an adiabatic shut off of the lattice potential leads to a homogeneously populated first Brillouin zone, which can be observed through absorption imaging after a time of flight expansion (b). If in addition higher Bloch bands were populated, higher Brillouin zones become populated as well (c).

allows one to adiabatically convert a deep optical lattice into a shallow one and eventually completely turn off the lattice potential. Under adiabatic transformation of the lattice depth the quasi-momentum is preserved and during the turn off process a Bloch wave in the nth energy band is mapped onto a corresponding free particle momentum p in the nth Brillouin zone (see Fig. 10) (Kastberg et al., 1995; Greiner et al., 2001; Köhl et al., 2004).

Such a behaviour has indeed been observed with both bosonic (Greiner et al., 2001) and fermionic (Köhl et al., 2004) atoms. For a situation where all atoms are confined to the lowest Bloch band and for a homogeneous filling of such an energy band, an adiabatic ramp down of the lattice potential leaves the central Brillouin zone—a square of width $2\hbar k$—fully occupied (see Fig. 11(b)). If on the other hand higher energy bands are populated, one also observes populations in higher Brillouin zones (see Fig. 11(c)). This method can be used to efficiently probe the distribution of the particles over different energy bands.

4. Bose–Hubbard Model of Interacting Bosons in Optical Lattices

The behavior of bosonic atoms with repulsive interactions in a periodic potential is fully captured by the Bose–Hubbard Hamiltonian of solid state physics (Fisher et al., 1989; Jaksch et al., 1998), which in the homogeneous case can be expressed through:

$$H = -J \sum_{\langle i,j \rangle} \hat{a}_i^\dagger \hat{a}_j + \frac{1}{2} U \sum_i \hat{n}_i \left(\hat{n}_i - 1 \right). \tag{34}$$

Here $\hat{a}_i^\dagger$ and $\hat{a}_i$ describe the creation and annihilation operators for a boson on the ith lattice site and $\hat{n}_i$ counts the number of bosons on the ith lattice site. The tunnel coupling between neighboring potential wells is characterized by the tunneling matrix element

$$J = -\int d^3x\, w(\mathbf{x} - \mathbf{x}_i) \left(-\hbar^2 \frac{\nabla^2}{2m} + V_{\text{lat}}(\mathbf{x}) \right) w(\mathbf{x} - \mathbf{x}_j), \tag{35}$$

where $w(\mathbf{x} - \mathbf{x}_i)$ is a single particle Wannier function localized to the ith lattice site and $V_{\text{lat}}(\mathbf{x})$ indicates the optical lattice potential. The repulsion between two atoms on a single lattice site is quantified by the on-site matrix element U

$$U = 4\pi \hbar^2 \frac{a}{m} \int \left| w(\mathbf{x}) \right|^4 d^3x, \tag{36}$$

with a being the scattering length associated with the interaction. Due to the short range of the interactions compared to the lattice spacing, the interaction energy is well described by the second term of Eq. (34) which characterizes a purely on-site interaction.

Both the tunnelling matrix element J and the on-site interaction matrix element can be calculated from a band structure calculation (see Fig. 12(b) for J). The tunnel matrix element J is related to the width of the lowest Bloch band through:

$$4J = \left| E_0 \left(q = \frac{\pi}{a} \right) - E_0(q = 0) \right|, \tag{37}$$

where a is the lattice period, such that $q = \pi/a$ corresponds to the quasi-momentum at the border of the first Brillouin zone. Care has to be taken to evaluate the tunnel matrix element through Eq. (35), when the Wannier function is approximated by the Gaussian ground state wave function on a single lattice site. This usually results in a severe underestimation of the tunnel coupling between lattice sites.

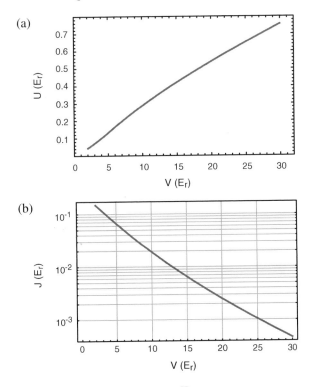

FIG. 12. On-site interaction matrix element U for ^{87}Rb (a) and tunnelling matrix element J (b) vs lattice depth. All values are given in units of the recoil energy E_r.

The interaction matrix element can be evaluated through the Wannier function with the help of Eq. (36). In this case, however, the approximation of the Wannier function through the Gaussian ground state wave function yields a very good approximation (see Fig. 12(a) for U).

Recently Zwerger (2003) has carried out a more sophisticated approximation of the tunneling matrix element and the on-site interaction matrix element; he finds (in units of the recoil energy):

$$J \approx \frac{4}{\sqrt{\pi}} \left(\frac{V_{\text{lat}}}{E_r} \right)^{3/4} e^{-2\sqrt{V_{\text{lat}}/E_r}} \tag{38}$$

and for the interaction matrix element

$$U \approx 4\sqrt{2\pi} \frac{a}{\lambda} \left(\frac{V_{\text{lat}}}{E_r} \right)^{3/4}. \tag{39}$$

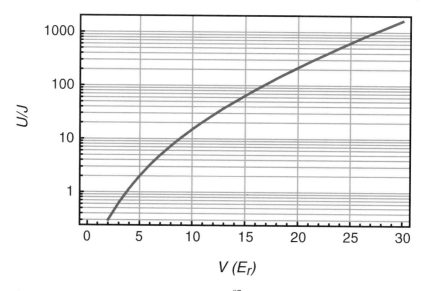

FIG. 13. U/J vs optical lattice potential depth for ^{87}Rb. By increasing the lattice depth one can tune the ratio U/J, which determines whether the system is strongly or weakly interacting.

The ratio U/J is crucial for determining whether one is a strongly interacting or a weakly interacting regime. It can be tuned continuously by simply changing the lattice potential depth (see Fig. 13).

4.1. GROUND STATES OF THE BOSE–HUBBARD HAMILTONIAN

The Bose–Hubbard Hamiltonian of Eq. (34) has two distinct ground states depending on the strength of the interaction U relative to the tunnel-coupling J. In order to gain insight into the two limiting ground-states, let us first consider the case of a double well system with only two interacting neutral atoms.

4.2. DOUBLE WELL CASE

In the double well system the two lowest lying states for noninteracting particles are the symmetric $|\varphi_S\rangle = 1/\sqrt{2}(|\varphi_L\rangle + |\varphi_R\rangle)$ and the antisymmetric $|\varphi_A\rangle = 1/\sqrt{2}(|\varphi_L\rangle - |\varphi_R\rangle)$ states, where $|\varphi_L\rangle$ and $|\varphi_R\rangle$ are the ground states of the left- and right-hand side of the double well potential. The energy difference between $|\varphi_S\rangle$ and $|\varphi_A\rangle$ will be denoted by $2J$, which characterizes the tunnel coupling between the two wells and depends strongly on the barrier height between the two potentials.

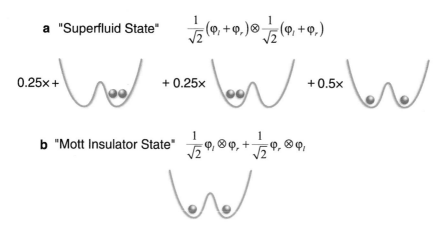

FIG. 14. Ground state of two interacting particles in a double well. For interaction energies U smaller than the tunnel coupling J the ground state of the two-body system is realized by the "superfluid" state (a). If on the other hand U is much larger than J, then the ground state of the two-body system is the Mott insulating state (b).

In the case of no interactions, the ground state of the two-body system is realized when each atom is in the symmetric ground state of the double well system (see Fig. 14(a)). Such a situation yields an average occupation of one atom per site with the single site many-body state actually being in a superposition of zero, one and two atoms. Let us now consider the effects due to a repulsive interaction between the atoms. If both atoms are again in the symmetric ground state of the double well, the total energy of such a state will increase due to the repulsive interactions between the atoms. This higher energy cost is a direct consequence of having contributions where both atoms occupy the same site of the double well. This leads to an interaction energy of $1/2U$ for this state.

If this energy cost is much greater than the splitting $2J$ between the symmetric and antisymmetric ground states of the noninteracting system, the system can minimize its energy when each atom is in a superposition of the symmetric and antisymmetric ground state of the double well $1/\sqrt{2}(|\varphi_S\rangle \pm |\varphi_A\rangle)$. The resulting many-body state can then be written as $|\Psi\rangle = 1/\sqrt{2}(|\varphi_L\rangle \otimes |\varphi_R\rangle + |\varphi_R\rangle \otimes |\varphi_L\rangle)$. Here exactly one atom occupies the left and right site of the double well. Now the interaction energy vanishes because both atoms never occupy the same lattice site. The system will choose this new "Mott insulating" ground state when the energy costs of populating the antisymmetric state of the double well system are outweighed by the energy reduction in the interaction energy. It is important to note that precisely the atom number fluctuations due to the delocalized single particle wave functions make the "superfluid" state unfavorable for large U.

Such a change can be induced by adiabatically increasing the barrier height in the double well system, such that J decreases exponentially and the energy cost for populating the antisymmetric state becomes smaller and smaller. Eventually it will then be favorable for the system to change from the "superfluid" ground state, where for $U/J \to \infty$ each atom is delocalized over the two wells, to the "Mott insulating" state, where each atom is localized to a single lattice site.

4.3. MULTIPLE WELL CASE

The above ideas can be extended readily to the multiple well case of the periodic potential of an optical lattice. For $U/J \ll 1$ the tunnelling term dominates the Hamiltonian and the ground-state of the many-body system with N atoms is given by a product of identical single particle Bloch waves, where each atom is spread out over the entire lattice with M lattice sites:

$$|\Psi_{SF}\rangle_{U/J \approx 0} \propto \left(\sum_{i=1}^{M} \hat{a}_i^\dagger\right)^N |0\rangle. \tag{40}$$

Since the many-body state is a product over identical single particle states, a macroscopic wave function can be used to describe the system. Here the single site many-body wave function $|\phi\rangle_i$ is almost equivalent to a coherent state. The atom number per lattice site then remains uncertain and follows a Poissonian distribution with a variance given by the average number of atoms on this lattice site $\mathrm{Var}(n_i) = \langle \hat{n}_i \rangle$. The nonvanishing expectation value of $\psi_i = \langle \phi_i | \hat{a}_i | \phi_i \rangle$ then characterizes the coherent matter wave field on the ith lattice site. This matter wave field has a fixed phase relative to all other coherent matter wave fields on different lattice sites.

If, on the other hand, interactions dominate the behavior of the Hamiltonian, such that $U/J \gg 1$, then fluctuations in the atom number on a single lattice site become energetically costly and the ground state of the system will instead consist of localized atomic wave functions that minimize the interaction energy. The many-body ground state is then a product of local Fock states for each lattice site. In this limit the ground state of the many-body system for a commensurate filling of n atoms per lattice site is given by:

$$|\Psi_{MI}\rangle_{J \approx 0} \propto \prod_{i=1}^{M} (\hat{a}_i^\dagger)^n |0\rangle. \tag{41}$$

Under such a situation the atom number on each lattice site is exactly determined but the phase of the coherent matter wave field on a lattice site has obtained a maximum uncertainty. This is characterized by a vanishing of the matter wave field on the ith lattice site $\psi_i = \langle \phi_i | \hat{a}_i | \phi_i \rangle \approx 0$.

In this regime of strong correlations, the interactions between the atoms dominate the behavior of the system and the many-body state is not amenable anymore to a description as a macroscopic matter wave, nor can the system be treated by the theories for a weakly interacting Bose gas of Gross, Pitaevskii and Bogoliubov (Pethick and Smith, 2001; Pitaevskii and Stringari, 2003).

For a 3D system, the transition to a Mott insulator occurs around $U/J \approx z \cdot 5.6$ (Sheshadri et al., 1993; Jaksch et al., 1998; van Oosten et al., 2001; Zwerger, 2003), where z is the number of next neighbors to a lattice site (for a simple cubic crystal $z = 6$).

4.4. SF–MI TRANSITION IN INHOMOGENEOUS POTENTIALS

Compared to experiments in condensed matter physics, where one typically works at a fixed density, experiments with ultracold atom clouds in an external potential are usually carried out at a fixed chemical potential. Density redistributions are therefore very common in inhomogeneous trapping potentials. How can one understand the basic effects of an external confining potential on top of the periodic potential? Let us assume an isotropic overall harmonic confinement of the form $V_{ext} = 1/2 m\omega^2 r^2$ and for simplicity discuss only the effects along one spatial direction of the trap. Using a local density approximation, one can then introduce a local chemical potential on the ith lattice site, which is related to the overall chemical potential, through:

$$\mu_i = \mu - \frac{1}{2} m\omega^2 i^2 \left(\frac{\lambda}{2}\right)^2, \quad \text{for } \mu_i \geq 0. \tag{42}$$

At the center of the trap, the local chemical potential equals the overall chemical potential, however as one increases the distance from the center, the local chemical potential continuously decreases until it vanishes at the border of the atom cloud. Using the local chemical potential in the SF–MI phase diagram (Fisher et al., 1989) (see Fig. 15), one finds that in the trap the system develops a shell structure, with alternating superfluid and Mott insulating regions. The superfluid rings, however, tend to vanish for deeper and deeper lattices, for which $J/U \to 0$. The border of the quantum gas, necessarily always has to be a superfluid, due to the vanishing occupation in this regime. Such a shell structure formation has indeed been predicted in more detailed mean-field calculations (Jaksch et al., 1998; van Oosten et al., 2001; Dickerscheid et al., 2003) and 2D and 3D quantum Monte Carlo simulations (Kashurnikov et al., 2002; Wessel et al., 2004) (see also Fig. 16). For one-dimensional situations a mean field approach is less valid and detailed theoretical analyses have been carried out using quantum Monte Carlo (Batrouni et al., 2002), Density Renormalization Group (Kollath et al., 2004) and adaptive Hilbert space methods (Daley et al., 2004;

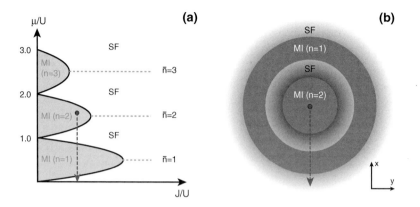

FIG. 15. Formation of a shell structure of alternating rings of Mott insulating and superfluid regions based on the SF–MI phase diagram (Fisher et al., 1989) (a). In an inhomogeneous trapping potential a local chemical potential can be introduced (see text), which decreases continuously as one propagates radially outward from the center of the atom cloud to its border (dashed line in (a)) and leads to a characteristic shell structure in the density distribution (b).

Clark and Jaksch, 2004) a variational ansatz (Garcia-Ripoll et al., 2004) and exact diagonalization approaches (Roth and Burnett, 2003b), also focussing on the detailed behaviour of the correlation functions in the superfluid and crossover regime to a Mott insulator.

4.5. SUPERFLUID TO MOTT INSULATOR TRANSITION

In the experiment the crucial parameter U/J that characterizes the strength of the interactions relative to the tunnel coupling between neighboring sites can be varied by simply changing the potential depth of the optical lattice potential. By increasing the lattice potential depth, U increases almost linearly due to the tighter localization of the atomic wave packets on each lattice site and J decreases exponentially due the decreasing tunnel coupling. The ratio U/J can therefore be varied over a large range from $U/J \approx 0$ up to values in our case of $U/J \approx 2000$.

In the superfluid regime (Cataliotti et al., 2001) phase coherence of the matter wave field across the lattice characterizes the many-body state. This can be observed by suddenly turning off all trapping fields, such that the individual matter wave fields on different lattice sites expand and interfere with each other. After a fixed time of flight period the atomic density distribution can then be measured by absorption imaging. Such an image directly reveals the momentum distribution of the trapped atoms. In Fig. 17(b) an interference pattern can be seen after releasing the atoms from a three-dimensional lattice potential.

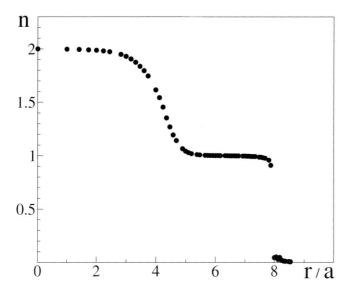

FIG. 16. Predicted shell structure of alternating Mott insulating and superfluid regions from a 3D quantum Monte Carlo simulation for $U/J = 80$. Reprinted figure with permission from (Kashurnikov et al., 2002).

© 2002 The American Physical Society.

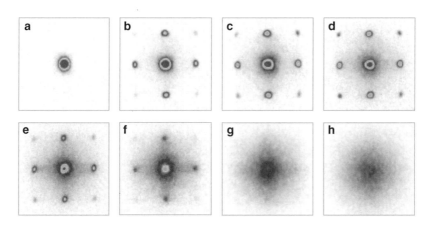

FIG. 17. Absorption images of multiple matter wave interference patterns after releasing the atoms from an optical lattice potential with a potential depth of (a) $0E_r$, (b) $3E_r$, (c) $7E_r$, (d) $10E_r$, (e) $13E_r$, (f) $14E_r$, (g) $16E_r$ and (h) $20E_r$. The ballistic expansion time was 15 ms. Adapted from (Greiner et al., 2002).

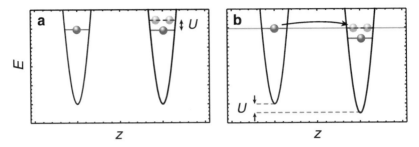

FIG. 18. (a) Excitation gap in the Mott insulator phase with exactly $n = 1$ atom on each lattice site. (b) If a correct potential gradient is added, atoms can tunnel again.

If on the other hand the optical lattice potential depth is increased such that the system is very deep in the Mott insulating regime ($U/J \rightarrow \infty$), phase coherence is lost between the matter wave fields on neighboring lattice sites due to the formation of Fock states (Jaksch et al., 1998; Fisher et al., 1989; Sachdev, 1999; Zwerger, 2003). In this case no interference pattern can be seen in the time of flight images (see Fig. 17(b)) (Greiner et al., 2002; Porto et al., 2003; Stöferle et al., 2004). For a Mott insulator at finite U/J one expects a residual visibility in the interference pattern (Kashurnikov et al., 2002; Roth and Burnett, 2004), which can be caused on the one hand by the residual superfluid shells and on the other hand through an admixture of coherent particle hole pairs to the ideal MI ground state (Gerbier et al., 2005).

In addition to the fundamentally different momentum distributions in the superfluid and Mott insulating regime, the excitation spectrum is markedly different as well in both cases. Whereas the excitation spectrum in the superfluid regime is gapless, it is gapped in the Mott insulating regime. This energy gap of order U (deep in the MI regime) can be attributed to the now localized atomic wave functions of the atoms (Fisher et al., 1989; Jaksch et al., 1998; van Oosten et al., 2001; Zwerger, 2003).

Let us consider, for example, a Mott insulating state with exactly one atom per lattice site. The lowest lying excitation to such a state is determined by removing an atom from a lattice site and placing it into the neighboring lattice site (see Fig. 18(a)). Due to the on-site repulsion between the atoms, however, such an excitation costs energy U which is usually not available to the system. Therefore these are only allowed in virtual processes and an atom in general has to remain immobile at its original position. If one adds a potential gradient such that the energy difference between neighboring lattice sites ΔE exactly matches the onsite energy cost U, then such an excitation becomes energetically possible and one is able to resonantly perturb the system (see Fig. 18(b) and Figs. 19, 20). It has been possible to measure this

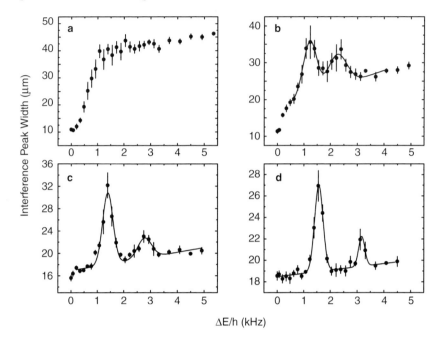

FIG. 19. Probing the excitation probability versus an applied vertical potential gradient. Width of interference peaks after adiabatic rampdown vs. the energy difference between neighboring lattice sites E, due to the potential gradient at different lattice depths: (a) $V_{max} = 10E_r$, (b) $V_{max} = 13E_r$, (c) $V_{max} = 16E_r$ and (d) $V_{max} = 20E_r$. The emergence of an energy gapped can be observed as the system is converted into a Mott insulator (adapted from (Greiner et al., 2002)).

change in the excitation spectrum by applying varying magnetic field gradients to the system for different lattice potential depths and detecting the response of the system to such perturbations (Greiner et al., 2002; Sachdev et al., 2002; Braun-Munzinger et al., 2004).

Recently, the excitation spectrum has also been probed via an intensity modulation of the lattice potential, which is effectively a Bragg spectroscopy for momentum transfer $q = 0$ (Stöferle et al., 2004). Here a similar behavior was predicted and observed: in the superfluid a broad excitation spectrum was detected (Stöferle et al., 2004; Schori et al., 2004; Büchler and Blatter, 2003), whereas the Mott insulating state displays sharp resonances at integer multiples of the on-site interaction matrix element U (Stöferle et al., 2004) (see also Fig. 20). Using Bragg spectroscopy it should also be possible to probe the excitation spectrum of a Mott insulator in more detail at various momentum transfers, which has been suggested and evaluated recently by (van Oosten et al., 2005).

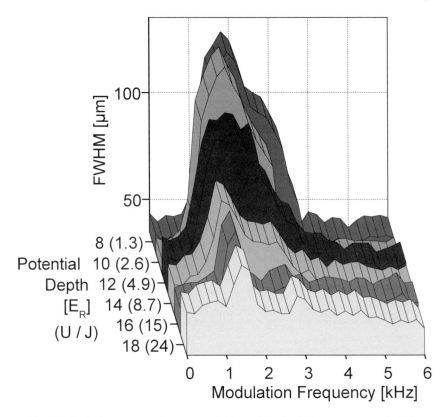

FIG. 20. Excitation spectrum from a superfluid to a Mott insulating state measured via Bragg spectroscopy with zero momentum transfer via intensity modulation of the lattice beams. Reprinted figure with permission from (Stöferle et al., 2004).
© 2004 The American Physical Society.

5. Collapse and Revival of a Macroscopic Quantum Field

A long standing question in interacting macroscopic quantum systems has been directed towards the problem of what happens to an initially well defined relative phase between two macroscopic quantum systems after they have been isolated from each other (Sols, 1994; Wright et al., 1996; Imamoglu et al., 1997; Castin and Dalibard, 1997; Dunningham et al., 1998). Equivalently one may ask: how do the individual macroscopic quantum fields evolve after they have been isolated from each other? Such a situation can be realized for example with a Bose–Einstein condensate that is split into two parts, such that a constant relative phase is initially established between the two subsystems BEC1 and BEC2 (see Fig. 21).

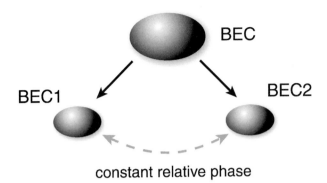

FIG. 21. A Bose–Einstein condensate is split into two parts with an initially constant phase between the two subsystems BEC1 and BEC2.

Whenever a condensate is split into two parts such that a fixed relative phase is established between those two parts, the many-body state in each of the BECs is in a superposition of different atom number states. Let us now consider the case of repulsive interactions between the atoms and determine how such superpositions of atom number states evolve over time, taking into account the collisions between the atoms. Let us first assume that all atoms in a subsystem occupy the ground state of its external confining potential. If the interaction energy is then small compared to the vibrational spacing in this potential well, the Hamiltonian governing the behavior of the atoms is given by:

$$H = \frac{1}{2} U \hat{n} (\hat{n} - 1). \tag{43}$$

The eigenstates of the above Hamiltonian are Fock states $|n\rangle$ in the atom number, with eigenenergies $E_n = Un(n - 1)/2$. The evolution with time of such an n-particle state is then simply given by $|n\rangle(t) = |n\rangle(0) \times \exp(-i E_n t/\hbar)$.

If the atoms in such a subsystem are brought into a superposition of atom number states $|n\rangle$, which always occurs whenever a fixed relative phase persists between the two subsystems, each subsystem is in a superposition of eigenstates $|n\rangle$ which results in a dynamical evolution of this state over time. Let us consider, for example, a coherent state $|\alpha\rangle = \exp(-|\alpha|^2/2) \sum_n \frac{\alpha^n}{\sqrt{n!}} |n\rangle$ in each subsystem. Here α is the amplitude of the coherent state with $|\alpha|^2$ corresponding to the average atom number in the subsystem. The evolution with time of such a coherent state can be evaluated by taking into account the time evolution of the different Fock states forming the coherent state:

$$|\alpha\rangle(t) = e^{-|\alpha|^2/2} \sum_n \frac{\alpha^n}{\sqrt{n!}} e^{-i(1/2)Un(n-1)t/\hbar} |n\rangle. \tag{44}$$

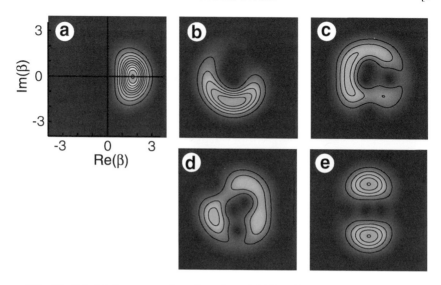

FIG. 22. Calculated quantum dynamics of an initially slightly number squeezed state with an average number of three atoms. Such a state can be parameterized as $|\alpha\rangle(t) \propto \sum_n g^{n(n-1)}(\alpha^n/\sqrt{n!}) \exp(-i(1/2)Un(n-1)t/\hbar)|n\rangle$, where $0 < g < 1$ characterizes the sub-Poissonian character of the many-body state. In the graph $g = 0.8$. The dynamical evolution of the quantum state is caused by the coherent cold collisions between the atoms. The graphs show the overlap of the dynamically evolved input state with an arbitrary coherent state of amplitude β. Evolution times are: (a) $0h/U$; (b) $\frac{1}{8}h/U$; (c) $\frac{1}{4}h/U$; (d) $\frac{3}{8}h/U$; (e) $\frac{1}{2}h/U$.

The coherent matter wave field ψ in each of the subsystems can then simply be evaluated through $\psi = \langle\alpha(t)|\hat{a}|\alpha(t)\rangle$, which exhibits an intriguing dynamical evolution (Narozhny et al., 1981; Yurke and Stoler, 1986; Buzek et al., 1992; Sols, 1994; Milburn et al., 1997; Wright et al., 1996, 1997; Imamoglu et al., 1997; Castin and Dalibard, 1997; Dunningham et al., 1998). At first, the different phase evolutions of the atom number states lead to a collapse of ψ. However, at integer multiples in time of h/U all phase factors in the above equation re-phase modulo 2π and thus lead to a revival of the initial coherent state (see also Fig. 22). The collapse and revival of the coherent matter wave field of a BEC is reminiscent to the collapse and revival of the Rabi oscillations in the interaction of a single atom with a single mode electromagnetic field in cavity quantum electrodynamics (Rempe et al., 1987; Brune et al., 1996). There, the nonlinear atom–field interaction induces the collapse and revival of the Rabi oscillations whereas here the nonlinearity due to the interactions between the atoms themselves leads to the series of collapse and revivals of the matter wave field. It should be pointed out that such a behavior has also been theoretically predicted to occur for a coherent light field propagating in a nonlinear medium (Yurke and Stoler, 1986) but to our knowledge has never been observed experimentally. Note also, that in related experiments the collapse and revival of the wave packet of a single atom in a superposition of vibrational motion states at a lattice site has been observed (Raithel et al., 1998). Here the collapse and revival in first quantization arises due to the nonlinearity in the harmonic oscillator potential at each lattice site.

In order to realize a coherent state in a potential well, one can again use the optical lattice potential and ramp it to a potential depth V_A, which is still completely in the superfluid regime. Then, for low lattice depths, the many-body state in each potential well is almost equal to that of a coherent state with a corresponding average atom number. In such a situation the phase of the matter wave field on the ith lattice site is fixed relative to the matter wave fields on the other lattice sites. In order to isolate the wells from each other, one rapidly increases the lattice potential depth to V_B with negligible tunnel coupling on a time scale that is fast compared to the tunnelling time h/J in the system. As a result, the atoms do not have time to redistribute themselves during the ramp-up of the optical potential and we preserve the initial atom number distribution on each lattice site. On the other hand, the time scale is slow compared to the oscillation frequencies on each lattice site such that no vibrational excitations are created in the ramp-up process and all atoms remain in the vibrational ground state of each well. Using this method one can freeze out the atom number distribution at a potential depth V_A and the dynamics of each of the matter wave fields on different lattice sites is now governed by the Hamiltonian of Eq. (43).

The dynamical evolution of the matter wave fields can subsequently be followed by holding the atoms at the lattice potential depth V_B for a variable time and then releasing them suddenly from the combined optical and magnetic trapping potentials. After a suitable time of flight period one then takes absorption images of the multiple matter wave interference pattern (see Fig. 23). Initially, directly after ramping up the lattice potential, the interference pattern is clearly visible, however after a time of ≈ 250 μs the interference pattern is completely lost. Here the vanishing of the interference pattern is caused by the collapse of the matter wave fields on each lattice site. But after a total hold time of 550 μs the original interference pattern is regained again, showing that the matter wave fields have revived. It is important to note that the atom number statistics in each of the wells remains constant throughout the dynamical evolution time. This is fundamentally different from the vanishing of the interference pattern in the Mott insulator case, where the atom number distribution changes, but no further dynamical evolution occurs.

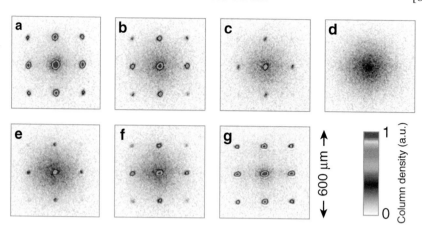

FIG. 23. Dynamical evolution of the multiple matter wave interference pattern after jumping from a potential depth $V_A = 8 \, E_r$ to a potential depth $V_B = 22 \, E_r$ and a subsequent hold time τ. Hold times τ: (a) 0 µs; (b) 100 µs; (c) 150 µs; (d) 250 µs; (e) 350 µs; (f) 400 µs and (g) 550 µs.

6. Quantum Gate Arrays via Controlled Collisions

6.1. SPIN-DEPENDENT TRANSPORT

So far the optical potentials used have been mostly independent of the internal ground state of the atom. However, it has been suggested that by using spin-dependent periodic potentials one could bring atoms on different lattice sites into contact and thereby realize fundamental quantum gates (Brennen et al., 1999, 2002; Raussendorf and Briegel, 2001; Briegel et al., 2000), create large scale entanglement (Jaksch et al., 1999; Briegel and Raussendorf, 2001), excite spin waves (Sorensen and Molmer, 1999), study quantum random walks (Dür et al., 2002) or form a universal quantum simulator to simulate fundamental condensed matter physics Hamiltonians (Jane et al., 2003). We show how the wave packet of an atom that is initially localized to a single lattice site can be split and delocalized in a controlled and coherent way over a defined number of lattice sites.

To accomplish this, the atoms are loaded into a three-dimensional lattice potential, on which spin-dependent transport is implemented along one direction. An experimental setup similar to Fig. 5 is used. In such a setup the polarization angle θ is initially set to a lin‖lin polarization configuration. Shortly before moving the atoms along this standing wave direction, the lattice potentials are adiabatically turned off along the y- and z-direction. This is done in order to reduce the interaction energy, which strongly depends on the confinement of the atoms at a single lattice site. One can thereby study the transport process itself, without having to take into account the phase shifts in the many-body state that result from a coherent collisional interaction between atoms.

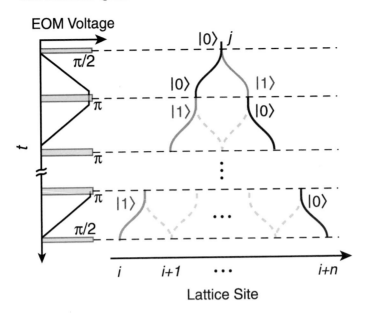

FIG. 24. General interferometer sequence used to delocalize an atom over an arbitrary number of lattice sites. Initially an atom is localized to the jth lattice site. The graph on the left indicates the EOM voltage and the sequence of $\pi/2$ and π microwave pulses that are applied over time (see text).

During the shifting process of the atoms it is crucial to avoid unwanted vibrational excitations, especially if the shifting process would be repeated frequently. Therefore care has to be taken not to expose the atoms to high accelerations, which could excite them to higher vibrational bands (Jaksch et al., 2000; Briegel et al., 2000; Calarco et al., 2000; Mandel et al., 2003).

In order to verify the coherence of the spin-dependent transport, the interferometer sequence of Fig. 24 has been used. Let us first consider the case of a single atom being initially localized to the jth lattice site. First, the atom is placed in a coherent superposition of the two internal states $|0\rangle_j$ and $|1\rangle_j$ with a $\pi/2$ microwave pulse (here the index denotes the position in the lattice). Then the polarization angle θ is rotated to 180°, such that the spatial wave packet of an atom in the $|0\rangle$ and the $|1\rangle$ state are transported in opposite directions. The final state after such a movement process is then given by $1/\sqrt{2}(|0\rangle_j + i \exp(i\beta)|1\rangle_{j+1})$, where the wave function of an atom has been delocalized over the jth and the $(j+1)$th lattice site. The phase β between the two wave packets depends on the accumulated kinetic and potential energy phases in the transport process and in general will be nonzero.

In order to reveal the coherence between the two wave packets, a final $\pi/2$ microwave pulse is applied, which erases the which-way information encoded

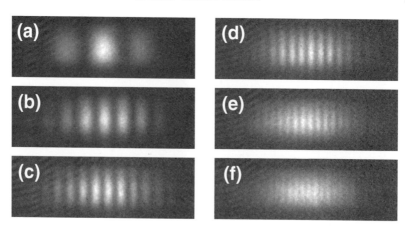

FIG. 25. Observed interference patterns in state $|1\rangle$ after initially localized atoms have been delocalized over (a) two, (b) three, (c) four, (d) five, (e) six and (f) seven lattice sites using the interferometer sequence of Fig. 24. The time of flight period before taking the images was 14 ms and the horizontal size of each image is 880 µm.

in the hyperfine states. The atoms are then released from the confining potential by suddenly turning off the standing wave optical potential and the momentum distribution of the trapped atoms in the $|1\rangle$ state is subsequently observed after a time of flight period. As a result of the above sequence, the spatial wave packet of an atom in the $|0\rangle$ ($|1\rangle$) state is delocalized over two lattice sites resulting in a double slit momentum distribution $w(p) \propto \exp(-p^2/(\hbar/\sigma_x)^2) \cdot \cos^2(p\delta x_0/2\hbar + \beta/2)$ (see Figs. 25(a) and 26) where δx_0 denotes the separation between the two wave packets and σ_x is the spatial extension of the Gaussian ground state wave function on each lattice site. In order to increase the separation between the two wave packets further, one could increase the polarization angle θ to further integer multiples of 180°. In practice, such an approach is however limited by the finite maximum voltage that can be applied to the EOM, however the use of spin-echo π-pulses, allows one to reverse the roles of both spin states and implement a transport over almost arbitrary distances.

With increasing separation between the two wave packets the fringe spacing of the interference pattern further decreases (see Fig. 25). Through these experiments it has become possible to demonstrate the coherent spin-dependent transport of neutral atoms in optical lattices, thereby showing an essential level of coherent control for many future applications. For example, the method also provides a simple way to continuously tune the interspecies interactions by controlling the overlap of the two ground state wave functions for the two spin states. Furthermore, if such a transport is carried out in a three-dimensional lattice, where the on-site interaction energy between atoms is large, one can induce interactions

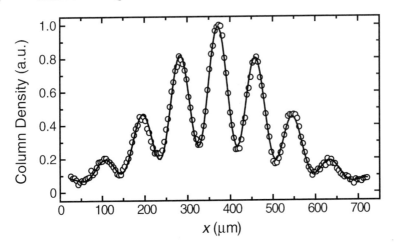

FIG. 26. Profile of the interference pattern obtained after delocalizing atoms over three lattice sites with a $\pi/2$-π-$\pi/2$ microwave pulse sequence. The solid line is a fit to the interference pattern with a sinusoidal modulation, a finite visibility ($\approx 60\,\%$) and a Gaussian envelope. The time of flight period was 15 ms.

between almost any two atoms on different lattice sites in a controlled way as will be shown in Section 6.2.

6.2. Controlled Collisions

In order to realize a controlled interaction between the particles on different lattice sites in a 3D Mott insulating quantum register, the above spin-dependent transport sequence can be used. This leads to collisions between neighboring atoms and can be described through an ensemble of quantum gates acting in parallel (Jaksch et al., 1999; Briegel et al., 2000). Alternatively, these quantum gates can be described as a controllable quantum Ising interaction (Briegel and Raussendorf, 2001):

$$H_{\text{int}} \propto g(t) \sum_j \frac{1 + \sigma_z^{(j)}}{2} \frac{1 - \sigma_z^{(j+1)}}{2}. \tag{45}$$

Here $g(t)$ denotes the time dependent coupling constant and $\sigma_z^{(j)}$ is the Pauli spin operator acting on an atom at the jth lattice site. For an interaction phase of $\varphi = \int_0^{t_{\text{hold}}} g(t)\,dt/\hbar = (2n+1)\pi$ one obtains a maximally entangled cluster state, whereas for $\varphi = 2\pi n$ one obtains a disentangled state (Briegel and Raussendorf, 2001). Here t_{hold} denotes the time for which the atoms are held together at a common site and n is an integer. Let us point out that the creation of such highly

(a)   **(b)**

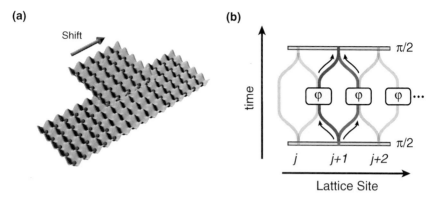

FIG. 27. (a) Controlled interactions between atoms on different lattice sites can be realized with the help of spin-dependent lattice potentials. In such spin-dependent potentials, atoms in a "blue" internal state experience a different lattice potential than atoms in a "red" internal state. These lattices can be moved relative to each other such that two initially separated atoms can be brought into controlled contact with each other. (b) This can be extended to form a massively parallel quantum gate array. Consider a string of atoms on different lattice sites. First the atoms are placed in a coherent superposition of the two internal states (red and blue). Then spin-dependent potentials are used to split each atom such that it simultaneously moves to the right and to the left and is brought into contact with the neighboring atoms. There both atoms interact and a controlled phase shift j is introduced. After such a controlled collision the atoms are again moved back to their original lattice sites.

entangled states can be achieved in a single lattice shift operational sequence described above and depicted in Fig. 27, independent of the number of atoms to be entangled (Jaksch et al., 1999; Briegel and Raussendorf, 2001).

A $\pi/2$ pulse first allows one to place the atom in a coherent superposition of the two states $|0\rangle$ and $|1\rangle$. After creating such a coherent superposition, a spin-dependent transfer is used to split and move the spatial wave function of the atom over half a lattice spacing in two opposite directions depending on its internal state (see Fig. 27). Atoms on neighboring sites interact for a variable amount of time t_{hold} that leads to a controlled conditional phase shift of the corresponding many-body state. After half of the hold time, a microwave π pulse is applied. This spin-echo type pulse is used mainly to cancel unwanted single particle phase shifts due, for example, to inhomogeneities in the trapping potentials. It does not however affect the nontrivial and crucial collisional phase shift due to the interactions between the atoms. After such a controlled collision, the atoms are moved back to their original site. Then a final $\pi/2$ microwave pulse with variable phase is applied and the atom number in state relative to the total atom number is recorded.

For short hold times, where no significant collisional phase shift is acquired, a Ramsey fringe with a high visibility of approx. 50% is recorded (see Fig. 28). For longer hold times one notices a strong reduction in the visibility of the Ramsey fringe, with an almost vanishing visibility of approximately 5% for a hold time of

210 µs. This hold time corresponds to an acquired collisional phase shift of $\varphi = \pi$ for which a minimum visibility is expected if the system is becoming entangled. For a two-particle system this can be understood by observing the resulting Bell state:

$$\frac{1}{\sqrt{2}}\left(|0\rangle_j|+\rangle^{\alpha}_{j+1} + |1\rangle_j|-\rangle^{\alpha}_{j+1}\right), \tag{46}$$

after the final $\pi/2$ pulse of the Ramsey sequence has been applied to the atoms. Here $|+\rangle^{\alpha}_{j+1}$ and $|-\rangle^{\alpha}_{j+1}$ represent two orthogonal superposition states of $|0\rangle$ and $|1\rangle$ for which $|\langle 1|+\rangle^{\alpha}|^2 + |\langle 1|-\rangle^{\alpha}|^2 = 0.5$. A measurement of atoms in state $|1\rangle$ therefore becomes independent of the phase corresponding to a vanishing Ramsey fringe. This indicates that no single particle operation can place all atoms in either spin-state when a maximally entangled state has been created. The disappearance of the Ramsey fringe has been shown to occur not only for a two-particle system, but is a general feature for an arbitrary N-particle array of atoms that have been highly entangled with the above experimental sequence (Jaksch, 1999; Briegel et al., 2000). For longer hold times, however, the visibility of the Ramsey fringe increases again reaching a maximum of 55% for a hold time of 450 µs (see Fig. 28). Here the system becomes disentangled again, as the collisional phase shift is close to $\varphi = 2\pi$ and the Ramsey fringe is restored with maximum visibility. The timescale of the observed collisional phase evolution is in good agreement with the measurements on the Mott insulator transition of the previous section and ab-initio calculations of the on-site matrix element U.

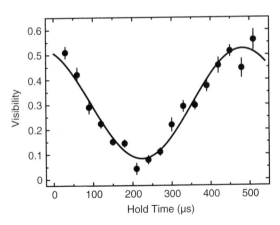

FIG. 28. Visibility of Ramsey fringes vs. hold times on neighboring lattice sites for the experimental sequence of Fig. 27. The solid line is a sinusoidal fit to the data including an offset and a finite amplitude. Such a sinusoidal behavior of the visibility vs. the collisional phase shift (determined by the hold time t_{hold}) is expected for a Mott insulating state with an occupancy of $n = 1$ atom per lattice site.

In a one-dimensional lattice shift the system is very susceptible to vacant lattice sites, as a defect will immediately limit the size of the cluster. However, the scheme can be extended to two or three dimensions by using two additional lattice shift operations along the remaining orthogonal lattice axes. As long as the filling factor of lattice sites would exceed the percolation threshold (31% for a 3D simple cubic lattice system) a large entangled cluster should be formed, making maximum entanglement of literally 100000 of atoms possible in only three operational steps. In addition, novel filtering schemes have been proposed, which would allow one create a Mott state with high fidelity through a coherent coupling to a BEC reservoir (Rabl et al., 2003).

6.3. USING CONTROLLED COLLISIONAL QUANTUM GATES

It has been proposed to use such controlled interactions of the Ising or Heisenberg type to simulate the behavior of quantum magnets (Sorensen and Molmer, 1999) and other quantum system (Jane et al., 2003), to realize quantum gates between different atoms (Brennen et al., 1999, 2002; Jaksch et al., 1999; Briegel et al., 2000; Raussendorf and Briegel, 2001), or to generate highly entangled cluster states $|\phi_c\rangle$ (Briegel et al., 2000). In a one-dimensional spin chain of length N, such a cluster state can be written to be of general form (Briegel and Raussendorf, 2001):

$$|\phi_c\rangle = \frac{1}{2^{N/2}} \bigotimes_{i=1}^{N} \left(|0\rangle_i \sigma_z^{i+1} + |1\rangle_i\right). \tag{47}$$

Up to local unitary transformations, for $N = 2$ the cluster state simply corresponds to the Bell state $1/\sqrt{2}(|0\rangle|0\rangle + |1\rangle|1\rangle)$ and for $N = 3$ to the three particle GHZ state $1/\sqrt{2}(|0\rangle|0\rangle|0\rangle + |1\rangle|1\rangle|1\rangle)$ (Greenberger et al., 1989); however, for larger N no such simple identification can be made.

Cluster states provide a high degree of persistent entanglement and a new form of quantum computing has been proposed based on such a state—the so called *one way quantum computer* (Raussendorf and Briegel, 2001; Raussendorf et al., 2003). At its heart lies the idea of first generating a massively entangled 2D cluster state. Then only single particle operations have to be implemented on this cluster state to realize quantum logical operations in the system and to perform a quantum computation (see Fig. 29).

7. Outlook

What are the topics that will be investigated in the future? We believe that one has just started to explore the road of strongly correlated quantum systems in optical lattices. One of the next natural steps is to load spin-mixtures into the lattice

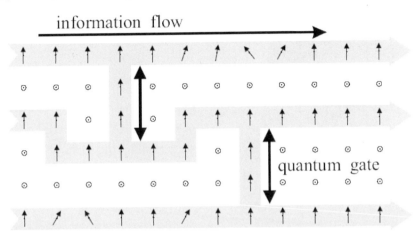

FIG. 29. Simulation of a quantum logic network by measuring two-state particles on a lattice. Circles ⊙ symbolize measurements of σ_z, vertical arrows are measurements of σ_x, while tilted arrows refer to measurements in the x–y-plane. Reprinted figure with permission from (Raussendorf and Briegel, 2001).
© 2001 The American Physical Society.

potential. For this case, fascinating novel quantum phases, such as a counterflow superfluid (Kuklov and Svistunov, 2003; Kuklov et al., 2004) and Cooper-type pairing (Paredes and Cirac, 2003) have been predicted, for which the total density of a two-component spin-mixture in a lattice is fixed, however the individual spin components remain completely superfluid. Moreover, by using spin-dependent lattice potentials one can map the Hamiltonian of a two-component Bose mixture onto a controlled quantum spin system Hamiltonian and therefore investigate fundamental quantum magnetic systems (Altman et al., 2003; Duan et al., 2003; Kuklov and Svistunov, 2003) in a highly controllable environment.

Another research effort will be directed towards disordered systems (Damski et al., 2003; Roth and Burnett, 2003a, 2003c; Sanpera et al., 2004). Strongly interacting quantum systems in random potentials are among one of the most difficult systems to analyze theoretically. As one of the highlights under discussion, (Fisher et al., 1989) have predicted the existence of a so called Bose-glass phase, an insulator without an energy gap, which should be observable in the experiments. In a most recent experiment in a one-dimensional optical lattice potential, random potentials were shown to be realizable via a laser speckle pattern and the transport properties of a BEC in this random potential have been investigated (Lye et al., 2004).

In addition to bosons, it is a natural next step to load fermions and into lattice potentials and investigate their quantum phases. In first experiments, the Fermi

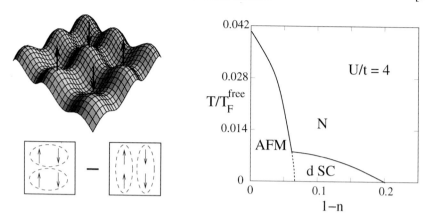

FIG. 30. In the case of repulsive interactions one finds either antiferromagnetic (AFM, upper left) or d-wave SF phases (lower left), depending upon the filling fraction n. Right: predicted phase diagram for repelling ^{6}Li atoms in a 2D lattice. Reprinted figure with permission from (Hofstetter et al., 2002). © 2002 The American Physical Society.

surface of a fermionic quantum gas in a three-dimensional lattice potential has been explored (Köhl et al., 2004) and the transport properties of fermions in the presence of bosonic collisional partners have been investigated (Ott et al., 2004). Fermions in optical lattices have also proven to be suitable for precision inter-ferometry measurements of the gravitational acceleration due to the absence of collisional shift in single species Fermi gases (Roati et al., 2004). Fermions in a three-dimensional optical lattice (Köhl et al., 2004) can be described by the fa-mous Hubbard Hamiltonian, which is believed to contain a possible explanation for High-Tc-Superconductivity (Hofstetter et al., 2002). Although this system has been investigated theoretically for decades, the exact form of the phase diagram remains unknown. Here fermions in optical lattices could be used to investigate this fundamental Hamiltonian and test, for example, where anti-ferromagnetic and superfluid phases of fermions with repulsive interactions in periodic potentials are located (Hofstetter et al., 2002) (see Fig. 30).

In addition to single species fermionic and bosonic gases, mixtures of bosons and fermionic quantum gases in optical lattice are predicted to contain a rich set of novel quantum phases (Albus et al., 2003; Büchler and Blatter, 2004; Lewenstein et al., 2004; Roth and Burnett, 2003c, 2004; Sanpera et al., 2004; Santos et al., 2004), which have never been observed in conventional condensed matter physics. Lewenstein et al. (2004), for example, found several novel quantum phases con-sisting of composite fermions in superfluid or metallic phases. The discovery of such novel quantum phases will shed more light on our understanding of complex quantum matter under extreme conditions, with many exciting developments just at our doorstep.

8. Acknowledgements

We would like to thank Ehud Altman, Hans Briegel, Ignacio Cirac, Tilman Esslinger, Fabrice Gerbier, Walter Hofstetter, Belén Paredes, Misha Lukin, Boris Svistunov and Peter Zoller for stimulating discussions and support with the figures of this work. Also, I.B. would like to acknowledge funding through DFG, a Marie Curie Excellence grant of the EU and AFOSR.

9. References

Albiez, M., Gati, R., Föelling, J., Hunsmann, S., Cristiani, M., Oberthaler, M. (2004). Direct observation of tunneling and nonlinear self-trapping in a single bosonic Josephson junction. cond-mat/0411757.

Albus, A., Illuminati, F., Eisert, J. (2003). Mixtures of bosonic and fermionic atoms in optical lattices. *Phys. Rev. A* **68**, 023606.

Altman, E., Hofstetter, W., Demler, E., Lukin, M. (2003). Phase diagram of two-component bosons on an optical lattice. *New J. Phys.* **5**, 113.

Anderson, B., Kasevich, M.A. (1998). Macroscopic quantum interference from atomic tunnel arrays. *Science* **282**, 1686–1689.

Anderson, M., Ensher, J., Matthews, M., Wieman, C.E., Cornell, E.A. (1995). Observation of Bose–Einstein condensation in a dilute atomic vapor. *Science* **269**, 198.

Ashcroft, N., Mermin, N. (1976). "Solid State Physics". Harcourt Brace College Publishers, Fort Worth.

Askar'yan, G. (1962). Effects of the gradient of a strong electromagnetic beam on electrons and atoms. *Soviet Phys. JETP* **15**, 1088.

Barrett, M., Sauer, J., Chapman, M. (2001). All-optical formation of an atomic Bose–Einstein condensate. *Phys. Rev. Lett.* **87**, 010404.

Batrouni, G., Rousseau, V., Scalettar, R., Rigol, M., Muramatsu, A., Denteneer, P., Troyer, M. (2002). Mott domains of bosons confined on optical lattices. *Phys. Rev. Lett.* **89**, 117203.

Bradley, C., Sackett, C., Tollett, J., Hulet, R. (1995). Evidence of Bose–Einstein condensation in an atomic gas with attractive interactions. *Phys. Rev. Lett.* **75**, 1687.

Braun-Munzinger, K., Dunningham, J.A., Burnett, K. (2004). Excitations of Bose–Einstein condensates in optical lattices. *Phys. Rev. A* **69**, 053613.

Brennen, G., Caves, C., Jessen, P., Deutsch, I.H. (1999). Quantum logic gates in optical lattices. *Phys. Rev. Lett.* **82**, 1060–1063.

Brennen, G., Deutsch, I.H., Williams, C. (2002). Quantum logic for trapped atoms via molecular hyperfine interactions. *Phys. Rev. A* **65**, 022313.

Briegel, H.J., Raussendorf, R. (2001). Persistent entanglement in arrays of interacting particles. *Phys. Rev. Lett.* **86** (5), 910–913.

Briegel, H.J., Calarco, T., Jaksch, D., Cirac, J.I., Zoller, P. (2000). Quantum computing with neutral atoms. *J. Mod. Opt.* **47** (2–3), 415–451.

Brune, M., Schmidt-Kaler, F., Maali, A., Dreyer, J., Hagley, E., Raimond, J.M., Haroche, S. (1996). Quantum Rabi oscillation: A direct test of field quantization in a cavity. *Phys. Rev. Lett.* **76** (11), 1800–1803.

Büchler, H.P., Blatter, G. (2003). Signature of quantum depletion in the dynamic structure factor of atomic gases. cond-mat/0312526.

Büchler, H.P., Blatter, G. (2004). Phase separation of atomic Bose–Fermi mixtures in an optical lattice. *Phys. Rev. A* **69**, 063603.

Buzek, V., Moya-Cessa, H., Knight, P.L., Phoenix, S.J.D. (1992). Schrödinger-cat states in the resonant Jaynes–Cummings model: Collapse and revival of oscillations of the photon-number distribution. *Phys. Rev. A* **45** (11), 8190–8203.

Calarco, T., Briegel, H.J., Jaksch, D., Cirac, J.I., Zoller, P. (2000). Entangling neutral atoms for quantum information processing. *J. Mod. Opt.* **47** (12), 2137–2149.

Castin, Y., Dalibard, J. (1997). Relative phase of two Bose–Einstein condensates. *Phys. Rev. A* **55** (6), 4330–4337.

Cataliotti, F.S., Burger, S., Fort, C., Maddaloni, P., Minardi, F., Trombettoni, A., Smerzi, A., Inguscio, M. (2001). Josephson junction arrays with Bose–Einstein condensates. *Science* **293**, 843–846. (5531).

Chin, C., Bartenstein, M., Altmeyer, A., Riedl, S., Jochim, S., Denschlag, J., Grimm, R. (2004). Observation of the pairing gap in a strongly interacting Fermi gas. *Science* **305**, 1128–1130. (5687).

Chu, S., Björkholm, J., Ashkin, A., Cable, A. (1986). Experimental observation of optically trapped atoms. *Phys. Rev. Lett.* **57**, 314.

Clark, S., Jaksch, D. (2004). Dynamics of the superfluid to Mott insulator transition in one dimension. *Phys. Rev. A* **70**, 043612.

Cohen-Tannoudji, C., Dupont-Roc, J., Grynberg, G. (1992). "Atom–Photon Interactions". Wiley-VCH, Berlin.

Cook, R. (1979). Atomic motion in resonant radiation: An application of Ehrenfest's theorem. *Phys. Rev. A* **20**, 224.

Courteille, P., Freeland, R., Heinzen, D., van Abeelen, F., Verhaar, B. (1998). Observation of a Feshbach resonance in cold atom scattering. *Phys. Rev. Lett.* **81**, 69.

Daley, A., Kollath, C., Schollwöck, U., Vidal, G. (2004). Time-dependent density-matrix renormalization-group using adaptive Hilbert spaces. *J. Stat. Mech.* P04005.

Damski, B., Santos, L., Tiemann, E., Lewenstein, M., Kotochigova, S., Julienne, P.S., Zoller, P. (2003). Creation of a dipolar superfluid in optical lattices. *Phys. Rev. Lett.* **90**, 110401.

Davis, K., Mewes, M., Andrews, M., van Druten, N., Durfee, D., Kurn, D., Ketterle, W. (1995). Bose–Einstein condensation in a gas of sodium atoms. *Phys. Rev. Lett.* **75**, 3969–3973.

deMarco, B., Jin, D. (1999). Onset of Fermi degeneracy in a trapped atomic gas. *Science* **285**, 1703–1706.

Dickerscheid, D., van Oosten, D., Denteneer, P.J.H., Stoof, H. (2003). Ultracold atoms in optical lattices. *Phys. Rev. A* **68**, 033606.

Duan, L.-M., Demler, E., Lukin, M. (2003). Controlling spin exchange interactions of ultracold atoms in an optical lattice. *Phys. Rev. Lett.* **91**, 090402.

Dunningham, J.A., Collett, M.J., Walls, D.F. (1998). Quantum state of a trapped Bose–Einstein condensate. *Phys. Lett. A* **245** (1–2), 49–54.

Dür, W., Raussendorf, R., Kendon, V., Briegel, H.J. (2002). Quantum random walks in optical lattices. *Phys. Rev. A* **66**, 052319.

Feynman, R. (1985). Quantum mechanical computers. *Opt. News* **11**, 11–20.

Feynman, R. (1986). Quantum mechanical computers. *Found. Phys.* **16**, 507–531.

Finkelstein, V., Berman, P., Guo, J. (1992). One-dimensional laser cooling below the Doppler limit. *Phys. Rev. A* **45**, 1829–1842.

Fisher, M.P.A., Weichman, P.B., Grinstein, G., Fisher, D.S. (1989). Boson localization and the superfluid–insulator transition. *Phys. Rev. B* **40** (1), 546–570.

Friebel, S., D'Andrea, C., Walz, J., Weitz, M. (1998). Co2-laser optical lattice with cold rubidium atoms. *Phys. Rev. A* **57**, R20.

Garcia-Ripoll, J., Cirac, J., Zoller, P., Kollath, C., Schollwock, U., von Delft, J. (2004). Variational ansatz for the superfluid Mott insulator transition in optical lattices. *Opt. Exp.* **12** (1), 42–54.

Gerbier, F., Widera, A., Fölling, S., Mandel, O., Gericke, T., Bloch, I. (2005). Phase coherence of an atomic Mott insulator. *Phys. Rev. Lett.* **95**, 050404.

Gordon, J., Ashkin, A. (1980). Motion of atoms in a radiation trap. *Phys. Rev. A* **21**, 1606.

Greenberger, D., Horne, M.A., Zeilinger, A. (1989). Going beyond Bell's theorem. In: Kafatos, M. (Ed.), "Bell's Theorem, Quantum Theory, and Conceptions of the Universe", Kluwer, Dordrecht, pp. 69–72.

Greiner, M., Bloch, I., Mandel, O., Hänsch, T.W., Esslinger, T. (2001). Exploring phase coherence in a 2d lattice of Bose–Einstein condensates. *Phys. Rev. Lett.* **87** (16), 160405.

Greiner, M., Mandel, O., Esslinger, T., Hänsch, T.W., Bloch, I. (2002). Quantum phase transition from a superfluid to a Mott insulator in a gas of ultracold atoms. *Nature* **415**, 39–44. (6867).

Grimm, R., Weidemüller, M., Ovchinnikov, Y. (2000). Optical dipole traps for neutral atoms. *Adv. At. Mol. Opt. Phys.* **42**, 95.

Hemmerich, A., Schropp, D., Esslinger, T., Hänsch, T. (1992). Elastic scattering of rubidium atoms by two crossed standing waves. *Europhys. Lett.* **18**, 391.

Hofstetter, W., Cirac, J.I., Zoller, P., Demler, E., Lukin, M.D. (2002). High-temperature superfluidity of fermionic atoms in optical lattices. *Phys. Rev. Lett.* **89**, 220407.

Imamoglu, A., Lewenstein, M., You, L. (1997). Inhibition of coherence in trapped Bose–Einstein condensates. *Phys. Rev. Lett.* **78** (13), 2511–2514.

Inouye, S., Andrews, M.R., Stenger, J., Miesner, H.-J., Stamper-Kurn, D.M., Ketterle, W. (1998). Observation of Feshbach resonances in a Bose–Einstein condensate. *Nature* **392**, 151–154.

Jaksch, D. (1999). Bose–Einstein condensation and applications. Ph.D. Thesis, Leopold-Franzens-University of Innsbruck.

Jaksch, D., Bruder, C., Cirac, J.I., Gardiner, C.W., Zoller, P. (1998). Cold bosonic atoms in optical lattices. *Phys. Rev. Lett.* **81** (15), 3108–3111.

Jaksch, D., Briegel, H.J., Cirac, J.I., Gardiner, C.W., Zoller, P. (1999). Entanglement of atoms via cold controlled collisions. *Phys. Rev. Lett.* **82** (9), 1975–1978.

Jaksch, D., Cirac, J., Zoller, P., Rolston, S., Côté, R., Lukin, M. (2000). Fast quantum gates for neutral atoms. *Phys. Rev. Lett.* **85**, 2208–2211.

Jane, E., Vidal, G., Dür, W., Zoller, P., Cirac, J.I. (2003). Simulation of quantum dynamics with quantum optical system. *Quantum Information and Computation* **3** (1), 15–37.

Jessen, P., Deutsch, I.H. (1996). Optical lattices. *Adv. At. Mol. Opt. Phys.* **37**, 95–139.

Kashurnikov, V.A., Prokof'ev, N.V., Svistunov, B. (2002). Revealing the superfluid–Mott insulator transition in an optical lattice. *Phys. Rev. A* **66**, 031601(R).

Kastberg, A., Phillips, W., Rolston, S.L., Spreeuw, R. (1995). Adiabatic cooling of cesium to 700 nK in an optical lattice. *Phys. Rev. Lett.* **74**, 1542–1545.

Kazantsev, A. (1973). Acceleration of atoms by a resonance field. *Soviet Phys. JETP* **36**, 861.

Kinoshita, T., Wenger, T., Weiss, D. (2004). Observation of a one-dimensional Tonks–Girardeau gas. *Science* **305**, 1125–1128.

Köhl, M., Moritz, H., Stöferle, T., Günter, K., Esslinger, T. (2004). Fermionic atoms in a 3d optical lattice: Observing Fermi-surfaces, dynamics and interactions. *Phys. Rev. Lett.* **94**, 080403.

Kollath, C., Schollwöck, U., von Delft, J., Zwerger, W. (2004). Spatial correlations of trapped one-dimensional bosons in an optical lattice. *Phys. Rev. A* **69** (3), 031601.

Kuklov, A., Svistunov, B. (2003). Counterflow superfluidity of two-species ultracold atoms in a commensurate optical lattice. *Phys. Rev. Lett.* **90**, 100401.

Kuklov, A., Prokof'ev, N.V., Svistunov, B. (2004). Commensurate two-component bosons in an optical lattice: Ground state phase diagram. *Phys. Rev. Lett.* **92**, 050402.

Laburthe-Tolra, B., O'Hara, K., Huckans, J., Phillips, W., Rolston, S., Porto, J. (2004). Observation of reduced three-body recombination in a correlated 1d degenerate Bose gas. *Phys. Rev. Lett.* **92** (19), 190401.

Lewenstein, M., Santos, L., Baranov, M.A., Fehrmann, H. (2004). Atomic Bose–Fermi mixtures in an optical lattice. *Phys. Rev. Lett.* **92**, 050401.

Lye, J., Fallani, L., Modugno, M., Wiersma, D., Fort, C., Inguscio, M. (2004). A Bose–Einstein condensate in a random potential. cond-mat/0412167.

Mandel, O., Greiner, M., Widera, A., Rom, T., Hänsch, T.W., Bloch, I. (2003). Coherent transport of neutral atoms in spin-dependent optical lattice potentials. *Phys. Rev. Lett.* **91**, 010407.

Metcalf, H., van der Straten, P. (1999). "Laser Cooling and Trapping". Springer, New York.

Milburn, G.J., Corney, J., Wright, E.M., Walls, D.F. (1997). Quantum dynamics of an atomic Bose–Einstein condensate in a double-well potential. *Phys. Rev. A* **55** (6), 4318–4324.

Moritz, H., Stöferle, T., Köhl, M., Esslinger, T. (2003). Exciting collective oscillations in a trapped 1d gas. *Phys. Rev. Lett.* **91**, 250402.

Morsch, O., Oberthaler, M. (2005). Rev. Mod. Phys., submitted for publication.

Narozhny, N.B., Sanchez-Mondragon, J.J., Eberly, J.H. (1981). Coherence versus incoherence: collapse and revival in a simple quantum model. *Phys. Rev. A* **23** (1), 236–247.

Ott, H., De Mirandes, E., Ferlaino, F., Roati, G., Modugno, G., Inguscio, M. (2004). Collisionally induced transport in periodic potentials. *Phys. Rev. Lett.* **92**. (160601).

Paredes, B., Cirac, J.I. (2003). From Cooper pairs to Luttinger liquids with bosonic atoms in optical lattices. *Phys. Rev. Lett.* **90**, 150402.

Paredes, B., Widera, A., Murg, V., Mandel, O., Fölling, S., Cirac, J.I., Shlyapnikov, G., Hänsch, T.W., Bloch, I. (2004). Tonks–Girardeau gas of ultracold atoms in an optical lattice. *Nature* **429**, 277–281.

Pethick, C., Smith, H. (2001). "Bose–Einstein Condensation in Dilute Gases". Cambridge Univ. Press, Cambridge.

Petsas, K., Coates, A., Grynberg, G. (1994). Crystallography of optical lattices. *Phys. Rev. A* **50**, 5173–5189.

Pitaevskii, L., Stringari, S. (2003). "Bose–Einstein Condensation". *International Series of Monographs on Physics.* Oxford Univ. Press, Oxford.

Porto, J., Rolston, S., Tolra, B., Williams, C., Phillips, W. (2003). Quantum information with neutral atoms as qubits. *Phil. Trans. Roy. Soc. A* **361**, 1417–1427. (1808).

Rabl, R., Daley, A., Fedichev, P., Cirac, J.I., Zoller, P. (2003). Defect-suppressed atomic crystals in an optical lattice. *Phys. Rev. Lett.* **91** (11), 110403.

Raithel, G., Phillips, W., Rolston, S. (1998). Collapse and revival of wave packets in optical lattices. *Phys. Rev. Lett.* **81**, 3615–3618.

Raussendorf, R., Briegel, H.J. (2001). A one-way quantum computer. *Phys. Rev. Lett.* **86** (22), 5188–5191.

Raussendorf, R., Browne, D., Briegel, H. (2003). Measurement-based quantum computation on cluster states. *Phys. Rev. A* **68**, 022312.

Regal, C., Greiner, M., Jin, D. (2004). Observation of resonance condensation of fermionic atom pairs. *Phys. Rev. Lett.* **92**, 040403.

Rempe, G., Walther, H., Klein, N. (1987). Observation of quantum collapse and revival in a one-atom maser. *Phys. Rev. Lett.* **58** (4), 353–356.

Roati, G., de Mirandes, E., Ferlaino, F., Ott, H., Modugno, G., Inguscio, M. (2004). Atom interferometry with trapped Fermi gases. *Phys. Rev. Lett.* **92**, 230402.

Roth, R., Burnett, K. (2003a). Phase diagram of bosonic atoms in two-color superlattices. *Phys. Rev. A* **68**, 023604.

Roth, R., Burnett, K. (2003b). Superfluidity and interference pattern of ultracold bosons in optical lattices. *Phys. Rev. A* **67**, 031602.

Roth, R., Burnett, K. (2003c). Ultracold bosonic atoms in two-color disordered optical superlattices. *J. Opt. B.*

Roth, R., Burnett, K. (2004). Quantum phases of atomic boson–fermion mixtures in optical lattices. *Phys. Rev. A* **69**, 021601(R).

Sachdev, S. (1999). "Quantum Phase Transitions". Cambridge Univ. Press, Cambridge.

Sachdev, S., Sengupta, K., Girvin, S. (2002). Mott insulators in strong electric fields. *Phys. Rev. B* **66**, 075128.

Saleh, B., Teich, M. (1991). "Fundamentals of Photonics". Wiley, New York.

Sanpera, A., Kantian, A., Sanchez-Palencia, L., Zakrzewski, J., Lewenstein, M. (2004). Atomic Fermi–Bose mixtures in inhomogeneous and random lattices: From Fermi glass to quantum spin glass and quantum percolation. *Phys. Rev. Lett.* **93**, 040401.

Santos, L., Baranov, M.A., Cirac, J.I., Everts, H., Fehrmann, H., Lewenstein, M. (2004). Atomic quantum gases in Kagomé lattices. *Phys. Rev. Lett.* **93**, 030601.

Schori, C., Stöferle, T., Moritz, H., Köhl, M., Esslinger, T. (2004). Excitations of a superfluid in a three-dimensional optical lattice. *Phys. Rev. Lett.* **93**, 240402.

Schreck, F., Khaykovich, L., Corwin, K., Ferrari, G., Bourdel, T., Cubizolles, J., Salomon, C. (2001). A quasipure Bose–Einstein condensate immersed in a Fermi sea. *Phys. Rev. Lett.* **87**, 080403.

Sheshadri, K., Krishnamurthy, H.R., Pandit, R., Ramakrishnan, T.V. (1993). Superfluid and insulating phases in an interacting-boson model: mean-field theory and the rpa. *Europhys. Lett.* **22** (4), 257–263.

Sols, F. (1994). Randomization of the phase after the suppression of the Josephson coupling. *Physica B* **194–196**, 1389–1390.

Sorensen, A., Molmer, K. (1999). Spin–spin interaction and spin squeezing in optical lattices. *Phys. Rev. Lett.* **83**, 2274–2277.

Stöferle, T., Moritz, H., Schori, C., Köhl, M., Esslinger, T. (2004). Transition from a strongly interacting 1d superfluid to a Mott insulator. *Phys. Rev. Lett.* **92**, 130403.

Takekoshi, T., Yeh, J., Knize, R. (1995). *Opt. Comm.* **114**, 421.

Truscott, A., Strecker, K., McAlexander, W., Partridge, G., Hulet, R. (2001). Observation of Fermi pressure in a gas of trapped atoms. *Science* **291**, 2570–2572.

van Oosten, D., van der Straten, P., Stoof, H. (2001). Quantum phases in an optical lattice. *Phys. Rev. A* **63**, 053601.

van Oosten, D., Dickerscheid, D., Farid, B., van der Straten, P., Stoof, H. (2005). Inelastic light scattering from a Mott insulator. *Phys. Rev. A* **71**, 021601.

Wessel, S., Alet, F., Troyer, M., Batrouni, G. (2004). Quantum Monte Carlo simulations of confined bosonic atoms in optical lattices. *Phys. Rev. A* **70**, 053615.

Wright, E.M., Walls, D.F., Garrison, J.C. (1996). Collapses and revivals of Bose–Einstein condensates formed in small atomic samples. *Phys. Rev. Lett.* **77** (11), 2158–2161.

Wright, E.M., Wong, T., Collett, M.J., Tan, S.M., Walls, D.F. (1997). Collapses and revivals in the interference between two Bose–Einstein condensates formed in small atomic samples. *Phys. Rev. A* **56** (1), 591–602.

Yurke, B., Stoler, D. (1986). Generating quantum mechanical superpositions of macroscopically distinguishable states via amplitude dispersion. *Phys. Rev. Lett.* **57** (1), 13–16.

Zwerger, W. (2003). Mott–Hubbard transition of cold gases in an optical lattice. *J. Opt. B* **5**, S9–S16.

Zwierlein, M., Stan, C., Schunck, C., Raupach, S., Kerman, A., Ketterle, W. (2004). Condensation of pairs of fermionic atoms near a Feshbach resonance. *Phys. Rev. Lett.* **92** (12), 120403.

THE KICKED RYDBERG ATOM

F.B. DUNNING[1], J.C. LANCASTER[1], C.O. REINHOLD[2,3], S. YOSHIDA[4] and
J. BURGDÖRFER[3,4]

[1]Department of Physics and Astronomy, and the Rice Quantum Institute, Rice University, MS 61,
6100 Main Street, Houston, TX 77005, USA

[2]Physics Division, Oak Ridge National Laboratory, Oak Ridge, TN 37831-6372, USA

[3]Department of Physics, University of Tennessee, Knoxville, TN 37996-1200, USA

[4]Institute for Theoretical Physics, Vienna University of Technology, A-1040 Vienna, Austria

Abstract

Recent developments in experimental technique allow study of the ionization and excitation of Rydberg atoms by pulsed unidirectional electric fields, termed half-cycle pulses (HCPs), with duration $T_p \ll T_n$, where T_n is the classical electron orbital period. In this limit, each HCP simply delivers an impulsive momentum transfer or "kick" to the Rydberg electron. The opportunities provided by this are discussed with the aid of both theory and experiment. Single HCPs can be used to

ISSN 1049-250X
DOI 10.1016/S1049-250X(05)52002-0

probe the momentum (and spatial) distribution of the excited electron. Atoms subject to a train of periodic HCPs provide an experimental realization of the kicked atom, a valuable test bed for the study of nonlinear dynamics in Hamiltonian systems. Studies of kicked atoms have furnished new insights into classical–quantum correspondence as well as a variety of phenomena such as dynamical stabilization and quantum localization which illuminate how the classical world emerges from the quantum world. Dynamical stabilization leads to strong transient periodic localization of the excited electron in phase-space and creation of periodic nondispersive wavepackets. The ability to manipulate the electron in phase-space can be extended by modulating the amplitude and frequency of the HCPs. Strong transient phase-space localization can also be generated using quasi-one-dimensional Rydberg atoms and a single HCP. Such localization provides new opportunities for control of the atomic wavefunction, i.e., for atomic engineering, by application of a carefully tailored sequence of HCPs to produce, for example, wavepackets whose behavior mimics that of a localized classical particle. Studies of the kicked Rydberg atom find application in a number of fields including fast ion–atom collisions, ion channeling in solids and exploring the behavior of atoms in ultra-fast, ultra-intense pulsed laser fields.

1. Introduction

1.1. IMPULSIVELY-DRIVEN OR "KICKED" SYSTEMS

Recent advances in experimental technique now allow study of the behavior of impulsively driven or "kicked" systems in which the duration of each impulse is short compared to the period of the unperturbed system. This has provided valuable new approaches for controlling and manipulating the state of a system as well as a test-bed for the study of nonlinear dynamics in Hamiltonian systems. Although systems subject to sinusoidal perturbations have been widely studied, for example, Rydberg atoms in a microwave field [1,2], experimental realizations of kicked systems are few. Two widely discussed impulsively-driven dynamical systems are the kicked rotor (or standard map) and the kicked hydrogen atom. Recently atoms cooled in a standing light wave have provided a realization of the one-dimensional standard map, demonstrating the existence of Anderson localization in the translational motion of a single atom [3–6]. The kicked atom has also been realized experimentally by exposing very-high-n Rydberg atoms to a sequence of unidirectional electric field pulses, termed half-cycle pulses (HCPs), whose duration T_p is short compared to the classical electron orbital period T_n [7–11]. In the limit that $T_p \ll T_n$, a single pulse $\vec{F}_{\mathrm{HCP}}(t)$ simply delivers an impulsive momentum transfer or "kick"

$$\Delta \vec{p} = -\int\limits_{0}^{\infty} \vec{F}_{\text{HCP}}(t)\, dt \qquad (1)$$

to the excited electron. (Unless otherwise noted atomic units are used throughout this article.) In the present review we focus on the kicked atom and the new insights it has provided into phenomena like dynamical stabilization and quantum localization as well as the new opportunities it affords for atomic engineering, i.e., for manipulating atomic wavefunctions, by application of carefully-tailored sequences of HCPs.

Kicked atoms provide a means to create atomic wavepackets whose behavior mimics that of a localized classical particle thereby providing an opportunity to examine classical–quantum correspondence. Typically, because of the nonlinearity of the atomic Hamiltonian, any localized wavepacket that initially follows the motion of the corresponding classical particle must spread out in time. This effect is classical in origin and results because the component states have different energies and must evolve at different rates. This spreading can be overcome by driving the system with a periodic external perturbation which results in a time-dependent Hamiltonian. For certain driving frequencies the internal motion in the atom becomes locked to the external drive. In general, for periodic driving, the Floquet theorem can be used which requires, by analogy to the Bloch theorem for potentials periodic in space, that the solution to the time-dependent Schrödinger equation be given by a linear combination of time-periodic functions, the Floquet eigenstates. If the system can be placed in a Floquet state that is well localized at a given phase of the driving field, this will be recovered every period of the drive leading to creation of a nondispersive localized wavepacket. As discussed in a recent review [12], such behavior has been studied extensively using hydrogen Rydberg atoms driven by linearly- or circularly-polarized microwave fields [13,14]. (In the latter case, in a frame rotating with the microwave field, the Hamiltonian becomes time-independent and the electron is trapped near the local minimum in the potential creating localized so-called Trojan states.) The periodically kicked atom provides new opportunities for examining the response of atoms to periodic perturbations and, through dynamical filtering, to obtain strong transient localization of the electron in phase-space. This ability to manipulate the electron in phase-space can be further extended by modulating the frequency and amplitude of the kicks. Interestingly, strong transient phase-space localization can also be generated using a single kick. Such localization provides a valuable starting point for further control of the atomic wavefunction.

1.2. RELATED PROBLEMS

Studies of the kicked atom provide new opportunities to address questions of interest in a variety of other areas. For example, during fast ion–atom collisions the atom is subject to an impulsive electric field produced by the ion that can lead to excitation and ionization. This field can be described as the superposition of a HCP directed perpendicular to the ion trajectory and a single cycle sine-like pulse directed parallel to the ion trajectory. In collision experiments it is not possible to control independently the relative collision velocity and the impact parameter. In contrast, it is possible to simulate such collisions in fine detail by applying electric field pulses of appropriately tailored duration, amplitude and direction [15,16]. During the passage of a fast ion/atom through a solid the outer electrons in the ion/atom are subject to a train of impulsive electric field pulses due to their interaction with electrons and heavy particles in the solid [17–21]. If the electron orbital period is much longer than a typical binary collision time inside the solid, the result of each elastic or inelastic collision can be reduced to an impulsive momentum transfer to the electron. Thus, the problem becomes equivalent to the simplified problem of an excited atomic state subject to a train of impulsive momentum transfers, i.e., a multiply kicked atom. For transmission through amorphous solids the train of impulses is indeterministic (noisy) and the electron undergoes a random walk through state space which ultimately leads to incoherent excitation and to ionization. For planar or axial channeling of fast ions/atoms through crystals the sequence of impulses is partially periodic and leads to coherent electronic transitions [22,23].

With the development of high-intensity ultrashort pulse laser systems and high harmonic generation techniques considerable interest is now focused on the behavior of atoms in ultrafast, ultraintense pulsed laser fields with timescales extending into the attosecond regime and field strengths sufficient to dominate the electron motion. Recently, it has been suggested that it might even be feasible to produce trains of attosecond-duration HCPs by appropriately combining harmonics generated through high harmonic conversion using a two-color laser field [24]. A fundamental difference exists between pulses obtained in high harmonic generation and those produced using pulse generators: high harmonic pulses are freely propagating electromagnetic waves and the time integral of the electric field must vanish [25]. This, however, does not preclude the possibility of achieving significant momentum transfer to an electron in an atom because it responds nonlinearly to strong external fields. For such electrons it is the large field near the peak of the HCPs that is important, rather than the integral over the entire temporal field distribution. As will be shown, it is possible using very-high-n Rydberg atoms ($n \sim 350$–400), to obtain detailed insights into the behavior of atoms when subject to such pulses using more modest fields and nanosecond timescales.

In the following we first discuss the experimental realization of the impulsive limit using Rydberg atoms. The response of an atom to a single HCP is then described together with the use of HCPs to characterize atomic states and to monitor the spatial and temporal evolution of wavepackets. The behavior of atoms subject to a train of impulsive HCPs is next addressed, focusing on effects such as dynamical stabilization and quantum localization, and the factors that govern these behaviors. Finally, the control and manipulation of atomic wavefunctions by carefully tailored sequences of HCPs is described with emphasis on the role of transient phase-space localization. Possible future research directions and opportunities are highlighted.

2. Realization of the Impulsive Limit

The approaches currently used to access the impulsive limit take advantage of the fact that the classical electron orbital period in an atom, given by $T_n = 2\pi n^3$, increases rapidly with principal quantum number n. For $n \sim 30$ the orbital period amounts to a few picoseconds but increases to a few nanoseconds by $n \sim 300$. Freely-propagating HCPs with durations of a picosecond or less can be generated using a photoconducting switch triggered by a femtosecond laser [26–29] allowing the impulsive limit to be reached using Rydberg atoms with relatively small values of n, $n \sim 30$. Such atoms are straightforward to produce and are not strongly perturbed by external stray fields. One difficulty in using freely-propagating HCPs is that they are not inherently truly unidirectional. Rather they comprise a short intense unipolar electric field pulse followed by a weaker pulse of opposite polarity and much longer duration. The resulting pulse shape has been referred to as an "asymmetric monocycle" and the effects of the reverse polarity pulse must be considered in analyzing data [25]. Significant momentum transfer to an excited Rydberg electron can be achieved, however, because of the very different timescales associated with the initial unipolar pulse and the ensuing opposite-polarity tail. True unipolar HCPs can be created by applying voltage pulses from a fast pulse generator(s) to an electrode. This approach has the advantage that by using a number of pulse generators in conjunction with pulse combiners and splitters it is possible to engineer complex pulse sequences that can be measured directly using a fast probe and sampling oscilloscope. The minimum pulse durations that can be generated, however, are typically limited to $\gtrsim 0.5$–1.0 ns meaning that the true impulsive limit can only be reached for atoms with $n \gtrsim 300$–400.

2.1. Experimental Apparatus

2.1.1. Freely-Propagating Half-Cycle Pulses

Apparatus typical of those used to study kicked atoms with freely-propagating HCPs is shown in Fig. 1 [30]. The HCPs are generated by illuminating a biased

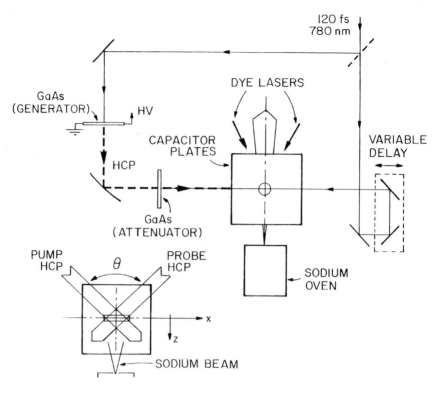

FIG. 1. Schematic diagram of apparatus used in studies with freely-propagating HCPs. The inset shows the arrangement employed in pump-probe experiments (see text).

photoconducting GaAs wafer with a 120 fs, 780 nm pulse from a mode locked Ti:Sapphire laser. The HCP generated by the resulting current pulse is transmitted through the wafer and collected with a parabolic mirror which directs it onto the Rydberg atoms. The peak field in the HCP increases linearly with the bias voltage across the wafer and can be varied continuously from 0 to ~50 kV cm^{-1}. The duration of the initial unipolar pulse is ~0.5 ps. Before reaching the interaction region, the HCP passes through a second GaAs wafer that is used as an optical gate. Illumination of this (unbiased) wafer with a fs-laser pulse transforms the wafer surface into a conductor greatly reducing its transmission. Thus by exciting the wafer at the appropriate time the amplitude of the opposite-polarity tail can be significantly attenuated with respect to the initial main HCP.

The HCP interacts with the Rydberg atoms in an interaction region defined by two parallel plates. The Rydberg atoms are created by photoexciting sodium atoms contained in a thermal-energy beam. The atoms are excited through the $3p_{1/2}$ level to an ns or nd state using a pair of pulsed dye lasers. Following exposure to the HCPs the number and excited state distribution of the surviving atoms is determined using selective field ionization (SFI). For this a slowly varying negative voltage ramp is applied to the lower interaction region electrode. Electrons produced by SFI are directed through a small hole in the upper interaction region plate and detected by a microchannel plate. Because atoms in different Rydberg states ionize at different applied fields, measurements of the field ionization signal as a function of time, i.e., of applied field, provide a measure of the excited-state distribution of those atoms present at the time of application of the ramp.

An interesting variant of the apparatus allows pump/probe experiments to monitor, for example, wavepacket evolution [31]. The principle of this technique is illustrated in the inset in Fig. 1. The pump and probe HCPs cross in the interaction region at some angle θ. Given the small longitudinal spatial extent of the HCP ($\sim$150 μm), the relative pump-probe delay varies linearly along the x axis indicated, which is perpendicular to the bisector of the angle between the propagating beams. The difference in the relative pulse delay between two points along this axis separated by a distance d is $\Delta t = 2d \sin(\theta/2)/c$. Thus by using a broad atom beam and creating a ribbon of Rydberg atoms aligned along the x axis, a range of time delays can be explored in a single shot, the exact time delay experienced by an atom depending on its location along the x axis. Ions created directly by the HCPs or by subsequent SFI are extracted through a slit cut in the upper interaction region electrode and imaged using a microchannel plate and phosphor screen. Analysis of the resulting image then provides the time evolution of the initial wavepacket. The time sweep corresponding to the width of the detector can be controlled by changing θ, the maximum being $\sim$100 ps for counterpropagating beams.

2.1.2. Studies at Very-High n

Measurements at very high n, $n \gtrsim 350$, present a number of challenges because the oscillator strengths for photoexcitation to high n states are small and because the Rydberg levels are closely spaced in energy. In consequence, excitation requires the use of narrow-linewidth, frequency-stabilized lasers. Furthermore, the excited electron is so far from the core ion that its motion can be strongly perturbed or even dominated by small stray fields requiring that these be reduced to very low values. These difficulties have been overcome using the apparatus in Fig. 2 [9]. Very-high-n Rydberg atoms are created by photoexciting potassium (or rubidium) atoms contained in a tightly-collimated thermal-energy beam to selected np states with $n \sim 350$–400 using the crossed output beam from an extracavity-doubled CW Rh6G dye laser. The output frequency is stabilized to $\lesssim$2 MHz day^{-1} using a scanning Fabry–Perot etalon and polarization-stabilized HeNe laser [32]. Excitation occurs near the center of an interaction region bounded by three pairs of large copper electrodes, each 10 cm × 10 cm. The use of

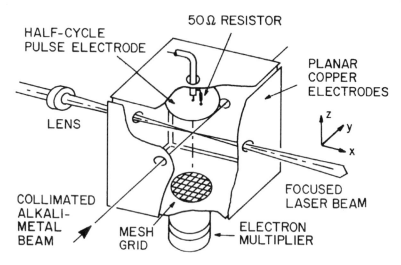

FIG. 2. Schematic diagram of apparatus employed in studies at very-high-*n*.

large electrodes well separated from the interaction region minimizes the effect of stray patch fields associated with nonuniformities in the electrode surfaces. Stray fields in the experimental volume are further reduced (to $\lesssim 50$ μV cm^{-1}) by application of small bias potentials to these electrodes which are determined using a technique based on the Stark effect [33]. (Mu-metal shields are used to reduce the stray magnetic field to $\lesssim 30$ mG.) The HCPs are generated by applying voltage pulses to a circular copper electrode mounted, together with a 50 Ω terminating resistor, on the end of a section of semi-rigid coaxial cable. This arrangement reduces the stray capacitance of the HCP electrode allowing fast pulse rise times (~ 200 ps) to be obtained. If required, small bias potentials can also be applied to this electrode to establish offset fields in the experimental volume.

Experiments are conducted in a pulsed mode. The laser output is formed into a train of pulses of ~ 1 μs duration and ~ 20 kHz repetition frequency using an acousto-optic modulator. (The probability that a Rydberg atom is formed during any laser pulse is small and data must be accumulated following many laser pulses.) Immediately following each laser pulse the atoms are subject to one or more HCPs. The surviving Rydberg atoms are analyzed using SFI. A positive voltage ramp is applied to the lower interaction region electrode and electrons resulting from field ionization are accelerated to, and detected by, a particle multiplier. Measurements in which no HCPs are applied are interspersed at routine intervals during data acquisition to monitor the number of Rydberg atoms initially created. The Rydberg atom survival probability is then determined by taking the ratio of the signals observed with and without application of the pulsed fields.

2.2. CREATION OF QUASI-ONE-DIMENSIONAL ATOMS

Photoexcitation of alkali atoms in zero field typically leads to creation of low-$|m|$ ns, np or nd states. However, much interest focuses on one-dimensional (1D) or at least quasi-1D atoms because their properties are straightforward to analyze theoretically, which facilitates comparison between the predictions of classical and quantum calculations. Furthermore, while they display much of the essential physics present in three dimensions they provide an opportunity to observe effects not seen with 3D atoms. As has been demonstrated by a number of workers [15, 28,34–36], quasi-1D atoms can be created by exciting the extreme members of a low-$|m|$, high-n Stark manifold in a small dc field. As illustrated in Fig. 3, which shows the calculated probability density distribution [37] for the extreme $n_2 = 0$ state in the hydrogenic $n = 100$, $m = 0$ manifold, such states have large permanent electric dipole moments and are approximately one-dimensional.

Figure 4 illustrates the essential features of the Stark energy level structure for potassium in the vicinity of $n = 50$ calculated by diagonalizing the Hamiltonian

$$H = H_{at} + F_{DC}z \tag{2}$$

where H_{at} is the field-free atomic Hamiltonian, F_{DC} is the strength of the applied field oriented along the z coordinate of the electron [36]. The Stark structure is complicated by the large quantum defects associated with the low-ℓ states and by the appearance of avoided crossings wherever neighboring levels approach. The

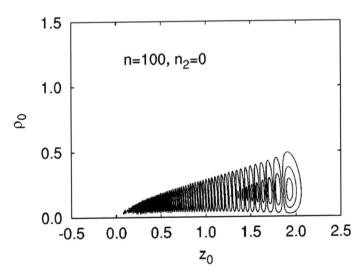

FIG. 3. Calculated probability density distribution in cylindrical polar coordinates for the extreme $n_2 = 0$ state in the H($n = 100$), $m = 0$ Stark manifold.

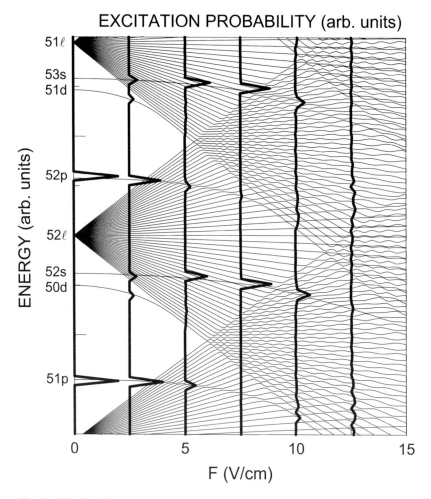

FIG. 4. Calculated Stark energy level structure for K($m = 0$) states in the vicinity of $n = 50$. Also shown is the probability for photoexcitation of these levels from the ground state (—) by a laser with a linewidth $\Delta\omega_L = 10^{-3}/n^2$ for several values of applied field.

local level structure evident in Fig. 4 is characteristic of that to be expected at all n. In particular, the positions of the low-ℓ s, p, and d states relative to their adjacent Stark manifolds remain invariant because their n-scaling parallels that of the n manifolds. The s and p states have large quantum defects ($\mu_s = 2.19$, $\mu_p = 1.71$) and, as the applied field is increased, initially encounter states in neighboring Stark manifolds of *different* n. In contrast, d states have a small quantum defect ($\mu_d = 0.25$) and first encounter Stark states of the *same* n. Given that the coupling is strongest between states of the same n, this leads to marked differences in behavior at the ensuing avoided crossings. As a d state approaches its neighboring manifold, a broad avoided crossing leads to strong mixing with the extreme lowest-energy, redshifted, "downhill" Stark state. In contrast, s and p states couple weakly to their neighboring Stark states resulting in a series of narrow avoided crossings. These states thus preserve more of their initial character. Figure 4 also includes calculated probabilities for photoexcitation of the various levels from the ground state. For $F_{DC} = 0$ excitation is limited exclusively to p states because of the dipole selection rule. As F_{DC} increases, other states, as well as the "fan" of quasi-hydrogenic Stark states, acquire p content. In particular, at fields where states from neighboring manifolds first begin to cross a significant part of the p dipole coupling has been transferred to states in the vicinity of the zero-field s and d levels. Since the latter contains a significant admixture of strongly polarized states from the lower edge of the neighboring Stark manifold, the photoexcitation of quasi-1D elongated states becomes possible.

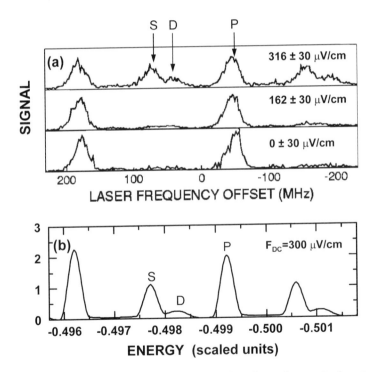

FIG. 5. (a) Measured excitation spectra for photoexcitation of ground-state potassium atoms into $m = 0$ Stark states in the vicinity of $n \sim 350$ for the values of applied dc field F_{dc} indicated. (b) Calculated excitation spectrum for $F_{dc} = 300\ \mu V\ cm^{-1}$ and an effective laser linewidth of 10 MHz.

Probabilities for the photoexcitation of ground-state potassium atoms into $m = 0$ Stark states in the vicinity of $n = 350$ measured using the apparatus in Fig. 2 are shown in Fig. 5 for three values of applied field [36]. (The 350d state merges with the $n = 350$ Stark manifold at a field $F_{DC} \sim 300 \ \mu V \ cm^{-1}$.) In the presence of a field three distinct peaks labeled S, P and D appear in the excitation spectrum that are associated with the zero-field s, p and d levels but represent Stark mixtures of several substates. Inspection of Fig. 4 shows that the peaks S and P result from excitation of a small number of adjacent states, each of which has a large oscillator strength, whereas peak D is associated with the excitation of a relatively broad range of states each having only a small oscillator strength. (The effective laser linewidth, ~ 10 MHz, spans a range of Stark states.) Calculations indicate that those states that make up the peaks S and P are not strongly polarized, whereas those that make up the peak D are strongly polarized, i.e., strongly quasi-1D, having dipole moments similar to those of the corresponding elongated hydrogenic states. The quasi-1D nature of these states has been investigated experimentally through measurements of their ionization in pulsed electric fields with durations $T_p \sim T_n$ applied parallel and antiparallel to the atomic axis [36]. Only small asymmetries in the ionization characteristics were observed with the laser tuned to the S and P features indicating that these states are only weakly polarized. Pronounced asymmetries, however, were evident with the laser tuned to the D feature confirming that its constituent states are indeed quasi-one-dimensional.

3. Effect of a Single HCP

3.1. Energy Transfer and Ionization

Consider initially the application of a single HCP to an atom that delivers an impulsive momentum transfer $\Delta \vec{p}$. Classically the application of such a HCP to an electron with momentum $\vec{p}_i$ and energy E_i results in an energy transfer

$$\Delta E = E_f - E_i = \left(\frac{\vec{p}_i + \Delta \vec{p}}{2} \right)^2 - \frac{p_i^2}{2} = \frac{\Delta p^2}{2} + \vec{p}_i \cdot \Delta \vec{p} \tag{3}$$

where E_f is the final electron energy. Taking the z axis to be parallel to the impulse delivered by the HCP, i.e., $\Delta \vec{p} = \Delta p_z \hat{k}$, this may be written

$$E_f = E_i + \frac{\Delta p_z^2}{2} + p_{zi} \Delta p_z. \tag{4}$$

If $E_f > 0$ ionization occurs. If $E_f < 0$, i.e., if

$$p_{zi} \Delta p_z < -E_i - \frac{\Delta p_z^2}{2} \tag{5}$$

the electron remains bound and the atom survives in a new final n state. Given the distribution of initial electron momenta associated with a classical orbit, application of a HCP leads to population of a range of n states centered about a final energy $E_f = E_i + \Delta p_z^2/2$, i.e., a final n state $n_f = 1/\sqrt{-2E_f}$. A typical final-state energy distribution, calculated using Classical Trajectory Monte Carlo (CTMC) techniques, is shown in Fig. 6 [38]. Here, and in many later figures, scaled units (denoted by the subscript 0) are used to explicitly exploit the scaling invariance of the classical dynamics of Rydberg Coulomb orbits. Classically, Rydberg atoms of different energy become equivalent to each other (frequently referred to as "mechanical similarity") when the following scaled units are employed: for coordinates $\vec{r}_0 \equiv \vec{r}/n^2$ ($z_0 \equiv z/n^2$), for momentum $\vec{p}_0 \equiv n\vec{p}$, for momentum transfer or impulse $\Delta \vec{p}_0 \equiv n\Delta\vec{p}$, for angular momentum $\vec{L}_0 = \vec{L}/n$, for time $t_0 = t/2\pi n^3$, for frequency $\nu_0 = 2\pi n^3\nu$, for energy $E_0 = n^2 E$, and for an applied field $\vec{F}_0 = n^4\vec{F}$. Using these scaled variables and a scaled Hamiltonian $H_0(\vec{r}_0, \vec{p}_0, t_0) \equiv n^2 H(\vec{r}, \vec{p}, t)$ the classical equations of motion are invariant and independent of n.

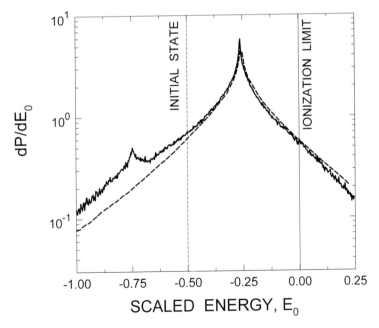

FIG. 6. Calculated final state energy distribution following application of a HCP to H(400p) atoms for (- - -) an ultrashort HCP ($T_p/T_n = 0$) and (—) a HCP of 1 ns duration ($T_p/T_n \sim 0.1$). The scaled momentum transfer is $\Delta p_0 = n\Delta p = 0.7$. The energy axis is expressed in scaled units $E_0 = n^2 E$.

The results in Fig. 6 pertain to application of a scaled impulse $\Delta p_0 = 0.7$ to a H(400p) atom, and calculations are included for both an ultrashort HCP ($T_0 \equiv T_p/T_n = 0$), i.e., a δ-function impulse, and for a HCP of 1 ns duration ($T_0 = T_p/T_n \sim 0.1$) more typical of those used experimentally. As expected, the final state distribution is peaked about the value $E_f = E_i + \Delta p^2/2$, which corresponds to a final n value $n_f \sim 560$. Its "width" encompasses a range of n values Δn that amounts to $\Delta n \sim 90$. The final distribution depends somewhat on the HCP duration. For HCPs of finite duration a small secondary peak is evident in the final-state distribution at energies corresponding to $n_f < n_i$ and the likelihood of direct ionization, i.e., production of states with $E_f > 0$, is somewhat reduced.

Initial studies of the kicked atom focused on their ionization [15,39–49]. The ionization of Rydberg atoms in "slow" pulsed electric fields whose risetime and width are much greater than electron orbital period has been extensively investigated. As shown in the inset in Fig. 7, application of a field $\vec{F} = F_z \hat{k}$ leads to the appearance of a saddle point, E_B, in the electron potential on the $-z$ axis. If the electron energy lies close to or above that of the saddle point ionization can occur from tunneling through the resulting potential barrier or from over-barrier escape [50–52]. The field required to induce such ionization depends on the initial electron energy and decreases rapidly with increasing n_i, scaling as $1/n_i^4$. In contrast, application of a HCP with $T_p \ll T_{n_i}$ lowers the Coulomb barrier only momentarily and ionization requires that the energy transfer to the electron (Eq. (3)) be greater than its initial binding energy E_{n_i} ($= -1/2n_i^2$). Consequently, the n_i scaling of the ionization threshold crosses over from the $1/n_i^4$ dependence characteristic of the adiabatic limit to a $1/n_i$ dependence in the limit of ultrashort HCPs. Initial experiments [39] using Na(np) Rydberg atoms with $13 \lesssim n_i \lesssim 35$ and very short ($T_p \sim 0.5$ ps) freely-propagating HCPs indicated that in the intermediate regime, $0.1 \lesssim T_p/T_n \lesssim 1$, the threshold field for ionization varied approximately as $1/n_i^2$. Classical and quantum calculations reproduced this behavior, but detailed comparisons between theory and experiment were difficult because of problems inherent in determining the absolute amplitude and shape of the HCPs [40]. Subsequent experiments using K(388p) and K(520p) atoms and HCPs with widths in the range $\sim(2–100)$ ns, which span the transition from the short pulse to the long-pulse regime and which could be accurately measured, yielded results that were in excellent agreement with 3D CTMC (and quantum) simulations validating the classical description of ionization [42,43]. One immediate consequence of this is that the experimental data satisfy classical scaling invariance. As shown in Fig. 7, when plotted as a function of scaled field strength $F_0 = n^4 F$, the survival probabilities displayed similar behavior governed only by the scaled duration $T_0 \equiv T_p/T_n$ of the pulse [43]. As expected, in the true impulsive limit, $T_0 \ll 1$, the field strength F_0 required for ionization scales as $\sim 1/T_0$, since the ionization probability is controlled by the impulse $\Delta p_0 = F_0 T_0$ delivered to the electron. The calculations also indicated that as T_p/T_n changes

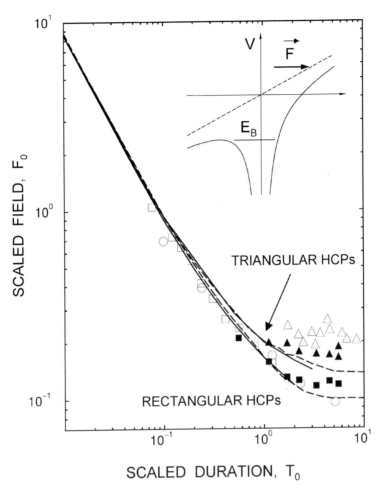

FIG. 7. Scaled peak field F_0 for 10% ionization of hydrogen np atoms as a function of scaled pulse duration. —, - - - -, results of quantum calculations and CTMC simulations, respectively, for rectangular and triangular HCPs of equal FWHM. Also included are measurements using near-rectangular HCPs taken from (■) [43], (○) [42] and (□) [39] (scaled by a factor 2.5), using "sawtooth" HCPs taken from (▲) [43], and using near triangular HCPs taken from (△) [43].

from the sudden to the adiabatic limit the impulsive energy transfer to the electron is decreased. This was verified experimentally through studies of the excited state distribution of the surviving atoms using selective field ionization. Interestingly, in the long-pulse regime, $T_0 \gtrsim 1$, the ionization probability was observed to depend markedly on HCP shape. This may be due to the different time dependence of the switch on of the field, dF/dt, for different HCP shapes. This, in turn, implies different degrees of nonadiabatic passage through the multitude of avoided crossing in the Stark map (see Fig. 4).

3.2. WAVEPACKET PRODUCTION AND EVOLUTION

Quantum-mechanically if an atom is initially in some stationary Rydberg state. $|\phi_i\rangle$, the electronic wavefunction immediately after application of a HCP may be written $|\Psi(t = 0)\rangle = |\phi_i^B\rangle = e^{i\Delta\vec{p}\cdot\vec{r}}|\phi_i\rangle$ and corresponds in the sudden limit to the initial state shifted, or "boosted," in momentum space by $\Delta\vec{p}$. The corresponding expectation values of the energy and momentum are

$$\langle E \rangle_{t=0} = \langle \phi_i^B | H_{at} | \phi_i^B \rangle = \langle \phi_i | H_{at} | \phi_i \rangle + \frac{\Delta p^2}{2} + \langle \phi_i | \vec{p} \cdot \Delta \vec{p} | \phi_i \rangle, \tag{6}$$

$$\langle \vec{p} \rangle_{t=0} = \langle \phi_i^B | \vec{p} | \phi_i^B \rangle = \Delta \vec{p} + \langle \phi_i | \vec{p} | \phi_i \rangle \tag{7}$$

where H_{at} is the atomic Hamiltonian. Since $\langle n\ell m | \vec{p} | n\ell m \rangle = 0$, $\langle E \rangle_{t=0} = E_i + \Delta p^2/2$, in agreement with the classical prediction. The "boosted" final electronic wavefunction can be expanded in terms of the field-free atomic basis as

$$|\Psi(t)\rangle \sim \sum_n e^{-iE_n t} \sum_\ell \langle n\ell m | \Psi(0) \rangle | n\ell m \rangle \tag{8}$$

and comprises a coherent superposition of many states having a broad distribution in n, ℓ that includes high-ℓ states. Equation (8) represents a nonstationary wavepacket whose evolution is governed by the atomic, i.e., field-free, Hamiltonian H_{at}.

The time evolution of this wavepacket can be understood classically by considering the time development of the expectation value of the z component, p_z, of the electron momentum following HCP application. Initially the HCP imparts a net momentum to the electron in the $+z$ direction, i.e., $\langle p_z \rangle > 0$. After approximately one half of the orbital period T_{n_f} associated with the final-state distribution, the "orbital" motion of the electron reverses direction, i.e., $\langle p_z \rangle < 0$. As the orbital motion continues the distribution once more becomes strongly peaked at positive p_z. This cycle then repeats causing $\langle p_z \rangle$ to oscillate between positive and negative values with period $\sim T_{n_f}$. The amplitude of the oscillation decreases steadily with time, however, because a (narrow) distribution of final states with a distribution of periods around the mean value $\langle T_{n_f} \rangle = T_{\langle n_f \rangle}$, where $\langle n_f \rangle$ is the mean final state quantum number, is excited that evolve differently in time leading to dephasing.

This evolution of the wavepacket can be monitored by applying a second probe HCP (again along the z axis) [27,31,38,53,54] after some time delay t_D. If the

impulse delivered by the probe pulse is in the same direction as the initial electron momentum, i.e., $\langle p_z \rangle$, at the moment the probe pulse is applied, the electron
momentum and energy will be increased. This can lead to ionization and a low
survival probability. If, however, the impulse and initial electron momentum are
in opposite directions (and $|p_{zi}| < |\Delta p_z|/2$, see Eq. (4)), the final electron energy
is decreased and the likelihood of survival is greater. Thus the time evolution of
the wavepacket, i.e., $\langle p_z(t_D) \rangle$, can be monitored by measuring the survival probability as a function of the time delay, t_D, between the initial and probe pulses.
This is illustrated in Fig. 8 which shows the survival probability for K($n = 351$)
quasi-1D atoms following application of two oppositely-directed HCPs as a function of time delay [55]. The first HCP delivers a scaled impulse $\Delta p_{z0} = -0.1$,

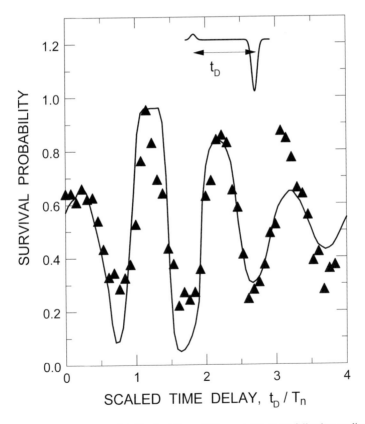

FIG. 8. Measured survival probability for K($n = 351$) quasi-1D atoms following application of
two oppositely directed HCPs as a function of scaled time delay, t_D/T_n. The HCP sequence is illustrated in the inset. The first HCP delivers an impulse $\Delta p_0 = -0.1$, the second probe pulse an impulse
$\Delta p = 0.9$. —, results of CTMC simulations.

the probe pulse an impulse $\Delta p_{z0} = 0.9$. Pronounced "quantum beats" with a frequency corresponding to the orbital period $\nu = 1/T_{\langle n_f \rangle}$ are evident that mirror the quasi-periodic evolution of the wavepacket. The experimental results are well reproduced by CTMC simulations showing that classical evolution can mimic quantum evolution for time intervals extending over several orbital periods. These beats are classical in nature and can be distinguished from modulations due to quantum effects occurring over longer time intervals.

The energy differences in the Rydberg series that determine the quantum beats are given to second order in $\delta n / \langle n_f \rangle$ by [38]

$$E_{\langle n_f \rangle + \delta n} - E_{\langle n_f \rangle} \simeq \delta \omega \omega_{\langle n_f \rangle} \left[1 - \frac{3}{2} \frac{\delta n}{\langle n_f \rangle} + 2 \left(\frac{\delta n}{\langle n_f \rangle} \right)^2 \right] \qquad (9)$$

where $|\delta n| \lesssim \Delta n / 2$. To leading order, the spectrum compares locally to that of a harmonic quantum oscillator, with $\omega_{\langle n_f \rangle} = 2\pi / T_{\langle n_f \rangle}$ being the classical orbital frequency. For a harmonic oscillator, quantum and classical expectation values agree which explains the presence in the classical simulation of beats with the mean orbital period. Damping of the beats is caused by the "anharmonic" correction. Dephasing (by π) or damping occurs over a time $t_D \sim 4n_f/3(\Delta n)^2 T_{\langle n_f \rangle}$. Classical–quantum correspondence will, however, break down for times approaching the revival time [56,57] $t_R \sim (n_f/3) T_{\langle n_f \rangle}$. Quantum theory (Eq. (9)) predicts the revival of quantum beats at this time due to the discreteness of the excitation spectrum, i.e., of n_f. In contrast, the classical action n_f (in units of $\hbar$) is continuously distributed.

The survival probability following application of a probe pulse Δp_z provides a measure of the fraction of electrons with values of p_{zi} such that $p_{zi} \Delta p_z < -E_i - \Delta p_z^2 / 2$ (see Eq. (5)). Thus, by measuring the survival probability as a function of Δp_z the distribution of p_{zi} can be determined and its time evolution monitored by varying the time of application of the probe pulse [31,53,58,59]. This technique, sometimes referred to as impulsive momentum retrieval (IMR), provides a valuable tool to study wavepacket evolution. The temporal evolution of the momentum distribution for a wavepacket created by exposing Na(25d) atoms to a strong 0.5 ps HCP is shown in Fig. 9. This was determined [53] via the IMR technique using a second 0.5 ps probe HCP. Clear quasi-periodic oscillations in the momentum distribution can be seen as the wavepacket evolves in time. Because of the anharmonicity represented by Eq. (9), the wavepacket also alternately disperses and revives as it oscillates.

While a probe HCP can be used to measure the electron momentum distribution, complementary information on the spatial distribution of the electron can be obtained by the sudden turn on of a quasi-dc field F_{dc}, i.e., by application of a "field step" with a risetime $t_R \ll T_n$ and duration $T_p \gg T_n$ [59]. Classically, the energy E_f of the electron immediately after the application of the field step is related to its initial energy E_i and its initial z coordinate z_i within a sudden

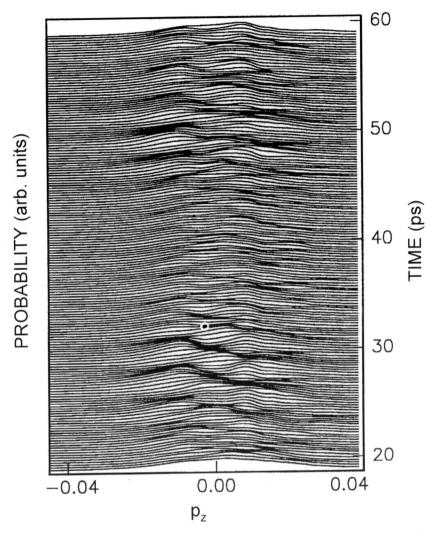

FIG. 9. Temporal evolution of the distribution of the z component of electron momentum p_z for the wavepacket created by exposing Na(25d) atoms to a strong 0.5 ps duration HCP. The time interval between adjacent profiles is $\sim$230 fs (from [53]).

approximation by

$$E_{\mathrm{f}} = E_{\mathrm{i}} + z_{\mathrm{i}} F_{\mathrm{dc}}. \tag{10}$$

If the field F_{dc} remains switched on for a sufficiently long period and if E_{f} lies above the energy E_{B} associated with the saddle point evident in Fig. 7, classical field ionization might be expected, whereupon the survival probability would provide a measure of those electrons with values of z_i such that

$$z_i F_{dc} < -E_i + E_B. \tag{11}$$

For $F_{dc} > 0$, the survival probability would therefore be larger (smaller) when $z_i < 0$ ($z_i > 0$). Equations (10) and (11) suggest that a field step might be used to monitor the spatial evolution of a wavepacket. Moreover, Eq. (11) offers hope that the distribution of z_i might be determined from measurements of the survival probability as a function of F_{dc}. While this is indeed true for a 1D atom, the situation is more complex for a 3D atom because the saddle point lies on the (negative) z axis whereas the electron is typically initially well removed from this axis and therefore has difficulty in locating the saddle. Detailed analysis shows that, in general, the explicit condition for survival is

$$2E_i z_i + F_{dc} r_i^2 > -\frac{E_i^2}{F_{dc}} + 2(1 + A_z) \tag{12}$$

where r_i is the initial radial coordinate of the electron and A_z is the z-component of the Runge–Lenz vector. The survival probability thus depends on two coordinates and, through A_z, on the orientation of the orbit meaning that only under special circumstances can a single coordinate be determined directly. Nonetheless, field steps have been used successfully to follow wavepacket evolution in coordinate space and, as expected from Eq. (12), the oscillations in survival probability do not mimic the time evolution of $\langle z \rangle$ but rather the quantity $\langle 2E_i z + F_{dc} r^2 \rangle$.

3.3. CHARACTERIZATION OF QUASI-1D ATOMS

Single HCPs can also be used to characterize quasi-1D atoms by applying them both parallel and perpendicular to the atomic axis [60]. Differences in the ionization characteristics provide a measure of the anisotropy of the momentum distribution, i.e., of the degree to which the state is quasi-1D. In the limiting case of a truly 1D atom oriented *perpendicular* to the z axis, the energy transfer produced by a kick Δp_z applied along the z axis is simply $\Delta E = \Delta p_z^2/2$ independent of the initial electron momentum because the kick is transverse to the atom, i.e., $p_{zi} = 0$. The survival probability will thus display a near step-like dependence on Δp_z, falling to zero as soon as $\Delta p_z^2/2 > |E_i|$. As the atomic state becomes less well localized along the perpendicular axis the transverse momentum distribution, i.e., the distribution of p_{zi}, broadens, leading to a range of energy transfers. The survival probability will then fall more gradually with increasing Δp_z, the rate of decrease providing a measure of the width of the momentum distribution (or Compton profile) along the direction of the impulse. The effect that the width of the (transverse) momentum distribution has on the ionization characteristics is illustrated by the results of a series of 3D CTMC simulations undertaken

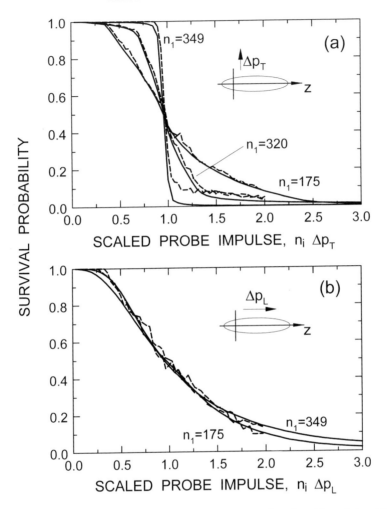

FIG. 10. Survival probabilities calculated using CTMC theory for selected lower lying states in the H($n = 350$), $m = 0$ Stark manifold as a function of the scaled impulse applied (a) transverse and (b) along the atomic axis. The solid and dashed lines show results for a δ-function impulse and a 600 ps-wide HCP, respectively.

for a variety of $n = 350$, $m = 0$ Stark states. The calculated survival probabilities are shown in Fig. 10 as a function of the scaled impulse Δp_{z0} for both a δ-function impulse and the 600 ps-wide HCPs typically used experimentally. Only very small differences in the ionization profiles are evident indicating that the experiment represents the impulsive limit. For the lowest member of the Stark manifold ($n_1 = 349$) the survival probability displays near steplike behavior due to the very narrow transverse momentum distribution. As n_1 decreases, the ionization profile broadens and the step disappears. The differences in ionization profiles between neighboring states are particularly pronounced for large values of n_1. This reflects the initial rapid rise in the width of the transverse momentum distribution with decreasing n_1 and shows that ionization by a transverse HCP can provide a sensitive test of the quasi-1D character of a state. For states near the center of the Stark manifold, the ionization profile is broad and relatively insensitive to n_1. Figure 10 also includes ionization profiles predicted for quasi-1D atoms aligned *along* the z axis, i.e., for kicks applied along the atomic axis. The profiles are broad and largely independent of n_1, as is the axial electron momentum distribution.

The response of states near the Stark-shifted d level in the $K(n = 350)$ Stark manifold to probe impulses applied both along and transverse to the atomic axis are shown in Fig. 11. With the probe kick applied along the atomic axis, the rate of decrease of the survival probability with increasing kick strength is significantly less than with the probe kick applied transverse to the atomic axis. This demonstrates that the width of the electron momentum distribution transverse to the atomic axis is substantially smaller than that parallel to the axis, confirming

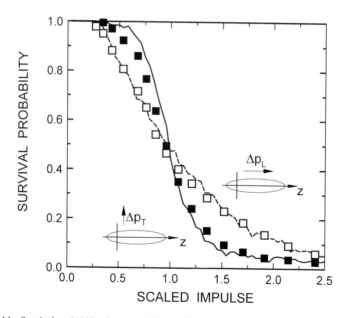

FIG. 11. Survival probability for quasi-1D potassium atoms in low-lying states in the $n = 350$ Stark manifold as a function of scaled impulse Δp_0. The experimental data are for probe HCPs applied transverse (■) and parallel (□) to the atomic axis, respectively. The corresponding lines show the results of CTMC simulations (see text).

the production of quasi-1D states. Even when using a transverse kick the survival probability does not display true steplike behavior indicating that the transverse electron momentum distribution has significant width. Figure 11 also includes the predictions of CTMC simulations. Calculations show that the oscillator strengths for excitation of Stark states in the vicinity of the Stark-shifted d level are small, with a broad maximum near $n_1 \sim 320$. The simulations thus assume creation of a statistical distribution of 36 neighboring Stark states centered at $n_1 = 320$. Experiment and theory are in good agreement and further demonstrate that it is indeed possible to selectively excite quasi-1D Stark states even at $n \sim 350$.

4. Effect of Multiple HCPs

4.1. DYNAMICAL STABILIZATION

4.1.1. 3D Atoms

The periodically-kicked Rydberg atom has been suggested as an excellent laboratory for studying nonlinear dynamics of simple driven systems [61–68]. Under the influence of repeated impulses the time evolution reduces to a sequence of discrete maps between adjacent kicks. This simplification permits detailed numerical studies of the long-term evolution using both classical and quantum dynamics and, hence, of the classical–quantum correspondence in microscopic systems that feature regular and chaotic dynamics. We consider here initially the response of an atom to a train of identical equispaced unipolar HCPs. Such a pulse train is sketched in the inset in Fig. 12 which itself shows the survival probability for K(351p) atoms following application of 20 identical HCPs each delivering a scaled impulse $\Delta p_0 = 0.3$ as a function of the scaled pulse repetition frequency $\nu_0 \equiv \nu_T / \nu_n$, where ν_T is the pulse repetition frequency in the HCP train and $\nu_n \ (= 1/T_n)$ is the classical electron orbital frequency [69]. Two features of the data are noteworthy: despite the strength and number of the kicks the survival probability is sizable, and the survival probability depends markedly on the pulse repetition frequency.

Insights into this behavior are provided by CTMC simulations that use the Hamiltonian [7–9,69]

$$H = H_{\text{at}} + zF(t) = H_{\text{at}} + \sum_{j=1}^{N} zF_{\text{HCP}}(t - j\nu_T^{-1}) \tag{13}$$

where N is the number of HCPs. The atomic Hamiltonian is given in cylindrical coordinates (z, ρ, ϕ) by

$$H_{\text{at}} = \frac{p_z^2 + p_\rho^2}{2} + \frac{L_z^2}{2\rho^2} + V_{\text{mod}}\left(\sqrt{z^2 + \rho^2}\right). \tag{14}$$

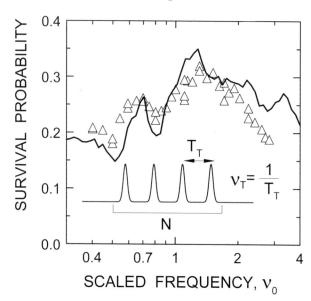

FIG. 12. Measured survival probabilities for K(351p) atoms following application of $N = 20$ HCPs each delivering a scaled impulse $\Delta p_0 = 0.3$ as a function of scaled pulse repetition frequency $\nu_0 \equiv \nu_T/\nu_n$. The solid line shows the results of CTMC simulations. The inset shows a schematic of the HCP train.

L_z denotes the projection of the angular momentum along the z axis, which is a constant of the motion. V_{mod} is a one-electron potential for the K^+ core potential that yields accurate quantum defects and reproduces the correct behavior of the potential in the limit of both large and small distances. An isotropic phase-space microcanonical ensemble with classical angular momenta $1 \leq L \leq 2$ is used to describe the laser-excited initial np state. In these calculations the ensemble of points in phase-space representing the initial state is propagated in 3D according to Hamilton's equation of motion. At the end of the propagation time the final energy E_f of the evolved electron is determined. The overall survival probability is determined by calculating the fraction of initial conditions for which $E_f < 0$.

As illustrated in Fig. 12, CTMC simulations using the experimental HCP profiles as input are in reasonable agreement with the experimental data and reproduce well the observed dependence on scaled frequency ν_0. This indicates that the increase in survival probability evident for $\nu_0 \gtrsim 1$ is classical in origin. (Equally good agreement is obtained if V_{mod} is replaced by the Coulomb potential, $-1/\sqrt{\rho^2 + z^2}$, showing that core effects are unimportant in the present context.) The origin of the broad maximum in the survival probability can be analyzed using the simplified model of a hydrogen atom subject to a sequence of

δ-function impulses. The corresponding Hamiltonian is

$$H = H_{\text{at}} - z\Delta p \sum_{j=1}^{N} \delta\left(t - j v_T^{-1}\right). \tag{15}$$

The single period evolution governed by this Hamiltonian is given by a map of phase-space coordinates [64]

$$M(\vec{r}, \vec{p}) = M_{\text{Coul}} \cdot M_{\Delta p}, \tag{16}$$

where $M_{\Delta p}$ describes the kick and M_{Coul} describes the evolution of the Coulomb orbit. The evolution of the phase-space coordinates after N kicks is given by $M^N(\vec{r}, \vec{p})$. For this simplified system it is computationally feasible to investigate the long term stability in the limit $N \to \infty$ (typically $N > 10^6$) and to perform a detailed analysis of the classical phase space. This model reproduces [9] the observed general structure in the survival probability in Fig. 12 suggesting that the stabilization mechanism can indeed be reasonably discussed using the simplified model. The predicted survival probabilities are slightly lower than those obtained when using the experimental HCP profile because the higher frequencies present in the δ-function perturbation provide stronger coupling to the continuum. The regions of phase-space that give rise to stabilization can be identified using Poincaré surfaces of section. The stroboscopic snapshots taken by M after each cycle lie on a four-dimensional manifold that can be sectioned by layers around fixed coordinates $(\rho \pm \Delta\rho, p_\rho \pm \Delta p_\rho)$. An example [7] of such a Poincaré surface of section is shown in Fig. 13 for a cut at $\rho \sim 0.1$, $p_\rho \sim 0$. Sizable stable islands enclosed in Kolmogorov–Arnold–Moser (KAM) tori are evident that are embedded in a chaotic sea pointing to the coexistence of regular and chaotic dynamics. The stable islands are responsible for both the size of the survival probability and its dependence on scaled frequency. If, at the start of the HCP train, the phase point of an atom resides within a stable island, it will remain trapped within the island giving rise to dynamical stabilization which allows the atom to survive a large number of kicks. On the other hand, if the initial phase point resides in the chaotic sea diffusive spreading occurs and the atom eventually ionizes.

The physical origin of the stability of the electron orbit against ionization can be understood by remembering that for an atom to survive its binding energy must be approximately unchanged following some integer number k of applied kicks. Thus after a large number of kicks the distribution of electron binding energies will be sharply peaked about its initial value. (If the first few kicks populate high energy levels these will typically be ionized by subsequent kicks.) For the period-one ($k = 1$) case this requires that the energy transfer $\Delta p^2/2 + p_{zi}\Delta p$ induced by each kick be (close to) zero, i.e., $p_{zi} = -\Delta p/2$, which agrees with the p_z coordinate of the center of the family of islands labeled A_0, A_1, A_2 in Fig. 13. The center of each island corresponds to a periodic orbit. Each kick maps the point

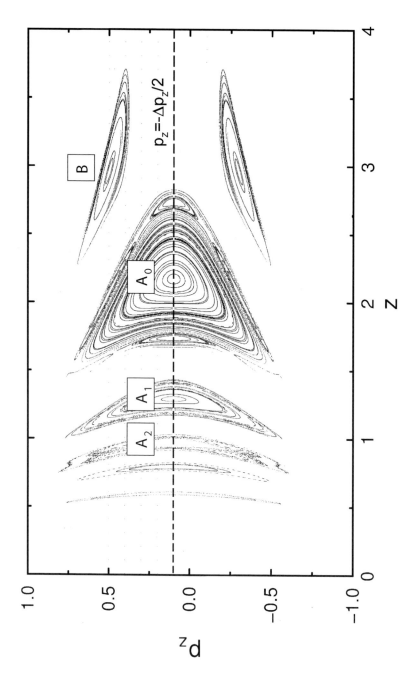

FIG. 13. Poincaré surface of section for the kicked hydrogen atom for $\nu_T = 1/2\pi$ and $\Delta p = -0.2$ and a cut at $\rho = 0.1$, $p_\rho = 0$. Each trajectory has been followed for up to one million kicks.

$(z, p_z = -\Delta p_z/2)$ on a given Coulomb orbit to the point $(z, p_z [= p_{zf}] = \Delta p_z/2)$ on a new orbit. The motion of the electron on this orbit must then be such that p_z evolves to a value $\sim(-\Delta p_z/2)$ at the time of the next kick. Points in phase-space where these conditions are approximately but not exactly met form quasi-periodic orbits and stable islands in the Poincaré surface of section. In consequence, the survival probability will depend markedly on the relationship between the pulse separation in the HCP train and the electron orbital period. The large island labeled A_0 in Fig. 13 corresponds to a scaled frequency $\nu_0 \sim 1.3$ for which the time separation between adjacent kicks is less than the Kepler period. The electron motion corresponds to a sequence of segments of a Coulomb orbit each cut short by a kick. The other islands $A_1, A_2, A_3, \ldots$ in Fig. 13 correspond to orbits that complete $1, 2, 3, \ldots$ full periods plus one segment prior to the next kick. The corresponding scaled frequencies are $\nu_0 = 0.57, 0.35, 0.26 \ldots$ and approximately match the positions of the peaks in the calculated survival probably evident in Fig. 12. (The actual positions of the peaks in Fig. 12 are slightly shifted because the time period between the first and second pulses in the train is slightly larger than the subsequent period of the train [64]).

More complex periodic orbits also exist that involve transfer back and forth between two (or more) energy levels. Figure 13 includes one such example that involves two energy levels, i.e., $k = 2$, and corresponds to the center of the regular island labeled B. In this case a kick excites the electron to a higher energy level and, subsequently, another kick returns it to its original level [8,70]. The electronic motion thus becomes trapped within these two levels.

The effects of dynamical stabilization are evident in Fig. 14(a) which shows the calculated survival probability for kicked H(390p) atoms as a function of the number N of (δ-shaped) kicks for a variety of different scaled frequencies, ν_0. For values of $\nu_0 = 1.9$ and 6.2 the initial states lie in the chaotic sea. The survival probability decreases monotonically with increasing N, a characteristic of chaotic ionization. For $\nu_0 = 0.6$ and 1.3 part of the initial state resides within a stable island and the survival probability, after initially decreasing, becomes essentially constant, even for $N \gtrsim 10^6$ kicks, providing clear evidence of dynamical stabilization.

4.1.2. 1D and Quasi-1D Atoms: Effect of Kick Direction

Many aspects of the behavior of 3D systems can be understood using 1D models [8,9]. These are less complex than 3D models making them simpler to implement and indeed the earliest studies of the dynamics of the kicked atom were undertaken using 1D models. Such models also permitted the first detailed quantum mechanical treatments of the problem thereby facilitating study of classical–quantum correspondence. In addition, 1D and quasi-1D atoms feature behavior that cannot be observed using, for example, np states. Consider therefore a 1D

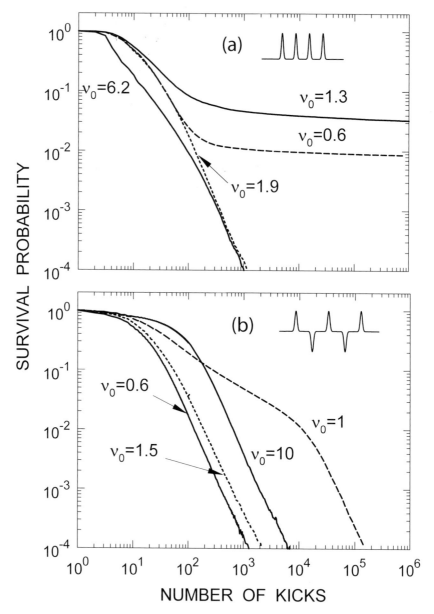

FIG. 14. Calculated survival probabilities for H(390p) atoms subject to a train of (a) unidirectional and (b) alternating δ-shaped kicks with various scaled frequencies ν_0 as a function of the number N of kicks.

hydrogen "atom" that is subject to a train of N δ-function kicks Δp. The dynamics of such an atom is governed by the Hamiltonian

$$H^{1\mathrm{D}}(q, p, t) = H_{\mathrm{at}}^{1\mathrm{D}}(q, p) + q\Delta p \sum_{k=1}^{N} \delta\left(t - \frac{k}{\nu_T}\right)$$

$$= \frac{p^2}{2} + \frac{\Lambda^2}{2q^2} - \frac{1}{q} + q\Delta p \sum_{k=1}^{N} \delta\left(t - \frac{k}{\nu_T}\right) \qquad (17)$$

where q and p denote the position and momentum of the electron and q is confined to positive values ($q > 0$). Within classical dynamics the latter constraint can be enforced by introducing an effective centrifugal potential with pseudo-angular momentum Λ into the unperturbed Hamiltonian H_{at}. The limit $\Lambda \to 0$ is taken and the only role of this potential is to provide an infinite barrier at $q = 0$ such that the electron motion is confined to $q > 0$. In quantum mechanics no such limiting procedure is needed. Instead, simply imposing the hard-wall boundary condition $\psi(q = 0) = 0$ suffices.

Figures 15(a) and (b) show Poincaré surfaces of section for kicks with scaled kick strengths $|\Delta p_0| = 0.3$ directed both towards ($\Delta p_0 < 0$) and away from ($\Delta p_0 > 0$) the origin, respectively [69]. For $\Delta p_0 < 0$ sizable islands of stability embedded in a chaotic sea are evident. This structure is similar to that shown in Fig. 13 for the kicked 3D atom. This is not surprising in that to obtain this figure phase-space was sliced near $\rho \sim 0.1$, $p_\rho \sim 0$ and thus $L_z \sim 0$. Setting $\rho = 0$, $p_\rho = 0$ and $L_z = 0$ in Eq. (14) yields a Hamiltonian that is equivalent to the 1D Hamiltonian, Eq. (17). Because of the higher-dimensional phase-space available, chaotic trajectories in 3D are unlikely to return to the original slice of the Poincaré section after a finite number of kicks. Even in 1D, however, ionization might not happen immediately. Simulations show that an electron can wander among various regions of phase-space associated with bound states ($E < 0$) before finally passing into the continuum. For a scaled frequency $\nu_0 = 1.3$, the initial state of the electron, denoted by the thick dashed line in Fig. 15(a), overlaps a large stable island. Those atoms whose initial phase points lie within this island remain trapped leading to dynamical stabilization. This is illustrated in Fig. 15(c) which shows the calculated survival probability as a function of the number N of kicks. The overall survival probability decreases only slowly with N and, because of the restricted phase-space in 1D, the fraction of atoms that undergo dynamical stabilization is greater than in 3D resulting in a larger overall survival probability. In contrast, for $\Delta p > 0$, the Poincaré section contains no islands of stability and the system is globally chaotic. Dynamical stabilization does not occur and as shown in Fig. 15(c) the survival probability falls quite rapidly as N increases.

This very different behavior is also evident in Fig. 16 which shows Rydberg atom survival probabilities measured as a function of scaled frequency ν_0 for

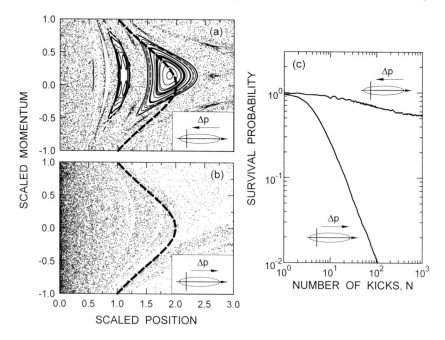

FIG. 15. Poincaré surfaces of section for a 1D hydrogen atom subject to kicks with scaled frequency $\nu_0 = 1.3$ and a scaled kick strength $|\Delta p_0| = 0.3$ directed both (a) towards ($\Delta p_0 < 0$) and (b) away from ($\Delta p_0 > 0$) the origin. The thick dashed line indicates the points of the energy shell (or torus) associated with the initial state. (c) Calculated survival probabilities as a function of the number N of kicks.

quasi-1D K($n = 350$) $m = 0$ Stark states subject to kicks applied parallel ($\Delta p > 0$) and antiparallel ($\Delta p < 0$) to the atomic axis [69]. The peaks in the measured survival probabilities at $\nu_0 \sim 1.3$ and 0.7 for $\Delta p < 0$ are again associated with dynamical stabilization. As predicted by the 1D simulations, the survival probability depends markedly on the direction of the kicks, being typically a factor of two to three larger for negative kicks. Figure 16 also includes the results of fully-3D CTMC simulations that use the theoretically-predicted initial Stark state distribution discussed earlier. These agree reasonably well with experiment, although the predicted asymmetry is larger than that observed. In particular, the survival probability calculated for negative kicks ($\Delta p < 0$) is much larger than that actually measured. One possible explanation for this discrepancy is noise resulting from fluctuations in the amplitudes and separation of pulses in the HCP train. Calculations show that for positive kicks the introduction of noise into the HCP train leads to only small changes in the calculated survival probability, which is not unexpected given that the system is globally chaotic. In contrast, for negative kicks, the survival probabilities are rather sensitive to the presence of noise

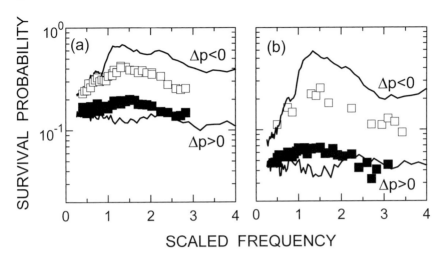

FIG. 16. Survival probabilities for quasi-1D $K(n = 350)$ $m = 0$ Stark states subject to (a) $N = 20$ and (b) $N = 40$ kicks of magnitude $|\Delta p_0| = 0.3$ directed both towards ($\Delta p_0 < 0$) and away from ($\Delta p_0 > 0$) the origin as a function scaled frequency. The symbols show experimental data, the lines the results of CTMC simulations.

indicating that it leads to a reduction in the size of the stable islands associated with dynamical stabilization.

4.2. CLASSICAL–QUANTUM CORRESPONDENCE

One-dimensional systems allow detailed studies of classical–quantum correspondence in impulsively driven systems. One motivation for this work is to study the correspondence between the stable classical islands observed in phase-space and their quantum counterparts. Another is to examine the apparent contradiction between the widespread occurrence of chaos in classical quantum dynamics and the lack thereof in quantum dynamics. Comparison of classical and quantum phase-space distributions renders a detailed test of classical–quantum correspondence [71]. For a comparison with the classical phase-space distribution, the Husimi distribution [72,73] is the quantum phase-space distribution of choice. The Husimi distribution can be understood as a convolution of the Wigner phase-space distribution with a minimum uncertainty Gaussian wavepacket. The effect of the Gaussian averaging is to remove the short-wavelength oscillatory structures associated with the de Broglie wavelength and to restore positive definiteness. The latter is of crucial importance for an interpretation in terms of a probability density. If quantum stabilization takes place, it originates from the localization of the stationary quantum states associated with the periodic perturbation, i.e., the

Floquet states $|\phi_k^F\rangle$ defined through the eigenvalue equation [74]

$$U(T_T)|\phi_k^F\rangle = \mathrm{e}^{-iE_kT}|\phi_k^F\rangle, \tag{18}$$

where $U(T_T)$ is the quantum period-one evolution operator. In practice $U(T_T)$ is evaluated in a finite Hilbert space with absorbing boundary conditions. Therefore the quasi-energies $E_k \equiv E_k^R - iE_k^I$ can be complex numbers [71]. After N kicks, the propagated wave function becomes

$$\big|\Psi(NT_T)\big\rangle = U^N(T_T)\big|\Psi(0)\big\rangle = \sum_{k=1}^{N} c_k \mathrm{e}^{-iE_k^R NT} \mathrm{e}^{-E_k^I NT}\big|\phi_k^F\big\rangle \tag{19}$$

where c_k are the expansion coefficients of $|\Psi(0)\rangle$ in the basis spanned by Floquet states. The key point for the stability analysis is that the imaginary part E_k^I corresponds to a decay probability within the finite Hilbert space and stable Floquet states can be identified as those whose E_k^I values are very close to zero (within numerical uncertainty).

Figure 17 illustrates [71] classical–quantum correspondence for dynamical stabilization of an $n = 50$ 1D atom subject to a train of kicks with strength $\Delta p_0 = -0.3$ and scaled frequency $\nu_0 = 1.09$. The Husimi distribution of the dominant Floquet state (Fig. 17(b)) associated with stabilization is seen to be localized in phase space around the dominant classical stable island visible in Fig. 17(a). The resolution of classical structures in quantum phase-space requires large quantum numbers and small uncertainty, i.e., that the size of the classical islands be bigger than the effective Planck constant $\hbar$. In the present case $\Delta p_0 \Delta q_0 \sim 1/n$ which, as illustrated in Fig. 17(a), corresponds to an area much smaller than the dominant classical island. For this reason, the Husimi distributions of several stable Floquet states are found to mimic the dominant classical island of stability. Quantum stabilization is particularly remarkable in view of the fact that direct "multiphoton" transitions to the continuum through high harmonics are present with the same strength as the fundamental frequency. Evidently, the system has become sufficiently classical that the KAM tori can efficiently confine phase-space flow, and "tunneling" through KAM barriers is inhibited.

Also remarkable is the fact that Husimi distributions of quantum Floquet states can be found that are localized around other classical islands in Fig. 17(a) even when their size is smaller than the size of the Planck cell, an example of which is given in Fig. 17(c). In this case, however, the quantum Floquet state is slightly unstable. An even more remarkable and quite universal feature is that partially stable Floquet states are found to be localized around classical periodic orbits of the system even in those cases where the periodic orbits are unstable, i.e., where the periodic orbits are surrounded by a chaotic sea. Quantum dynamics (characterized by a finite size of the Planck cell) allows building on the skeleton of classical dynamics such that unstable periodic orbits become the carrier of a significant

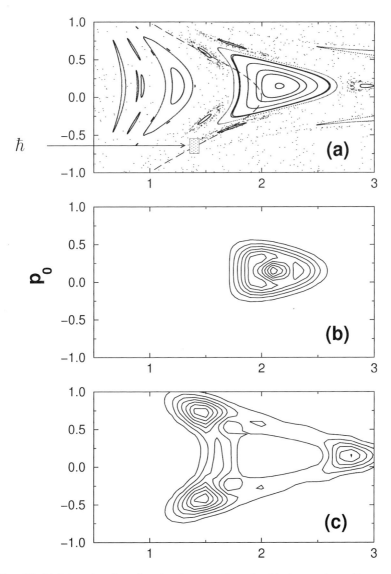

FIG. 17. (a) Poincaré surface of section for a one-dimensional hydrogen atom with $n_i = 50$ subject to a train of kicks delivering a scaled impulse $\Delta p_0 = -0.3$ with scaled frequency $\nu_0 = 1.09$. The thick dashed line indicates the unperturbed torus associated with the initial state, the shaded box the size of the Planck cell. (b) and (c) Quantum Husimi distributions of (b) the Floquet state that has the largest overlap with the initial state and (c) a slightly unstable Floquet state. The contours of the quantum calculations were obtained using a linear scale.

fraction of the localized probability density. Unstable periodic orbits can be identified as scars in the quantum phase-space and the corresponding Floquet states are thereby called scarred states [75,76]. This is a truly quantum phenomenon and considered to be one hallmark of quantum localization or quantum suppression of classically chaotic diffusion.

Quantum localization was first predicted for the kicked rotor [77] and it was shown that it could be mapped onto the well-known Anderson or strong localization in solid state physics [78,79]. (In cases where quantum localization is structurally equivalent to strong localization, it is sometimes referred to as dynamical localization.) Subsequently quantum localization was predicted for Rydberg atoms subject to microwave fields [1,2,80–82] where it manifested itself through enhanced stabilization of the atom against ionization, an effect that has now been confirmed experimentally. This suggests that, even in the limit of very high n, the correspondence between classical and quantum dynamics cannot hold for arbitrary periods of time. A more detailed discussion of quantum localization in the kicked atom is provided elsewhere [83,84], but has yet to be observed experimentally.

4.3. FREELY-PROPAGATING ATTOSECOND HCP TRAINS

As noted earlier, it has been suggested that it might be possible to generate trains of unidirectional attosecond-duration HCPs using high harmonic conversion techniques opening the possibility that these might be employed to manipulate and shape wavepackets in low-lying atomic states [24]. For such a freely-propagating pulse train, however, the average electric field experienced by the atom must be zero, or, equivalently, the net momentum transfer Δp_{free} to a free electron must be zero. To investigate the effect of this constraint on the shape of the HCP train on the dynamics of the kicked atom measurements [85] have been undertaken at very high n in which the average field experienced by an atom during a unidirectional HCP train is varied by using a second pulse generator to superpose an offset field F_{offset} during the train resulting in the waveform shown in the inset in Fig. 18. The average field thus becomes

$$F_{\text{av}} = F_{\text{offset}} - \overline{F} \tag{20}$$

where $\overline{F} = \nu_T \Delta p$ is the average field associated with the HCP train alone. By setting $F_{\text{offset}} = \overline{F}$ the behavior of atoms subject to freely-propagating HCP trains can be simulated.

The survival probability for K(351p) atoms subject to $N = 20$ and $N = 40$ HCPs is shown in Fig. 18(a) as a function of the offset field for a scaled HCP repetition frequency $\nu_0 \sim 2.7$. The HCPs used to obtain these (and later) data had a peak amplitude of $\sim 130 \text{ mV cm}^{-1}$ and a full width at half maximum duration

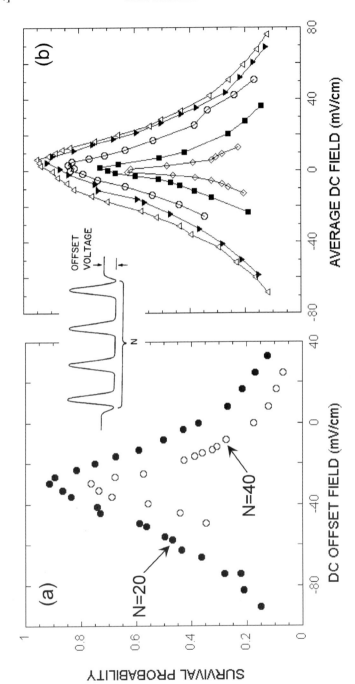

FIG. 18. Measured survival probabilities for K(351p) atoms subject to (a) $N = 20$ and $N = 40$ HCPs with scaled repetition frequency $\nu_0 = 2.7$ and delivering a scaled momentum transfer $\Delta p_0 = 0.3$ as a function of the offset field F_{offset}, and (b) $N = 20$ HCPs with values of ν_0 ($\diamond$) 0.33, ($\blacksquare$) 0.8, ($\bigcirc$) 1.3, ($\blacktriangledown$) 2.7 and ($\triangle$) 3.3 expressed as a function of the net average field F_{av} experienced by the atoms. The inset shows a schematic of the applied HCP train.

of $\sim$600 ps. The survival probability depends strongly on the size and direction of the offset field, peaking at an offset field of $\sim(-30)$ mV cm^{-1} for which the average field F_{av} experienced by the atoms is close to zero. These data, and data recorded at scaled frequencies ranging from $\nu_0 \sim 0.33$ to $\nu_0 \sim 3.3$, are presented in Fig. 18(b) expressed as a function of F_{av}. In each case, the survival probability peaks when F_{av} is, to within the experimental error, equal to zero. This behavior, and the reduction in peak survival probability at low scaled frequencies, is well reproduced by CTMC simulations.

For high scaled frequencies, $\nu_0 \sim 3.3$, the survival probability is close to one when $F_{av} = 0$. This results because the electron responds as a quasi-free particle in the limit that the external driving frequency is fast compared to the electron orbital motion. Given the short time interval between successive HCPs, the off-set field itself may be viewed as delivering a series of impulsive kicks that are in the opposite direction to those delivered by the HCPs. When these are of equal magnitude, i.e., when $F_{av} = 0$, their effects cancel leading to a maximum in survival probability. In the limit of very high scaled frequencies no ionization is expected for $F_{av} = 0$. For lower frequencies the maximum survival proba-bility drops well below one but still peaks at $F_{av} = 0$. Stabilization for kick frequencies comparable to or less than the orbital frequency of the initial state is at first sight surprising but results from the trapping of the electron in states near the ionization threshold. Such suppression of ionization has been observed in other systems such as the microwave driven Rydberg atom [86] and in con-voy electron transport through solids [87]. It can be understood using Fig. 19 which shows, for a low scaled frequency $\nu_0 \sim 0.33$, the evolution of the elec-tron energy distribution as a function of the number N of kicks. With no offset field applied, i.e., when $\overline{F} = \nu_T \Delta p \neq 0$, the distribution moves rapidly up in energy and across the ionization threshold ($E = 0$), the electron continuing to gain energy from successive kicks. A small part of the initial distribution sur-vives below threshold as a result of dynamical stabilization. In marked contrast, when $F_{av} = 0$, a significant pile up of probability is observed near thresh-old. The energy distribution initially moves rapidly towards higher energies as N increases but ultimately becomes sharply peaked near $E = 0$. Those elec-trons with $E \leq 0$ are in states with n values typically $\gtrsim 800$ and contribute to the large overall survival probability. The classical electron orbital periods asso-ciated with such states are more than an order of magnitude larger than those associated with the parent 351p state. The scaled frequency of the applied HCPs relative to these states is thus large allowing the offset field and the HCPs to be viewed as providing high-frequency alternating kicks. These high-lying states remain localized close to the continuum for extended periods because the elec-tron is transported into a region in phase-space where its response to the ex-ternal pulses closely resembles that of a free electron, i.e., the electron can no longer absorb momentum or energy from the pulse train. This picture is con-

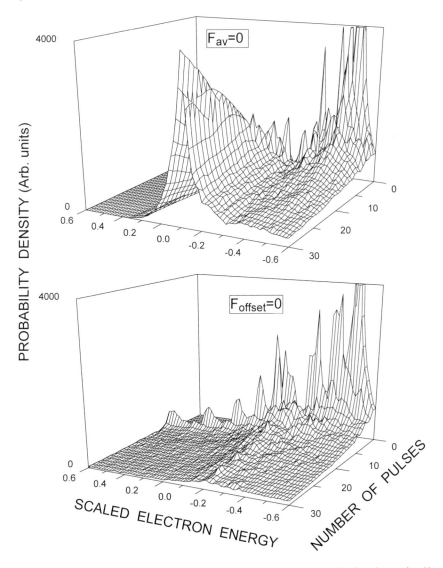

FIG. 19. Calculated time development of the distribution of electron energies E as the number N of HCPs (with $\nu_0 = 0.33$) is increased when no offset field is applied ($F_{\text{offset}} = 0$) and when the average net field experienced by the atoms is zero ($F_{\text{av}} = 0$).

firmed by SFI spectra which show that for $F_{\text{av}} = 0$ the great majority of the surviving atoms are in very high n states whereas for $F_{\text{offset}} = 0$ the final state distribution is broad, peaking at $n \sim 400$, and few very-high-n states are evident.

It is reasonable to expect that, if F_{offset} is small, its presence should not strongly affect the dynamics or the existence of stable islands. As discussed earlier, for $F_{offset} = 0$ the most stable periodic orbit that contributes to dynamical stabilization is a Coulomb orbit cut short by a kick. For small values of F_{offset}, dynamical stabilization is still expected, the most stable orbit being a Stark orbit associated with the Stark Hamiltonian $H_{at} + zF_{offset}$ (which corresponds to a slightly deformed Coulomb orbit) that is cut short by a kick. Figure 19, however, suggests that when F_{offset} is large and $F_{av} = 0$ significant changes in dynamics occur. If the offset field F_{offset} itself is large and is sufficient to ionize a substantial fraction of the parent Rydberg atoms, the presence of the HCP train is observed to result in an increase in survival probability. The HCPs suppress field ionization by periodically directing the electron back towards the atom as it attempts to escape under the influence of F_{offset}. The dynamics of such "pulse-frustrated field ionization" differ substantially from those occurring when F_{offset} is small and might provide a new model system in which to examine dynamical stabilization.

4.4. ALTERNATING KICKS

Atoms subject to a train of HCPs such as illustrated in the inset of Fig. 14(b) that provides kicks that alternate in direction also experience zero average field, i.e., $F_{av} = 0$. Again, cancellation of the net field leads to dramatic changes in the dynamics of the atom [9]. The survival probability for K(351p) atoms following application of a total of $N = 20$ alternating kicks of magnitude $|\Delta p_0| = 0.33$ as a function of scaled pulse repetition frequency ν_0 is shown in Fig. 20. The survival probability depends strongly on ν_0 but, in contrast to the case for unidirectional HCPs, no broad peak in the survival probability is evident. Rather the survival probability tends to increase steadily with increasing ν_0. This behavior is not unexpected. At high scaled frequencies the time delay between successive pulses becomes very short and the effect of one kick is, in essence, immediately reversed by the next (equal but opposite) kick. Figure 20 also includes the results of CTMC simulations which reproduce well the general features of the experimental data. A peak is observed for scaled frequencies at $\nu_0 \sim 1$ which is suggestive of dynamical stabilization.

Poincaré surfaces of section for 1D and 3D hydrogen atoms subject to a series of alternating δ-function kicks with $|\Delta p_0| = 0.3$ are shown in Fig. 21. The time cuts are taken immediately before the first, third, fifth ... kicks. The 3D results are again obtained using a cut at $\rho_0 \sim 0.1$, $p_\rho = 0$. The 1D system features a number of islands bounded by KAM tori that are structurally stable and confining. Those atoms whose initial phase points lie within them remain trapped and

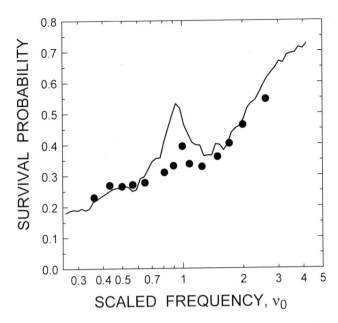

FIG. 20. Measured survival probabilities for K(351p) atoms subject to $N = 20$ kicks of magnitude $|\Delta p_0| = 0.33$ that alternate in sign as a function of scaled frequency ν_0. The symbols show experimental data, the line the results of CTMC simulations. The experimental data are normalized to the simulations.

suffer only small fluctuations in energy. In the 3D case the islands that resemble tori are in fact "leaky" tori or cantori. The trajectories that form the closed loops only intersect the slice of the figure for a finite period of time after which they drift away and eventually become ionized. Leaky tori can slow down diffusion but not completely suppress it. Starting on a leaky torus energy diffusion will initially be slow. As the trajectory wanders towards the edge of the torus it will eventually reach the interconnected chaotic sea, which is accompanied by a transition to fast diffusion and rapid ionization. The narrow maximum in survival probability evident in Fig. 20 at $\nu_0 = 1$, which is associated with this feature, is thus a signature of the delayed onset of chaotic diffusion rather than a signature of true dynamical stabilization. This is further evident from Fig. 14(b), which includes calculated survival probabilities for alternating kicks. These show that, in contrast to the case for unidirectional kicks, the survival probability simply decreases steadily as the number of kicks N is increased for all scaled frequencies, even those in the neighborhood of $\nu_0 \sim 1$. The rate of decrease is, however, significantly lower in this regime leading to a local maximum in the survival probability.

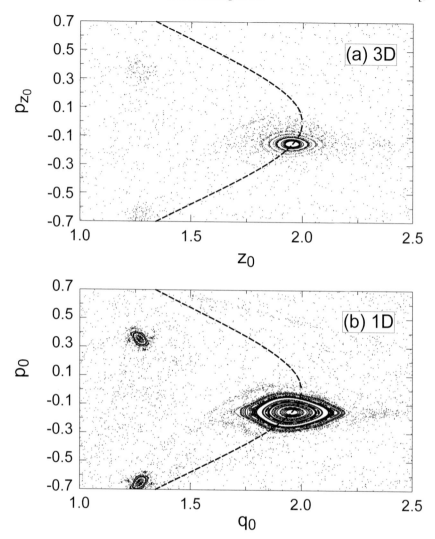

FIG. 21. Poincaré surfaces of section for (a) the 3D and (b) the 1D hydrogen atom subject to a series of alternating δ-function kicks with $|\Delta p_0| = 0.3$. The 3D slice is taken at $\rho_0 = 0.1$, $p_{\rho 0} = 0$. The time cuts are immediately before the $(2N + 1)$th kick ($N = 0, 1, \ldots$).

5. Phase-Space Localization

5.1. DYNAMICAL FILTERING

In recent years there has been increasing interest in the control and manipulation of atomic wavefunctions to generate, for example, wavepackets that mimic classical behavior or that are tailored to specific applications such as data storage [88]. Here we discuss how such control can be achieved by use of a carefully tailored sequence of HCPs [89,90]. The ease with which some targeted final state can be produced is governed by the initial position and momentum distributions of the electron. The more tightly the initial state is localized in phase-space the more straightforward it is to generate some selected final state [28]. One approach to obtaining such phase-space localization takes advantage of the dynamical properties of the kicked atom. As noted previously, the phase-space for this system comprises a number of stable islands enclosed by KAM tori that are embedded in a chaotic sea. After a large number N of kicks those atoms whose initial phase points reside in the chaotic sea are ionized by diffusion to the continuum. That fraction of the initial states that lie in one (or more) of the stable islands remain trapped and survive as a superposition of selected bound states which, immediately before each kick, are strongly localized in phase-space [70]. This "dynamical filtering" technique requires a sizable number of kicks and has the disadvantage that only that part of the initial wavefunction positioned within the stable island survives meaning that a large number of the initial Rydberg atoms are discarded. However, since the wavepacket is associated with dynamical stabilization, it can, in principle, be "trapped" for extended periods by continuing the HCPs and then "released" at a specified time simply by turning off the HCP train [70].

An experimental example of such a dynamically filtered wavepacket is shown in Fig. 22(a). The wavepacket is localized using dynamical filtering by a train of 100 HCPs. Following the pulse train the evolution of the wavepacket is examined using a delayed probe pulse (see inset). Figure 22 shows the measured survival probabilities as a function of the time delay before application of the probe pulse, together with the results of a CTMC simulation. Pronounced oscillations in the survival probability are evident pointing to creation of a well-localized wavepacket. The width Δn of the energy distribution within the stable island where the wavepacket is localized causes dephasing leading to a damping of the oscillations. Figure 22 also includes data recorded under conditions where the initial state is embedded in the chaotic sea. No oscillations are observed indicating the absence of any transient localization.

5.2. NAVIGATING IN PHASE-SPACE

HCP trains can be used to extend control beyond simple dynamical filtering. For example, if the initial state overlaps a period-two island successive kicks will cycle the atom back and forth between upper and lower levels. Control can be exercised over which of these states remains by applying either an even or odd number of kicks [70]. Such a state might thus, in principle, be used to represent a

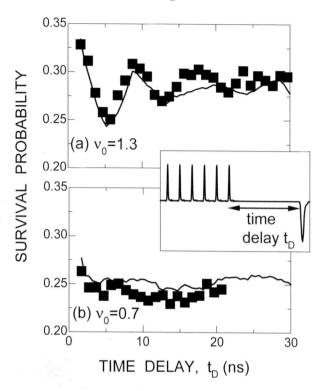

FIG. 22. Rb(390p) Rydberg atom survival probabilities following application of a train of $N = 100$ unidirectional HCPs with scaled frequencies of (a) $\nu_0 = 1.3$ and (b) $\nu_0 = 0.7$ followed by a probe pulse as a function of the time delay t_D (see inset) between the end of the HCP train and the probe pulse. The results of 3D CTMC simulations using the experimental pulse profiles are also included (—).

qbit. Protocols have also been developed to steer Rydberg wavepackets to selected regions of phase-space using chirped HCP trains [91]. The development of such protocols is aided by classical phase-space portraits. The classical phase-space changes its structure as the frequency and strength of the HCPs is modulated. The quantum wavepacket can be made to follow such changes and, therefore, by modulating the HCP train it is possible to direct the wavepacket to targeted regions of phase-space. (Final state control has also been demonstrated in studies using chirped microwave pulses [92].)

Such control can be understood using Fig. 23 which compares the classical Poincaré surfaces of section for a 1D atom with $n_i = 50$ subject to a train of HCPs delivering kicks of strength $\Delta p_0 = -0.1$ and scaled frequencies of $\nu_0 \sim 1.1$ and $\nu_0 \sim 10$. Clearly, changes in scaled frequency lead to dramatic

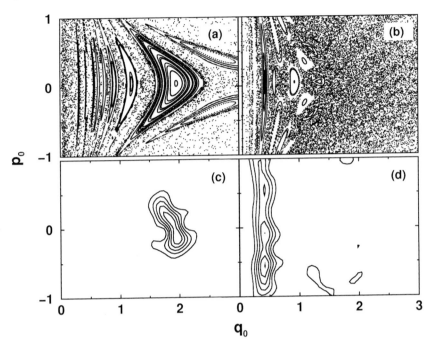

FIG. 23. Poincaré surfaces of section for a 1D hydrogen atom with $n_i = 50$ for (a) $\nu_0 = 1.1$ and (b) $\nu_0 = 10$ for a kick strength $\Delta p_0 = n_i \Delta p = -0.1$. The time cuts are taken immediately before each kick. Also included are the corresponding quantum Husimi distributions (c) before and (d) after application of a chirped train of HCPs.

changes in the phase-space structure, the stable islands being displaced towards the nucleus as ν_0 increases. Thus by chirping the HCP frequency the phase-space structure can be smoothly changed from that in Fig. 23(a) to that in Fig. 23(b). If the frequency chirping is sufficiently slow a wavepacket will be able to follow these changes when a wavepacket initially trapped in the island centered at $(q_0, p_0) = (2.0, 0.05)$ in Fig. 23(a) will be transferred to the island centered at $(q_0, p_0) = (0.4, 0.05)$ in Fig. 23(b), i.e., the wavepacket will have been steered towards the nucleus. This behavior is illustrated by the Husimi distributions included in Fig. 22. That in Fig. 23(c) is for a wavepacket initially produced by applying 14 HCPs with $\Delta p_0 = -0.1$, $\nu_0 \sim 1.1$ to a 1D $n_i = 50$ atom. The HCP train is then chirped such that the scaled frequency is incremented by $\Delta \nu_0 = 0.1$ between successive pulses until (after 89 further pulses) a scaled frequency $\nu_0 = 10$ is reached, whereupon ν_0 is held constant for a further 100 kicks. The Husimi distribution corresponding to the final state is shown in Fig. 23(d). The shift of the probability density towards the nucleus is clearly evident as is the good correspondence between the large stable island at $(q_0, p_0) = (0.4, 0.05)$ in Fig. 23(b) and the final wavefunction, Fig. 23(d).

If the frequency modulation is too rapid (i.e., in the sudden limit) the electron will be unable to follow the shifts in position of the stable islands and will be stranded in the chaotic sea. Classically, the fraction of the probability density left behind in the chaotic sea will eventually ionize. In contrast, quantum mechanical analysis shows that part of the wavepacket left in the chaotic sea does not completely ionize but rather forms a scarred wavefunction which manifests localization around unstable periodic orbits. Fast modulation therefore might provide another opportunity to unravel quantum effects and control and enhance breakdown of classical–quantum correspondence.

Modulating the kick strength rather than frequency shifts both the position and momentum coordinates of the stable islands. However, by simultaneously changing both the kick strength and frequency the shifts in spatial coordinate can be eliminated allowing the momentum coordinate of the wavepacket to be separately controlled. As illustrated in Fig. 24, shifting the main island towards larger values of p requires the use of larger kicks. Increasing the size of the kicks also reduces the size of the stable islands due to resonant island overlap. This decreases the overall survival probability but leads to better localization because the extremities of the initial wavepacket are trimmed off as the island shrinks.

Sudden changes in the strength and frequency of a HCP train can be used to transfer a wavepacket between islands associated with different periodic orbits. Such "island hopping" can be understood by reference to the phase portraits shown in Fig. 25 for an atom subject to HCP trains with $\Delta p_0 = -0.41$, $\nu_0 = 1.1$ and with $\Delta p_0 = -0.07$ and $\nu_0 = 1.3$. A wavepacket trapped inside the main island centered at $(q_0, p_0) = (2, 0.205)$ in Fig. 25(a) can be transferred to either the island at $(q_0, p_0) = (1.7, 0.05)$ in Fig. 25(b) or the coupled period-two islands at $(q_0, p_0) = (2.3, 0.4)$ and $(2.3, -0.3)$ by suddenly changing the period and amplitude of the HCP train. Even further control can be achieved by carefully preparing and positioning the initial wavepacket within the main island.

5.3. TRANSIENT PHASE-SPACE LOCALIZATION

Strong transient phase-space localization can also be obtained by taking advantage of wavepacket propagation in a 1D, or quasi-1D, atom following application of a single HCP [28,55,93]. This is illustrated in Fig. 26 which shows classical and quantum calculations of the time evolution of the phase-space distribution $\rho(q, p, t)$ for a 1D hydrogen atom as a function of time after application of a single HCP of strength $\Delta p_0 = -0.054$. The first distribution $(t_0 = 0)$ corresponds to the phase-space distribution of the initial state prior to application of the kick. The classical distribution is somewhat tighter than its quantum counterpart because the

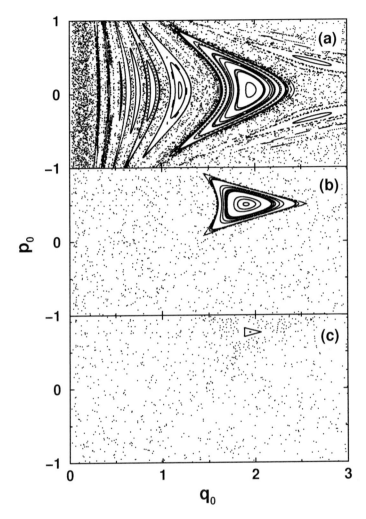

FIG. 24. Poincaré surfaces of section for a 1D hydrogen atom with $n_i = 50$ for (a) $\Delta p_0 = -0.1$, $v_0 = 1.1$; (b) $\Delta p_0 = -1$, $v_0 = 1.77$; and (c) $\Delta p_0 = -1.53$, $v_0 = 2.1$.

former is initially represented by a microcanonical distribution, i.e., a line in phase space, while the latter is represented by a Husimi distribution. Application of the kick leads to a very strong time dependence in the phase-space distribution. The distribution initially broadens until at a scaled time $t_0 = t/T_n \sim 1.45$ the phase-space density is spread over a broad range of coordinates in phase-space. (Since Δp_0 is small, the average period of the electron orbits after the kick remains close to T_n.) The distribution, however, then begins to narrow until at $t_0 \sim 3$ it becomes

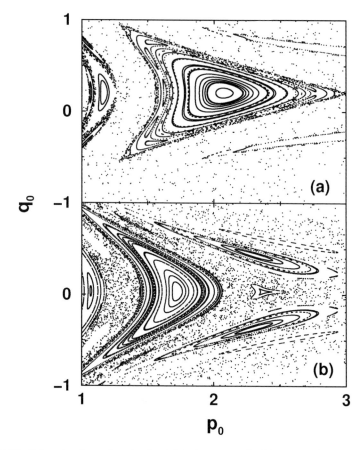

FIG. 25. Poincaré surfaces of section for a 1D hydrogen atom for (a) $\Delta p_0 = -0.41$ and $\nu_0 = 1.1$, and (b) $\Delta p_0 = -0.07$ and $\nu_0 = 1.3$.

strongly localized, the probability density being concentrated in a small region of phase-space near the outer classical turning point. The distribution then once again starts to broaden and the system undergoes a series of periodic variations in localization. As expected, at late times classical–quantum correspondence breaks down. This is illustrated by the results shown for $t_0 = 21.5$, which corresponds to a quantum revival of phase-space localization. At this time the classical distribution has relaxed to an equilibrium phase-space distribution and no localization is evident.

The periodic variation in phase-space localization can also be seen by considering the time development of the widths of the coordinate and momentum distributions $\sigma_q^2 = \langle q^2 \rangle - \langle q \rangle^2$ and $\sigma_p^2 = \langle p^2 \rangle - \langle p \rangle^2$. Figure 27 shows the time

QUANTUM   CLASSICAL

(a) t_0=0

(b) t_0=1.45

(c) t_0=3

(d) t_0=21.5

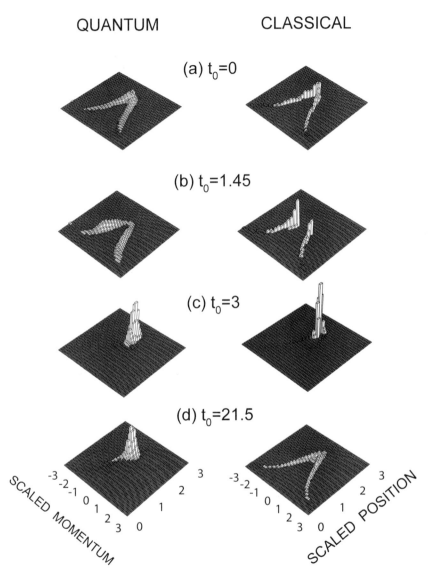

FIG. 26. Quantum (left) and classical (right) phase-space distributions for a 1D hydrogen atom with $n_i = 50$ ($t = 0$) and for selected scaled times ($t_0 = t/T_n = 1.45$, 3 and 21.5) following application of a kick $\Delta p_0 = -0.05$.

evolution of σ_q and σ_p following application of a kick of strength $\Delta p_0 = -0.1$. Remarkably both σ_q and σ_p minimize simultaneously at a scaled time $t_0 \sim 1$. The associated increase in the localization of the system can be evaluated by dividing phase-space into rectangular cells of area $\delta q \delta p$ centered at points (q_i, p_j), $i, j = 1, 2, \ldots$, and using the coarse-grained Renyi entropy [94] defined by

$$S_c(t) = -\ln\left[\delta q \delta p \sum_{i,j=1}^{\infty} \rho_{i,j}^2(t)\right] \tag{21}$$

where $\rho_{i,j}(t)$ is the average value of $\rho(q, p, t)$ in the cell centered at (q_i, p_j). As illustrated in Fig. 27, $S_c(t)$ oscillates with time, its minima coinciding with the

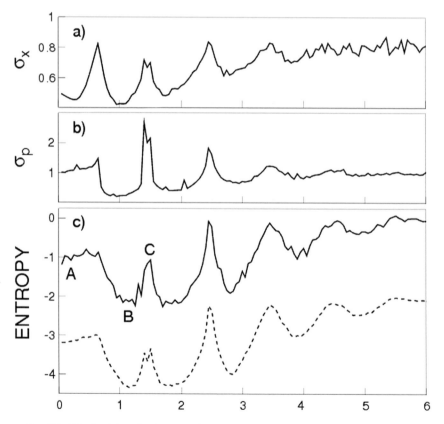

FIG. 27. Calculated time development of the widths of (a) the coordinate and (b) the momentum distributions following application of a kick of strength $\Delta p_0 = -0.1$. The corresponding evolution of the coarse-grained entropy $S_c(t)$ is shown in (c) for bin sizes $\delta p_0 = 0.1$, $\delta q_0 = 0.1$ (—) and $\delta p_0 = 0.0125$, $\delta q_0 = 0.125$ (- - -).

minima for $\sigma_q(t)$ and $\sigma_p(t)$. Detailed analysis shows that phase-space localization is a consequence of phase-space focusing in classical Coulomb systems that is analogous to rainbow scattering [55]. Remarkably, the essential features of this classical analysis remain valid in quantum dynamics.

The time dependence in the localization can be examined experimentally by applying a probe HCP at selected time delays following the initial kick and measuring the survival probability as a function of the size of the probe HCP. Figure 28(a) shows survival probabilities measured with the probe HCP applied directly to the parent quasi-1D K($n = 350$) states [55]. The survival probability varies relatively slowly with increasing probe strength pointing to a broad range of electron momenta. Figure 28(b) shows survival probabilities measured following application of a kick of strength $\Delta p_0 = -0.085$ after a scaled time delay $t_0 \sim 1$. The behavior of the survival probability resembles a step function, confirming that, as suggested by Fig. 27, transient momentum localization does indeed oc-

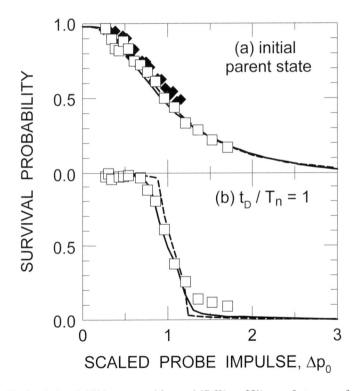

FIG. 28. Survival probabilities measured for quasi-1D K($n = 351$), $m = 0$ atoms as a function of scaled probe impulse Δp_0 when applied (a) directly to the parent state and (b) following a pump kick of scaled strength $\Delta p_0 = -0.085$ and scaled time delay $t_D/T_n = 1$. The results of 1D (- - -) and 3D (—) CTMC simulations are also included.

cur at this time. Because the electron energy distribution is narrow, momentum localization also implies localization of the position coordinate. Thus the data in Fig. 28 provide a clear signature of phase-space localization. Figure. 28 also includes the results of CTMC simulations which are in excellent agreement with the experimental data. Although the degree of localization already achieved is excellent it might be possible to improve it in the future by controlling the detailed shape of the HCP or by application of two (or more) sequential pulses. Theory can be used to evaluate different trial HCP shapes and sequences which can be refined by use of learning algorithms and feedback techniques [95].

Once transient phase-space localization is achieved it can be recovered periodically by positioning the localized state atop a stable island associated with a subsequent HCP train. In this manner transient phase space localization can be maintained for an extended period. Preliminary measurements confirm this and demonstrate stable trapping for microsecond timescales. Furthermore, because of dynamical filtering, the degree of transient phase-space localization is observed to improve with time. Overall, using this approach, it is possible to trap $\gtrsim 50\%$ of the stationary Rydberg atoms initially created by laser excitation in the (principal) stable island following $N > 100$ kicks. This is much higher than the number that can be trapped using the same HCP train relying simply on dynamical filtering and points up the advantage of "prelocalization" using an initial HCP.

6. Outlook

6.1. ATOMIC ENGINEERING

Recent experimental advances now allow the unprecedented manipulation of the internal state of a Rydberg atom opening up a wealth of new opportunities for atomic engineering. Once phase-space localization is achieved it is, in principle, straightforward to transfer the electron to some other final targeted state. For example, for a quasi-1D state localized near the outer classical turning point application of an impulse in the $+z$ direction can be used to create even more highly-elongated very-high-n states. These will allow direct study of the behavior of quasi-1D atoms at very high scaled frequencies $\nu_0 \gtrsim 20$ where it is predicted, for example, that the survival probability following application of a train of N HCPs will not simply decrease monotonically with increasing N [76]. Alternately, if an impulse is applied perpendicular to the z axis the electron can be launched into a near-circular orbit allowing production of high-ℓ (and high-m) states. (The production of high-ℓ states using a field step has also been demonstrated but affords less control [96,97].) Simulations suggest that the centrifugal barrier present in the atomic Hamiltonian for high-ℓ states can profoundly influence their dynamical properties. Since the near-circular state is initially created at a well defined time, the subsequent position of the electron in its "orbit" is known—at least at early times. This enables exploration of the dependence of dynamical stabilization on the position of the electron when the first HCP in the train is applied and whether it is oriented along or perpendicular to the electron path. Clearly, the initial impulse used to generate highly-elongated or near-circular states will lead to population of a distribution of final n states. If required, this distribution can be narrowed substantially by use of dynamical filtering.

6.2. CLASSICAL LIMIT OF QUANTUM MECHANICS

Other questions of interest concern the classical limit of quantum mechanics. While quantum localization has been observed for the kicked rotor and microwave driven Rydberg atom it remains to be seen with the kicked atom. This is perhaps not surprising given that the majority of studies of the kicked atom to date have concerned atoms with $n \gtrsim 350$, well into the (semi)classical regime. However, recent theoretical work suggests that it might be possible to see the effects of the suppression of (diffusive) classical ionization by quantum localization even at $n \sim 350$ by using quasi-1D atoms and applying $N \sim 200\text{--}300$ kicks with scaled frequencies $\nu_0 > 4$ if conditions are chosen such that the ensemble of initial phase points lies in the chaotic sea [87]. This effect should become apparent after a time $t \sim 50T_n$ which is significantly earlier than the time $t \sim nT_n/3$ of the first full revival following application of a single kick [98] thereby placing less stringent demands on phase coherence and on noise in the apparatus.

The study of how the classical world emerges from the quantum world is not only of fundamental interest but could also be of practical importance in quantum information storage and processing. True quantum information is destroyed by the time a system becomes classical. This "decoherence" of the quantum system results from the coupling of the quantum system with its environment. Recent work suggests decoherence rates can be quantified by studying the dephasing of Rydberg wavepackets due to their interaction with "controlled" environments. Such environments might include a gas of atoms or molecules (whose density can be varied) [99] or well characterized noise [100] produced, for example, using a random sequence of electric field pulses. Clearly the goal would be to control and modify decoherence with the view to the design and construction of decoherence-free subspaces within which qbits are embedded.

6.3. FURTHER APPLICATIONS

Another avenue for future investigation is the study of pulse-driven recombination. Bound-free transitions induced by stochastic sequences of momentum transfers play an important role in atom transport through solids [18] and in plasmas [101,102]. No systematic investigations of recombination induced by multiple pulses have been reported although recombination driven by a single HCP has been examined in connection with recombination in plasmas and discussed as an efficient means to recombine positrons and antiprotons to form antihydrogen [102–104].

The extension from Rydberg atoms to "designer atoms," i.e., quantum dots in semiconductor heterostructures, is also of interest. HCPs with durations in the (sub)picosecond regime represent an ultrashort impulsive perturbation for the electronic ground state of quantum dots and quantum wells [105]. Such HCPs could therefore be used to perform momentum spectroscopy of the quantum dot wavefunction in close analogy to the Compton profile mapping for Rydberg atoms (Fig. 28). Such momentum imaging would allow the effective (self-consistent) potential that confines the electronic quasi-particle inside the nano-structure to be determined. HCP-induced momentum imaging could provide much more detailed information on the shape and anisotropy of the confining potential than is available from the excitation spectrum.

As mentioned previously, one of the original motivations to explore the dynamics of a "kicked" Rydberg electron was the classical [18] and quantum transport [20,21] of energetic electrons. One very recent and topical extension of this is the study of the transport of (cold) electrons through mesoscopic structures in the diffusive multiple-scattering limit. In the limit of short-ranged disorder where the correlation length is short compared to the de Broglie wavelength of the electron, transport coefficients such as conductance and shot noise can be modelled by an indeterministic random walk under the influence of s-wave scattering [106]. It is thus tempting to speculate that the simple model of an electron subject to a random (i.e., noisy) isotropic sequence of HCPs could capture the essential features of this fundamental transport problem.

To summarize, the kicked atom provides a novel laboratory in which to examine nonlinear dynamics and the control and manipulation of atomic wavefunctions that can be exploited further in the future as advances in experimental and theoretical techniques continue.

7. Acknowledgements

We are grateful to M.T. Frey, B.E. Tannian, C.L. Stokely, W. Zhao, E. Persson and D. Arbó for their valuable contributions to the work highlighted in this review. FBD acknowledges support from the National Science Foundation under grants Nos. PHY0096392 and PHY0353424 and from the Robert A. Welch Foundation. JCL also received support from the Office of Naval Research. COR acknowledges support from the OBES, U. S. DoE to ORNL which is managed by the UT-Batelle

LLC under Contract No. DE-AC0500OR22725. SY and JB received support from SFB016ADLIS of the FWF (Austria).

8. References

[1] P.M. Koch, K.A.H. van Leeuwen, *Phys. Rep.* **255** (1995) 289;
    M.R.W. Bellerman, P.M. Koch, D. Richards, *Phys. Rev. Lett.* **78** (1997) 3840;
    L. Sirko, P.M. Koch, *Phys. Rev. Lett.* **89** (2002) 274101.
[2] R.V. Jensen, *Nature (London)* **355** (1992) 311;
    M.M. Sanders, R.V. Jensen, *Amer. J. Phys.* **64** (1996) 21.
[3] F. Moore, J. Robinson, C. Bharucha, P. Williams, M.G. Raizen, *Phys. Rev. Lett.* **73** (1994) 2974.
[4] B.G. Klappauf, W.H. Oskay, D.A. Steck, M.G. Raizen, *Phys. Rev. Lett.* **79** (1997) 4790.
[5] D.A. Steck, V. Milner, W.H. Oskay, M.G. Raizen, *Phys. Rev. E* **62** (2000) 3461.
[6] S.A. Gardiner, J.I. Cirac, P. Zoller, *Phys. Rev. Lett.* **79** (1997) 4790.
[7] C.O. Reinhold, J. Burgdörfer, M.T. Frey, F.B. Dunning, *Phys. Rev. Lett.* **79** (1997) 5226.
[8] M.T. Frey, F.B. Dunning, C.O. Reinhold, S. Yoshida, J. Burgdörfer, *Phys. Rev. A* **59** (1999) 1434.
[9] B.E. Tannian, C.L. Stokely, F.B. Dunning, C.O. Reinhold, S. Yoshida, J. Burgdörfer, *Phys. Rev. A* **62** (2000) 043402.
[10] C.O. Reinhold, S. Yoshida, J. Burgdörfer, B.E. Tannian, C.L. Stokely, F.B. Dunning, *J. Phys. B: At. Mol. Opt. Phys.* **34** (2001) L551.
[11] F.B. Dunning, C.O. Reinhold, J. Burgdörfer, *Phys. Scripta* **68** (2003) C44.
[12] A. Buchleitner, D. Delande, J. Zakrzewski, *Phys. Rep.* **368** (2002) 409.
[13] I. Bialynicki-Birula, M. Kalinski, J.H. Eberly, *Phys. Rev. Lett.* **73** (1994) 1777.
[14] H. Maeda, T.F. Gallagher, *Phys. Rev. Lett.* **92** (2004) 133004.
[15] R.R. Jones, N.E. Tielking, D. You, C. Raman, P.H. Bucksbaum, *Phys. Rev. A* **51** (1995) R2687.
[16] T.J. Bensky, G. Haeffler, R.R. Jones, *Phys. Rev. Lett.* **79** (1997) 2018.
[17] J. Burgdörfer, C. Bottcher, *Phys. Rev. Lett.* **61** (1988) 2917.
[18] J. Burgdörfer, J. Gibbons, *Phys. Rev. A* **42** (1990) 1206.
[19] C.O. Reinhold, J. Burgdörfer, J. Kemmler, P. Koschar, *Phys. Rev. A* **45** (1992) R2655.
[20] D.G. Arbó, C.O. Reinhold, P. Kurpick, S. Yoshida, J. Burgdörfer, *Phys. Rev. A* **60** (1999) 109.
[21] T. Minami, C.O. Reinhold, J. Burgdörfer, *Phys. Rev. A* **67** (2003) 022902.
[22] S. Datz, C.D. Moak, O.H. Crawford, H.F. Krause, P.F. Dittner, J.G. del Campo, J.A. Biggerstaff, P.D. Miller, P. Hvelplund, H. Knudsen, *Phys. Rev. Lett.* **40** (1978) 843.
[23] T. Azuma, T. Ito, K. Komaki, Y. Yamazaki, M. Sano, M. Torikoshi, A. Kitagawa, E. Takada, T. Murakami, *Phys. Rev. Lett.* **83** (1999) 528.
[24] E. Persson, S. Puschkarski, X.-M. Tong, J. Burgdörfer, in: F. Krausz (Ed.), *Ultrafast Optics*, Springer, 2004.
[25] C. Wesdorp, F. Robicheaux, L.D. Noordam, *Phys. Rev. Lett.* **87** (2001) 083001.
[26] D. You, R.R. Jones, P.H. Bucksbaum, D.R. Dykaar, *Opt. Lett.* **18** (1993) 290.
[27] N.E. Tielking, R.R. Jones, *Phys. Rev. A* **52** (1995) 1371.
[28] J. Bromage, C.R. Stroud, *Phys. Rev. Lett.* **83** (1999) 4963.
[29] A. Wetzels, A. Gurtler, L.D. Noordam, F. Robicheaux, C. Dinu, H.G. Muller, M.J.J. Vrakking, W.J. van der Zande, *Phys. Rev. Lett.* **89** (2002) 273003.
[30] N.E. Tielking, T.J. Bensky, R.R. Jones, *Phys. Rev. A* **51** (1995) 3370.
[31] M.B. Campbell, T.J. Bensky, R.R. Jones, *Opt. Express* **1** (1997) 197.
[32] B.G. Lindsay, K.A. Smith, F.B. Dunning, *Rev. Sci. Instrum.* **62** (1991) 1656.
[33] M.T. Frey, X. Ling, B.G. Lindsay, K.A. Smith, F.B. Dunning, *Rev. Sci. Instrum.* **64** (1993) 3649.

[34] M.L. Zimmerman, M.G. Littman, M.M. Kash, D. Kleppner, *Phys. Rev. A* **20** (1979) 2251.
[35] T.F. Gallagher, *Rep. Prog. Phys.* **51** (1988) 143;
     T.F. Gallagher, Rydberg Atoms, Cambridge Univ. Press, New York, 1992.
[36] C.L. Stokely, J.C. Lancaster, F.B. Dunning, D.G. Arbó, C.O. Reinhold, J. Burgdörfer, *Phys. Rev. A* **67** (2003) 013403.
[37] E. Persson, S. Yoshida, X.-M. Tong, C.O. Reinhold, J. Burgdörfer, *Phys. Rev. A* **68** (2003) 063406.
[38] C.O. Reinhold, J. Burgdörfer, M.T. Frey, F.B. Dunning, *Phys. Rev. A* **54** (1996) R33.
[39] R.R. Jones, D. You, P.H. Bucksbaum, *Phys. Rev. Lett.* **70** (1993) 1236.
[40] C.O. Reinhold, M. Melles, H. Shao, J. Burgdörfer, *J. Phys. B: At. Mol. Opt. Phys.* **26** (1993) L659.
[41] C.O. Reinhold, J. Burgdörfer, R.R. Jones, C. Raman, P.H. Bucksbaum, *J. Phys. B: At. Mol. Opt. Phys.* **28** (1995) L457.
[42] M.T. Frey, F.B. Dunning, C.O. Reinhold, J. Burgdörfer, *Phys. Rev. A* **53** (1996) R2929;
     P. Kristensen, G.M. Lankhuijzen, L.D. Noordam, *J. Phys. B* **30** (1997) 1481.
[43] S. Yoshida, C.O. Reinhold, J. Burgdörfer, B.E. Tannian, R.A. Popple, F.B. Dunning, *Phys. Rev. A* **58** (1998) 2229;
     B.E. Tannian, R.A. Popple, F.B. Dunning, S. Yoshida, C.O. Reinhold, J. Burgdörfer, *J. Phys. B: At. Mol. Opt. Phys.* **31** (1998) L455.
[44] V. Enss, V. Kostrykin, R. Schrader, *Phys. Rev. A* **50** (1994) 1578.
[45] K.J. LaGuttuta, P. Lerner, *Phys. Rev. A* **49** (1994) R1547.
[46] C.D. Schwieters, J.B. Delos, *Phys. Rev. A* **51** (1995) 1023.
[47] P. Krstic, Y. Hahn, *Phys. Rev. A* **50** (1994) 4629.
[48] A. Bugacov, B. Piraux, M. Pont, R. Shakeshaft, *Phys. Rev. A* **51** (1995) 1490.
[49] C.O. Reinhold, J. Burgdörfer, *Phys. Rev. A* **51** (1995) R3410.
[50] T.F. Gallagher, L.M. Humphrey, R.M. Hill, S.A. Edelstein, *Phys. Rev. Lett.* **37** (1976) 1465.
[51] M.G. Littman, M.M. Kash, D. Kleppner, *Phys. Rev. Lett.* **41** (1978) 103.
[52] T.H. Jeys, G.W. Folz, K.A. Smith, E.J. Beiting, F.G. Kellert, F.B. Dunning, R.F. Stebbings, *Phys. Rev. Lett.* **44** (1980) 390.
[53] R.R. Jones, *Phys. Rev. Lett.* **76** (1996) 3927.
[54] A. Wetzels, A. Gürtler, H.G. Muller, L.D. Noordam, *Europhys. J. D* **14** (2001) 157.
[55] D.G. Arbó, C.O. Reinhold, J. Burgdörfer, A.K. Pattanayak, C.L. Stokely, W. Zhao, J.C. Lancaster, F.B. Dunning, *Phys. Rev. A* **67** (2003) 063401.
[56] J.A. Yeazell, C.R. Stroud, *Phys. Rev. Lett.* **60** (1988) 1494.
[57] M. Mallalieu, C.R. Stroud, *Phys. Rev. A* **49** (1994) 2329.
[58] M.T. Frey, F.B. Dunning, C.O. Reinhold, J. Burgdörfer, *Phys. Rev. A* **55** (1997) R865.
[59] B.E. Tannian, C.L. Stokely, F.B. Dunning, C.O. Reinhold, J. Burgdörfer, *Phys. Rev. A* **64** (2001) 021404(R).
[60] W. Zhao, J.C. Lancaster, F.B. Dunning, C.O. Reinhold, J. Burgdörfer, *Phys. Rev. A* **69** (2004) 041401(R).
[61] A.K. Dhar, M.A. Nagarajan, F.M. Izrailev, R.R. Whitehead, *J. Phys. B* **16** (1983) L17.
[62] A. Carnegie, *J. Phys. B* **17** (1984) 3435.
[63] T. Grozdanov, H.S. Taylor, *J. Phys. B: At. Mol. Opt. Phys.* **20** (1987) 3683.
[64] J. Burgdörfer, *Nucl. Instrum. Meth. Phys. Res. B* **42** (1989) 500.
[65] C.F. Hillermeir, R. Blumel, U. Smilansky, *Phys. Rev. A* **45** (1992) 3486.
[66] M. Melles, C.O. Reinhold, J. Burgdörfer, *Nucl. Instrum. Meth. Phys. Res. B* **79** (1993) 109.
[67] G. Casati, I. Guarneri, G. Mantica, *Phys. Rev. A* **50** (1994) 5018.
[68] H. Wiedemann, J. Mostowski, F. Haake, *Phys. Rev. A* **49** (1994) 1171.
[69] C.O. Reinhold, W. Zhao, J.C. Lancaster, F.B. Dunning, E. Persson, D.G. Arbó, S. Yoshida, J. Burgdörfer, *Phys. Rev. A* **70** (2004) 033402.

[70] C.O. Reinhold, S. Yoshida, J. Burgdörfer, B.E. Tannian, C.L. Stokely, F.B. Dunning, *J. Phys. B: At. Mol. Opt. Phys.* **34** (2001) L551.

[71] S. Yoshida, C.O. Reinhold, P. Krisföfel, J. Burgdörfer, S. Watanabe, F.B. Dunning, *Phys. Rev. A* **59** (1999) R4121.

[72] K. Husimi, *Proc. Phys. Math. Soc. Japan* **22** (1940) 264.

[73] M. Hillery, R.F. O'Connell, M.O. Scully, E.P. Wigner, *Phys. Rep.* **106** (1984) 121.

[74] J.H. Shirley, *Phys. Rev.* **138** (1965) B979.

[75] E.J. Heller, *Phys. Rev. Lett.* **53** (1984) 1515.

[76] S. Yoshida, C.O. Reinhold, J. Burgdörfer, *Phys. Rev. Lett.* **84** (2000) 2602.

[77] G. Casti, B.V. Chirikov, F.M. Izraelev, J. Ford, in: *Lecture Notes in Phys.*, vol. 93, 1979, p. 334.

[78] S. Fishman, D.R. Grempel, R.E. Prange, *Phys. Rev. Lett.* **49** (1982) 509.

[79] P.W. Anderson, *Phys. Rev.* **109** (1958) 1492, *Rev. Mod. Phys.* **50** (1978) 191.

[80] G. Casati, B.V. Chirikov, D.L. Shepelyansky, I. Guarneri, *Phys. Rep.* **154** (1987) 77.

[81] R.V. Jensen, S.M. Susskind, M.M. Sanders, *Phys. Rep.* **201** (1991) 1.

[82] A. Buchleitner, D. Delande, *Chaos Solitons Fractals* **5** (1995) 1125.

[83] S. Yoshida, C.O. Reinhold, P. Kristöfel, J. Burgdörfer, *Phys. Rev. A* **62** (2000) 023408.

[84] E. Persson, S. Yoshida, X.-M. Tong, C.O. Reinhold, J. Burgdörfer, *Phys. Rev. A* **66** (2002) 043407.

[85] W. Zhao, J.C. Lancaster, F.B. Dunning, C.O. Reinhold, J. Burgdörfer, *J. Phys. B: At. Mol. Opt. Phys.* **38** (2005) S191.

[86] M.W. Noel, W.M. Griffith, T.F. Gallagher, *Phys. Rev. Lett.* **83** (1999) 1747.

[87] J. Burgdörfer, in: D. Berenyi, G. Hock (Eds.), *Lecture Notes in Phys.*, vol. 294, Springer, New York, 1988, p. 344.

[88] J. Ahn, T.C. Weinacht, P.H. Bucksbaum, *Science* **287** (2000) 463.

[89] T.C. Weinacht, J. Ahn, P.H. Bucksbaum, *Nature (London)* **397** (1999) 233.

[90] R.R. Jones, L.D. Noordam, *Adv. At. Mol. Opt. Phys.* **38** (1998) 1.

[91] S. Yoshida, C.O. Reinhold, E. Persson, J. Burgdörfer, F.B. Dunning, *J. Phys. B: At. Mol. Opt. Phys.* **28** (2005) S209.

[92] J. Lambert, M.W. Noel, T.F. Gallagher, *Phys. Rev. A* **66** (2002) 053413.

[93] C.L. Stokely, F.B. Dunning, C.O. Reinhold, J. Burgdörfer, *Phys. Rev. A* **65** (2002) 021405.

[94] A.K. Pattanayak, P. Brumer, *Phys. Rev. E* **56** (1977) 5174; A.K. Pattanayak, *Physica D* **148** (2001) 1.

[95] T.C. Weinacht, P.H. Bucksbaum, *J. Opt. B: Quantum Semiclass. Opt.* **4** (2002) R35.

[96] B.E. Tannian, C.L. Stokely, F.B. Dunning, C.O. Reinhold, J. Burgdörfer, *J. Phys. B: At. Mol. Opt. Phys.* **32** (1999) L517.

[97] M.B. Campbell, T.J. Bensky, R.R. Jones, *Phys. Rev. A* **59** (1999) R4117.

[98] Z.D. Gaeta, C.R. Stroud, *Phys. Rev. A* **42** (1990) 6308.

[99] C.O. Reinhold, J. Burgdörfer, F.B. Dunning, *Nucl. Instrum. Meth. Phys. Res. B* (in press).

[100] V. Milner, D.A. Steck, W.H. Oskay, M.G. Raizen, *Phys. Rev. E* **61** (2000) 7223; D.A. Steck, V. Milner, W.H. Oskay, M.G. Raizen, *Phys. Rev. E* **62** (2000) 3461.

[101] J.G. Zeibel, R.R. Jones, *Phys. Rev. Lett.* **89** (2002) 093204.

[102] T.J. Bensky, M.B. Campbell, R.R. Jones, *Phys. Rev. Lett.* **81** (1998) 3112.

[103] C. Wesdorp, F. Robicheaux, L.D. Noordam, *Phys. Rev. Lett.* **84** (2000) 3799.

[104] C. Wesdorp, F. Robicheaux, L.D. Noordam, *Phys. Rev. A* **64** (2001) 033414.

[105] L. Rebohle, F. Schrey, S. Hofer, G. Strasser, K. Unterrainer, *Appl. Phys. Lett.* **81** (2002) 2079.

[106] S. Oberholzer, E. Sukhorukov, C. Schönberger, *Nature* **415** (2002) 765.

ADVANCES IN ATOMIC, MOLECULAR AND OPTICAL PHYSICS, VOL. 52

PHOTONIC STATE TOMOGRAPHY

J.B. ALTEPETER, E.R. JEFFREY and P.G. KWIAT

Department of Physics, University of Illinois at Urbana-Champaign, Urbana, IL 61801, USA

Abstract

   Quantum state tomography is the process by which an identical ensemble of unknown quantum states is completely characterized. A sequence of identical measurements within a series of different bases allow the reconstruction of a complete quantum wavefunction. This article reviews state representation and notation, lays out the theory of ideal tomography, and details the full experimental realization (measurement, electronics, error correction, numerical analysis, measurement choice, and estimation of uncertainties) of a tomographic system applied to polarized photonic qubits.

ISSN 1049-250X
DOI 10.1016/S1049-250X(05)52003-2

Unlike their classical counterparts, quantum states are notoriously difficult to measure. In one sense, the spin of an electron can be in only one of two states, up or down. A simple experiment can discover which state the electron occupies, and further measurements on the same electron will always confirm this answer. However, the simplicity of this picture belies the complex, complete nature of an electron which always appears in one of exactly two states—states which *change* depending on how it is measured.

Quantum state tomography is the process by which any quantum system, including the spin of an electron, can be characterized completely using an ensemble of many identical particles. Measurements of multiple types reconstruct a quantum state from different eigenbases, just as classical tomography can image a three-dimensional object by scanning it from different physical directions. Additional measurements in any single basis bring that dimension into sharper relief.

This article is structured into two major partitions:[1] the theory of tomography (Sections 1 and 2) and the experimental tomography of photonic systems (Sections 3–6). The theoretical sections provide a foundation for quantum state tomography, and should be applicable to any system, including photons (White et al., 1999; Sanaka et al., 2001; Mair et al., 2001; Nambu et al., 2002; Giorgi et al., 2003; Yamamoto et al., 2003; Sergienko et al., 2003; Pittman et al., 2003; O'Brien et al., 2003; Marcikic et al., 2003), spin-$\frac{1}{2}$ particles (as, e.g., are used in NMR quantum computing (Cory et al., 1997; Jones et al., 1997; Weinstein et al., 2001; Laflamme et al., 2002)), and (effectively) 2-level atoms (Monroe, 2002; Schmidt-Kaler et al., 2003). Section 1 provides an introduction to state representation and the notation of this article. Section 2 describes the theory of tomographic reconstruction assuming error-free, exact measurements. The second part of the article contains not only information specific to the experimental measurement of photon polarization (e.g., how to deal with imperfect waveplates), but extensive information on how to deal with real, error-prone systems; information useful to anyone implementing a real tomography system. Section 3 concerns the collection of experimental data (projectors, electronics, systematic error correction) and Section 4 deals with its analysis (numerical techniques for reconstructing states). Sections 5 and 6 describe how to choose which measurements to make and how to estimate the uncertainty in a tomography, respectively.

---

[1] This manuscript is based on a shorter article (Altepeter et al., 2004) which appeared in the special volume *Quantum State Estimation*; here we have rewritten that article in order to be specific to polarization-based photonic tomography and extended the results to include qudits, imperfect waveplates, a new type of maximum likelihood techniques, and information on the choice of measurements. Because the conceptual background is identical, some of the text and figures have been borrowed from that earlier work.

In order to facilitate the use of these techniques by groups and individuals working in any field, a website is available which provides both further details about these techniques and working, documented code for implementing them.[2]

1. State Representation

Before states can be analyzed, it is necessary to understand their representation. In particular, the reconstruction of an unknown state is often simplified by a specific state parametrization.

1.1. REPRESENTATION OF SINGLE-QUBIT STATES

Rather than begin with a general treatment of tomography for an arbitrary number of qubits, throughout this chapter the single-qubit case is investigated initially. This provides the opportunity to strengthen an intuitive grasp of the fundamentals of state representation and tomography before moving on to the more complex (and more useful) general case. In pursuance of this goal, we use graphical representations available only at the single-qubit level.

1.1.1. Pure States, Mixed States, and Diagonal Representations

In general, any single qubit in a pure state can be represented by

$$|\psi\rangle = \alpha|0\rangle + \beta|1\rangle, \tag{1}$$

where α and β are complex and $|\alpha|^2 + |\beta|^2 = 1$ (Nielsen and Chuang, 2000). If the normalization is written implicitly and the global phase is ignored, this can be rewritten as

$$|\psi\rangle = \cos\left(\frac{\theta}{2}\right)|0\rangle + \sin\left(\frac{\theta}{2}\right)e^{i\phi}|1\rangle. \tag{2}$$

EXAMPLE 1 (*Pure states*). Throughout this chapter, examples will be provided using qubits encoded into the electric field polarization of photons. For a single photon, this system has two levels, e.g., horizontal ($|H\rangle \equiv |0\rangle$) and vertical ($|V\rangle \equiv |1\rangle$), with all possible pure polarization states constructed from coherent superpositions of these two states. For example, diagonal, antidiagonal, right-circular and left-circular light are respectively represented by

---

[2] http://www.physics.uiuc.edu/research/QuantumPhotonics/Tomography/

$$|D\rangle \equiv \frac{|H\rangle + |V\rangle}{\sqrt{2}},$$

$$|A\rangle \equiv \frac{|H\rangle - |V\rangle}{\sqrt{2}},$$

$$|R\rangle \equiv \frac{|H\rangle + \mathrm{i}|V\rangle}{\sqrt{2}}, \quad \text{and} \tag{3}$$

$$|L\rangle \equiv \frac{|H\rangle - \mathrm{i}|V\rangle}{\sqrt{2}}.$$

This representation enables the tomography of an ensemble of *identical* pure states, but is insufficient to describe either an ensemble containing a variety of *different* pure states or an ensemble whose members are not pure (perhaps because they are entangled to unobserved degrees of freedom). In this case the overall state is *mixed*.

In general, these mixed states may be described by a probabilistically weighted incoherent sum of pure states, i.e., they behave as if any particle in the ensemble has a specific probability of being in a given pure state, and this state is distinguishably labelled in some way. If it were not distinguishable, the total state's constituent pure states would add coherently (with a definite relative phase), yielding a single pure state.

A mixed state can be represented by a density matrix $\hat{\rho}$, where

$$\hat{\rho} = \sum_i P_i |\psi_i\rangle\langle\psi_i| = \begin{array}{c} \\ |0\rangle \\ |1\rangle \end{array} \overset{\langle 0| \qquad \langle 1|}{\begin{pmatrix} A & Ce^{\mathrm{i}\phi} \\ Ce^{-\mathrm{i}\phi} & B \end{pmatrix}}. \tag{4}$$

P_i is the probabilistic weighting ($\sum_i P_i = 1$), A, B and C are all real and non-negative, $A + B = 1$, and $C \leq \sqrt{AB}$ (Nielsen and Chuang, 2000).

While any ensemble of pure states can be represented in this way, it is also true that *any* ensemble of single-qubit states can be represented by an ensemble of only *two* orthogonal pure states. (Two pure states $|\psi_i\rangle$ and $|\psi_j\rangle$ are orthogonal if $|\langle\psi_i|\psi_j\rangle| = 0$.) For example, if the matrix from Eq. (4) were diagonal, then it would clearly be a probabilistic combination of two orthogonal states, as

$$\begin{array}{c} \\ |0\rangle \\ |1\rangle \end{array} \overset{\langle 0| \quad \langle 1|}{\begin{pmatrix} A & 0 \\ 0 & B \end{pmatrix}} \equiv A|0\rangle\langle 0| + B|1\rangle\langle 1|. \tag{5}$$

However, *any* physical density matrix can be diagonalized, such that

$$
\hat{\rho} = \begin{array}{c} |\psi\rangle \\ |\psi^\perp\rangle \end{array} \overset{\langle\psi| \ \langle\psi^\perp|}{\begin{pmatrix} E_1 & 0 \\ 0 & E_2 \end{pmatrix}} = E_1|\psi\rangle\langle\psi| + E_2|\psi^\perp\rangle\langle\psi^\perp|, \tag{6}
$$

where $\{E_1, E_2\}$ are the eigenvalues of $\hat{\rho}$, and $\{|\psi\rangle, |\psi^\perp\rangle\}$ are the eigenvectors (recall that these eigenvectors are always mutually orthogonal, denoted here by the $\perp$ symbol). Thus the representation of any quantum state, no matter how it is constructed, is identical to that of an ensemble of two orthogonal pure states.[3]

EXAMPLE 2 (*A mixed state*). Now consider measuring a source of photons which emits a one-photon wave packet each second, but alternates—perhaps randomly—between horizontal, vertical, and diagonal polarizations. Their emission time labels these states (in principle) as distinguishable, and so if we ignore that timing information when they are measured, we must represent their state as a density matrix $\hat{\rho}$:

$$
\hat{\rho} = \frac{1}{3}\left(|H\rangle\langle H| + |V\rangle\langle V| + |D\rangle\langle D|\right)
$$

$$
= \frac{1}{3}\left(\begin{array}{c} |H\rangle \\ |V\rangle \end{array} \overset{\langle H| \ \langle V|}{\begin{pmatrix} 1 & 0 \\ 0 & 0 \end{pmatrix}} + \begin{array}{c} |H\rangle \\ |V\rangle \end{array} \overset{\langle H| \ \langle V|}{\begin{pmatrix} 0 & 0 \\ 0 & 1 \end{pmatrix}} + \begin{array}{c} |H\rangle \\ |V\rangle \end{array} \overset{\langle H| \ \langle V|}{\begin{pmatrix} \frac{1}{2} & \frac{1}{2} \\ \frac{1}{2} & \frac{1}{2} \end{pmatrix}} \right)
$$

$$
= \frac{1}{6}\left(\begin{array}{c} |H\rangle \\ |V\rangle \end{array} \overset{\langle H| \ \langle V|}{\begin{pmatrix} 3 & 1 \\ 1 & 3 \end{pmatrix}} \right). \tag{7}
$$

---

[3] It is an interesting question whether all physical states described by a mixed state—e.g., Eq. (6)—are indeed completely equivalent. For example, Lehner, Leonhardt, and Paul discussed the notion that two types of unpolarized light could be considered, depending on whether the incoherence between polarization components arose purely due to an averaging over rapidly varying phases, or from an entanglement with another quantum system altogether (Lehner et al., 1996). This line of thought can even be pushed further, by asking whether all mixed states necessarily arise only from tracing over some unobserved degrees of freedom with which the quantum system has become entangled, or if indeed such entanglement may 'collapse' when the systems involved approach macroscopic size (Kwiat and Englert, 2004). If the latter were true, then there would exist mixed states that could *not* be seen as pure in some larger Hilbert space. In any event, these subtleties of interpretation do not in any way affect experimental results, at least insofar as state tomography is concerned.

When diagonalized,

$$\hat{\rho} = \frac{1}{3} \begin{pmatrix} & & \langle D| & \langle A| \\ |D\rangle & & \begin{pmatrix} 2 & 0 \\ 0 & 1 \end{pmatrix} \\ |A\rangle & & \end{pmatrix} = \frac{2}{3}|D\rangle\langle D| + \frac{1}{3}|A\rangle\langle A|, \tag{8}$$

which, as predicted in Eq. (6), is a sum of only *two* orthogonal states.

Henceforth, the 'bra' and 'ket' labels will be suppressed from written density matrices where the basis is $\{|0\rangle, |1\rangle\}$ or $\{|H\rangle, |V\rangle\}$.

1.1.2. The Stokes Parameters and the Poincaré Sphere

Any single-qubit density matrix $\hat{\rho}$ can be represented uniquely by three parameters $\{S_1, S_2, S_3\}$:

$$\hat{\rho} = \frac{1}{2} \sum_{i=0}^{3} S_i \hat{\sigma}_i. \tag{9}$$

The $\hat{\sigma}_i$ matrices are

$$\hat{\sigma}_0 \equiv \begin{pmatrix} 1 & 0 \\ 0 & 1 \end{pmatrix}, \qquad \hat{\sigma}_1 \equiv \begin{pmatrix} 0 & 1 \\ 1 & 0 \end{pmatrix},$$

$$\hat{\sigma}_2 \equiv \begin{pmatrix} 0 & -i \\ i & 0 \end{pmatrix}, \qquad \hat{\sigma}_3 \equiv \begin{pmatrix} 1 & 0 \\ 0 & -1 \end{pmatrix}, \tag{10}$$

and the S_i values are given by

$$S_i \equiv \mathrm{Tr}\{\hat{\sigma}_i \hat{\rho}\}. \tag{11}$$

For all pure states, $\sum_{i=1}^{3} S_i^2 = 1$; for mixed states, $\sum_{i=1}^{3} S_i^2 < 1$; for the completely mixed state, $\sum_{i=1}^{3} S_i^2 = 0$. Due to normalization, S_0 will always equal one.

Physically, each of these parameters corresponds directly to the outcome of a specific pair of projective measurements:

$$\begin{aligned}
S_0 &= P_{|0\rangle} + P_{|1\rangle}, \\
S_1 &= P_{1/\sqrt{2}(|0\rangle+|1\rangle)} - P_{1/\sqrt{2}(|0\rangle-|1\rangle)}, \\
S_2 &= P_{1/\sqrt{2}(|0\rangle+i|1\rangle)} - P_{1/\sqrt{2}(|0\rangle-i|1\rangle)}, \\
S_3 &= P_{|0\rangle} - P_{|1\rangle},
\end{aligned} \tag{12}$$

where $P_{|\psi\rangle}$ is the probability to measure the state $|\psi\rangle$. As we shall see below, these relationships between probabilities and S parameters are extremely useful in understanding more general operators. Because $P_{|\psi\rangle} + P_{|\psi^\perp\rangle} = 1$, these can be simplified in the single-qubit case, and

$$P_{|\psi\rangle} - P_{|\psi^\perp\rangle} = 2P_{|\psi\rangle} - 1. \tag{13}$$

The probability of projecting a given state $\hat{\rho}$ into the state $|\psi\rangle$ (the probability of measuring $|\psi\rangle$) is given by (Gasiorowitz, 1996):

$$P_{|\psi\rangle} = \langle\psi|\hat{\rho}|\psi\rangle = \text{Tr}\{|\psi\rangle\langle\psi|\hat{\rho}\}. \tag{14}$$

In Eqs. (12) above, the S_i are defined with respect to three states, $|\phi\rangle_i$:

$$
\begin{aligned}
|\phi\rangle_1 &= \frac{1}{\sqrt{2}}\left(|0\rangle + |1\rangle\right) \\
|\phi\rangle_2 &= \frac{1}{\sqrt{2}}\left(|0\rangle + i|1\rangle\right) \\
|\phi\rangle_3 &= |0\rangle,
\end{aligned}
\tag{15}
$$

and their orthogonal complements, $|\phi^\perp\rangle$. Parameters similar to these and serving the same function can be defined with respect to any three arbitrary states, $|\psi_i\rangle$, as long as the matrices $|\psi_i\rangle\langle\psi_i|$ along with the identity are linearly independent. Operators analogous to the $\hat{\sigma}$ operators can be defined relative to these states:

$$\hat{\tau}_i \equiv |\psi_i\rangle\langle\psi_i| - |\psi_i^\perp\rangle\langle\psi_i^\perp|. \tag{16}$$

We can further define an 'S-like' parameter T, given by:

$$T_i \equiv \text{Tr}\{\hat{\tau}_i\hat{\rho}\}. \tag{17}$$

Continuing the previous convention and to complete the set, we define $\hat{\tau}_0 \equiv \hat{\sigma}_0$, which then requires that $T_0 = 1$. Note that the S_i parameters are simply a special case of the T_i, for the case when $\hat{\tau}_i = \hat{\sigma}_i$.

Unlike the specific case of the S parameters which describe *mutually unbiased*[4] (MUB) measurement bases, for biased measurements

$$\hat{\rho} \neq \frac{1}{2}\sum_{i=0}^{3} T_i\hat{\tau}_i. \tag{18}$$

In order to reconstruct the density matrix, the T parameters must first be transformed into the S parameters using Eq. (21).

---

[4] Two measurement bases, $\{\langle\psi_i|\}$ and $\{\langle\psi_j|\}$, are mutually unbiased if $\forall_{i,j} |\langle\psi_i|\psi_j\rangle|^2 = 1/d$, where d is the dimension of the system (for a system of n qubits, $d = 2^n$). A set of measurement bases are mutually unbiased if each basis in the set is mutually unbiased with respect to every other basis in the set. In single-qubit Poincaré space, the axes indicating mutually unbiased measurement bases are at right angles (Lawrence et al., 2002).

EXAMPLE 3 (*The Stokes parameters*). For photon polarization, the S_i are the famous Stokes parameters (though normalized), and correspond to measurements in the D/A, R/L, and H/V bases (Stokes, 1852). In terms of the $\hat{\tau}$ matrices just introduced, we would define a set of basis states $|\psi_1\rangle \equiv |D\rangle$, $|\psi_2\rangle \equiv |R\rangle$, and $|\psi_3\rangle \equiv |H\rangle$. For these analysis bases, $\hat{\tau}_1 = \hat{\sigma}_1$, $\hat{\tau}_2 = \hat{\sigma}_2$, and $\hat{\tau}_3 = \hat{\sigma}_3$ (and therefore $T_i = S_i$ for this specific choice of analysis bases).

As the simplest example, consider the input state $|H\rangle$. Applying Eq. (11), we find that

$$
\begin{aligned}
S_0 &= \mathrm{Tr}\{\hat{\sigma}_0 \hat{\rho}_H\} = 1, \\
S_1 &= \mathrm{Tr}\{\hat{\sigma}_1 \hat{\rho}_H\} = 0, \\
S_2 &= \mathrm{Tr}\{\hat{\sigma}_2 \hat{\rho}_H\} = 0, \\
S_3 &= \mathrm{Tr}\{\hat{\sigma}_3 \hat{\rho}_H\} = 1,
\end{aligned}
\tag{19}
$$

which from Eq. (9) implies that

$$
\hat{\rho}_H = \frac{1}{2}(\hat{\sigma}_0 + \hat{\sigma}_3) = \begin{pmatrix} 1 & 0 \\ 0 & 0 \end{pmatrix}.
\tag{20}
$$

When the Stokes parameters (S_i) are used as coordinates in 3-space, all physically possible states fall within a sphere of radius one (the Poincaré sphere for polarization, the Bloch sphere for electron spin or other two-level systems; see Born and Wolf (1987)). The pure states are found on the surface, states of linear polarization on the equator, circular states at the poles, mixed states within, and the totally mixed state—corresponding to completely unpolarized photons—at the center of the sphere. This provides a very convenient way to visualize one-qubit states (see Fig. 1). The θ and ϕ values from Eq. (2) allow any pure state to be easily mapped onto the sphere surface. These values are the polar coordinates of the pure state they represent on the Poincaré sphere.[5] In addition to mapping states, the sphere can be used to represent any unitary operation as a rotation about an arbitrary axis. For example, waveplates implement rotations about an axis that passes through the equator.

Any state $|\psi_0\rangle$ and its orthogonal partner, $|\psi_0^\perp\rangle$, are found on opposite points of the Poincaré sphere. The line connecting these two points forms an axis of the sphere, useful for visualizing the outcome of a measurement in the $|\psi_0\rangle/|\psi_0^\perp\rangle$ basis. The projection of any state $\hat{\rho}$ (through a line perpendicular to the $|\psi_0\rangle/|\psi_0^\perp\rangle$

---

[5] These polar coordinates are by convention rotated by 90°, so that $\theta = 0$ is on the equator corresponding to the state $|H\rangle$ and $\theta = 90°$, $\phi = 90°$ is at the North Pole corresponding to the state $|R\rangle$. This 90° rotation is particular to the Poincaré representation of photon polarization (Peters et al., 2003); representations of two-level systems on the Bloch sphere do not introduce it.

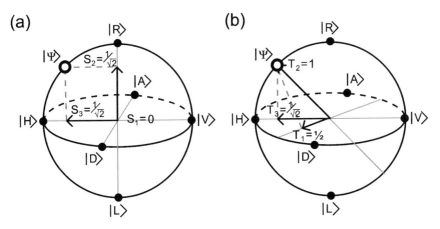

FIG. 1. The Poincaré (or Bloch) sphere. Any single-qubit quantum state $\hat{\rho}$ can be represented by three parameters $T_i = \mathrm{Tr}\{\hat{\tau}_i \hat{\rho}\}$, as long as the operators $\hat{\tau}_i$ in addition to the identity are linearly independent. Physically, the T_i parameters directly correspond to the outcome of a specific projective measurement: $T_i = 2P_i - 1$, where P_i is the probability of success for the measurement. The T_i may be used as coordinates in 3-space. Then all 1-qubit quantum states fall on or within a sphere of radius one. The surface of the sphere corresponds to pure states, the interior to mixed states, and the origin to the totally mixed state. Shown is a particular pure state $|\psi\rangle$, which is completely specified by its projection onto a set of nonparallel basis vectors. (a) When $\hat{\tau}_i = \hat{\sigma}_i$ (the Pauli matrices), the basis vectors are orthogonal, and in this particular case the T_i are equal to the S_i, the well known Stokes parameters, corresponding to measurements of diagonal (S_1), right-circular (S_2), and horizontal (S_3) polarizations. (b) A nonorthogonal coordinate system in Poincaré space. It is possible to represent a state using its projection onto nonorthogonal axes in Poincaré space. This is of particular use when attempting to reconstruct a quantum state from mutually biased measurements. Shown here are the axes corresponding to measurements of 22.5° linear (T_1), elliptical light rotated 22.5° from H towards R (T_2), and horizontal (T_3). Taken from Altepeter et al. (2004).

axis), will lie a distance along this axis corresponding to the relevant Stokes-like parameter ($T = \langle\psi_0|\hat{\rho}|\psi_0\rangle - \langle\psi_0^{\perp}|\hat{\rho}|\psi_0^{\perp}\rangle$).

Thus, just as any point in three-dimensional space can be specified by its projection onto three linearly independent axes, any quantum state can be specified by the three parameters $T_i = \mathrm{Tr}\{\hat{\tau}_i \hat{\rho}\}$, where $\hat{\tau}_{i=1,2,3}$ are linearly independent matrices equal to $|\psi_i\rangle\langle\psi_i| - |\psi_i^{\perp}\rangle\langle\psi_i^{\perp}|$. The $\hat{\tau}_i$ correspond to general Stokes-like parameters for any three linearly independent axes on the Poincaré sphere. However, they can differ from the canonical Stokes axes and need not even be orthogonal. See Fig. 1(b) for an example of state representation using nonorthogonal axes.

In order to use these mutually biased Stokes-like parameters, it is necessary to be able to transform a state from the mutually biased representation to the Stokes representation and vice-versa. In general, for any two representations $S_i =$

$\text{Tr}\{\hat{\sigma}_i \hat{\rho}\}$ and $T_i = \text{Tr}\{\hat{\tau}_i \hat{\rho}\}$ it is possible to transform between them by using

$$\begin{pmatrix} T_0 \\ T_1 \\ T_2 \\ T_3 \end{pmatrix} = \frac{1}{2} \begin{pmatrix} \text{Tr}\{\hat{\tau}_0\hat{\sigma}_0\} & \text{Tr}\{\hat{\tau}_0\hat{\sigma}_1\} & \text{Tr}\{\hat{\tau}_0\hat{\sigma}_2\} & \text{Tr}\{\hat{\tau}_0\hat{\sigma}_3\} \\ \text{Tr}\{\hat{\tau}_1\hat{\sigma}_0\} & \text{Tr}\{\hat{\tau}_1\hat{\sigma}_1\} & \text{Tr}\{\hat{\tau}_1\hat{\sigma}_2\} & \text{Tr}\{\hat{\tau}_1\hat{\sigma}_3\} \\ \text{Tr}\{\hat{\tau}_2\hat{\sigma}_0\} & \text{Tr}\{\hat{\tau}_2\hat{\sigma}_1\} & \text{Tr}\{\hat{\tau}_2\hat{\sigma}_2\} & \text{Tr}\{\hat{\tau}_2\hat{\sigma}_3\} \\ \text{Tr}\{\hat{\tau}_3\hat{\sigma}_0\} & \text{Tr}\{\hat{\tau}_3\hat{\sigma}_1\} & \text{Tr}\{\hat{\tau}_3\hat{\sigma}_2\} & \text{Tr}\{\hat{\tau}_3\hat{\sigma}_3\} \end{pmatrix} \begin{pmatrix} S_0 \\ S_1 \\ S_2 \\ S_3 \end{pmatrix}. \quad (21)$$

This relation allows S parameters to be transformed into any set of T parameters. In order to transform from T to S, we can invert the 4 by 4 matrix in Eq. (21) and multiply both sides by this new matrix. This inversion is possible because we have chosen the $\hat{\tau}_i$ operators to be linearly independent; otherwise the T_i parameters would not specify a single point in Hilbert space.

1.2. REPRESENTATION OF MULTIPLE QUBITS

With the extension of these ideas to cover multiple qubits, it becomes possible to investigate nonclassical features, including the quintessentially quantum mechanical phenomenon of entanglement.

1.2.1. Pure States, Mixed States, and Diagonal Representations

As the name implies, multiple-qubit states are constructed out of individual qubits. As such, the Hilbert space of a many qubit system is spanned by state vectors which are the tensor product of single-qubit state vectors. A general n-qubit system can be written as

$$|\psi\rangle = \sum_{i_1,i_2,\ldots,i_n=0,1} \alpha_{i_1,i_2,\ldots,i_n} |i_1\rangle \otimes |i_2\rangle \otimes \cdots \otimes |i_n\rangle. \quad (22)$$

Here the α_i are complex, $\sum_i |\alpha_i|^2 = 1$, and $\otimes$ denotes a tensor product, used to join component Hilbert spaces. For example, a general two-qubit pure state can be written as

$$|\psi\rangle = \alpha|00\rangle + \beta|01\rangle + \gamma|10\rangle + \delta|11\rangle, \quad (23)$$

where $|00\rangle$ is shorthand for $|0\rangle_1 \otimes |0\rangle_2$.

As before, we represent a general mixed state through an incoherent sum of pure states:

$$\hat{\rho} = \sum_i P_i |\psi_i\rangle\langle\psi_i|. \quad (24)$$

And, as before, this 2^n-by-2^n density matrix representing the n-qubit state may always be diagonalized, allowing any state to be written as

$$\hat{\rho} = \sum_{i=1}^{2^n} P_i |\phi_i\rangle\langle\phi_i|. \quad (25)$$

(24) differs from (25) in that the ϕ_i are necessarily orthogonal ($\langle \phi_i | \phi_j \rangle = \delta_{ij}$), and there are at most 2^n of them; in (24) there could be an arbitrary number of $|\psi_i\rangle$.

EXAMPLE 4 (*A general two-qubit polarization state*). Any two-qubit polarization state can be written as

$$
\hat{\rho} = \begin{array}{c}
\\
|HH\rangle \\
|HV\rangle \\
|VH\rangle \\
|VV\rangle
\end{array}
\begin{array}{cccc}
\langle HH| & \langle HV| & \langle VH| & \langle VV| \\
\left(\begin{array}{cccc}
A_1 & B_1 e^{i\phi_1} & B_2 e^{i\phi_2} & B_3 e^{i\phi_3} \\
B_1 e^{-i\phi_1} & A_2 & B_4 e^{i\phi_4} & B_5 e^{i\phi_5} \\
B_2 e^{-i\phi_2} & B_4 e^{-i\phi_4} & A_3 & B_6 e^{i\phi_6} \\
B_3 e^{-i\phi_3} & B_5 e^{-i\phi_5} & B_6 e^{-i\phi_6} & A_4
\end{array} \right)
\end{array},
\tag{26}
$$

where $\hat{\rho}$ is positive and Hermitian with unit trace. Henceforth, the 'bra' and 'ket' labels will be omitted from density matrices presented in this standard basis.

EXAMPLE 5 (*The Bell states*). Perhaps the most famous examples of pure two-qubit states are the Bell states (Bell, 1964):

$$
|\phi^{\pm}\rangle = \frac{1}{\sqrt{2}}(|HH\rangle \pm |VV\rangle),
$$
$$
|\psi^{\pm}\rangle = \frac{1}{\sqrt{2}}(|HV\rangle \pm |VH\rangle).
\tag{27}
$$

Mixed states of note include the Werner states (Werner, 1989),

$$
\hat{\rho}_W = P|\gamma\rangle\langle\gamma| + (1-P)\frac{1}{4}I,
\tag{28}
$$

where $|\gamma\rangle$ is a maximally entangled state and $\frac{1}{4}I$ is the totally mixed state, and the maximally entangled mixed states (MEMS), which possess the maximum amount of entanglement for a given amount of mixture (Munro et al., 2001).

Measures of entanglement and mixture may be derived from the density matrix; for reference, we now describe several such measures used to characterize a quantum state.

Fidelity. Fidelity is a measure of state overlap:

$$
F(\rho_1, \rho_2) = \left(\mathrm{Tr}\left\{ \sqrt{\sqrt{\rho_1}\rho_2\sqrt{\rho_1}} \right\} \right)^2,
\tag{29}
$$

which—for ρ_1 and ρ_2 pure—simplifies to $\mathrm{Tr}\{\rho_1\rho_2\} = |\langle\psi_1|\psi_2\rangle|^2$ (Jozsa, 1994).[6]

---

[6] Note that some groups use an alternative definition of fidelity, equal to the square root of the formula presented here.

Tangle. The concurrence and tangle are measures of the nonclassical properties of a quantum state (Wooters, 1998; Coffman et al., 2000). For two qubits,[7]
concurrence is defined as follows: consider the non-Hermitian matrix $\widehat{R} =$
$\hat{\rho}\widehat{\Sigma}\hat{\rho}^T\widehat{\Sigma}$ where the superscript T denotes transpose and the 'spin flip matrix' $\widehat{\Sigma}$ is
defined by

$$\hat{\Sigma} \equiv \begin{pmatrix} 0 & 0 & 0 & -1 \\ 0 & 0 & 1 & 0 \\ 0 & 1 & 0 & 0 \\ -1 & 0 & 0 & 0 \end{pmatrix}. \tag{30}$$

If the eigenvalues of $\widehat{R}$, arranged in decreasing order, are given by $r_1 \geq r_2 \geq r_3$
$\geq r_4$, then the concurrence is defined by

$$C = \text{Max}\{0, \sqrt{r_1} - \sqrt{r_2} - \sqrt{r_3} - \sqrt{r_4}\}. \tag{31}$$

The tangle is calculated directly from the concurrence:

$$T \equiv C^2. \tag{32}$$

The tangle (and the concurrence) range from 0 for product states (or, more generally, any incoherent mixture of product states) to a maximum value of 1 for Bell
states.

Entropy and the linear entropy. The Von Neuman entropy quantifies the degree of mixture in a quantum state, and is given by

$$S \equiv -\text{Tr}\{\hat{\rho}\ln[\hat{\rho}]\} = -\sum_i p_i \ln\{p_i\}, \tag{33}$$

where the p_i are the eigenvalues of ρ. The linear entropy (White et al., 1999) is a
more analytically convenient description of state mixture. The linear entropy for
a two-qubit system is defined by

$$S_L = \frac{4}{3}(1 - \text{Tr}\{\hat{\rho}^2\}) = \frac{4}{3}\left(1 - \sum_{a=1}^{4} p_a^2\right), \tag{34}$$

where p_a are the eigenvalues of ρ. Note that for pure states, $\hat{\rho}^2 = \hat{\rho}$, and $\text{Tr}[\hat{\rho}]$ is
always 1, so that S_L ranges from 0 for pure states to 1 for the completely mixed
state.

---

[7] The analysis in this subsection applies to the two-qubit case only. Measures of entanglement for
mixed n-qubit systems are a subject of on-going research: see, for example, (Terhal, 2001) for a recent
survey. In some restricted cases it may be possible to measure entanglement directly, without quantum state tomography; this possibility was investigated in Sancho and Huelga (2000). Also, one can
detect the presence of nonzero entanglement, without quantifying it, using so-called "entanglement
witnesses" (Lewenstein et al., 2000). Elsewhere we describe the trade-offs associated with these other
entanglement characterization schemes (Altepeter et al., 2005).

1.2.2. Multiple Qubit Stokes Parameters

Extending the single-qubit density matrix representation of Eq. (9), any n-qubit state $\hat{\rho}$ may be represented as

$$\hat{\rho} = \frac{1}{2^n} \sum_{i_1,i_2,\dots,i_n=0}^{3} S_{i_1,i_2,\dots,i_n} \hat{\sigma}_{i_1} \otimes \hat{\sigma}_{i_2} \otimes \cdots \otimes \hat{\sigma}_{i_n}. \tag{35}$$

Normalization requires that $S_{0,0,\dots,0} = 1$, leaving $4^n - 1$ real parameters (the multiple-qubit analog of the single-qubit Stokes parameters) to identify any point in Hilbert space, just as three parameters determined the exact position of a one-qubit state in the Bloch/Poincaré sphere. Already for two qubits, the state space is much larger, requiring 15 independent real parameters to describe it. For this reason, there is no convenient graphical picture of this space, as there was in the single-qubit case (see, however, the interesting approaches made by Zyczkowski (2000, 2001)).

For multiple qubits the link between the multiple-qubit Stokes parameters (James et al., 2001; Abouraddy et al., 2002) and measurement probabilities still exists. The formalism of $\hat{\tau}$ operators also still holds for larger qubit systems, so that

$$T = \text{Tr}\{\hat{\tau}\hat{\rho}\}. \tag{36}$$

For 'local' measurements (a local measurement is the tensor product of a number of single-qubit measurements: the first projecting qubit one along $\hat{\tau}_{i_1}$, the second qubit two along $\hat{\tau}_{i_2}$, etc.), $\hat{\tau} = \hat{\tau}_{i_1} \otimes \hat{\tau}_{i_2} \otimes \cdots \otimes \hat{\tau}_{i_n}$. Combining Eqs. (35) and (36),

$$\begin{aligned}
T_{i_1,i_2,\dots,i_n} &= \text{Tr}\{(\hat{\tau}_{i_1} \otimes \hat{\tau}_{i_2} \otimes \cdots \otimes \hat{\tau}_{i_n})\hat{\rho}\} \\
&= \frac{1}{2^n} \sum_{j_1,j_2,\dots,j_n=0}^{3} \text{Tr}\{\hat{\tau}_{i_1}\hat{\sigma}_{j_1}\} \text{Tr}\{\hat{\tau}_{i_2}\hat{\sigma}_{j_2}\} \cdots \text{Tr}\{\hat{\tau}_{i_n}\hat{\sigma}_{j_n}\} S_{j_1,j_2,\dots,j_n}.
\end{aligned} \tag{37}$$

Recall that for single qubits,

$$\begin{aligned}
T_{i=1,2,3} &= P_{|\psi_i\rangle} - P_{|\psi_i^\perp\rangle}, \\
T_0 &= P_{|\psi\rangle} + P_{|\psi^\perp\rangle} = 1, \quad \forall \psi.
\end{aligned} \tag{38}$$

Therefore, for an n-qubit system,

$$\begin{aligned}
T_{i_1,i_2,\dots,i_n} = \left(P_{|\psi_{i_1}\rangle} \pm P_{|\psi_{i_1}^\perp\rangle}\right) \otimes \left(P_{|\psi_{i_2}\rangle} \pm P_{|\psi_{i_2}^\perp\rangle}\right) \otimes \cdots \\
\otimes \left(P_{|\psi_{i_n}\rangle} \pm P_{|\psi_{i_n}^\perp\rangle}\right),
\end{aligned} \tag{39}$$

where the plus sign is used for a zero index and the minus sign is used for a nonzero index. For a two-qubit system where $i_1 \neq 0$ and $i_2 \neq 0$, T_{i_1,i_2} simplifies dramatically, giving

$$T_{i_1,i_2} = (P_{|\psi_{i_1}\rangle} - P_{|\psi_{i_1}^\perp\rangle}) \otimes (P_{|\psi_{i_2}\rangle} - P_{|\psi_{i_2}^\perp\rangle})$$

$$= P_{|\psi_{i_1}\rangle|\psi_{i_2}\rangle} - P_{|\psi_{i_1}\rangle|\psi_{i_2}^\perp\rangle} - P_{|\psi_{i_1}^\perp\rangle|\psi_{i_2}\rangle} + P_{|\psi_{i_1}^\perp\rangle|\psi_{i_2}^\perp\rangle}. \qquad (40)$$

This relation will be crucial for rebuilding a two-qubit state from local measurements.

As before, we are not restricted to multiple-qubit Stokes parameters based only on mutually unbiased operators. Extending Eq. (21) to multiple qubits, and again assuming two representations $S_{i_1,i_2,\ldots,i_n} = \text{Tr}\{(\hat\sigma_{i_1} \otimes \hat\sigma_{i_2} \otimes \cdots \otimes \hat\sigma_{i_n})\hat\rho\}$, and $T_{i_1,i_2,\ldots,i_n} = \text{Tr}\{(\hat\tau_{i_1} \otimes \hat\tau_{i_2} \otimes \cdots \otimes \hat\tau_{i_n})\hat\rho\}$,

$$T_{i_1,i_2,\ldots,i_n} = \frac{1}{2^n} \sum_{j_1,j_2,\ldots,j_n=0}^{3} \text{Tr}\Big\{\big(\hat\tau_{i_1} \otimes \hat\tau_{i_2} \otimes \cdots \otimes \hat\tau_{i_n}\big)$$

$$\times \big(\hat\sigma_{j_1} \otimes \hat\sigma_{j_2} \otimes \cdots \otimes \hat\sigma_{j_n}\big)\Big\} S_{j_1,j_2,\ldots,j_n}. \qquad (41)$$

In general, a given $\hat\tau$ operator is not uniquely mapped to a single pair of analysis states. For example, consider measurements of $|H\rangle$ and $|V\rangle$ corresponding to $\hat\tau_H = |H\rangle\langle H| - |V\rangle\langle V| = \hat\sigma_3$ and $\hat\tau_V = |V\rangle\langle V| - |H\rangle\langle H| = -\hat\sigma_3$. Therefore, $\hat\tau_{H,H} \equiv \hat\sigma_3 \otimes \hat\sigma_3 = -\hat\sigma_3 \otimes -\hat\sigma_3 \equiv \hat\tau_{V,V}$. This artifact of the mathematics does not in practice affect the results of a tomography.

EXAMPLE 6 (*A separable two-qubit polarization state*). Consider the state $|HH\rangle$. Following the example in Eqs. (20),

$$\hat\rho_{HH} = |HH\rangle\langle HH|$$

$$= \frac{1}{2}(\hat\sigma_0 + \hat\sigma_3) \otimes \frac{1}{2}(\hat\sigma_0 + \hat\sigma_3)$$

$$= \frac{1}{4}(\hat\sigma_0 \otimes \hat\sigma_0 + \hat\sigma_0 \otimes \hat\sigma_3 + \hat\sigma_3 \otimes \hat\sigma_0 + \hat\sigma_3 \otimes \hat\sigma_3). \qquad (42)$$

This implies that for this state there are exactly four nonzero two-qubit Stokes parameters: $S_{0,0}$, $S_{0,3}$, $S_{3,0}$, and $S_{3,3}$—all of which are equal to one. (As earlier, for the special case when $\hat\tau_{i,j} = \hat\sigma_{i,j}$, we relabel the $T_{i,j}$ as $S_{i,j}$, the two-qubit Stokes parameters (James et al., 2001; Abouraddy et al., 2002).) The separable nature of this state makes it easy to calculate the two-qubit Stokes decomposition.

EXAMPLE 7 (*The singlet state*). If instead we investigate the entangled state $|\psi^-\rangle \equiv (|HV\rangle - |VH\rangle)/\sqrt{2}$, it will be necessary to calculate each two-qubit Stokes parameter from the $\hat\sigma$ matrices. As an example, consider $\hat\sigma_{3,3} \equiv \hat\sigma_3 \otimes \hat\sigma_3$, for which

$$S_{3,3} = \text{Tr}\{\hat{\sigma}_{3,3}|\psi^-\rangle\langle\psi^-|\} = -1. \tag{43}$$

We could instead calculate $S_{3,3}$ directly from probability outcomes of measurements on $|\psi^-\rangle$:

$$\begin{aligned}
S_{3,3} &= (P_H - P_V) \otimes (P_H - P_V) \\
&= P_{HH} - P_{HV} - P_{VH} + P_{VV} \\
&= 0 - \frac{1}{2} - \frac{1}{2} + 0 = -1.
\end{aligned} \tag{44}$$

Continuing on, we measure $S_{0,3}$:

$$\begin{aligned}
S_{0,3} &= (P_H + P_V) \otimes (P_H - P_V) \\
&= P_{HH} - P_{HV} + P_{VH} - P_{VV} \\
&= 0 - \frac{1}{2} + \frac{1}{2} - 0 = 0.
\end{aligned} \tag{45}$$

Here the signs of the probabilities changed due to the zero index in $S_{0,3}$. These results would have been the same even if the analysis bases of the first qubit had been shifted to any other orthogonal basis, i.e., $S_{0,3} = (P_\psi + P_{\psi\perp}) \otimes (P_H - P_V)$.

If the method above is continued for all the Stokes parameters, one concludes that

$$\begin{aligned}
\hat{\rho}_{\psi^-} &= \frac{1}{2}\big(|HV\rangle - |VH\rangle\big)\big(\langle HV| - \langle VH|\big) \\
&= \frac{1}{4}\big(\hat{\sigma}_0 \otimes \hat{\sigma}_0 - \hat{\sigma}_1 \otimes \hat{\sigma}_1 - \hat{\sigma}_2 \otimes \hat{\sigma}_2 - \hat{\sigma}_3 \otimes \hat{\sigma}_3\big).
\end{aligned} \tag{46}$$

1.3. REPRESENTATION OF NONQUBIT SYSTEMS

Although most interest within the field of quantum information and computation has focused on two-level systems (qubits) due to their simplicity, availability, and similarity to classical bits, nature contains a multitude of many-level systems, both discrete and continuous. A discussion of continuous systems is beyond the scope of this work—see Leonhardt (1997), but we will briefly address here the representation and tomography of discrete, d-level systems ("qudits"). For a more detailed description of qudit tomography, see Thew et al. (2002).

1.3.1. Pure, Mixed, and Diagonal Representations

Directly extending Eqs. (1) and (2), a d-level qudit can be represented as

$$|\psi\rangle = \alpha_0|0\rangle + \alpha_1|1\rangle + \cdots + \alpha_{d-1}|d-1\rangle, \tag{47}$$

where $\sum_i |\alpha_i|^2 = 1$. Mixed qudit states can likewise be represented by generalizing Eqs. (4) and (6):

$$\rho = \sum_k P_k |\phi_k\rangle\langle\phi_k| \tag{48}$$

$$= \sum_{i=0}^{d-1} P_i |\psi_i\rangle\langle\psi_i|. \tag{49}$$

Here $\{|\phi_k\rangle\}$ is completely unrestricted while $|\langle\psi_i|\psi_j\rangle| = \delta_{ij}$. In other words, while any mixed state is an incoherent superposition of an undetermined number of pure states, any mixed state can be represented by an incoherent superposition of only n orthogonal states (the diagonal representation).

EXAMPLE 8 (*Orbital angular momentum modes*). The Laguerre–Gaussian modes of an optical field propagating in the z direction possess z components of orbital angular momentum that serve as the quantum numbers of a multiple-level photonic system that has recently been studied for quantum information (Mair et al., 2001; Langford et al., 2004; Arnaut and Barbosa, 2000). Consider a qudit system with an infinite number of levels representing the quantization of orbital angular momentum. A superposition of the three lowest angular momentum levels would look like

$$|\psi\rangle = |+1\rangle + |0\rangle + |-1\rangle, \tag{50}$$

where $|+1\rangle$ ($|-1\rangle$) corresponds to a mode where each photon has $+\hbar$ ($-\hbar$) z-component of orbital angular momentum, and $|0\rangle$ corresponds to a zero angular momentum mode, e.g., a mode having a Gaussian transverse profile. Using specially designed holograms, these states can be measured and interconverted (Allen et al., 2004).

1.3.2. Qudit Stokes Parameters

In order to completely generalize the qubit mathematics laid out previously to the qudit case, it is necessary to find Stokes-like parameters which satisfy the following conditions:

$$\hat{\rho} = \sum_{i=0}^{n} S_i \hat{\sigma}_i, \tag{51}$$

$$S_i \equiv \text{Tr}\{\hat{\sigma}_i \hat{\rho}\}. \tag{52}$$

In addition, in order to easily generalize the tomographic techniques of the next section, it will be necessary to find S_i as a function of measurable probabilities:

$$S_i = \mathcal{F}(\{P_{|\psi\rangle}\}). \tag{53}$$

Obviously, it would be ideal to find a simple form similar to the qubit $\hat{\sigma}$ matrices. Conveniently, the general qudit sigma matrices and corresponding S_i parameters can be divided into three groups ($\{\hat{\sigma}_i^X, \hat{\sigma}_i^Y, \hat{\sigma}_i^Z\}$ and $\{S_i^X, S_i^Y, S_i^Z\}$), according to their similarity to $\hat{\sigma}_x = \hat{\sigma}_1$, $\hat{\sigma}_y = \hat{\sigma}_2$, and $\hat{\sigma}_z = \hat{\sigma}_3$, respectively (Thew et al., 2002). Using these divisions, we can expand Eq. (51):

$$\hat{\rho} = S_0\hat{\sigma}_0 + \sum_{\substack{j,k\in\{0,1,\ldots,n-1\} \\ j\neq k}} \left(S_{j,k}^X\hat{\sigma}_{j,k}^X + S_{j,k}^Y\hat{\sigma}_{j,k}^Y\right) + \sum_{r=1}^{n-1} S_r^Z\hat{\sigma}_r^Z. \tag{54}$$

Investigating the simplest group first, it is unsurprising that

$$\hat{\sigma}_0 = I, \qquad S_0 = 1, \tag{55}$$

continuing the previous qubit convention. The X and Y related variables are defined almost identically to their predecessors:

$$\hat{\sigma}_{j,k}^X = |j\rangle\langle k| + |k\rangle\langle j|, \tag{56}$$

$$S_{j,k}^X = P_{1/\sqrt{2}(|j\rangle+|k\rangle)} - P_{1/\sqrt{2}(|j\rangle-|k\rangle)}, \tag{57}$$

$$\hat{\sigma}_{j,k}^Y = -i\left(|j\rangle\langle k| - |k\rangle\langle j|\right), \tag{58}$$

$$S_{j,k}^Y = P_{1/\sqrt{2}(|j\rangle+i|k\rangle)} - P_{1/\sqrt{2}(|j\rangle-i|k\rangle)}. \tag{59}$$

The definitions for $\hat{\sigma}_i^Z$ and S_i^Z are slightly more complicated:

$$\hat{\sigma}_r^Z = \sqrt{\frac{2}{r(r+1)}}\left[\left(\sum_{j=0}^{r-1}|j\rangle\langle j|\right) - r|r\rangle\langle r|\right], \tag{60}$$

$$S_r^Z = \sqrt{\frac{2}{r(r+1)}}\left[\left(\sum_{j=0}^{r-1}P_{|j\rangle}\right) - r P_{|r\rangle}\right]. \tag{61}$$

These definitions complete the set of n^2 sigma matrices, and have a slightly more complex form in order to satisfy $\mathrm{Tr}[\hat{\sigma}_i] = 0$ and $\mathrm{Tr}[\hat{\sigma}_i\hat{\sigma}_j] = 2\delta_{ij}$ (these definitions apply to all $\hat{\sigma}_i$ except $\hat{\sigma}_0$).

EXAMPLE 9 (*The qutrit*). For a 3-level system ($|0\rangle$, $|1\rangle$, and $|2\rangle$), the $\hat{\sigma}$ matrices can be defined as:

$$\hat{\sigma}_0 = \begin{pmatrix} 1 & 0 & 0 \\ 0 & 1 & 0 \\ 0 & 0 & 1 \end{pmatrix}, \qquad \hat{\sigma}_1^Z = \begin{pmatrix} 1 & 0 & 0 \\ 0 & -1 & 0 \\ 0 & 0 & 0 \end{pmatrix},$$

$$\hat{\sigma}_2^Z = \sqrt{\frac{1}{3}} \begin{pmatrix} 1 & 0 & 0 \\ 0 & 1 & 0 \\ 0 & 0 & -2 \end{pmatrix}, \qquad \hat{\sigma}_{1,2}^X = \begin{pmatrix} 0 & 1 & 0 \\ 1 & 0 & 0 \\ 0 & 0 & 0 \end{pmatrix},$$

$$\hat{\sigma}_{1,3}^X = \begin{pmatrix} 0 & 0 & 1 \\ 0 & 0 & 0 \\ 1 & 0 & 0 \end{pmatrix}, \qquad \hat{\sigma}_{2,3}^X = \begin{pmatrix} 0 & 0 & 0 \\ 0 & 0 & 1 \\ 0 & 1 & 0 \end{pmatrix},$$

$$\hat{\sigma}_{1,2}^Y = \begin{pmatrix} 0 & -i & 0 \\ i & 0 & 0 \\ 0 & 0 & 0 \end{pmatrix}, \qquad \hat{\sigma}_{1,3}^Y = \begin{pmatrix} 0 & 0 & -i \\ 0 & 0 & 0 \\ i & 0 & 0 \end{pmatrix},$$

$$\hat{\sigma}_{2,3}^Y = \begin{pmatrix} 0 & 0 & 0 \\ 0 & 0 & -i \\ 0 & i & 0 \end{pmatrix}.$$

Expanding the S_i parameters in terms of probabilities, we find that:

$$S_0 = 1,$$
$$S_{1,2}^X = P_{1/\sqrt{2}(|0\rangle+|1\rangle)} - P_{1/\sqrt{2}(|0\rangle-|1\rangle)},$$
$$S_{1,3}^X = P_{1/\sqrt{2}(|0\rangle+|2\rangle)} - P_{1/\sqrt{2}(|0\rangle-|2\rangle)},$$
$$S_{2,3}^X = P_{1/\sqrt{2}(|1\rangle+|2\rangle)} - P_{1/\sqrt{2}(|1\rangle-|2\rangle)},$$
$$S_{1,2}^Y = P_{1/\sqrt{2}(|0\rangle+i|1\rangle)} - P_{1/\sqrt{2}(|0\rangle-i|1\rangle)},$$
$$S_{1,3}^Y = P_{1/\sqrt{2}(|0\rangle+i|2\rangle)} - P_{1/\sqrt{2}(|0\rangle-i|2\rangle)},$$
$$S_{2,3}^Y = P_{1/\sqrt{2}(|1\rangle+i|2\rangle)} - P_{1/\sqrt{2}(|1\rangle-i|2\rangle)},$$
$$S_1^Z = P_{|0\rangle} - P_{|1\rangle},$$
$$S_2^Z = \frac{1}{\sqrt{3}}\left(P_{|0\rangle} + P_{|1\rangle} - 2P_{|2\rangle}\right).$$

2. Tomography of Ideal Systems

The goal of tomography is to reconstruct the density matrix of an ensemble of particles through a series of measurements. In practice, this can never be performed exactly, as an infinite number of particles would be required to eliminate statistical error. If exact measurements were taken on an infinite number of identically prepared systems, each measurement would yield an exact probability of success, which could then be used to reconstruct a density matrix. Though unrealistic, it is highly illustrative to examine this exact tomography before considering the more general treatment. Hence, this section will treat all measurements as yielding exact probabilities, and ignore all sources of error in those measurements.

2.1. SINGLE-QUBIT TOMOGRAPHY

Although reconstructive tomography of any size system follows the same general procedure, beginning with tomography of a single qubit allows the visualization of each step using the Poincaré sphere, in addition to providing a simpler mathematical introduction.

2.1.1. Visualization of Single-Qubit Tomography

Exact single-qubit tomography requires a sequence of three linearly independent measurements. Each measurement exactly specifies one degree of freedom for the measured state, reducing the free parameters of the unknown state's possible Hilbert space by one.

As an example, consider measuring R, D, and H on the partially mixed state

$$\hat{\rho} = \begin{pmatrix} \frac{5}{8} & \frac{-i}{2\sqrt{2}} \\ \frac{i}{2\sqrt{2}} & \frac{3}{8} \end{pmatrix}. \tag{62}$$

Rewriting the state using Eq. (9) as

$$\hat{\rho} = \frac{1}{2}\left(\hat{\sigma}_0 + \frac{1}{\sqrt{2}}\hat{\sigma}_2 + \frac{1}{4}\hat{\sigma}_3\right) \tag{63}$$

allows us to read off the normalized Stokes parameters corresponding to these measurements:

$$S_1 = 0, \quad S_2 = \frac{1}{\sqrt{2}}, \quad \text{and} \quad S_3 = \frac{1}{4}. \tag{64}$$

As always, $S_0 = 1$ due to normalization. Measuring R (which determines S_2) first, and looking to the Poincaré sphere, we determine that the unknown state must lie in the $z = 1/\sqrt{2}$ plane (as $S_2 = 1/\sqrt{2}$). A measurement in the D basis (with the result $P_D = P_A = \frac{1}{2}$) further constrains the state to the $y = 0$ plane, resulting in a total confinement to a line parallel to and directly above the x axis. The final measurement of H pinpoints the state. This process is illustrated in Fig. 2(a). Obviously the order of the measurements is irrelevant: it is the intersection point of three orthogonal planes that defines the location of the state.

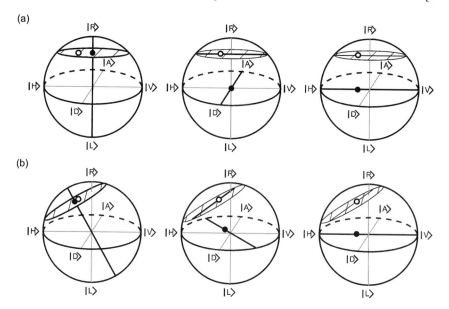

FIG. 2. A sequence of three linearly independent measurements isolates a single quantum state in Hilbert space (shown here as an open circle in the Poincaré sphere representation). The first measurement isolates the unknown state to a plane perpendicular to the measurement basis. Further measurements isolate the state to the intersections of nonparallel planes, which for the second and third measurements correspond to a line and finally a point. The black dots shown correspond to the projection of the unknown state onto the measurement axes, which determines the position of the aforementioned planes. (a) A sequence of measurements along the right-circular, diagonal, and horizontal axes. (b) A sequence of measurements on the same state taken using nonorthogonal projections: elliptical light rotated 30° from *H* towards *R*, 22.5° linear, and horizontal. Taken from Altepeter et al. (2004).

If instead measurements are made along nonorthogonal axes, a very similar picture develops, as indicated in Fig. 2(b). The first measurement always isolates the unknown state to a plane, the second to a line, and the third to a point.

Of course, in practice, the experimenter has no knowledge of the unknown state before a tomography. The set of the measured probabilities, transformed into the Stokes parameters as above, allows a state to be directly reconstructed.

2.1.2. A Mathematical Look at Single-Qubit Tomography

Using the tools developed in the first section of this chapter, single-qubit tomography is relatively straightforward. Recall Eq. (9), $\hat{\rho} = \frac{1}{2} \sum_{i=0}^{3} S_i \hat{\sigma}_i$. Considering that S_1, S_2, and S_3 completely determine the state, we need only measure them to complete the tomography. From Eq. (13), $S_{j>0} = 2P_{|\psi\rangle} - 1$, therefore three measurements in the $|0\rangle$, $\frac{1}{\sqrt{2}}(|0\rangle + |1\rangle)$, and $\frac{1}{\sqrt{2}}(|0\rangle + i|1\rangle)$ bases will completely specify the unknown state. If instead measurements are made in another basis, even a nonorthogonal one, they can be easily related back to the S_i parameters, and therefore the density matrix, by means of Eq. (21).

While this procedure is straightforward, there is one subtlety which will become important in the multiple-qubit case. Projective measurements generally refer to the measurement of a single basis state and return a single value between zero and one. This corresponds, for example, to an electron beam passing through a Stern–Gerlach apparatus with a detector placed at one output. While a single detector and knowledge of the input particle intensity will—in the one-qubit case—completely determine a single Stokes parameter, one could collect data from both outputs of the Stern–Gerlach device. This would measure the probability of projecting not only onto the state $|\psi\rangle$, but also onto $|\psi^{\perp}\rangle$, without needing to know the input intensity. All physical measurements on single qubits, regardless of implementation, can in principle be measured this way (though in practice measurements of some qubit systems may typically detect a population in only *one* of the states, as in Kielpinski et al. (2001)). We will see below that although one detector functions as well as two in the single-qubit case, this situation will not persist into higher dimensions.

2.2. MULTIPLE-QUBIT TOMOGRAPHY

The same methods used to reconstruct an unknown single-qubit state can be applied to multiple-qubit systems. Just as each single-qubit Stokes vector can be expressed in terms of measurable probabilities—Eq. (12), each multiple-qubit Stokes vector can be measured in terms of the probabilities of projecting the multiple-qubit state into a sequence of separable bases—Eq. (39).

Using the most naive method, an n-qubit system, represented by 4^n Stokes parameters, would require $4^n \times 2^n$ probabilities to be reconstructed (2^n probabilities for each of 4^n Stokes parameters).

Of course, because an n-qubit density matrix contains $4^n - 1$ free parameters, the $4^n \times 2^n$ measured probabilities must be linearly dependent. As expected, by using the extra information that measurements of complete orthogonal bases must sum to one (e.g., $P_{HH} + P_{HV} + P_{VH} + P_{VV} = 1$, $P_{HH} + P_{HV} = P_{HD} + P_{HA}$), we find that only $4^n - 1$ probability measurements are necessary to reconstruct a density matrix.

While we can easily construct a minimum measurement set for an n-qubit system by measuring every combination of $\{H, V, D, R\}$ at each qubit, i.e.,

$$\{M\} = \{H, V, D, R\}_1 \otimes \{H, V, D, R\}_2 \otimes \cdots \{H, V, D, R\}_n, \qquad (65)$$

this is almost never optimal (see Section 5). See Section 2.4 for a formal method for testing whether a specific set of measurements is sufficient for tomography.

EXAMPLE 10 (*An ideal 2-qubit tomography of photon pairs*). Consider measuring a state in nine complete four-element bases, for a total of 36 measurement results. These results are compiled below, with each row representing a single basis, and therefore a single two-qubit Stokes parameter.

$$S_{1,1} = \frac{+P_{DD}}{\frac{1}{3}} \quad \frac{-P_{DA}}{\frac{1}{6}} \quad \frac{-P_{AD}}{\frac{1}{6}} \quad \frac{+P_{AA}}{\frac{1}{3}} = \frac{1}{3},$$

$$S_{1,2} = \frac{+P_{DR}}{\frac{1}{4}} \quad \frac{-P_{DL}}{\frac{1}{4}} \quad \frac{-P_{AR}}{\frac{1}{4}} \quad \frac{+P_{AL}}{\frac{1}{4}} = 0,$$

$$S_{1,3} = \frac{+P_{DH}}{\frac{1}{4}} \quad \frac{-P_{DV}}{\frac{1}{4}} \quad \frac{-P_{AH}}{\frac{1}{4}} \quad \frac{+P_{AV}}{\frac{1}{4}} = 0,$$

$$S_{2,1} = \frac{+P_{RD}}{\frac{1}{4}} \quad \frac{-P_{RA}}{\frac{1}{4}} \quad \frac{-P_{LD}}{\frac{1}{4}} \quad \frac{+P_{LA}}{\frac{1}{4}} = 0,$$

$$S_{2,2} = \frac{+P_{RR}}{\frac{1}{6}} \quad \frac{-P_{RL}}{\frac{1}{3}} \quad \frac{-P_{LR}}{\frac{1}{3}} \quad \frac{+P_{LL}}{\frac{1}{6}} = -\frac{1}{3}, \qquad (66)$$

$$S_{2,3} = \frac{+P_{RH}}{\frac{1}{4}} \quad \frac{-P_{RV}}{\frac{1}{4}} \quad \frac{-P_{LH}}{\frac{1}{4}} \quad \frac{+P_{LV}}{\frac{1}{4}} = 0,$$

$$S_{3,1} = \frac{+P_{HD}}{\frac{1}{4}} \quad \frac{-P_{HA}}{\frac{1}{4}} \quad \frac{-P_{VD}}{\frac{1}{4}} \quad \frac{+P_{VA}}{\frac{1}{4}} = 0,$$

$$S_{3,2} = \frac{+P_{HR}}{\frac{1}{4}} \quad \frac{-P_{HL}}{\frac{1}{4}} \quad \frac{-P_{VR}}{\frac{1}{4}} \quad \frac{+P_{VL}}{\frac{1}{4}} = 0,$$

$$S_{3,3} = \frac{+P_{HH}}{\frac{1}{3}} \quad \frac{-P_{HV}}{\frac{1}{6}} \quad \frac{-P_{VH}}{\frac{1}{6}} \quad \frac{+P_{VV}}{\frac{1}{3}} = \frac{1}{3}.$$

Measurements are taken in each of these nine bases, determining the above nine two-qubit Stokes parameters. The six remaining required parameters, listed below, are dependent upon the same measurements.

$$S_{0,1} = \frac{+P_{DD}}{\frac{1}{3}} \quad \frac{-P_{DA}}{\frac{1}{6}} \quad \frac{+P_{AD}}{\frac{1}{6}} \quad \frac{-P_{AA}}{\frac{1}{3}} = 0,$$

$$S_{0,2} = \frac{+P_{RR}}{\frac{1}{6}} \quad \frac{-P_{LR}}{\frac{1}{3}} \quad \frac{+P_{RL}}{\frac{1}{3}} \quad \frac{-P_{LL}}{\frac{1}{6}} = 0,$$

$$S_{0,3} = \frac{+P_{HH}}{\frac{1}{3}} \quad \frac{-P_{HV}}{\frac{1}{6}} \quad \frac{+P_{VH}}{\frac{1}{6}} \quad \frac{-P_{VV}}{\frac{1}{3}} = 0,$$

$$S_{1,0} = \frac{+P_{DD}}{\frac{1}{3}} \quad \frac{+P_{DA}}{\frac{1}{6}} \quad \frac{-P_{AD}}{\frac{1}{6}} \quad \frac{-P_{AA}}{\frac{1}{3}} = 0, \qquad (67)$$

$$S_{2,0} = \frac{+P_{RR}}{\frac{1}{6}} \quad \frac{+P_{LR}}{\frac{1}{3}} \quad \frac{-P_{RL}}{\frac{1}{3}} \quad \frac{-P_{LL}}{\frac{1}{6}} = 0,$$

$$S_{3,0} = \frac{+P_{HH}}{\frac{1}{3}} \quad \frac{+P_{HV}}{\frac{1}{6}} \quad \frac{-P_{VH}}{\frac{1}{6}} \quad \frac{-P_{VV}}{\frac{1}{3}} = 0.$$

These terms will not in general be zero. Recall—cf. Eq. (42)—that for $|HH\rangle$, $S_{0,3} = S_{3,0} = 1$. Of course, $S_{0,0} = 1$. Taken together, these two-qubit Stokes parameters determine the density matrix:

$$\hat{\rho} = \frac{1}{4}\left(\hat{\sigma}_0 \otimes \hat{\sigma}_0 + \frac{1}{3}\hat{\sigma}_1 \otimes \hat{\sigma}_1 - \frac{1}{3}\hat{\sigma}_2 \otimes \hat{\sigma}_2 + \frac{1}{3}\hat{\sigma}_3 \otimes \hat{\sigma}_3\right)$$

$$= \frac{1}{6} \begin{pmatrix} 2 & 0 & 0 & 1 \\ 0 & 1 & 0 & 0 \\ 0 & 0 & 1 & 0 \\ 1 & 0 & 0 & 2 \end{pmatrix} = \frac{1}{6} \begin{pmatrix} 1 & 0 & 0 & 1 \\ 0 & 0 & 0 & 0 \\ 0 & 0 & 0 & 0 \\ 1 & 0 & 0 & 1 \end{pmatrix} + \frac{1}{6} \begin{pmatrix} 1 & 0 & 0 & 0 \\ 0 & 1 & 0 & 0 \\ 0 & 0 & 1 & 0 \\ 0 & 0 & 0 & 1 \end{pmatrix}.$$

(68)

This is the final density matrix, a Werner state, as defined in Eq. (28).

2.3. TOMOGRAPHY OF NONQUBIT SYSTEMS

By making use of the qudit extensions to the Stokes parameter formalism—Eqs. (57)–(61), we can reconstruct any qudit system in exactly the same manner as qubit systems. For a single particle d-level system, a single Stokes parameter is dependent on $d - 1$ independent probabilities, and $d + 1$ Stokes parameters are necessary to reconstruct the density matrix.

Multiple-qudit systems can be reconstructed by using separable projectors (Thew et al., 2002) upon which the multiple qudit Stokes parameters are dependent (these dependencies were laid out in Section 1.3.2). Likewise, the following section on general tomography, while specific to qubits, can be easily adapted to qudit systems.

2.4. GENERAL QUBIT TOMOGRAPHY

As discussed earlier, qubit tomography will require $4^n - 1$ probabilities in order to define a complete set of T_i parameters. In practice, this will mean that 4^n measurements are necessary in order to normalize counts to probabilities. By making projective measurements on each qubit and only taking into account those results where a definite result is obtained (e.g., the photon was transmitted by the polarizer), it is possible to reconstruct a state using the results of 4^n measurements.

Our first task is to represent the density matrix in a useful form. To this end, define a set of $2^n \times 2^n$ matrices which have the following properties:

$$\mathrm{Tr}\{\widehat{\Gamma}_v \cdot \widehat{\Gamma}_\mu\} = \delta_{v,\mu},$$
$$\hat{A} = \sum_v \widehat{\Gamma}_v \mathrm{Tr}\{\widehat{\Gamma}_v \cdot \hat{A}\} \quad \forall \hat{A},$$

(69)

where $\hat{A}$ is an arbitrary $2^n \times 2^n$ matrix. A convenient set of $\widehat{\Gamma}$ matrices to use are tensor-products of the $\hat{\sigma}$ matrices used throughout this paper:

$$\widehat{\Gamma}_v = \hat{\sigma}_{i_1} \otimes \hat{\sigma}_{i_2} \otimes \cdots \otimes \hat{\sigma}_{i_n},$$

(70)

where v is simply a short-hand index by which to label the Γ matrices (there are 4^n of them) which is more concise than $i_1, i_2, \ldots, i_n$. Transforming Eq. (35) into this notation, we find that

$$\hat{\rho} = \frac{1}{2^n} \sum_{\nu=1}^{4^n} \hat{\Gamma}_\nu S_\nu. \tag{71}$$

Next, it is necessary to consider exactly which measurements to use. In particular, we now wish to determine the necessary and sufficient conditions on the 4^n measurements to allow reconstruction of any state.[8] Let $|\psi_\mu\rangle$ ($\mu = 1$ to 4^n) be the measurement bases, and define the probability of the μth measurement as $P_\mu \equiv \langle \psi_\mu | \hat{\rho} | \psi_\mu \rangle$.

Combining this with Eq. (71),

$$P_\mu = \langle \psi_\mu | \frac{1}{2^n} \sum_{\nu=1}^{4^n} \hat{\Gamma}_\nu S_\nu | \psi_\mu \rangle = \frac{1}{2^n} \sum_{\nu=1}^{4^n} B_{\mu,\nu} S_\nu, \tag{72}$$

where the $4^n \times 4^n$ matrix $B_{\mu,\nu}$ is given by

$$B_{\mu,\nu} = \langle \psi_\mu | \hat{\Gamma}_\nu | \psi_\mu \rangle. \tag{73}$$

Immediately we find a necessary and sufficient condition for the completeness of the set of tomographic states $\{|\psi_\mu\rangle\}$: if the matrix $B_{\mu,\nu}$ is nonsingular, then Eq. (72) can be inverted to give

$$S_\nu = 2^n \sum_{\mu=1}^{4^n} \left(B^{-1}\right)_{\mu,\nu} P_\mu. \tag{74}$$

While this provides an exact solution if exact probabilities are known, it leads to a number of difficulties in real systems. First, it is possible for statistical errors to cause a set of measurements to lead to an illegal density matrix. Second, if more than the minimum number of measurements are taken and they contain any error, they will overdefine the problem, eliminating the possibility of a single analytically calculated answer. To solve these problems it is necessary to analyze the data in a fundamentally different way, in which statistically varying probabilities are assumed from the beginning and optimization algorithms find the state most likely to have resulted in the measured data (Section 4.2).

---

[8] If exact probabilities are known, only $4^n - 1$ measurements are necessary. However, often only numbers of counts (successful measurements) are known, with no information about the number of counts which would have been measured by detectors in orthogonal bases. In this case an extra measurement is necessary to normalize the inferred probabilities.

3. Collecting Tomographic Measurements

Before discussing the analysis of real experimental data, it is necessary to understand how that experimental data is collected. This chapter outlines the experimental implementation of tomography on polarization entangled qubits generated from spontaneous parametric downconversion (Kwiat et al., 1999), though the techniques for projection and particularly systematic error correction will be applicable to many systems. We filter these photon pairs using both spatial filters (irises used to isolate a specific k-vector, necessary because our states are angle-dependent) and frequency filters (interference filters, typically 5–10 nm wide, FWHM).

After this initial filtering, measurement collection involves two central issues: projection (into an ideally arbitrary range of states) and systematic error correction (to compensate for any number of experimental problems ranging from imperfect optics to accidental coincidences).

3.1. PROJECTION

Any tomography, in fact any measurement on a quantum system, depends on state projection; for the purposes of this chapter, these projections will be *separable*. While tomography could be simplified by using arbitrary projectors (e.g., joint measurements on two qubits), this is experimentally difficult. We therefore focus on the ability to create arbitrary single-qubit projectors which will then be easily chained together to create any separable projector.

3.1.1. Arbitrary Single-Qubit Projection

An arbitrary polarization measurement and its orthogonal complement can be realized using, in order, a quarter-wave plate, a half-waveplate, and a polarizing beam splitter. Waveplates implement unitary operations, and in the Poincaré sphere picture, act as rotations about an axis lying within the linear polarization plane (the equator) (Born and Wolf, 1987). Specifically, a waveplate whose optic axis is oriented at angle θ with respect to the horizontal induces a rotation on the Poincaré sphere about an axis 2θ from horizontal, in the linear plane. The magnitude of this rotation is equal to the waveplate's retardance (90° for quarter-wave plates and 180° for half-wave plates). For the remainder of this chapter we adopt the convention that polarizing beam splitters transmit horizontally polarized light and reflect vertically polarized light—though for some types the roles are reversed.

This analysis, while framed in terms of waveplates acting on photon polarization, is directly applicable to other systems, e.g., spin-$\frac{1}{2}$ particles (Cory et al.,

1997; Jones et al., 1997; Weinstein et al., 2001; Laflamme et al., 2002) or two-level atoms (Monroe, 2002; Schmidt-Kaler et al., 2003). In these systems, measurements in arbitrary bases are obtained using suitably phased π- and $\frac{\pi}{2}$-pulses (externally applied electromagnetic fields) to rotate the state to be measured into the desired analysis basis.

To derive the settings for these waveplates as a function of the projection state desired, we use the Poincaré sphere (see Fig. 3). For any state on the surface of the sphere, a 90° rotation about a linear axis directly below it will rotate that state into a linear polarization (see Fig. 3(b)). Assume the desired projection state is

$$|\psi_P\rangle = \cos\left(\frac{\theta}{2}\right)|H\rangle + \sin\left(\frac{\theta}{2}\right)e^{i\phi}|V\rangle. \tag{75}$$

Simple coordinate transforms from spherical to Cartesian coordinates reveal that a quarter-waveplate at $\theta_{QWP} = \frac{1}{2}\cos^{-1}\{\sin(\theta)\tan(\phi)\}$ will rotate the projection state (75) into a linear state

$$|\psi_P'\rangle = \cos\left(\frac{\theta'}{2}\right)|H\rangle + \sin\left(\frac{\theta'}{2}\right)|V\rangle. \tag{76}$$

A half-waveplate at $\frac{1}{4}\theta'$ (with respect to horizontal orientation) will then rotate this state to $|H\rangle$.[9] Finally, the PBS will transmit the projected state and reflect its orthogonal complement.

Mathematically, this process of rotation and projection can be described using unitary transformations. The unitary transformations for half- and quarter-waveplates in the H/V basis are

$$U_{\text{HWP}}(\theta) = \begin{bmatrix} \cos^2(\theta) - \sin^2(\theta) & 2\cos(\theta)\sin(\theta) \\ 2\cos(\theta)\sin(\theta) & \sin^2(\theta) - \cos^2(\theta) \end{bmatrix},$$

$$U_{\text{QWP}}(\theta) = \begin{bmatrix} \cos^2(\theta) + i\sin^2(\theta) & (1-i)\cos(\theta)\sin(\theta) \\ (1-i)\cos(\theta)\sin(\theta) & \sin^2(\theta) + i\cos^2(\theta) \end{bmatrix}, \tag{77}$$

with θ denoting the rotation angle of the waveplate with respect to horizontal. Assume that during the course of a tomography, the νth measurement setting requires that the QWP be set to $\theta_{\text{QWP},\nu}$ and the HWP to $\theta_{\text{HWP},\nu}$. Therefore, the total unitary[10] for the νth measurement setting will be

$$U_\nu = U_{\text{HWP}}(\theta_{\text{HWP},\nu})U_{\text{QWP}}(\theta_{\text{QWP},\nu}). \tag{78}$$

---

[9] $\theta' = \cos^{-1}/\cos^{-1}\{\sin(\theta)\tan(\phi)\} - \cos^{-1}/\cos^{-1}\{\cot(\theta)\cot(\phi)\}$. In practice, care must be taken that consistent conventions are used (e.g., right- vs. left-circular polarization), and it may be easier to calculate this angle directly from waveplate operators and the initial state.

[10] Note the order of the unitary matrices for the HWP and QWP. Incoming light encounters the QWP first, and therefore U_{QWP} is last when defining U_ν.

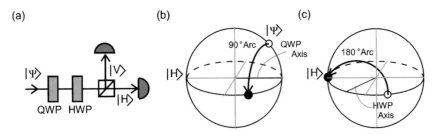

FIG. 3. A quarter-waveplate (QWP), half-waveplate (HWP), and polarizing beam splitter (PBS) are used to make an arbitrary polarization measurement. Both a diagram of the experimental apparatus (a) and the step-by-step evolution of the state on the Poincaré sphere are shown. (b) The quarter-waveplate rotates the projection state (the state we are projecting into, *not* the incoming unknown state) into the linear polarization plane (the equator). (c) The half-waveplate rotates this linear state to horizontal. The PBS transmits the projection state (now $|H\rangle$) and reflects its orthogonal complement (now $|V\rangle$), which can then both be measured.

For multiple qubits, we can directly combine these unitaries such that

$$U_\nu = {}^1U_\nu \otimes {}^2U_\nu \otimes \cdots \otimes {}^nU_\nu, \tag{79}$$

where $^qU_\nu$ denotes the qth qubit's unitary transformation due to waveplates. The total projection operator for this system is therefore $\langle 0|U_\nu$, where $|0\rangle$ is the first computational basis state (the state which passes through the beamsplitters—most likely $|H\rangle$ for each qubit). The measurement state (the state which will pass through the measurement apparatus and be measured every time) is therefore $U_\nu^\dagger|0\rangle$.

Of course, these calculations assume that we are using waveplates with retardances equal to exactly π or $\frac{\pi}{2}$ (or Rabi pulses producing perfect phase differences). Imperfect yet well characterized waveplates will lead to measurements in slightly different, yet known, bases. This can still yield an accurate tomography, but first these results must be transformed from a biased basis into the canonical Stokes parameters using Eq. (21). As discussed below (see Section 3), the maximum likelihood technique provides a different but equally effective way to accommodate for imperfect measurements.

3.1.2. Compensating for Imperfect Waveplates

While the previous section shows that it is possible using a quarter- and half-waveplate to project into an arbitrary single qubit state, perfect quarter- and half-waveplates are experimentally impossible to obtain. More likely, the experimenter will have access to waveplates with known retardances slightly different than the ideal values of π (HWP) and $\frac{\pi}{2}$ (QWP). Even in this case, it is often possible to obtain arbitrary single-qubit projections. (Note that this is the second solution to the problem of imperfect waveplates. Imperfect waveplates could be used at virtually any angles during a tomography—such as the same angles at which perfect waveplates would be used to measure in the canonical basis—resulting in a set of biased bases. The tomography mathematics have already been shown to function for either mutually biased or unbiased bases, as long as the set of bases is complete. In contrast, this section describes how—*even using imperfect waveplates*—one can still measure in the canonical, mutually unbiased bases.)

Analytically finding the angles where this is possible proves to be inconvenient and, for some waveplates, impossible. Rather than solve a system of equations based on the unitary waveplate matrices, we will examine the effect of these waveplates graphically using the Poincaré sphere. For the remainder of this discussion, we will assume that the experimenter has access to two waveplates, WP_1 and WP_2, which will respectively take the place of the QWP and HWP normally present in the experimental setup. We constrain the retardances of these waveplates to be $0 \leq \phi_1 \leq \phi_2 \leq \pi$.

In order to project into an arbitrary state $|\psi\rangle$, WP_1 and WP_2 must together rotate the state $|\psi\rangle$ into the state $|H\rangle$ (assuming a horizontal polarizer is used after the waveplates—any linear polarizer is equivalent). Taking a piecewise approach, first consider which states are possible after acting on the input state $|\psi\rangle$ with WP_1. Figure 4(a) shows several example cases on the Poincaré sphere, each resulting in a curved band of possible states that can be reached by varying the orientation of WP_1. Next consider which states could be rotated by WP_2 into the target state $|H\rangle$. Figure 4(b) shows several examples of these states, which also take the form of a curved band, traversed by varying the orientation of WP_2. In order for state $|\psi\rangle$ to be rotatable into state $|H\rangle$, these two bands of potential states (shown in Figs. 4(a) and (b)) must overlap.

Briefly examining the geometry of this system, it appears that for most states this will be possible as long as the waveplate phases do not differ too much from the ideal HWP and QWP. Further consideration reveals that it is sufficient to be able to project into the states on the H-R-V-L great circle. There are two conditions under which this will *not* occur. First, if WP_2 is too close to a HWP, with WP_1 far from a QWP, the states at the poles (close to $|R\rangle$ and $|L\rangle$) will be unreachable from $|H\rangle$ (see Fig. 4(c)). Quantifying this condition, we require that

$$2\left|\frac{\pi}{2} - \phi_1\right| \leq \pi - \phi_2. \tag{80}$$

Put another way, the error in the QWP must be less than half the error in the HWP. Second, the combined retardances from both waveplates can be insufficient to reach $|V\rangle$ (see Fig. 4(c)):

$$\phi_1 + \phi_2 \geq \pi. \tag{81}$$

Given these two conditions, numerical simulations confirm that arbitrary single-qubit projectors can be constructed with two waveplates.

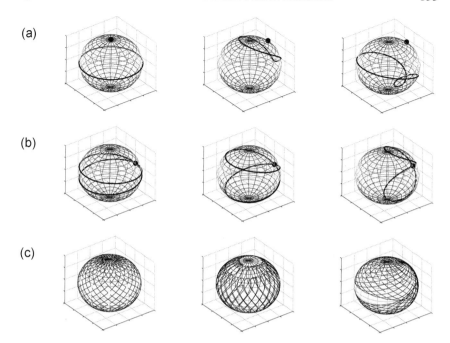

FIG. 4. Possible projectors simulated by waveplates and a stationary polarizer, graphically shown on Poincaré spheres. (a) WP$_1$, depending on its orientation, can rotate an incoming state into a variety of possible output states. Shown here on three Poincaré spheres are an initial incoming state (represented by a solid dot) and the set of all output states that WP$_1$ can rotate it into (represented by a dark band on the surface of the sphere). From left to right, the spheres depict $|R\rangle$ transformed by a $(\pi/2)$-waveplate, $|\gamma\rangle = \cos(\pi/8)|H\rangle + \mathrm{i}\sin(\pi/8)|V\rangle$ transformed by a $(\pi/3)$-waveplate, and $|\gamma\rangle$ transformed by a $(2\pi/3)$-waveplate. (b) WP$_2$, depending on its orientation, can rotate a variety of states into the target state $|H\rangle$. Shown here from left to right are the states able to be rotated into $|H\rangle$ by a $(11\pi/12)$-waveplate, a $(3\pi/4)$-waveplate, and a $(\pi/2)$-waveplate. (c) The possible projectors able to be produced by two waveplates and a horizontal polarizer. A series of arcs blanketing the Poincaré sphere show the areas of the sphere representing achievable projectors for each waveplate combination. From left to right, the spheres show the states (in this case, all of them) accessible from an ideal QWP and HWP, the states accessible using $(\pi/3)$- and $(11\pi/12)$-waveplates (groups of states near the poles are inaccessible), and the states accessible using $(\pi/3)$- and $(3\pi/5)$-waveplates (states on the equator are inaccessible). Note that the spheres shown in (c) are *not* simply combinations of the spheres above it, but include retardance values chosen to illustrate the possible failure modes of imperfect waveplates.

To clarify, as discussed in the previous section, one does not *require* arbitrary single-qubit projectors, since an accurate tomography can be obtained with *any* set of linearly independent projectors as long as they are known. In fact, one advantage to this approach is that the exact same tomography measurement system can be used on photons with different wavelengths (on which the waveplates' birefringent phase retardances depend), simply by entering in the analysis program what the actual phase retardances are at the new wavelengths (Peters et al., 2005).

Wedged waveplates It is an experimental reality that all commercially available waveplates have some degree of *wedge* (i.e., the surfaces of the waveplate are not parallel). This leads to a number of insidious difficulties which the experimenter must confront, grouped into two categories: (1) The thickness of the waveplate will change along its surface, providing a corresponding change in the phase retardance of the waveplate. This means that during a tomography when the waveplate is routinely rotated to different orientations, its total phase after rotation will change according to a much more complex—and often very difficult to calculate—formula. (In fact, if a large collection aperture is used, then different parts of the beam will experience different phase shifts.) (2) The direction (k-vector) of a beam will be deflected after passing through a wedged waveplate. This deflection will again depend on waveplate orientation, therefore changing throughout a tomography. This can have the effect of changing detector efficiencies (if, as in our case, a lens is used to focus to a portion of a very small detector area, different pieces of which have different efficiencies). This deflection will also affect any interferometric effects that depend on the beam direction being stable under waveplate rotation. Some of these problems can be mitigated (e.g., by taking care to pass through the exact center of the waveplate), but in general the best solution is to select waveplates with faces very close to parallel.

3.1.3. Multiple-Qubit Projections and Measurement Ordering

For multiple-qubit systems, separable projectors can be implemented by using in parallel the single-qubit projectors described above. This, by construction, allows the implementation of arbitrary separable projectors.

In practice, depending on the details of a specific tomography (see Section 5 for a discussion of how to choose which and how many measurements to use), multiple-qubit tomographies can require a large number of measurements. If the time to switch from one measurement to another varies depending on which measurements are switched between (as is the case with waveplates switching to different values for each projector), minimizing the time spent switching is a problem equivalent to the travelling salesman problem (Cormen et al., 2001). A great deal of time can be saved by implementing a simple, partial solution to this canonical problem (e.g., a genetic algorithm which is not guaranteed to find the optimal solution but likely to find a comparably good solution).

3.2. n vs. $2n$ Detectors

Until now, this chapter has discussed the use of an array of n detectors to measure a single separable projector at a time. While this is conceptually simple, there is an extension to this technique which can dramatically improve the efficiency and accuracy of a tomography: using an array of $2n$ detectors, project every incoming n-qubit state into one of 2^n basis states. This is the generalization of simultaneously measuring both outputs in the single-qubit case (the two detectors used for single-qubit measurement are shown in Fig. 3(a)), or all four basis states (HH, HV, VH, and VV) in the two-qubit case; in the general case $2n$ detectors will measure in n-fold coincidence with 2^n possible outcomes.

It should be emphasized that these additional detectors are not some 'trick', effectively masking a number of sequential settings of n detectors. If only n detectors are used, then over the course of a tomography most members comprising the input ensemble will never be measured. For example, consider measuring the projection of an unknown state into the $|00\rangle$ basis using two detectors. While this will give some number of counts, unmeasured coincidences will be routed into the $|01\rangle$, $|10\rangle$, and $|11\rangle$ modes. The information of how many coincidences are routed to which mode will be lost, unless another two detectors are in place in the '1' modes to measure it.

Returning to the notation of Section 3.1, recall that the state which passes through every beamsplitter is $U_\nu^\dagger|0\rangle$, but when $2n$ detectors are employed, the states $U_\nu^\dagger|r\rangle$ can all be measured, where r ranges from 0 to $2^n - 1$ and $|r\rangle$ denotes the rth element of the canonical basis (the canonical basis is chosen/enforced by the beamsplitters themselves).

EXAMPLE 11 (*The $|r\rangle$ notation for two qubits*). For two qubits, each incident on separate beamsplitters which transmit $|H\rangle$ and reflect $|V\rangle$, we can define the following values of $|r\rangle$, the canonical basis:

$$|0\rangle \equiv |HH\rangle, \qquad |1\rangle \equiv |HV\rangle, \qquad |2\rangle \equiv |VH\rangle, \qquad |3\rangle \equiv |VV\rangle. \quad (82)$$

The usefulness of this notation will become apparent during the discussion of the Maximum Likelihood algorithm in Section 4.2.

The primary advantage to using $2n$ detectors is that every setting of the analysis system (every group of the projector and its orthogonal complements) generates exactly enough information to determine a single multiple-qubit Stokes vector. Expanding out the probabilities that a multiple-qubit Stokes vector (which for now we limit to those with only nonzero indices) is based on,

$$
\begin{aligned}
S_{i_1,i_2,\ldots,i_n} &= \left(P_{\psi_1} - P_{\psi_1^\perp}\right) \otimes \left(P_{\psi_2} - P_{\psi_2^\perp}\right) \otimes \cdots \otimes \left(P_{\psi_n} - P_{\psi_n^\perp}\right) \\
&= P_{\psi_1,\psi_2,\ldots,\psi_n} - P_{\psi_1,\psi_2,\ldots,\psi_n^\perp} - \cdots \pm P_{\psi_1^\perp,\psi_2^\perp,\ldots,\psi_n^\perp},
\end{aligned} \quad (83)
$$

where the sign of each term on the last line is determined by the parity of the number of orthogonal ($\perp$) terms.

These probabilities are precisely those measured by a single setting of the entire analysis system followed by a $2n$ detector array. Returning to our primary decomposition of the density matrix from Eq. (35),

$$\hat{\rho} = \frac{1}{2^n} \sum_{i_1,i_2,\dots,i_n=0}^{3} S_{i_1,i_2,\dots,i_n} \hat{\sigma}_{i_1} \otimes \hat{\sigma}_{i_2} \otimes \cdots \otimes \hat{\sigma}_{i_n},$$

we once again need only determine all of the multiple-qubit Stokes parameters to exactly characterize the density matrix. At first glance this might seem to imply that we need to use $4^n - 1$ settings of the analysis system, in order to find all of the multiple-qubit Stokes parameters save $S_{0,0,\dots,0}$, which is always one.

While this is certainly sufficient to solve for $\hat{\rho}$, many of these measurements are redundant. In order to choose the smallest possible number of settings, note that the probabilities that constitute some multiple-qubit Stokes parameters overlap exactly with the probabilities for other multiple-qubit Stokes parameters. Specifically, any multiple-qubit Stokes parameter with at least one 0 subscript is derived from a set of probabilities that at least one other multiple-qubit Stokes vector (with no 0 subscripts) is also derived from. As an example, consider that

$$S_{0,3} = P_{|00\rangle} - P_{|01\rangle} + P_{|10\rangle} - P_{|11\rangle}, \tag{84}$$

while

$$S_{3,3} = P_{|00\rangle} - P_{|01\rangle} - P_{|10\rangle} + P_{|11\rangle}. \tag{85}$$

These four probabilities, measured simultaneously, will provide enough information to determine both values. This dependent relationship between multiple-qubit Stokes vectors is true, in general, as can be seen by returning to Eq. (83). Each subscript with nonzero value for S contributes a term to the tensor product on the right that looks like $(P_{\psi_i} - P_{\psi_i^\perp})$. Had there been subscripts with value zero, however, they each would have contributed a $(P_{\psi_i} + P_{\psi_i^\perp})$ term; as an aside, terms with zero subscripts are always dependent on terms will all positive subscripts. This reduces the minimum number of analysis settings to 3^n, a huge improvement in multiple qubit systems (e.g., 9 vs. 15 settings for 2-qubit tomography, 81 vs. 255 for 4-qubit tomography, etc.). Note that, as discussed earlier, this benefit is only possible if one employs $2n$ detectors, leading to a total of 6^n measurements (2^n measurements for each of 3^n analysis settings).[11]

Because Eq. (41) can be used to transform any set of nonorthogonal multiple-qubit Stokes parameters into the canonical form, orthogonal measurement sets

---

[11] These measurements, even though they result from the minimum number of analysis settings for $2n$ detectors, are overcomplete. A density matrix has only $4^n - 1$ free parameters, which implies that only $4^n - 1$ measurements are necessary to specify it (see n-detector tomography). Because the overcomplete set of 6^n measurements is not linearly independent, it can be reduced to a $4^n - 1$ element subset and still completely specify an unknown state.

need not be used. One advantage of the option to use nonorthogonal measurement sets is that an orthogonal set may not be experimentally achievable, for instance, due to waveplate imperfections, as discussed in Section 3.1.2.

3.3. ELECTRONICS AND DETECTORS

Single photon detectors and their supporting electronics are crucial to any photonic tomography. Figure 5 shows a simple diagram of the electronics used to count in coincidence from a pair of Si-avalanche photodiodes. An electrical pulse from a single-photon generated avalanche in the silicon photodiode sends a signal to a discriminator, which, after receiving a pulse of the appropriate amplitude and width, produces in a fan-out configuration several TTL (transistor–transistor logic) signals which are fed into the coincidence circuitry. In order to avoid pulse reelections, a fan-out configuration is used in preference to repeatedly splitting one signal.

The signals from these discriminators represent physical counts, with the number of discriminator signals sent to a detector equal to the singles counts for that detector. A copy of this signal, after travelling through a variable length delay line, is input into an AND gate with a similar pulse (with a static delay) from a complimentary detector. The pulses sent from the discriminators are variable width, typically about 2 ns, producing a 4-ns window in which the AND gate can produce a signal. (The coincidence window is chosen to be as small as conveniently possible, in order to reduce the number of "accidental" coincidences, discussed below.) This signal is also sent to the counters and is recorded as a coincidence between its two parent detectors.

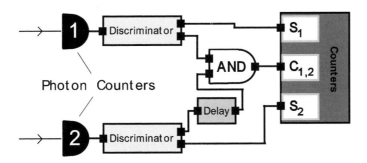

FIG. 5. A simple diagram of the electronics necessary to operate a coincidence-based photon counting circuit. While this diagram depicts a two-detector counting circuit, it is easily extendible to multiple detectors; by adding additional detectors each fed into a discriminator and a fan-out, we gain the signals necessary for one singles counter per detector and an AND gate for each pair of detectors capable of recording a coincidence.

As with any system of this sort, the experimenter must be wary of reflected pulses generating false counts, delay lines being properly matched for correct AND gate operation, and system saturation for high count rates.

3.4. Collecting Data and Systematic Error Correction

The projection optics and electronics described above will result in a list of co-incidence counts each tied to a single projective measurement. Incorporating the projectors defined earlier in this section, we can now make a first estimate on the number of counts we expect to receive for a given measurement of the state $\hat{\rho}$:

$$\bar{n}_{v,r} = I_0 \text{Tr}\{\widetilde{M}_{v,r}\hat{\rho}\},$$
$$\widetilde{M}_{v,r} = U_v^\dagger |r\rangle\langle r|U_v. \tag{86}$$

Our eventual strategy (see Section 4.2) will be to vary $\hat{\rho}$ until our expectations optimally match our actual measured counts. Here $\bar{n}_{v,r}$ is the expectation value of the number of counts recorded for the vth measurement setting on the rth pair of detectors (this is the pair of detectors which projects into the canonical basis state $|r\rangle$). The density matrix to be measured is denoted by $\hat{\rho}$ and I_0 is a constant scaling factor which takes into account the duration of a measurement and the rate of state production. Note that regardless of whether n or $2n$ detectors are used, each distinct measurement setting will be indexed by v. For n detectors, there will be a single value of r for each value of v, as each measurement setting projects into a single state. For $2n$ detectors, there will be 2^n values of r, one for each pair of detectors capable of registering coincidences.

Throughout this section we will modify Eq. (86) to give a more complete estimate of the expected count rates, taking into account real errors and statistical deviations. In particular, without adjustment, the expected coincidence counts will likely be inaccurate without adjustment due to experimental factors including accidental coincidences, imperfect optics, mismatched detector efficiencies, and drifts in state intensity. Below we will discuss each of these in turn.

3.4.1. Accidental Coincidences

In general, the spontaneous generation of photon pairs from downconversion processes can result in several pairs of photons being generated at the same time. These multiple-pair generation events can lead to two uncorrelated photons being detected as a coincidence, which will tend to raise all measured counts and lead to state tomographies resulting in states closer to the maximally mixed state.[12]

---

[12] There is also a similar but generally smaller contribution from one real photon and a detector noise count, and a smaller contribution still from two detector noise counts.

We can model these accidental coincidences for the two-qubit case by considering the probability that *any* given singles count will be detected during the coincidence window of a conjugate photon. This model implies that the accidental coincidences for the vth measurement setting on the rth detector pair ($n_{v,r}^{\text{accid}}$) will be dependent on the singles totals in each channel ($^1S_{v,r}$ and $^2S_{v,r}$), the total coincidence window (Δt_r, approximately equal to twice the pulse width produced by the discriminators[13]), and the total measurement time (T_v). When the singles channels are far from saturation ($^{1,2}S_{v,r}\Delta t_{\text{dead}} \ll T_v$, where Δt_{dead} is the dead time of the detectors, i.e., the time it takes after a detector registers a singles count before it can register another), the percentage of time that a channel is triggered (able to produce a coincidence) is approximated by $^{1,2}S_{v,r}\Delta t_r/T_v$. The probability that the other channel will produce a coincidence within this time (again in the unsaturated regime) is proportional to the singles counts on that channel. This allows us to approximate the total accidentals as

$$n_{v,r}^{\text{accid}} \simeq \frac{{}^1S_{v,r}\,{}^2S_{v,r}\Delta t_r}{T_v}, \tag{87}$$

implying that

$$\bar{n}_{v,r} = I_0\text{Tr}\{\widetilde{M}_{v,r}\hat{\rho}\} + n_{v,r}^{\text{accid}}. \tag{88}$$

Because the accidental rate will be necessary for analyzing the data, these expected accidental counts will need to be calculated from the singles rates for each measurement and recorded along with the actual measured coincidence counts.[14]

3.4.2. Beamsplitter Crosstalk

In most experimental implementations, particularly those involving $2n$ detectors, the polarizer used for single-qubit projection will be a beamsplitter, either based on dielectric stacks, or crystal birefringence. In practice, all beamsplitters function with some levels of crosstalk and absorption, i.e., some probability of reflecting or absorbing the polarization which should be transmitted and vice versa. By measuring these crosstalk probabilities and adjusting the measured counts accordingly, it is possible to recreate the approximate measurement values that would have resulted from a crosstalk-free system.

We can characterize a beamsplitter using four numbers $C_{r'\to r}$ which represent the probability that state r' will be measured as state r.

---

[13] If the pulses are not square, or the AND logic has speed limitations, this approximation may become inaccurate.

[14] It is advisable to initially experimentally determine Δt_r by directly measuring the accidental coincidence rate (by introducing an extra large time delay into the variable time delay before the AND gate, shown in Figure 5), and using Eq. (87) to solve for Δt_r. This should be done for every pair of detectors, and ideally at several count rates, in case there are nonlinear effects in the detectors.

EXAMPLE 12 (*A faulty beamsplitter*). Assume that we have measured a beam-splitter which transmits 90% and absorbs 10% of incident horizontal light (state 0), while reflecting 80% and transmitting 10% of vertical light (state 1). We would therefore use

$$\begin{aligned}
C_{0 \to 0} &= 0.9, \\
C_{0 \to 1} &= 0, \\
C_{1 \to 0} &= 0.1, \\
C_{1 \to 1} &= 0.8,
\end{aligned}$$

(89)

to characterize the behavior of this beamsplitter.

EXAMPLE 13 (*Two-qubit crosstalk*). Consider two faulty beamsplitters identical to the one presented in Example 12, with crosstalk coefficients $C^A_{r' \to r}$ and $C^B_{r' \to r}$. Assuming that we label the two qubit canonical basis $|r\rangle$ as $|0\rangle \equiv |HH\rangle$, $|1\rangle \equiv |HV\rangle$, $|2\rangle \equiv |VH\rangle$, and $|3\rangle \equiv |VV\rangle$, we can derive the general two-qubit crosstalk coefficients $C_{r' \to r}$ by multiplying the single-qubit crosstalk coefficients, according to the rule:

$$C_{(2r'_A + r'_B) \to (2r_A + r_B)} \equiv C^A_{r'_A \to r_A} C^B_{r'_B \to r_B}.$$

(90)

Thus, the total crosstalk matrix will be

$$
C_{r' \to r} \equiv
\begin{array}{c}
\\
\to 0 \\
\to 1 \\
\to 2 \\
\to 3
\end{array}
\begin{array}{cccc}
0' & 1' & 2' & 3' \\
\left(\begin{array}{cccc}
0.81 & 0.09 & 0.09 & 0.01 \\
0 & 0.72 & 0 & 0.08 \\
0 & 0 & 0.72 & 0.08 \\
0 & 0 & 0 & 0.64
\end{array}\right)
\end{array}.
$$

(91)

If we use this notation to modify Eq. (88) for predicted counts, we find that

$$\begin{aligned}
\bar{n}_{v,r} &= I_0 \mathrm{Tr}\{\widehat{M}_{v,r}\hat{\rho}\} + n^{\mathrm{accid}}_{v,r}, \\
\widehat{M}_{v,r} &\equiv \sum_{r'} (C_{r' \to r}) \tilde{M}_{v,r'}.
\end{aligned}$$

(92)

3.4.3. Detector-Pair Efficiency Calibration

Because single-photon detectors will in general have different efficiencies, it may be necessary to measure the relative efficiencies of any detector pairs used in the course of a tomography. For the n-detector case, this is unnecessary, as all recorded counts will be taken with the same detectors and scaled equally. For the $2n$-detector configuration, this can be a noticeable problem, with each of the n^2

measurement bases using a different combination of detectors, with a different total coincidence efficiency. By measuring the relative efficiencies of each combination, it is possible to correct the measured counts by dividing them by the appropriate relative efficiency.

Note that it is not necessary to know the *absolute* efficiency of each detector combination, but only the *relative* efficiencies. Knowing only the relative efficiencies leaves a single scaling factor that is applied to all counts, but as the error on a set of counts is dependent on the *measured* counts, rather than the total number of incident states, this ambiguity does not affect the tomography results.

The tomography process itself may be used to conveniently determine the full set of relevant relative efficiencies. By performing enough measurements to perform an n-detector tomography while using $2n$ detectors, it is possible to perform a tomography for *each* detector combination, using only the results of that detector combination's measurements. Each of these sets will be sufficient to perform a tomography, and the tomography algorithm (see Section 4) will necessarily determine the total state intensity. The ratios between these state intensities (one for each detector combination) will provide the relative efficiencies of each detector combination. In the two-qubit case, this means using four detectors and 36 measurement settings, for a total of 144 measurements to calibrate the relative efficiencies.

In order to continue to update our equation for $\bar{n}_{v,r}$, we define an efficiency $\widehat{E}_r$ which describes the relative efficiency of the rth detector combination. This allows us to correct our previous equation to

$$\bar{n}_{v,r} = I_0 E_0 E_r \text{Tr}\{\widehat{M}_{v,r}\hat{\rho}\} + n_{v,r}^{\text{accid}}, \tag{93}$$

where E_0 is a constant scalar, which combined with the easier to measure relative efficiency E_r, gives the absolute efficiency of each detector pair.

3.4.4. Intensity Drift

In polarization experiments based on downconversion sources, a major cause of error can be drift in the intensity (or direction) of the pump, which causes a drift in the rate of downconversion and therefore state production. If this intensity drift is recorded, then the prediction of the expected number of counts can be adjusted to account for this additional information. Alternatively, if $2n$ detectors are used, the sum of the counts from each of the detectors will automatically give the normalized intensity for each measurement setting, since the sum of the counts in orthonormal bases must add up to the total counts (assuming no state-dependent losses, e.g., in the polarizing beamsplitters). However, when summing the counts from a complete basis like this, the measurements *must be taken at the same time*, and the summed counts must take other sources of error, like detector inefficiency, accidental counts, and beamsplitter crosstalk into account.

By whatever method it is measured, assume that the relative size of the ensemble subject to the νth measurement setting is given by I_ν. Then

$$\bar{n}_{\nu,r} = I_0 E_0 I_\nu E_r \text{Tr}\{\widehat{M}_{\nu,r}\hat{\rho}\} + n_{\nu,r}^{\text{accid}}. \tag{94}$$

Now it becomes clear that I_0 is the factor (not necessarily the total number of pairs produced) which, combined with the relative efficiency I_ν, gives the total number of incident states for the νth measurement setting.

4. Analyzing Experimental Data

As discussed earlier, any real experiment will contain statistical and systematic errors which preclude the use of the ideal tomography described in Section 2. Instead, it is necessary to use an algorithm (the Maximum Likelihood technique) which assumes some uncertainty or error in measurement results, and returns a state which is the most likely to have produced the measured results.

In order to describe real tomography, we will first discuss the types of errors which are present in an experiment, the Maximum Likelihood algorithm, and some details of the optimization of the entire process using numerical search techniques. We first list the information

U_ν—measurement settings,
$n_{\nu,r}$—counts recorded,
$n_{\nu,r}^{\text{accid}}$—accidental counts,
$C_{r'\rightarrow r}$—crosstalk coefficients,
E_r—relative efficiencies,
I_ν—relative intensities (not used with $2n$ detectors)

that should have been gathered during the experimental phase of the experiment, followed by the formulae

$$\bar{n}_{\nu,r} \equiv I_0 E_0 I_\nu E_r \text{Tr}\{\widehat{M}_{\nu,r}\hat{\rho}\} + n_{\nu,r}^{\text{accid}},$$
$$\widehat{M}_{\nu,r} \equiv \sum_{r'}(C_{r'\rightarrow r})\widetilde{M}_{\nu,r'},$$
$$\widetilde{M}_{\nu,r} \equiv U_\nu^\dagger |r\rangle\langle r|U_\nu$$

used to determine the expected number of measured counts for the νth measurement setting on the rth detector combination.

Given this information, we are able to numerically estimate which state was most likely to return the measured results. Note that the relative intensities I_ν are optional, but can be included for an n-detector tomography. For a $2n$ detector tomography, the I_ν parameters are varied as part of the optimization algorithm, and do not need to be provided as part of the experimental data.

4.1. Types of Errors and State Estimation

Errors in the measurement of a density matrix fall into three main categories: errors in the measurement basis, errors from counting statistics, and errors from experimental stability. The first problem can be addressed by increasing the accuracy of the measurement apparatus (e.g., obtaining higher tolerance waveplates, better controlling the Rabi pulses, etc.) while the second problem is reduced by performing each measurement on a larger ensemble (counting for a longer time). The final difficulty is drift which occurs over the course of the tomography.[15] This drift occurs either in the state produced or the efficiency of the detection system, and can constrain the data-collection time. Figure 6(a) shows what a basis error looks like on the Poincaré sphere and how that error affects the ability to isolate a state in Poincaré space. This picture indicates that a basis error is more pronounced when measuring a pure state, but actually has no effect when measuring a totally mixed state (because all bases give the same answer).

Figure 6(b) shows the same analysis of errors in counting statistics. Any real measurement can be carried out only on a limited size ensemble. Though the details of the statistics will be dealt with later, the detection events are accurately

(a)   (b)

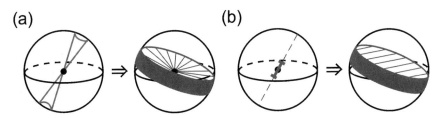

FIG. 6. Graphical representation of errors in a single-qubit tomography. (a) Basis errors. Errors in the setting of measurement apparatus can result in an accurate measurement being taken in an unintended basis. Shown graphically is the effect that an uncertainty in the measurement basis can have on the reconstruction of a state. Instead of a single axis on the Poincaré sphere, the possible measurement axes form uncertainty cones touching at the center, since all possible measurement axes pass through the origin. This uncertainty in axis is then translated into an uncertainty in the state (shown on the right). Instead of isolating the state to a plane, all possible measurement axes trace out a volume with large uncertainty near the surface of the sphere and low uncertainty near the center. (b) Counting errors. Even if the measurement basis is exactly known, only a limited number of qubits can be measured to gain an estimate of a state's projection onto this axis (taken directly from the probability of a successful measurement). This uncertainty results in an unknown state being isolated to a one-dimensional Gaussian (approximately) in three-dimensional space, rather than to a plane.

---

[15] These are the main sources of error that are likely to be present to some degree in *any* qubit implementation. In addition, each implementation may have its own unique errors, such as the wedged waveplates described earlier or accidental background counts from noisy detectors. Here we neglect such system-specific difficulties.

described by a Poissonian distribution, which for large numbers of counts is well approximated by a Gaussian distribution. This will cause the resultant knowledge about the unknown state to change from a plane (in the exact case) to a thick disk (uniformly thick for pure and mixed states), a one-dimensional Gaussian distribution plotted in three-dimensional space.

After all sources of error are taken into account, a single measurement results in a distribution over all possible states describing the experimenter's knowledge of the unknown state. This distribution represents the likelihood that a particular state would give the measured results, relative to another state. When independent measurements are combined, these distributions are multiplied, and ideally the knowledge of the unknown state is restricted to a small ball in Poincaré space, similar to a three-dimensional Gaussian (as a large uncertainty in any one direction will lead to a large uncertainty in the state). State isolation occurs regardless of which measurements are taken, as long as they are linearly independent, and is shown graphically in Fig. 7 for a set of orthogonal measurements.

In contrast to the ideal case in the previous section, for which the accuracy of a reconstructed state did not depend on whether mutually unbiased measurements were made, with real measurements the advantage of mutually unbiased measurement bases becomes clear. In contrast to the measurements shown in Fig. 7, mutually biased measurements result in a nonsymmetric error ball, increasing the error in state estimation in one direction in Hilbert space.

Even after tomography returns a distribution of likelihood over Poincaré space, one final problem remains. It is very possible, especially with low counts or with

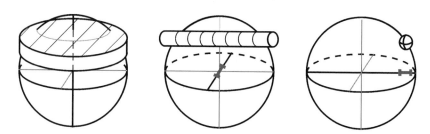

FIG. 7. Isolation of a quantum state through inexact measurements. Although a series of real measurements (those with uncertainties) will never be able to exactly isolate an unknown quantum state, they can isolate it to a region of Hilbert space that is far more likely than any other region to contain the unknown state. Consider a series of three measurements, each containing counting errors, along orthogonal axes. From left to right, the area of Hilbert space containing the unknown state is truncated from a one-dimensional Gaussian probability distribution (the disk in the left figure) to a two-dimensional Gaussian (the cylinder in the middle figure) and finally to a three-dimensional Gaussian (the ball in the right figure). This results in an 'error ball' which approximates the position of the unknown state. The global maximum, however, can often be outside allowed Hilbert space (outside the Poincaré sphere), which is one reason a maximum likelihood technique must be used to search over only allowed quantum states.

the measurement of very pure states, that state estimation will return an "illegal" state. For example, in Fig. 7, the measurements seem to place the error ball just on the edge of the sphere and slightly outside it. As all legal states have a radius of less than or equal to one in Poincaré space, it is necessary to find a way to return the most likely *legitimate* state reconstructed from a set of measurements.

4.2. The Maximum Likelihood Technique

The problem of reconstructing illegal density matrices is resolved by selecting the legitimate state most likely to have returned the measured counts (James et al., 2001; Hradil and Rehacek, 2001). In practice, analytically calculating this maximally likely state is prohibitively difficult, and a numerical search is necessary. Three elements are required: a manifestly legal parametrization of a density matrix, a likelihood function which can be maximized, and a technique for numerically finding this maximum over a search of the density matrix's parameters.

The Stokes parameters are an unacceptable parametrization for this search, as there are clearly combinations of these parameters which result in an illegal state (e.g., $S_1 = S_2 = S_3 = 1$). In this context, a legitimate state refers to a nonnegative definite Hermitian density matrix of trace one. The property of nonnegative definiteness for any matrix $\widehat{G}$ is written mathematically as

$$\langle \psi | \widehat{G} | \psi \rangle \geq 0 \quad \forall | \psi \rangle. \tag{95}$$

Any matrix that can be written in the form $\widehat{G} = \widehat{T}^\dagger \widehat{T}$ must be nonnegative definite. To see that this is the case, substitute into Eq. (95):

$$\langle \psi | \widehat{T}^\dagger \widehat{T} | \psi \rangle = \langle \psi' | \psi' \rangle \geq 0, \tag{96}$$

where we have defined $| \psi' \rangle = \widehat{T} | \psi \rangle$. Furthermore $(\widehat{T}^\dagger \widehat{T})^\dagger = \widehat{T}^\dagger (\widehat{T}^\dagger)^\dagger = \widehat{T}^\dagger \widehat{T}$, i.e., $\widehat{G} = \widehat{T}^\dagger \widehat{T}$ must be Hermitian. To ensure normalization, one can simply divide by the trace. Thus the matrix $\hat{g}$ given by the formula

$$\hat{g} = \frac{\widehat{T}^\dagger \widehat{T}}{\mathrm{Tr}\{\widehat{T}^\dagger \widehat{T}\}} \tag{97}$$

has all three of the mathematical properties required for density matrices.

For the one-qubit system, we have a 2×2 density matrix with 3 independent real parameters (although we will search over 4 in order to fit the intensity of the data). Since it will be useful to be able to invert relation (97), it is convenient to choose a tri-diagonal form for $\widehat{T}$:

$$\widehat{T}(\vec{t}) = \begin{pmatrix} t_1 & 0 \\ t_3 + it_4 & t_2 \end{pmatrix} \tag{98}$$

where $\vec{t}$ is a vector containing each t_i. The multiple-qubit form of the same equation is given by:

$$\widehat{T}(\vec{t}) = \begin{pmatrix} t_1 & 0 & \cdots & 0 \\ t_{2^n+1} + it_{2^n+2} & t_2 & \cdots & 0 \\ \cdots & \cdots & \cdots & 0 \\ t_{4^n-1} + it_{4^n} & t_{4^n-3} + it_{4^n-2} & t_{4^n-5} + it_{4^n-4} & t_{2^n} \end{pmatrix}. \quad (99)$$

The manifestly 'physical' density matrix $\hat{\rho}_p$ is then given by the formula

$$\hat{\rho}_p(\vec{t}) = \frac{\widehat{T}^\dagger(\vec{t})\widehat{T}(\vec{t})}{\text{Tr}\{\widehat{T}^\dagger(\vec{t})\widehat{T}(\vec{t})\}}. \quad (100)$$

This satisfies the first criterion for a successful maximum likelihood search, by providing an explicitly physical parametrization for $\hat{\rho}$. The second criterion, a likelihood function, will in general depend on the specific measurement apparatus used and the physical implementation of the qubit (as these will determine the statistical distributions of counts, and therefore their relative weightings). If we assume Gaussian counting statistics, then we can easily provide a suitable likelihood function.

Let $n_{v,r}$ be the result for the vth measurement setting on the rth detector combination. Let $\bar{n}_{v,r}$ be the counts that would be expected from the state $\hat{\rho}$, given all information about the system:

$$\bar{n}_{v,r} \equiv I_0 E_0 I_v E_r \text{Tr}\{\widehat{M}_{v,r}\hat{\rho}\} + n_{v,r}^{\text{accid}} \quad (101)$$

$$\widehat{M}_{v,r} \equiv \sum_{r'} (C_{r'\to r})\widetilde{M}_{v,r'} \quad (102)$$

$$\widetilde{M}_{v,r} \equiv U_v^\dagger |r\rangle \langle r| U_v. \quad (103)$$

Given that we wish to search over the parameters of $\vec{t}$, rather than $\hat{\rho}$, we will rewrite this equation as

$$\bar{n}_{v,r} = I_v E_r \text{Tr}\{\widehat{M}_{v,r}\widehat{T}^\dagger(\vec{t})\widehat{T}(\vec{t})\} + n_{v,r}^{\text{accid}}. \quad (104)$$

Notice that the unknown scalars I_0 and E_0 have been absorbed into the unnormalized $\widehat{T}^\dagger(\vec{t})\widehat{T}(\vec{t})$, allowing our numerical search to discover what their combined effect is without ever knowing their individual values.

Given these definitions, the probability of obtaining the vth measurement on the rth set of detectors, $n_{v,r}$, from the search parameters $\vec{t}$ is proportional to

$$\exp\left[-\frac{(\bar{n}_{v,r} - n_{v,r})^2}{2\hat{\sigma}_{v,r}^2}\right], \quad (105)$$

where $\hat{\sigma}_{v,r}$ is the standard deviation of the vth measurement (given approximately by $\sqrt{\bar{n}_{v,r}}$). Therefore, the total probability of $\hat{\rho}$ yielding the counts $\{n_{v,r}\}$ is given by:

$$P(n_{v,r}) = \frac{1}{Norm} \prod_{v,r} \exp\left[-\frac{(\bar{n}_{v,r} - n_{v,r})^2}{2\bar{n}_{v,r}} \right], \tag{106}$$

where *Norm* is the normalization constant. In order to find the ideal $\vec{t}$, and therefore the ideal $\hat{\rho}$, we need to maximize the probability function above. This is equivalent to maximizing the log of the same function, or equivalently, minimizing its negation, giving us our final likelihood function (notice that the normalization constant is ignored for this function, as it will not affect the minimum):

$$\mathcal{L}(\vec{t}) = \sum_{v,r} \frac{(\bar{n}_{v,r} - n_{v,r})^2}{2\bar{n}_{v,r}}. \tag{107}$$

The final piece in the maximum likelihood technique is an optimization routine, of which there are many available. The authors' implementation will be discussed in the next subsection.[16] After a minimum is found, $\hat{\rho}$ can be reconstructed from the values of $\vec{t}$.

EXAMPLE 14 (*A single-qubit tomography*). Photon pairs generated via spontaneous parametric downconversion from a nonlinear crystal can be used to generate single-photon states. Measuring a photon in one arm collapses the state of its partner to a single-qubit Fock state (Hong and Mandel, 1986). An ensemble of these photons can be characterized using the maximum likelihood technique. The following data was taken from an experiment in "Remote State Preparation" (Peters et al., 2005):

$$H = 6237, \quad D = 5793,$$
$$V = 8333, \quad R = 6202.$$

For this first example we will assume that no intensity normalization or crosstalk compensation needs to occur (see Example 15 for a more thorough example). After minimizing the likelihood function, we obtain the following $\widehat{T}$ matrix

$$\widehat{T} = \begin{pmatrix} 73.4 & 0 \\ -29.0 - 1.2i & 77.1 \end{pmatrix}, \tag{108}$$

from which we can derive the density matrix,

$$\hat{\rho} = \frac{\widehat{T}^\dagger \widehat{T}}{\mathrm{Tr}\{\widehat{T}^\dagger \widehat{T}\}} = \begin{pmatrix} 0.5121 & 0.1837 + 0.0075i \\ 0.1837 - 0.0075i & 0.4879 \end{pmatrix}. \tag{109}$$

---

[16] For freely available code and further examples, see: http://www.physics.uiuc.edu/research/ QuantumPhotonics/Tomography/

Note that the maximum likelihood technique easily adapts to measurements in mutually biased bases (e.g., due to imperfect yet well characterized waveplates) and overcomplete measurements (taking more measurements than is necessary). In the first case the set of $|\psi\rangle$ is mutually biased (i.e., not in the canonical bases), though still governed by the mathematics of tomography we have laid out; in the second case the sum in Eq. (107) is extended beyond the minimum number of measurement settings.

4.3. Optimization Algorithms and Derivatives of the Fitness Function

In order to complete a tomography, the likelihood function $\mathcal{L}(\vec{t})$ must be minimized. A number of optimization programs exist which can search over a large number of parameters (e.g., $\vec{t}$) in order to minimize a complex function. The authors use the *Matlab 7.0* function lsqnonlin, which is optimized to minimize a sum of squares. This type of optimized algorithm is more efficient than a generic search, such as the *Matlab* function fminunc. In order for this minimization to work most effectively, it takes as parameters $f(\vec{t})$ and $\partial f(\vec{t})/\partial t_i$, where $\mathcal{L}(\vec{t})$ is of the form

$$\mathcal{L}(\vec{t}) = \sum_x \left[f_x(\vec{t}) \right]^2. \tag{110}$$

For the problem of tomography, we can write

$$f_{v,r} \equiv \frac{\bar{n}_{v,r} - n_{v,r}}{\sqrt{2\bar{n}_{v,r}}} \tag{111}$$

$$= \frac{I_v E_r \mathrm{Tr}\{\widehat{M}_{v,r}\widehat{T}^\dagger(\vec{t})\widehat{T}(\vec{t})\} + n_{v,r}^{\mathrm{accid}} - n_{v,r}}{\sqrt{2(I_v E_r \mathrm{Tr}\{\widehat{M}_{v,r}\widehat{T}^\dagger(\vec{t})\widehat{T}(\vec{t})\} + n_{v,r}^{\mathrm{accid}})}}. \tag{112}$$

With some effort, we can analytically derive the partial derivatives of these terms, allowing the optimization algorithm to not only run much faster, but to converge quickly regardless of the initial search condition:

$$\begin{aligned}
\frac{\partial f_{v,r}}{\partial t_i} &= \frac{\left[\frac{\partial}{\partial t_i}(\bar{n}_{v,r} - n_{v,r})\right]\sqrt{\bar{n}_{v,r}} - (\bar{n}_{v,r} - n_{v,r})\left(\frac{\partial}{\partial t_i}\sqrt{\bar{n}_{v,r}}\right)}{\sqrt{2}\bar{n}_{v,r}} \\[2mm]
&= \frac{\left(\frac{\partial \bar{n}_{v,r}}{\partial t_i}\right)(\bar{n}_{v,r})^{1/2} - (\bar{n}_{v,r} - n_{v,r})\left[\frac{1}{2}(\bar{n}_{v,r})^{-1/2}\frac{\partial \bar{n}_{v,r}}{\partial t_i}\right]}{\sqrt{2}\bar{n}_{v,r}} \\[2mm]
&= \frac{1}{2\sqrt{2\bar{n}_{v,r}}}\frac{\partial \bar{n}_{v,r}}{\partial t_i}\left(1 + \frac{n_{v,r}}{\bar{n}_{v,r}}\right).
\end{aligned} \tag{113}$$

Note that it is impossible for this function to go to zero unless $\partial \bar{n}_{v,r}/\partial t_i$ goes to zero, important when considering whether or not the maximum likelihood function will have several local minima. Because $\widehat{T}$ is a linear function, we can easily write down

$$\frac{\partial \bar{n}_{v,r}}{\partial t_i} = I_v E_r \frac{\partial}{\partial t_i} \mathrm{Tr}\left\{\widehat{M}_{v,r} \widehat{T}^{\dagger}(\vec{t}) \widehat{T}(\vec{t})\right\}$$

$$= I_v E_r \mathrm{Tr}\left\{\widehat{M}_{v,r}\left[\widehat{T}^{\dagger}(\vec{t})\widehat{T}(\vec{\delta}_{ij}) + \widehat{T}^{\dagger}(\vec{\delta}_{ij})\widehat{T}(\vec{t})\right]\right\}, \tag{114}$$

where $\vec{\delta}_{ij}$ is a j-element vector whose ith element is equal to one. All other elements of $\vec{\delta}_{ij}$ are equal to zero. (Here j is the length of $\vec{t}$.)

Even using these derivatives (and especially if they are not used), it is important to choose an initial condition for the search which is as close as possible to the correct answer. This amounts to making the best analytic guess possible using the ideal tomographic techniques presented in Section 2. It is possible that those ideal techniques will result in an illegal density matrix, i.e., some of its eigenvalues will be negative. If this happens (indeed, this happening is the reason we *need* the Maximum Likelihood technique), we simply set those negative eigenvalues to zero, renormalize the positive eigenvalues, and use this truncated state as the starting condition for the search.

5. Choice of Measurements

After describing how to take measurements and how to analyze them, it is necessary to discuss how to choose which measurements to take. This general problem includes several choices, from whether to measure projectors independently or simultaneously as part of a complete basis (n versus $2n$ detectors) to how many measurements to take (the minimum set or an overcomplete set full of redundancy).

5.1. How Many Measurements?

The choice of how many measurements to perform depends on the details of the experimental setup and the goals of the experiment. For most applications, the speed of a tomography is paramount. At first it might appear obvious that the optimal number of measurements to perform would also be the *minimum* number of measurements to perform, seemingly implying the minimum time necessary to finish a tomography. This intuitive assumption is in fact false; it is often true that taking *more* measurements can in fact *reduce* the total time to achieve a specified level of accuracy for a complete tomography.

If changing between measurements requires little time or effort, as is the case when using fast, automated waveplates, then the primary consideration will be the total number of state copies necessary to run an accurate tomography. If, however, changing bases is time-intensive or error-prone, such as when moving waveplates by hand, it may be desirable to minimize the number of measurement settings.

As discussed in Section 2, minimizing the number of measurements will result in 4^n distinct measurement settings for n detectors, or 3^n settings for $2n$ detectors. In order to minimize state copies necessary, even for n detectors, every measurement should be accompanied by its complementary orthogonal measurements, in effect simulating a $2n$ detector measurement with only n detectors. For example, when measuring a two-qubit system, one should use 9 measurement settings (with 4 detectors) or 36 measurement settings (with only 2 detectors).

The reason that making more measurements can be so beneficial is simple: in order to transform the measured counts into probabilities, complete bases are necessary. If no complete bases are measured, then there is no scaling information about the rate at which state copies are being measured. Taking this information redundantly within every measured basis very precisely determines the state intensity, *allowing every other measurement to become more accurate.* Alternatively, this can function as a detector of systematic errors, allowing the experimenter to notice when the sum of the measured counts in one basis differs from the sum in another basis, e.g., due to a drift in the production rate of the source.

5.2. How Many Counts per Measurement?

In order to quantify the conclusions of the previous section, Monte Carlo simulations were used to estimate the number of state copies per measurement that would be necessary to achieve an average of 99% fidelity (between the tomography result and the "unknown" state) for a variety of two-qubit states, using both 2- and 4-detector measurements. Keep in mind that the 4-detector results, each of which require nine measurements, could be achieved by using 36 measurements with two detectors, which in many cases is still superior to the 16-measurement case. (For example, the Bell state requires 150 counts per measurement setting for 16 measurements on two detectors—a total of 150×16 counts, while at the same time 36 measurements on two detectors requires a total of 50×36 counts— a factor of two improvement.) Table I shows the state copies *per measurement* necessary to achieve a 99% fidelity for five families of states.

- The Werner state

$$\hat{\rho}_1(\lambda) = (1 - \lambda)\hat{\rho}\frac{1}{\sqrt{2}}(|HH\rangle + |VV\rangle) + \lambda\frac{I}{4},$$

Table I

Counts per measurement setting necessary to achieve, on average, a 99% fidelity state. To calculate the total counts necessary for a 2- (4-)detector tomography, multiply the listed number by 16 (9).

	λ	0	0.01	0.1	0.25	0.5	1.0
$\hat{\rho}_1$	2 det.	150	900	1300	4900	2400	1600
	4 det.	50	150	1200	600	350	350
$\hat{\rho}_2$	2 det.	150	200	250	300	350	350
	4 det.	50	100	150	150	150	150
$\hat{\rho}_3$	2 det.	75	200	350	350	300	350
	4 det.	50	100	150	150	150	150
$\hat{\rho}_4$	2 det.	150	1100	1600	1500	1000	700
	4 det.	50	100	300	500	250	200
$\hat{\rho}_5$	2 det.	75	300	6800	3500	2100	1600
	4 det.	50	100	1000	450	350	350

- The Bell state $\longrightarrow$ The half-mixed state

$$\hat{\rho}_2(\lambda) = \frac{1}{2}\big(|HH\rangle\langle HH| + |VV\rangle\langle VV|\big)$$
$$+ \frac{1-\lambda}{2}\big(|HH\rangle\langle VV| + |VV\rangle\langle HH|\big),$$

- HH $\longrightarrow$ The half-mixed state

$$\hat{\rho}_3(\lambda) = \left(1 - \frac{\lambda}{2}\right)|HH\rangle\langle HH| + \frac{\lambda}{2}|VV\rangle\langle VV|,$$

- Maximally Entangled Mixed States (MEMS)

$$\hat{\rho}_4\left(\lambda < \frac{1}{3}\right) = \left(\frac{1-\lambda}{2}\right)\big(|HH\rangle + |VV\rangle\big)\big(\langle HH| + \langle VV|\big)$$
$$+ \lambda|HV\rangle\langle HV|$$
$$\hat{\rho}_4\left(\lambda \geq \frac{1}{3}\right) = \left(\frac{1-\lambda}{2}\right)\big(|HH\rangle\langle VV| + |VV\rangle\langle HH|\big)$$
$$+ \frac{1}{3}\big(|HH\rangle\langle HH| + |HV\rangle\langle HV| + |VV\rangle\langle VV|\big),$$

- HH $\longrightarrow$ The totally mixed state

$$\hat{\rho}_5(\lambda) = (1 - \lambda)|HH\rangle\langle HH| + \lambda\frac{I}{4}.$$

In order to understand these numbers, consider just the Werner state's behavior as it transitions from a maximally entangled state ($\lambda = 0$) to a totally mixed state ($\lambda = 1$). Figure 8 shows the counts necessary to achieve either a 99% (dotted line) or 99.9% (solid line) fidelity for the full range of λ.

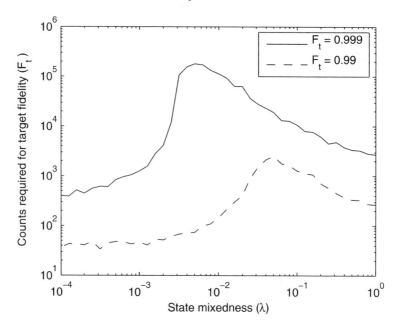

FIG. 8. This plot shows the counts per measurement for a 4-detector system using the minimum nine measurement settings to achieve either a 99% (dotted line) or a 99.9% (solid line) target fidelity (F_t) with the ideal state. These results used Monte Carlo simulations on the Werner state $(1 - \lambda)\hat{\rho}_{1/\sqrt{2}|HH+VV\rangle} + \lambda I/4$, for a range of values of λ.

For very low λ (very pure states), almost no state copies are necessary. When measuring these states, several measurements return a value close to zero counts. This value of zero probability guarantees that the state will be near the border of allowable states (e.g., near the surface of the Poincaré sphere for one-qubit states). This one measurement, because it isolates the allowable states to a very small area, can very quickly lead to a high fidelity. For example, if after measuring 100 copies in the H–V basis, one has recorded 100 H counts and 0 V counts, then there is a *very* high probability that the state $|H\rangle$ is the target state, because every other axis in Hilbert space has also been isolated by this measurement (in this specific example, the results of measurements in the D–A and R–L basis—50 counts in each—can be inferred from this one $H-V$ measurement).

As the state becomes more mixed, it moves away from the border in Hilbert space and now many counts are necessary to isolate the position of the state in those other directions. Continuing the example above, if our first measurement was instead 90 H and 10 V, we cannot infer the results of a D–A measurement. If, for instance, it results in 50 D and 50 A, these counts have a great deal more uncertainty than a 90/10 split, due to the nature of the fitness function. This accounts for the sharp transition in Fig. 8, a transition which moves closer to the pure states for a higher fidelity cutoff, as expected.

The last behavior shown in the graph, the gradual decrease in necessary counts for increasing mixedness, occurs because fidelity becomes less sensitive for states of greater mixedness (Peters et al., 2004).

6. Error Analysis

Error analysis of reconstructed density matrices is in practice a nontrivial process. The traditional method of error analysis involves analytically solving for the error in each measurement due to each source of error, then propagating these errors through a calculation of any derived quantity. In the photon case, for example, errors in counting statistics and waveplate settings were analyzed in some detail in reference (James et al., 2001), giving errors in both density matrices and commonly derived quantities, such as the tangle and the linear entropy. In practice, however, these errors appear to be too large: We have experimentally repeated some of our measurements many times, and observed a spread in the value of derived quantities which is approximately an order of magnitude smaller that the spread predicted from an analytic calculation of the uncertainty. Obviously, the correctness of the analytic calculation is questionable. Thus it is worthwhile to discuss alternate methods of error analysis.

One promising numerical method is the 'Monte Carlo' technique, whereby additional numerically simulated data is used to provide a statistical distribution over any derived quantity. Once an error distribution is understood over a single measurement (e.g., Gaussian for waveplate setting errors or Poissonian over count statistics), a set of 'simulated' results can be generated. These results are simulated using the known error distributions in such a way as to produce a full set of numerically generated data which could feasibly have come from the same system. These data are numerically generated (at the measured counts level), and each set is used to calculate a density matrix via the maximum likelihood technique. This set of density matrices is then used to calculate the standard error on any quantity implicit in or derived from the density matrix.

As an example, consider the application of the Monte Carlo technique to the downconversion results from Example 15. Two polarization-encoded qubits are generated within ensembles that obey Poissonian statistics, and these ensembles are used to generate a density matrix using the maximum likelihood technique. In order to find the error on a quantity derived from this density matrix (e.g., the tangle), 36 new measurement results are numerically generated, each drawn randomly from a Poissonian distribution with mean equal to the original number of counts. These 36 numerically generated results are then fed into the maximum likelihood technique, in order to generate a new density matrix, from which, e.g., the tangle may be calculated. This process is repeated many times, generating many density matrices and a distribution of tangle values, from which the error in the initial tangle may be determined. In practice, additional sets of simulated data must be generated until the error on the quantity of interest converges to a single value. For the data in Examples 14 and 15, a total of 100 simulations were used.

7. A Complete Example of Tomography

In order to demonstrate how all of the concepts presented in this chapter are actually applied, we have included an example which from start to finish uses laboratory parameters and data, taken from a two-qubit entangled photon source. Throughout this example we will use our usual convention for the canonical basis: $|0\rangle \equiv |HH\rangle$, $|1\rangle \equiv |HV\rangle$, $|2\rangle \equiv |VH\rangle$, and $|3\rangle \equiv |VV\rangle$.

EXAMPLE 15 (*A complete two-qubit tomography*). Before collecting tomography data, there are several measurement parameters that must be measured. After experimentally determining that each of our beamsplitters has negligible absorption, a 0.8% chance to reflect $|H\rangle$, and a 0.5% chance to transmit $|V\rangle$, we can determine that

$$
C_{r'\to r} \equiv
\begin{array}{c}
\to 0 \\
\to 1 \\
\to 2 \\
\to 3
\end{array}
\overset{\begin{array}{cccc} 0' & 1' & 2' & 3' \end{array}}{
\begin{pmatrix}
0.9842 & 00049 & 0.0049 & 0.0000 \\
0.0079 & 0.9871 & 0.0000 & 0.0050 \\
0.0079 & 0.0000 & 0.9871 & 0.0050 \\
0.0001 & 0.0079 & 0.0079 & 0.9901
\end{pmatrix}}. \tag{115}
$$

Rather than measuring intensity fluctuations by picking off a part of the pump laser, we will choose during this tomography to fit the intensity parameters I_ν as part of the maximum likelihood technique (we use four detectors, which will allow us to fit a relative intensity to each measurement setting by using the measured counts from each of four orthogonal projectors).

Because this particular tomography will use a total of nine measurement settings (the minimum number required), there will not be enough information to fit for the detector-pair efficiencies. A previous tomography (using 36 measurement settings and not shown here) was used to solve for the E_r, using a two-detector tomography applied to the 36 measurement results from *each* of the four pairs of detectors:

$$E_1 = 0.9998, \qquad E_3 = 0.9195,$$
$$E_2 = 1.0146, \qquad E_4 = 0.9265.$$

To simplify the example, we will make all measurements in the canonical bases (this could be accomplished using either ideal waveplates or, in some cases, imperfect waveplates—see Section 3.1.2).

With these parameters recorded, we can now take the data. The following counts were recorded for a slightly mixed Bell state (close to $\frac{1}{\sqrt{2}}(|HH\rangle+i|VV\rangle)$):

$n_{1,r}$: $HH = 3708$ $HV = 77$ $VH = 51$ $VV = 3642$
$n_{2,r}$: $HD = 1791$ $HA = 1987$ $VD = 2096$ $VA = 3642$
$n_{3,r}$: $HR = 2048$ $HL = 1854$ $VR = 1926$ $VL = 1892$
$n_{4,r}$: $DH = 1766$ $DV = 1914$ $AH = 2153$ $AV = 1741$
$n_{5,r}$: $DD = 1713$ $DA = 1945$ $AD = 2208$ $AA = 1647$
$n_{6,r}$: $DR = 3729$ $DL = 91$ $AR = 102$ $AL = 3662$
$n_{7,r}$: $RH = 2017$ $RV = 1709$ $LH = 1917$ $LV = 1955$
$n_{8,r}$: $RD = 3686$ $RA = 102$ $LD = 109$ $LA = 3651$
$n_{9,r}$: $RR = 2404$ $RL = 1474$ $LR = 1712$ $LL = 2209,$

with the corresponding accidental counts (calculated using the measured singles rates and the previously determined coincidence window Δt_r (c.f., Eq. (87)).

$n_{1,1}^{\text{accid}} = 5.4$ $n_{1,2}^{\text{accid}} = 5.6$ $n_{1,3}^{\text{accid}} = 5.9$ $n_{1,4}^{\text{accid}} = 6.0$
$n_{2,1}^{\text{accid}} = 5.2$ $n_{2,2}^{\text{accid}} = 5.5$ $n_{2,3}^{\text{accid}} = 5.6$ $n_{2,4}^{\text{accid}} = 6.0$
$n_{3,1}^{\text{accid}} = 5.3$ $n_{3,2}^{\text{accid}} = 5.5$ $n_{3,3}^{\text{accid}} = 5.6$ $n_{3,4}^{\text{accid}} = 5.9$
$n_{4,1}^{\text{accid}} = 5.2$ $n_{4,2}^{\text{accid}} = 5.3$ $n_{4,3}^{\text{accid}} = 6.0$ $n_{4,4}^{\text{accid}} = 6.1$
$n_{5,1}^{\text{accid}} = 5.2$ $n_{5,2}^{\text{accid}} = 5.4$ $n_{5,3}^{\text{accid}} = 5.9$ $n_{5,4}^{\text{accid}} = 6.2$
$n_{6,1}^{\text{accid}} = 5.2$ $n_{6,2}^{\text{accid}} = 5.3$ $n_{6,3}^{\text{accid}} = 5.9$ $n_{6,4}^{\text{accid}} = 6.1$
$n_{7,1}^{\text{accid}} = 5.3$ $n_{7,2}^{\text{accid}} = 5.4$ $n_{7,3}^{\text{accid}} = 5.9$ $n_{7,4}^{\text{accid}} = 6.1$
$n_{8,1}^{\text{accid}} = 5.4$ $n_{8,2}^{\text{accid}} = 5.9$ $n_{8,3}^{\text{accid}} = 6.1$ $n_{8,4}^{\text{accid}} = 6.6$
$n_{9,1}^{\text{accid}} = 5.3$ $n_{9,2}^{\text{accid}} = 5.4$ $n_{9,3}^{\text{accid}} = 6.0$ $n_{9,4}^{\text{accid}} = 6.2.$

After minimizing the likelihood function, we obtain the following $\widehat{T}$ matrix

$$\widehat{T} = \begin{pmatrix} 0 & 0 & 0 & 0 \\ 2.401 + 3.167i & 2.372 & 0 & 0 \\ -6.381 - 2.649i & 3.919 - 0.897i & 2.674 & 0 \\ -8.975 + 58.630i & 1.356 - 2.106i & 1.685 - 1.514i & 60.08 \end{pmatrix},$$

(116)

from which we can derive the density matrix,

$$\hat{\rho} = \frac{\widehat{T}^{\dagger}\widehat{T}}{\text{Tr}\{\widehat{T}^{\dagger}\widehat{T}\}}$$

$$= \begin{pmatrix} 0.50 & -0.02 - 0.01i & -0.02 - 0.01i & -0.07 - 0.49i \\ -0.02 + 0.01i & 0.00 & 0.00 + 0.00i & 0.01 + 0.02i \\ -0.02 + 0.01i & 0.00 - 0.01i & 0.00 & 0.01 + 0.01i \\ -0.07 + 0.49i & 0.01 - 0.02i & 0.01 - 0.01i & 0.50 \end{pmatrix}.$$

(117)

Our search algorithm returned this density matrix because it minimized not only
the main search parameters $\vec{t}$, but the intensities I_ν:

$$I_1 = 7647 \quad I_2 = 7745 \quad I_3 = 7879$$
$$I_4 = 7725 \quad I_5 = 7669 \quad I_6 = 7754$$
$$I_7 = 7751 \quad I_8 = 7716 \quad I_9 = 7967,$$

allowing us to calculate the expected counts $\bar{n}_{\nu,r}$ (for the final density matrix):

$$\bar{n}_{1,1} = 3792 \quad \bar{n}_{1,2} = 81 \quad \bar{n}_{1,3} = 67 \quad \bar{n}_{1,4} = 3544$$
$$\bar{n}_{2,1} = 1794 \quad \bar{n}_{2,2} = 1956 \quad \bar{n}_{2,3} = 2106 \quad \bar{n}_{2,4} = 1735$$
$$\bar{n}_{3,1} = 2046 \quad \bar{n}_{3,2} = 1787 \quad \bar{n}_{3,3} = 1933 \quad \bar{n}_{3,4} = 1956$$
$$\bar{n}_{4,1} = 1815 \quad \bar{n}_{4,2} = 1895 \quad \bar{n}_{4,3} = 2108 \quad \bar{n}_{4,4} = 1758$$
$$\bar{n}_{5,1} = 1618 \quad \bar{n}_{5,2} = 2050 \quad \bar{n}_{5,3} = 2247 \quad \bar{n}_{5,4} = 1604$$
$$\bar{n}_{6,1} = 3792 \quad \bar{n}_{6,2} = 97 \quad \bar{n}_{6,3} = 103 \quad \bar{n}_{6,4} = 3594$$
$$\bar{n}_{7,1} = 2032 \quad \bar{n}_{7,2} = 1699 \quad \bar{n}_{7,3} = 1901 \quad \bar{n}_{7,4} = 1966$$
$$\bar{n}_{8,1} = 3758 \quad \bar{n}_{8,2} = 103 \quad \bar{n}_{8,3} = 105 \quad \bar{n}_{8,4} = 3583$$
$$\bar{n}_{9,1} = 2271 \quad \bar{n}_{9,2} = 1580 \quad \bar{n}_{9,3} = 1751 \quad \bar{n}_{9,4} = 2206.$$

Using the error analysis techniques presented in Section 6, we can estimate this
state's fidelity with the Bell state $\frac{1}{\sqrt{2}}(|HH\rangle + i|VV\rangle)$ to be $98.4 \pm 0.2\%$.

8. Outlook

This chapter represents our best efforts to date to experimentally optimize quantum tomography of discrete systems, but fails to address several key areas that future research will need to include: (1) As discussed earlier, tomographic error analysis is still in the nascent stages of development, and to date the only acceptable error estimates have come from Monte Carlo simulations. Analytic solutions to the problem of error estimation could greatly speed up this computationally intensive task. (2) The study of *adaptive* tomography has motivated a number of the improvements presented in this chapter, notably the use of more measurement settings to achieve a more efficient and accurate tomography in less time. A general theory of how to adapt measurement settings and measurement times based on the data that has already been collected—which can be experimentally applied in real time—has yet to be fully realized (see, however, D'Ariano et al. (2004)). (3) The number of measurements necessary to perform tomography grows exponentially with the number of qubits; it will eventually be necessary to partially characterize states using fewer measurements. This will be particularly important for error— as the analysis of large systems takes more and more time, it will become less and less feasible to use Monte Carlo simulations to estimate the error. (4) Each distinct qubit implementation provides a unique challenges, which will need to be explored by the experimental groups specializing in those systems; hopefully, the study of each system's differences will illuminate new areas for tomographic improvement.

9. Acknowledgements

We would like to thank Daniel James and Andrew White for their assistance in the development of the theory of tomography and for helpful discussions throughout the preparation of this manuscript. We would like to thank Rob Thew for helpful conversations concerning qudit tomography. This work was supported by the National Science Foundation (Grant No. EIA-0121568) and the MURI Center for Photonic Quantum Information Systems (ARO/ARDA program DAAD19-03-1-0199).

10. References

Abouraddy, A.F., Sergienko, A.V., Saleh, B.E.A., Teich, M.C. (2002). Quantum entanglement and the two-photon Stokes parameters. *Opt. Commun.* **201**, 93.

Allen, L., Barnett, S.M., Padgett, M.J. (2004). "Optical Angular Momentum". Institute of Physics Publishing, London.

Altepeter, J.B., James, D.F.V., Kwiat, P.G. (2004). "Quantum State Estimation". *Lecture Notes in Phys.* Springer, Berlin.

Altepeter, J.B., Jeffrey, E.R., Kwiat, P.G., Tanzilli, S., Gisin, N., Acin, A. (2005). Experimental methods for detecting entanglement. Submitted to *Phys. Rev. Lett.*

Arnaut, H.H., Barbosa, G.A. (2000). Orbital and intrinsic angular momentum of single photons and entangled pairs of photons generated by parametric down-conversion. *Phys. Rev. Lett.* **85**, 286–289.

Bell, J.S. (1964). On the Einstein–Podolsky–Rosen paradox. *Physics* **1**, 195–200.

Born, M., Wolf, E. (1987). "Principles of Optics". Pergamon Press, Oxford, UK.

Coffman, V., Kundu, J., Wootters, W.K. (2000). Distributed entanglement. *Phys. Rev. A* **61**, 052306.

Cormen, T.H., Leiserson, C.E., Rivest, R.L., Stein, C. (2001). "Introduction to Algorithms", 2nd Edn. MIT Press, London, UK.

Cory, D.G., Fahmy, A.F., Havel, T.F. (1997). Ensemble quantum computing by NMR spectroscopy. *Proc. Nat. Acad. Sci. USA* **94**, 1634.

D'Ariano, G.M., Paris, M.G.A., Sacchi, M.F. (2004). "Advances in Imaging and Electron Physics", vol. 128. Academic Press, New York.

Gasiorowitz, S. (1996). "Quantum Physics". Wiley, New York.

Giorgi, G., Nepi, G.D., Mataloni, P., Martini, F.D. (2003). A high brightness parametric source of entangled photon states. *Laser Phys.* **13**, 350.

Hong, C.K., Mandel, L. (1986). Experimental realization of a localized one-photon state. *Phys. Rev. Lett.* **56**, 58.

Hradil, Z., Rehacek, J. (2001). Efficiency of maximum-likelihood reconstruction of quantum states. *Fortschr. Phys.* **49**, 1083.

James, D.F.V., Kwiat, P.G., Munro, W.J., White, A.G. (2001). Measurement of qubits. *Phys. Rev. A* **64**, 052312.

Jones, J.A., Mosca, M., Hansen, R.H. (1997). Implementation of a quantum search algorithm on a quantum computer. *Nature* **392**, 344.

Jozsa, R. (1994). Fidelity for mixed quantum states. *J. Modern Opt.* **41**, 12.

Kielpinski, D., Meyer, V., Rowe, M.A., Sackett, C.A., Itano, W.M., Monroe, C., Wineland, D.J. (2001). A decoherence-free quantum memory using trapped ions. *Science* **291**, 1013.

Kwiat, P.G., Englert, B.-G. (2004). "Science and Ultimate Reality: Quantum Theory, Cosmology and Complexity". Cambridge Univ. Press, Cambridge, UK.

Kwiat, P.G., Waks, E., White, A.G., Appelbaum, I., Eberhard, P.H. (1999). Ultrabright source of polarization-entangled photons. *Phys. Rev. A* **60**, R773.

Laflamme, R., Knill, E., Cory, D., Fortunato, E., Havel, T., Miquel, C., Martinez, R., Negrevergne, C., Ortiz, G., Pravia, M., Sharf, Y., Sinha, S., Somma, R., Viola, L. (2002). Introduction to NMR quantum information processing. quant-ph/0207172.

Langford, N.K., Dalton, R.B., Harvey, M.D., O'Brien, J.L., Pryde, G.J., Gilchrist, A., Bartlett, S.D., White, A.G. (2004). Measuring entangled qutrits and their use for quantum bit commitment. *Phys. Rev. Lett.* **93**, 053601.

Lawrence, J., Brukner, C., Zeilinger, A. (2002). Mutually unbiased binary observable sets on n qubits. *Phys. Rev. A* **65**, 032320.

Lehner, J., Leonhardt, U., Paul, H. (1996). Unpolarized light: Classical and quantum states. *Phys. Rev. A* **53**, 2727.

Leonhardt, U. (Ed.) (1997). "Measuring the Quantum State of Light". Cambridge Univ. Press, Cambridge, UK.

Lewenstein, M., Kraus, B., Cirac, J.I., Horodecki, P. (2000). Optimization of entanglement witnesses. *Phys. Rev. A* **62**, 052310.

Mair, A., Vaziri, A., Weihs, G., Zeilinger, A. (2001). Entanglement of the orbital angular momentum states of photons. *Nature* **412**, 313.

Marcikic, I., de Riedmatten, H., Tittel, W., Zbinden, H., Gisin, N. (2003). Long-distance teleportation of qubits at telecommunication wavelengths. *Nature* **421**, 509.

Monroe, C. (2002). Quantum information processing with atoms and photons. *Nature* **416**, 238.

Munro, W.J., James, D.F.V., White, A.G., Kwiat, P.G. (2001). Maximizing the entanglement of two mixed qubits. *Phys. Rev. A* **64**, 030302.

Nambu, Y., Usami, K., Tsuda, Y., Matsumoto, K., Nakamura, K. (2002). Generation of polarization-entangled photon pairs in a cascade of two type-i crystals pumped by femtosecond pulses. *Phys. Rev. A* **66**, 033816.

Nielsen, M.A., Chuang, I.L. (2000). "Quantum Computation and Quantum Information". Cambridge Univ. Press, Cambridge, UK.

O'Brien, J.L., Pryde, G.J., White, A.G., Ralph, T.C., Branning, D. (2003). Demonstration of an all-optical quantum controlled-not gate. *Nature* **426**, 264.

Peters, N., Altepeter, J.B., Jeffrey, E.R., Branning, D., Kwiat, P.G. (2003). Precise creation, characterization and manipulation of single optical qubits. *Quantum Information and Computation* **3**, 503.

Peters, N., Wei, T.-C., Kwiat, P.G. (2004). Mixed-state sensitivity of several quantum-information benchmarks. *Phys. Rev. A* **70**, 052309.

Peters, N.A., Barreiro, J.T., Goggin, M.E., Wei, T.-C., Kwiat, P.G. (2005). Remote state preparation: Arbitrary remote control of photon polarization. To appear in *Phys. Rev. Lett.*

Pittman, T.B., Fitch, M.J., Jacobs, B.C., Franson, J.D. (2003). Experimental controlled-not logic gate for single photons in the coincidence basis. *Phys. Rev. A* **68**, 032316.

Sanaka, K., Kawahara, K., Kuga, T. (2001). New high-efficiency source of photon pairs for engineering quantum entanglement. *Phys. Rev. Lett.* **86**, 5620.

Sancho, J.M.G., Huelga, S.F. (2000). Measuring the entanglement of bipartite pure states. *Phys. Rev. A* **61**, 042303.

Schmidt-Kaler, F., Häffner, H., Riebe, M., Gulde, S., Lancaster, G.P.T., Deuschle, T., Becher, C., Roos, C.F., Eschner, J., Blatt, R. (2003). Realization of the ciraczoller controlled-not quantum gate. *Nature* **422**, 408.

Sergienko, A.V., Giuseppe, G.D., Atatüre, M., Saleh, B.E.A., Teich, M.C. (2003). Entangled-photon state engineering. In: Shapiro, J.H., Hirota, O. (Eds.), "Proceedings of the Sixth International Conference on Quantum Communication, Measurement and Computing (QCMC)", Rinton, Princeton, p. 147.

Stokes, G.C. (1852). On the composition and resolution of streams of polarized light from different sources. *Trans. Cambridge Phil. Soc.* **9**, 399.

Terhal, B.M. (2001). Detecting quantum entanglement. quant-ph/0101032.

Thew, R.T., Nemoto, K., White, A.G., Munro, W.J. (2002). Qudit quantum-state tomography. *Phys. Rev. A* **66**, 012303.

Weinstein, Y.S., Pravia, M.A., Fortunato, E.M., Lloyd, S., Cory, D.G. (2001). Implementation of the quantum Fourier transform. *Phys. Rev. Lett.* **86**, 1889.

Werner, R. (1989). Quantum states with Einstein–Podolsky–Rosen correlations admitting a hidden-variable model. *Phys. Rev. A* **40**, 4277.

White, A.G., James, D.F.V., Eberhard, P.H., Kwiat, P.G. (1999). Nonmaximally entangled states: Production, characterization, and utilization. *Phys. Rev. Lett.* **83**, 3103.

Wooters, W.K. (1998). Entanglement of formation of an arbitrary state of two qubits. *Phys. Rev. Lett.* **80**, 2245.

Yamamoto, T., Koashi, M., Özdemir, S.K., Imoto, N. (2003). Experimental extraction of an entangled photon pair from two identically decohered pairs. *Nature* **421**, 343.

Zyczkowski, K. (2000). Geometry of entangled states. quant-ph/0006068.

Zyczkowski, K. (2001). cp^n, or, entanglement illustrated. quant-ph/0108064.

FINE STRUCTURE IN HIGH-L RYDBERG STATES: A PATH TO PROPERTIES OF POSITIVE IONS

STEPHEN R. LUNDEEN

Dept. of Physics, Colorado State University, USA

Abstract

Rydberg states of atoms, molecules, and ions with large values of orbital angular momentum ($L \geq 4$) represent a unique class of excited states with properties quite different from lower-L Rydberg states. Because special techniques, both experimental and theoretical, are required to study them effectively, they are still relatively unexplored. As experimental knowledge and theoretical understanding of these states improves, they represent an important source of information about the positive ions that form their cores. This note will describe the special experimental and theoretical techniques that have been developed to study these states, review the progress made to date, and discuss some of the remaining challenges.

ISSN 1049-250X
DOI 10.1016/S1049-250X(05)52004-4

1. Introduction

Rydberg states of atoms, molecules, and ions are states where one electron is highly excited, but still weakly bound to the positively charged ion which represents the remainder of the system. To a first approximation, the spectroscopy of Rydberg states is very simple. The study of Rydberg states could be said to date from the discovery by J.R. Rydberg that the wave numbers of different members of a series of atomic emission lines could be related by the formula:

$$\frac{1}{\lambda} = N_0 - \frac{R}{(n - \delta)^2} \qquad (1)$$

where R is a universal constant, N_0 and δ are constants characteristic of a particular series, and n is a running integer identifying members of the series (Rydberg, 1890). This discovery was also, of course, a critical step in the pre-history of quantum mechanics (White, 1934). We now recognize that Rydberg's observation reflects the fact that the energies of successive members of one Rydberg series are given approximately by the simple formula:

$$E = -\frac{R}{(n - \delta)^2} \qquad (2)$$

where R is now known as the Rydberg constant and δ as the "quantum defect" of the series. In the intervening years, there have been a large number of experimental studies involving Rydberg atoms, many of which have been reviewed elsewhere (Stebbings and Dunning, 1983; Gallagher, 1994). Most of these studies have been concerned with the special properties of these highly excited states, and not particularly focused on their spectroscopy. Because their other properties, their large size ($r \sim n^2 a_0$), low binding energy ($E_B \sim 13.6$ eV$/n^2$), and extreme sensitivity to external fields, are in dramatic contrast with the usual properties of atoms, this is understandable. One aspect of Rydberg state spectroscopy that has been widely studied is the complex behavior that occurs when two or more Rydberg series interact strongly, resulting in substantial shifts in the binding energies (i.e., quantum defects) in each series. This has led to the development of Multichannel Quantum Defect Theory (MQDT) and related theoretical methods (Aymar et al., 1996). This note discusses progress in a very different area of Rydberg spectroscopy, the spectroscopy of nonpenetrating, high-L Rydberg states. These states have very small quantum defects and, since they are usually not strongly perturbed by other Rydberg series, MQDT is not normally necessary in their description. Even though the quantum defects of high-L Rydberg states may be small, they are certainly not zero, and they give rise to definite differences between their actual binding energies and the simple hydrogenic result with $\delta = 0$. These differences, which we will refer to here as "fine structure", occur in characteristic patterns. The scale of these fine structure patterns is related to the long-range interactions between the Rydberg electron and the core ion that are present in addition to the simple Coulomb interaction binding it in its orbit. Careful measurement of these fine structure patterns, combined with a good theoretical understanding of the factors that influence them, can be used to obtain precise measurements of certain properties of the positive ions that form the cores of these Rydberg systems.

Figure 1 shows a simple view of a nonpenetrating Rydberg electron bound to a positively charged ion core in a classical elliptical orbit. The Rydberg electron is primarily sensitive to the total charge of the core, which binds it in its orbit. Each principal quantum number, n, supports a whole family of such Rydberg states, analogous to elliptical orbits sharing the length of their major axis but ranging in shape from linear ($L = 0$) to nearly circular ($L = n - 1$). The classical turning points of radial motion for these hydrogenic states are given by:

$$r_{\pm} = n^2 a_0 \pm n^2 a_0 \sqrt{1 - \frac{L(L+1)}{n^2}}. \tag{3}$$

The sum of r_+ and r_- corresponds to the major axis of the classical ellipse, $2n^2 a_0$. The inner turning point, the distance of closest approach of the Rydberg electron to the core ion is given by:

$$r_- \cong \frac{L(L+1)a_0}{2}\left[1 + \frac{L(L+1)}{4n^2} + \cdots\right] \geq \frac{L(L+1)a_0}{2}. \tag{4}$$

Consequently, an excited electron with $L \geq 4$ is blocked by the centrifugal barrier from approaching within $10a_0$ of the ion core, and higher-L electrons are kept at an even greater distance. For the purposes of this note, we may consider $L \geq 4$ as

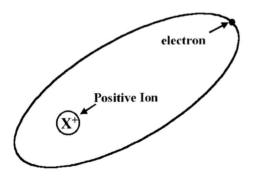

FIG. 1. Classical electron orbit analogous to a high-L Rydberg electron bound to the positively charged core ion. The quantum wavefunctions faithfully reflect the radial probability distributions of such classical orbits. In high-L orbits, the Rydberg electron never approaches the core ion, and so is primarily sensitive to its total charge. Other long-range interactions, much weaker by comparison, produce the Rydberg fine structure.

a working definition of a "nonpenetrating", "high-L" Rydberg state. Because of the lack of penetration of the core ion, the wave function of such a high-L Rydberg electron is close to a purely hydrogenic wave function. Whatever interactions are present between the Rydberg electron and the core ion beyond the dominant Coulomb attraction are weak by comparison and are exclusively long-range. This leads to a spectroscopy that is distinctly different from the familiar spectroscopy of lower-L states.

A useful, though certainly over-simplified, picture which represents the basic features common to high-L Rydberg states is to consider that the Rydberg electron moves in an "effective potential" which is the sum of the dominant Coulomb potential binding it to the core ion and an additional interaction potential due to its other long-range interactions with the core ion. These other interactions are primarily due to electrical forces of two types: (1) the multipole fields from the permanent electric moments of the core ion, if any; and (2) the fields due to electric moments induced in the core ion by the field of the Rydberg electron. While there is a whole series of multipole orders of each type, the lowest order terms will generally dominate, making the permanent quadrupole moment and the induced dipole moment the primary factors in the Rydberg fine structure. Thus, the most significant terms of this "effective potential" may be written as:

$$V_{\text{eff}}(r) = \left[-\frac{1}{r} \right] + \left[-\frac{Q}{r^3} T^{[2]}(\vec{J_c}) \cdot T^{[2]}(\vec{L}) - \frac{\alpha_1}{2r^4} + \cdots \right]. \tag{5}$$

Here, Q is the permanent electric quadrupole moment of the core ion, J_c is its angular momentum, and α_1 is the dipole polarizability of the core. $T^{[2]}$ denotes an appropriate second rank tensor in each space. If the core ion has a quadrupole moment, which requires $J_c \geq 1$, the quadrupole term will usually dominate the fine structure pattern, producing a "tensor fine structure" pattern consisting of $2J_c + 1$ fine structure levels for each value of L. These states are eigenstates of J_c, L, and the vector sum, $\vec{K} = \vec{J_c} + \vec{L}$. In the absence of a quadrupole moment, the dipole polarization term, proportional to α_1 the core ion's dipole polarizability, would usually dominate, giving rise to a "scalar fine structure" pattern consisting of a single fine structure level for each value of L. In general, both types of structure will be present, as illustrated for a typical case in Fig. 2, which shows the fine structure pattern observed in high-L Rydberg states of atomic neon (Ward et al., 1996). For comparison to Eq. (2), in $n = 10$ a fine structure energy of 2000 MHz corresponds to a quantum defect of about 0.0003.

Figure 2 also illustrates another general feature of these high-L fine structure patterns: the pattern varies smoothly in scale as L increases, approaching a purely hydrogenic limit as L nears its maximum value. The details of this variation with L can be used to isolate the leading contributions to the structure, identified in Eq. (5), from higher order corrections that become relatively less important as L increases. Thus, when measured carefully across a sufficient range of L, these

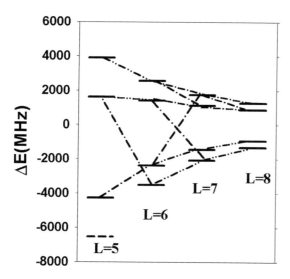

FIG. 2. Fine structure pattern measured in high-L, $n = 10$ Rydberg states of neon (Ward et al., 1996). Since $J_c = 3/2$, there are four eigenstates of $\vec{K} = \vec{J}_c + \vec{L}$ for each value of L, corresponding to four possible orientations of the Rydberg orbit plane with respect to the core angular momentum. The energy of each eigenstate, shown by a solid horizontal line, is referred to the zeroth order hydrogenic energy. This fine structure pattern is a superposition of a tensor pattern, due to the electric quadrupole moment of the Ne$^+$ core, and a scalar pattern due to the dipole polarizability of the core. The dash–dot lines illustrate the microwave transitions that were measured to determine the relative positions of these levels. The position of the lowest energy $L = 5$ level was not determined, so it is shown as a dashed line.

high-L fine structure patterns can provide precise measurements of the properties of the core ions that control the size of these long-range interactions, such as polarizabilities and electric moments. In essence, this application takes advantage of the known wave function of a high-L Rydberg electron to use that electron as a sensitive probe of the weak, but nonzero long-range interactions with the core ion.

The idea of using measured fine structure patterns in Rydberg states to determine properties of the positive ion core has a long history, but it has been applied with much greater success since the development of experimental techniques to measure high-L fine structure patterns precisely. The idea was first suggested in a 1933 paper titled "The Polarizabilities of Ions from Spectra" (Mayer and Goeppert Mayer, 1933). Initially, only atoms with S-state cores were considered, so the chief object of this exercise was to determine the dipole polarizability of the core ions. Although it was recognized from the outset that this program was best carried out in states of high angular momentum, the lack of experimental data about such states meant that the early studies were usually concerned with states of, at most, $L = 3$. Since these states were only marginally qualified as "nonpenetrating", complications in the structure made these studies only partially successful. The possibility of parameterizing Rydberg state binding energies with a small number of core parameters, even in relatively low-L Rydberg series, was discussed by Edlen in an widely cited article (Edlen, 1964). The suggestion that Rydberg systems with P-state cores could yield measurements of core ion quadrupole moments was advanced in 1982 (Chang and Sakai, 1982), but again without access to data on high-L levels the procedure provided only limited precision. The experimental techniques described in Section 2 have now provided a rich base of data about high-L fine structure patterns. This, in turn, has helped to stimulate progress in the theoretical techniques that use long-range models to describe the high-L Rydberg fine structure. Together, these developments have made it possible to extract core properties from high-L Rydberg fine structure patterns with precision that is at least comparable to the best methods available for the study of similar properties of neutral atoms.

An important theme in understanding the properties of high-L Rydberg states is the interplay between special theoretical methods, designed specifically to describe these nonpenetrating states, and more traditional methods of atomic and molecular structure theory. On the one hand, the characteristics of high-L states are so dramatically different from lower-L states that it is quite inefficient to describe them with the usual theoretical methods. Special methods are much more efficient, and also provide the connection to core ion properties that can be an important motivation for experimental studies. On the other hand, traditional theoretical methods form an important complement to the special techniques that can help to clarify their limitations and range of applicability.

Section 2 of this paper discusses the experimental methods that have been used to study Rydberg states of high-L ($L \geq 4$). Section 3 discusses the theoretical methods developed to describe these states and the types of fine structure patterns they predict. Section 4 reviews the conclusions drawn from the existing experimental studies with the aid of theoretical models. Section 5 points to some questions and experimental opportunities that remain open for future studies.

2. Experimental Methods

2.1. EARLY STUDIES

It may at first seem surprising that the spectroscopy of high-L Rydberg states of atoms is relatively unexplored. After all, the Rydberg states with $L \geq 4$ form the vast majority of atomic excited states, and an enormous quantity of work has been reported cataloging the optical spectra of atoms. The fact is, however, that almost none of these reports concern excited states with $L \geq 4$, since these states are quite difficult to examine with conventional spectroscopic methods. Absorption spectroscopy from the ground states of atoms or molecules, which tend to be states of relatively low angular momentum, can only directly access other relatively low angular momentum excited states because or the selection rule $\Delta L = \pm 1$, which applies to dipole excitation. Similarly, the emission spectra of atoms, molecules, and ions is dominated by the emission of $L < 3$ excited states, because only these can decay directly to low-lying low-L states. Higher L excited states, while they may contribute to the emission spectrum, radiate at much lower rates, emitting photons of much lower energy, perhaps in the infrared or far-infrared portion of the optical spectrum. Even if these weak emissions could be seen, it would generally require optical spectrometers of very high resolution to distinguish the high-L emission energies from hydrogenic energies. So, despite extensive optical spectroscopy of Rydberg series of S, P, D, and F states, very little direct spectroscopic data exists for states with $L \geq 4$. For example, an examination of the online NIST Atomic Spectra Database for the first row elements shows that of the over 600 excited state level positions recorded there, only 8 correspond to states with $L = 4$, and none with $L > 4$ (Martin et al., 1999).

Despite the difficulties, some of the earliest reported studies of high-L Rydberg spectroscopy were reported with traditional methods. One early study reported the optical emission spectrum from ng and nh Rydberg states of Mg$^+$, with $n = 5$–9 (Risberg, 1955). Another early example of optical emission spectroscopy is the study of V^{4+} emission lines by Van Deurzen, who reported positions of ng, nh and ni states with $n = 5$–8 (Van Deurzen, 1977). Both Mg$^+$ and V^{4+} are alkali-like ions showing only scalar fine structure. The precision of these early measurements, made with grating spectrometers, was limited to about 0.01 cm^{-1}. Several later reports were based on high resolution studies of emission spectra, obtained with Fourier Transform Spectrometers (FTS). Sansonetti et al. reported observations of the nG to 4F ($n = 5$–11) emission lines in Cs with a resolution of about $\lambda/\Delta\lambda = 400,000$, leading to determination of the positions of the nG levels with precision of about 0.001 cm^{-1} (Sansonetti et al., 1981). A second example is the identification of 5g–4f (Herzberg and Jungen, 1982), and later the 6h–5g and 6g–5f (Jungen et al., 1989) emission lines in H$_2$ and D$_2$, also obtained with FTS, although at somewhat lower resolution. A third example is the discovery of narrow emission lines near 12 μm in the natural emission spectrum of our sun (Breault and Noyes, 1983), which were subsequently identified as high-L ($L = 4, 5, 6$) emission from $n = 7$ Rydberg states of Mg, Al, and Si (Chang and Noyes, 1983; Chang, 1984).

Other early studies relied on high magnetic fields to reveal indirectly the fine structure of high-L states of helium. Beyer and Kolath studied the effects of anti-crossings on the nD–2P emission lines in helium in a magnetic field, and deduced the G–H, H–I, and I–K fine structure intervals in $n = 7$–10 of helium (Beyer and Kollath, 1978). Pannock et al. studied transitions between 7^{1}S and 9L ($L = 1$–8) states of helium driven with a CO_2 laser in a high magnetic field and determined the fine structure intervals between all $n = 9$ levels of helium (Panock et al., 1980). The precision of these measurements, however, was somewhat less than that obtained in the FTS emission studies, typically 0.002 cm^{-1}.

Still other studies began to explore the use of microwave spectroscopy to reveal the high-L fine structures. Because it is much less subject to Doppler broadening, microwave spectroscopy is able to reveal the small fine structure intervals much better than direct optical spectroscopy. One relatively simple approach which was used to detect such microwave transitions and to obtain spectroscopic information about states with $L \geq 4$ is to observe the cascade decay products arising from the decay of the high-L states, rather than the directly emitted infrared photons. This approach was used by Farley et al. to measure intervals between D, F, and G states in $n = 5$–11 levels of helium (Farley et al., 1979). In this case the excited states were initially produced by electron beam excitation of a low pressure cell. Another early experiment using the cascade decays was the observation of F–G, F–H, and F–I intervals in $n = 7, 8$ of helium (Cok and Lundeen, 1981). In this case the high-L states were initially populated by charge transfer from an accelerated He$^+$ beam, producing a fast beam of helium Rydberg atoms. In both experiments, single and multiphoton microwave resonances were used to measure the fine structure intervals separating states with common n. The use of microwave resonance methods led to measurement precision on the order of 0.3 MHz, or 10^{-5} cm^{-1}, about two orders of magnitude better than obtained previously with other methods. While this illustrated the advantages of microwave spectroscopy in studies of high-L fine structure, both the excitation and the detection of the high-L states were quite inefficient with these methods.

2.2. Stepwise Excitation Microwave Studies

Much more efficient excitation of high-L states was obtained by using stepwise laser excitation to populate a Rydberg D or F state, and then using microwave fields to drive transitions to higher-L levels, while still relying on the cascade fluorescence to detect the presence of transitions. This technique was used to measure d–f–g–h fine structure intervals in Na ($n = 13$–17) (Gallagher et al., 1976), d–f and d–g intervals in Li ($n = 7$–11) (Cooke et al., 1977), and f–g intervals in Cs ($n = 15$–17) (Ruff et al., 1980). These measurements led to increased interest in high-L spectroscopy, and in the possibility of using such spectroscopy to make precise measurements of dipole polarizabilities of the positive ion cores (Freeman and Kleppner, 1976).

A further improvement in the measurements based on stepwise laser excitation was the introduction of a more efficient method of detection of transitions based on Stark ionization. By taking advantage of the longer lifetime of the high-L levels produced in microwave transitions, Safinya et al. were able to efficiently detect transitions to high-*L* states by Stark ionizing the Rydberg population after a fixed delay time. This Delayed Field Ionization (DFI) method was used to measure f–h and f–i fine structure intervals in Cs ($n = 16$–18) (Safinya et al., 1980), with the Cs sample in the form of a thermal atomic beam. Later a similar technique was used to measure g–h–i–k intervals in Ba Rydberg levels around $n = 20$ (Gallagher et al., 1982). The combination of a thermal atomic beam source, selective laser excitation to a state of moderate *L* (2 or 3), followed by microwave excitation to a higher-*L* level which is then detected by time delayed selective field ionization provides a powerful and general method for exploring the spectroscopy of high-*L* Rydberg systems. The 1980 and 1982 studies of Cs and Ba Rydberg levels by Gallagher and co-workers represent the first instances of measurements of this kind. Figure 3 illustrates the apparatus used for the Ba Rydberg study, which typifies this approach. Barium atoms in a thermal beam are excited by three tunable lasers to the 6s20g $^1G_4$ level. One element of this excitation scheme is the decay of the 6s6p $^1P_1$ state excited in the first laser to the 6s5d $^1D_2$ level, which occurs 10% of the time. This allows the excitation of an $L = 4$ level while using only

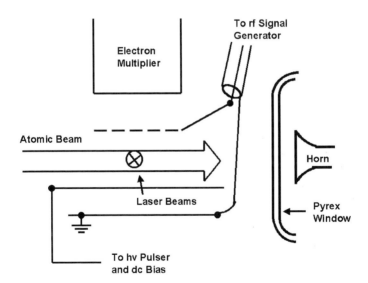

FIG. 3. Apparatus used to measure the fine structure in high-*L* Rydberg states of barium (Gallagher et al., 1982). Barium atoms in a thermal atomic beam were excited by three pulsed lasers to the 6sng state with $18 \leq n \leq 23$. Transitions to higher-*L* levels were then induced by application of microwave and rf electric fields. After a time delay that discriminated in favor of longer lived high-*L* states produced in such transitions, remaining Rydberg atoms were Stark ionized and the resulting electrons were collected and counted. Fine structure intervals were measured by observing resonances in the electron count rate as a function of applied microwave frequency. Similar experimental schemes have been used in number of subsequent studies of high-*L* Rydberg fine structure.

three lasers. Similar 6sng states with $18 \leq n \leq 23$ were excited by retuning the third excitation laser. Transitions to Rydberg states of higher L were induced with microwave fields and detected by ionizing high-n levels after a delay time that discriminated in favor of the longer-lived higher-L level. Resonances corresponding to g–h, g–i, and g–k transitions were observed, the later two requiring more than one microwave photon. The measurements determined the relative positions of $L = 4, 5, 6$, and 7 levels in Rydberg Ba, with high precision.

In the intervening years, similar methods have been used to study high-L states in a range of alkali and alkali-earth atoms. Similar techniques were used to study nf–ng intervals in Ca (Vaidyanathan et al., 1982) and a wide range of d–f–g–h intervals in Na (Tran et al., 1984; Sun and MacAdam, 1994) and Mg (Lyons and Gallagher, 1998). More recently, similar techniques have been used to study d–f and d–g intervals in Rydberg states of Al (Dyubko et al., 2003).

Studies of this type, using stepwise laser excitation from the atomic ground state in a thermal atomic beam, followed by single and multiphoton microwave transitions to high-L Rydberg levels detected by selective field ionization, have proved to be a versatile and powerful method of studying Rydberg states with $L \geq 4$. The atomic beam environment is relatively free of stray fields, which can easily perturb the fine structure patterns under study. In the most recent applications of this technique, very high signal-to-noise studies of high-L levels have been demonstrated. However, in only one case have states with L as large as 7 been observed, that being the original study of Ba Rydberg states (Gallagher et al., 1982). Reaching higher-L levels would be possible in principle with this method, but it would require additional laser excitation steps or microwave transitions with more than three photons. Either of these approaches would pose additional experimental difficulties, and increasing L by additional units would become progressively more difficult. Another limitation of this technique is that it can only be applied in cases where laser excitation from the atomic ground state is possible. This explains the prominence of alkali and alkali-earth atoms in studies of this type. It also helps to explain why, to date, no high-L atomic Rydberg system with tensor fine structure has been explored with this method, since none of the atoms which display this type of Rydberg structure, such as neon, are easily excited from their ground states.

2.3. RESONANT EXCITATION STARK IONIZATION SPECTROSCOPY (RESIS)

An alternative experimental method, which has also been quite productive in exploring high-L fine structure, is the Resonant Excitation Stark Ionization Spectroscopy (RESIS) method. This method does not rely on laser excitation from the ground state, but instead works with a sample of Rydberg atoms or molecules that is prepared by charge exchange from an accelerated beam of positive ions.

A very wide variety of excited levels is formed in this way, including Rydberg states of high-L. While these states may not be formed efficiently, they may still provide an adequate sample for useful spectroscopy if a sufficiently sensitive and selective method of detection is used. Early studies using cascade detection confirmed that useful levels of high-L products were formed in the charge exchange process (Brandenberger et al., 1976; Cok and Lundeen, 1981). However, the most productive use of these fast Rydberg beams awaited the development of a more efficient detection method. With the RESIS method, atoms in particular high-L Rydberg states are detected with near 100% efficiency by exciting them with a laser to a very weakly bound Rydberg level, and then Stark ionizing this state and collecting the ions produced.

The first use of the RESIS detection method for high-L Rydberg states in a fast beam was the study by Palfrey of high-L $n = 10$ Rydberg states of helium (Palfrey and Lundeen, 1984). Figure 4 shows a schematic diagram of the apparatus used for that study. A beam of He$^+$ ions, of about 10 keV energy, was partially neutralized in a differentially pumped charge exchange cell, forming a fast beam of helium Rydberg states. The residual charged beam, and any neutral atoms formed in very high n states ($n > 15$) were deflected out of the beam by an electric field sufficiently strong to ionize $n = 15$ levels. The atoms in the beam in an $n = 10$ state were detected in a two step process. First, those in a specific $10L$ level, such as the $10G$ level, were efficiently excited to a much higher level, such as the $30H$ level, using a Doppler-tuned CW CO$_2$ laser. Then any atoms so excited were efficiently Stark ionized and the resulting ions collected and counted. This Res-

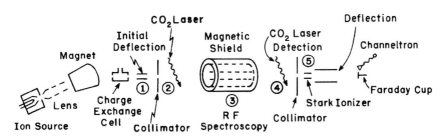

FIG. 4. Apparatus used to measure the fine structure of high-L states of helium (Palfrey and Lundeen, 1984). The method is typical of what has become known as Resonant Excitation Stark Ionization Spectroscopy (RESIS). A fast beam of Rydberg atoms is created by when He$^+$ ions in an accelerated beam capture an electron in a charge exchange cell. Electrons initially captured into very weakly bound states ($n > 15$) are immediately ionized in a field that also deflects remaining ions out of the beam. Specific high-L Rydberg states in the beam, e.g., the $10G$ state, are selectively detected by exciting them to a much higher state, e.g., the $27H$ states, using a Doppler-tuned CO$_2$ laser at (4). Any atoms so excited are Stark ionized at (5) and the resulting ions collected and counted. Microwave transitions, induced in section of transmission line, are detected by the resulting change in population of the selectively excited level. The microwave signal size is enhanced by deliberately creating a population difference between the two coupled levels, using a second CO$_2$ laser excitation at (2).

onant Excitation and Stark Ionization detection results in efficient and selective detection of atoms in a specific Rydberg level, such as in this case the 10G level. Because of the prior Stark ionization of atoms initially formed in high n levels, very little background signal is present. This L-selective detection provides the basis for detection of microwave induced transitions between different $n = 10$ levels. First, an initial CO_2 laser excitation is used to deplete the population of a particular (10L) level, such as the 10G. Then a microwave field drives transitions between this level and a nearby level, such as the 10H. When the frequency and strength of the microwave field is correct, this results in repopulation of the 10G level, and an increase in the ion current detected following the second 10G–30H laser excitation. Palfrey used this technique to measure the G–H, H–I, and I–K intervals in helium with precision of about than $+/- 10$ kHz (Palfrey and Lundeen, 1984).

The RESIS method is well suited to detecting $n = 9$ or $n = 10$ Rydberg levels, using a CO_2 laser. This laser can be discretely tuned to about 50 different single frequency lines covering the approximate range of 900 to 1100 cm^{-1}. The frequency separation between adjacent CO_2 lines is about 1 or 2 cm^{-1}, but Doppler tuning provides a convenient way to achieve fine tuning, by varying the angle of intersection between the CO_2 laser and the fast Rydberg beam. As long as the beam velocity is at least 0.001c, varying the angle provides continuous coverage of the frequency intervals between discrete lines, and insures that a desired transition can be tuned into exact resonance. The range of laser frequencies available in this way includes transitions exciting $n = 9$ or $n = 10$ states to higher levels. Typical transitions are from $n = 10$ to states near $n = 30$ or from $n = 9$ to states near $n = 20$. The linewidth of these excitation transitions is limited by combination of factors: (a) the transit time through the CO_2 beam, (b) the angular spread of the fast beam, and (c) the velocity spread of the fast beam, and (d) possible stray electric fields which Stark broaden the upper state of the transition. In the first study, the observed linewidths were on the order of 100 MHz. This is narrow enough that $n = 10$ helium Rydberg states with $L = 4, 5$, or 6 could be selectively detected.

A key advantage of the RESIS method of forming and detecting high-L states is the fact that all high-L levels are formed in charge exchange collisions and can be detected with this method. Although in this first study, the laser frequency resolution prevented selective detection of states with $L > 6$, no selection rule prevents excitation and detection of states with the highest possible L. On the other hand, a disadvantage of this method is that it is convenient only for a limited number of principal quantum numbers, and therefore a limited range of L. The $n = 9$ and 10 levels can be excited directly. The $n = 7$ levels can be detected with two step CO_2 excitation, first from 7 to 9, and then from 9–20 (Stevens and Lundeen, 1999). It is also possible to observe microwave transitions in the upper state of the RESIS transition (Arcuni et al., 1990a, 1990b). Still compared with methods based on laser excitation from the ground state, the range of states that can be studied is limited.

Subsequent RESIS studies in helium (Hessels et al., 1992; Storry et al., 1997b; Stevens and Lundeen, 1999) and H_2 (Sturrus et al., 1991; Jacobson et al., 2000) used the same basic technique, but incorporated modifications that extended its scope and improved its precision. One of the primary limitations in the precision of the microwave studies is the presence of stray electric fields in the microwave interaction region. Even after motional electric fields due to the earth's magnetic field are eliminated with magnetic shielding, fields on the order of 10–100 mV/cm are observed to build up over time, presumably due to charging of nominally conducting surfaces. The effect of such fields was most critical in the helium studies, where they were ultimately reduced to about the 5 mV/cm level by heating the microwave interaction region (Stevens and Lundeen, 1999). Other improvements included substituting a Cs vapor charge exchange cell for the gas charge exchange cell used in the earliest experiments, and improvements in the design of the Stark ionization detector to further reduce the background (Jacobson et al., 2000).

The most significant technical improvement in the RESIS technique, since the first experiments, is the introduction of a method of selectively populating the specific Rydberg levels in the fast beam that can be detected with the RESIS method. This is based on the use of a selectively excited Rydberg charge exchange target, which replaces the gas or Cs vapor cell. The electron transferred from the Rydberg target to the projectile ion will tend to have a similar binding energy to the projectile ion as it had to the Rb^+ ion in the target. Therefore an $n = 10$ or $n = 9$ Rydberg target should produce a fast Rydberg population peaked at $n = 10$ or $n = 9$, in contrast to the Cs vapor cell which would favor more tightly bound states. Many of the features of this resonant charge transfer process were studied experimentally by MacAdam and co-workers (MacAdam et al., 1990), and have been found to be in approximate agreement with predictions obtained from Classical Trajectory Monte Carlo (CTMC) calculations (Pascale et al., 1990). The first application of the Rydberg charge exchange target to a RESIS experiment was a study of sulfur Rydberg levels using a Rydberg target formed by selective excitation of either 10F or 8F levels of Rb (Deck et al., 1993). In that study it was found that the ratio of RESIS signal to total neutral beam was 10,000 times larger when the charge exchange occurred in the Rydberg target than when it took place in an Argon gas cell. This is consistent with predictions of a much higher fraction of the neutral population being in the $n = 9$ and $n = 10$ states of interest when capture occurs in the Rydberg target. One disadvantage of the Rydberg target is that the charge capture fraction was much smaller because of the limited density of the Rydberg target. Capture fractions of 0.1–1.0% are typical for a singly charged beam, much lower than the 30% capture fractions that were routinely obtained with a gas cell or Cs vapor cell. Nevertheless, the high concentration of Rydberg products in the states that may be detected with the RESIS method results in larger RESIS signals and lower background. In a typical case, capture from an $n = 10$ Rydberg target by ions with velocity of about 0.10 a.u., CTMC predicts that about 10% of the capture products will be formed in $n = 10$, with a similar fraction occupying all states with $n > 15$.

The combination of Rydberg target and RESIS detection makes for a versatile and powerful method for studying high-L Rydberg levels. In principle, any Rydberg system whose positive ion core can be formed into an accelerated ion beam could be studied with this technique. The formation of the Rydberg states through resonant charge exchange from the Rydberg target should not depend on the identity of the positive ion, nor should the approximate frequencies of the RESIS excitation transitions or the fields necessary to Stark ionize the upper states. The only feature of the method which depends sensitively on the actual identity of the core ion is the Rydberg fine structure, which determines the details of the excitation frequencies. To date, studies have been reported in He (Stevens and Lundeen, 1999 and references therein), S (Deck et al., 1993), Ne (Ward et al., 1996), C and O (Ward, 1994), N (Jacobson et al., 1996), Li (Storry et al., 1997a), and Ba (Snow et al., 2005), but many other atoms could be studied with the same technique. The same technique has also been applied to study high-L Rydberg states of H_2 and D_2 (Jacobson et al., 2000 and references therein). These microwave studies had much higher resolution than the earlier optical studies in H_2 and D_2 (Herzberg and Jungen, 1982). Another difference from these earlier studies was that only a subset of the possible H_2 Rydberg levels were seen with the RESIS technique. Rydberg states with $n = 9$ or 10 bound to $\nu > 0$ states of H_2^+ were not seen because they autoionize in a time shorter than a μsec, the time required to travel from the charge exchange cell to the first CO_2 laser. In addition, since some of the rotationally excited upper states of the RESIS transition are unstable to rotational autoionization, some of the transitions involving rotationally excited H_2^+ levels are not detected. Nevertheless, within these limitations, the RESIS/microwave technique has provided some very high precision studies of high-L Rydberg states of H_2. To date, no other Rydberg molecules have yet been studied with this technique.

The most recent application of the RESIS technique is to the study of Rydberg ions. When a multiply-charged ion beam captures a single electron from a Rydberg target, the resulting Rydberg ion will tend to have a binding energy to the multiply-charged ion core that is still approximately the same as it had to the Rb^+ ion in the target. In other words, the Rydberg ion formed by capture on an $n = 10$ Rydberg target will tend to be formed in states whose binding energy is approximately one CO_2 photon, and which can therefore be efficiently excited and ionized using the RESIS technique. RESIS detection of specific Rydberg ion states was demonstrated in a study of resonant charge transfer by multiply charged ions on Rydberg atoms (Fisher et al., 2001) In this study, positive ions with charges of 1, 2, 3, 4, 6, 8, and 11 captured single electrons from a tunable Rydberg target, and products formed in specific principal quantum numbers were monitored using the RESIS detection method. By noting the variation in the production of one product state as a function of the choice of principal quantum number of the Rydberg target, which could be varied between $n = 7$ and $n = 26$, the capture resonance was studied in detail. More recently, the first microwave resonance studies of high-*L* Rydberg fine structure were carried out using the RESIS method. Studies of Si^+, $n = 19$ included states with $9 \leq L \leq 16$ (Komara et al., 2005). Studies of Si^{2+}, $n = 29$ included states with $8 \leq L \leq 14$ (Komara et al., 2003). These studies demonstrate the potential of the RESIS/microwave method to reveal details of high-*L* Rydberg fine structure across a range of ion charge states. The only modification of the method, compared with the studies of neutral atoms and molecules, is that the initial ionization region cannot be a simple deflection field, since this would remove both the primary ion beam and the charge transfer products. Instead, the initial ionization and beam selection is performed with an einzellens, adjusted to focus the charge transfer beam while defocusing or even reflecting the primary beam. This is possible because the potential barrier posed by the positive potential of the einzellens is higher for the primary beam than for the charge transfer beam. With suitable choice of the lens geometry, the electric field encountered is still sufficient to ionize Rydberg levels that would otherwise contribute to a background.

2.4. OTHER EXPERIMENTAL METHODS

In addition to the two main approaches discussed above, which have yielded virtually all the high resolution studies of high-*L* Rydberg fine structure, three other approaches that have received more limited use should also be mentioned. One is the use of the Stark-switching technique first suggested by Freeman and Kleppner (1976) to selectively populate high-*L* Rydberg levels by laser excitation. With this approach, a single Rydberg *Stark level* is selectively excited with a laser, and then the electric field is adiabatically reduced to zero, resulting in population of a single high-*L* level. This technique can be used to populate very high-*L* levels if the electric field is turned off slowly enough. By itself, it yields no direct information about the spectroscopy of the high-*L* levels. However, this technique has been widely used to study doubly excited Rydberg levels, beginning with the study of Jones and Gallagher (1988). In these studies, after the one electron is excited into a high-*L* level using the Stark-switching technique, a second electron is also excited, making a doubly excited atom. The main focus of these studies has been the *L*-dependence of the autoionization rates of these doubly excited Rydberg levels. In some cases, however, when the second electron is excited into a state with enough angular momentum to support a quadrupole moment, the tensor fine structure of the doubly-excited Rydberg levels has been resolved in the optical excitation of the inner electron. The best example of this is the study of $6p_{3/2}nl$

and $6d_{5/2}nl$ states of Barium (Pruvost et al., 1991), which revealed the partially resolved tensor fine structures of these states with $5 < L < 12$.

A second technique that has also showed recent success in studying high-L states is the use of sensitive frequency-modulation absorption spectroscopy with diode lasers, in which the sample is prepared in a plasma discharge. Studies of this type have been reported of Mg (Lemoine et al., 1990) and H_2 (Basterrechea et al., 1994) Rydberg levels.

A third technique which has been used to spectrally resolve high-L fine structures in Rydberg ions is the use of stimulated recombination between highly-charged ions and electrons in an ion storage ring. An example is the study of recombination of electrons and O^{5+} ions at the heavy ion storage ring in Heidelberg (Schussler et al., 1995). In this study, Be-like O^{4+} ions are formed by two step stimulated recombination into high-L $n = 9$ and 10 levels. The highly relativistic Li-like O^{5+} ions combine with the electrons in the electron cooler section of the storage ring. First a NdYAG laser stimulates recombination into $n = 16$ levels of O^{4+}, and then a tunable dye laser stimulates deexcitation from 16 to either 9 or 10. In this second stage individual high-L $n = 9$ or 10 levels with $L \geq 5$ are resolved. Later studies showed similar resolution of high-L levels of N^{3+}, $n = 8$ (Wolf et al., 2000). This technique seems capable of studying a wide range of highly charged high-L Rydberg ions, but to date the resolution is rather limited.

3. Theoretical Methods

A variety of theoretical models have been used to describe high-L Rydberg fine structures. Despite the relative simplicity of these high-L systems, there are still many open questions about the range of validity and the possible corrections to these models. However, even in its current state, theoretical understanding of the spectroscopy of high-L Rydberg states is sufficient so that measurements of high-L fine structure patterns can be used to extract precise information about the ions that bind the Rydberg electron. In some cases, the level of understanding of the fine structure patterns is such that relativistic and radiative corrections must be included to achieve agreement with experimental data. In order to support the discussion of the conclusions drawn from high-L studies and of the remaining theoretical questions, this section aims to outline the main ideas contained in the existing theoretical models.

3.1. LONG-RANGE MODEL FOR ATOMS

The two fundamental simplifying assumptions that motivate special treatment of high-L Rydberg states are that, for sufficiently high-L, the Rydberg electron may be considered to be (1) distinguishable from the electrons in the core ion, and (2) nonpenetrating. Incorporating the first assumption into a theoretical model means that it is unnecessary to fully symmetrize the multielectron wavefunction and that exchange effects involving the high-L electron can be neglected. In a nonrelativistic treatment, the usual starting point is to write:

$$H = \left[H_C^0 + H_R^0 \right] + V,$$

$$H_C^0 = \sum_{i=1}^{N-1} \left[\frac{|\vec{p}_i|^2}{2} - \frac{Z}{r_i} \right] + \sum_{\substack{i,j=1 \\ j>i}}^{N-1} \frac{1}{|\vec{r}_i - \vec{r}_j|},$$

$$H_R^0 = \frac{|\vec{p}_N|^2}{2} - \frac{(Z - N + 1)}{r_N}, \tag{6}$$

$$V = \sum_{i=1}^{N-1} \frac{1}{|\vec{r}_N - \vec{r}_i|} - \frac{(N - 1)}{r_N}.$$

Here Z is the nuclear charge and N is the number of electrons. The Rydberg electron is taken to be the distinguishable Nth electron. For simplicity, the electron spin is neglected in this heuristic discussion. The zeroth order solutions are unsymmetrized product wavefunctions:

$$\Psi^{[0]}(\gamma, nL) = \Psi_C^{[0]}(\gamma) \Psi_R^{[0]}(nL) \tag{7}$$

with zeroth order energy

$$E^{[0]} = E_C^{[0]}(\gamma) + E_R^{[0]}(n) \tag{8}$$

where γ stands for the quantum numbers of a zeroth order core state. In general, the zeroth order core states are known only formally, while the zeroth order Rydberg states' hydrogenic wave functions are known analytically. The usual case for a Rydberg state is that the core ion is in its ground state, which we denote by "g", and its angular momentum by L_g. The assumption that the Rydberg electron is nonpenetrating implies that if r_N is the coordinate of the Rydberg electron,

$$|\vec{r}_N| \geq |\vec{r}_i| \quad i = 1, N - 1. \tag{9}$$

This, in turn, implies that the scalar term in V vanishes identically, leaving

$$V = \sum_{\kappa=1}^{\infty} \sum_{i=1}^{N-1} \frac{r_i^\kappa}{r_N^{\kappa+1}} C^{[\kappa]}(\hat{r}_i) \cdot C^{[\kappa]}(\hat{r}_N) \tag{10}$$

where $C^{[\kappa]}$ is a spherical tensor operator of rank κ. For atoms with nonzero core angular momentum, L_g, the first order perturbation energy in V gives a series of terms in the *even* permanent electric moments of the core, including moments with

$\kappa \leq 2L_g$. The odd moments vanish by parity. This leads to tensor fine structure, discussed further below.

For atoms with S-state cores, the first order perturbation energy vanishes. This leaves the second-order perturbation energy as the lowest order correction to the energies of high-L Rydberg states. This can be formally written as:

$$E^{[2]}\big(g, (n, L)\big)$$
$$= \sum \frac{|\langle (g), (n, L); L| \sum_{i=1}^{N-1} \sum_{\kappa=1}^{\infty} (r_i^{\kappa}/r_N^{\kappa+1}) C^{[\kappa]}(\hat{r}_i) \cdot C^{[\kappa]}(\hat{r}_N)|(\gamma, L_c), (n', L'); L\rangle|^2}{(E_C^{[0]}(g) - E_C^{[0]}(\gamma)) + (E_R^{[0]}(n) - E_R^{[0]}(n'))}.$$
$$(11)$$

The treatment up to this point is common to most theoretical models of high-L fine structure. From this point on, however, several different approaches have been used. The main difference between them is their treatment of the "nonadiabatic" corrections, i.e., the influence of the Rydberg energy differences, $\Delta E_R \equiv E_R^{[0]}(n) - E_R^{[0]}(n')$, in the denominator of Eq. (11).

3.1.1. Adiabatic Model with S-State Cores

The simplest approach, by far, is to neglect the contributions of ΔE_R to the energy denominator of Eq. (11) (Mayer and Goeppert Mayer, 1933; Edlen, 1964). Then, using the completeness of Rydberg radial functions, Eq. (11) reduces to the sum of adiabatic polarization energies of each multipole order.

$$E_{\text{Adiabatic}}^{[2]} = \sum_{\kappa=1}^{\infty} -\frac{1}{2} \alpha_\kappa \left\langle \frac{1}{r_N^{2\kappa+2}} \right\rangle_{nL}, \qquad (12)$$

$$\alpha_\kappa \equiv 2 \sum \frac{|\langle g| \sum_{i=1}^{N-1} r_i^{\kappa} C_0^{[\kappa]}(\hat{r}_i)|\gamma, L_c = \kappa \rangle|^2}{E_C^{[0]}(\gamma) - E_C^{[0]}(g)}. \qquad (13)$$

Including only dipole and quadrupole polarizabilities, this predicts that the fine structure energies are expectation values of the effective potential:

$$V_{\text{Adiabatic}}(r) = -\frac{1}{2}\frac{\alpha_1}{r^4} - \frac{1}{2}\frac{\alpha_2}{r^6} + \cdots. \qquad (14)$$

The expectation value is over the zeroth order hydrogenic wavefunctions can be computed using (Bockasten, 1974):

$$\langle r^{-4} \rangle_{nL} = \frac{3n^2 - L(L+1)}{2n^5(L-\frac{1}{2})(L)(L+\frac{1}{2})(L+1)(L+\frac{3}{2})} \simeq \frac{1.5}{n^3 L^5}, \quad L \ll n,$$

$$\langle r^{-6} \rangle_{nL} = -\frac{35n^4 - 5n^2(6L(L+1)-5) + 3(L-1)L(L+1)(L+2)}{8n^7(L-\frac{3}{2})(L-1)(L-\frac{1}{2})L(L+\frac{1}{2})(L+1)(L+\frac{3}{2})(L+2)(L+\frac{5}{2})} \qquad (15)$$

$$\simeq \frac{35}{8n^3 L^9}, \quad L \ll n.$$

The variation of these expectation values with L is traceable to the inner turning point of the radial wavefunctions, which increase approximately quadratically with L. Since the expectation values of inverse powers of r are primarily sensitive to the smallest values of r in the radial wavefunction, this leads to a ratio between the $\langle r^{-6} \rangle$ and $\langle r^{-4} \rangle$ expectation values of approximately L^{-4}. For high-L states ($L \geq 4$), this makes the second term in Eq. (14) a small correction to the first, which becomes increasingly dominant as L increases.

3.1.2. Nonadiabatic Corrections

Correction for nonadiabatic effects were obtained by Eissa and Opik (1967). Starting again from the same zeroth order Hamiltonian and perturbation V, they found corrections to the zeroth order Rydberg energies given by:

$$\Delta E_{\text{NonAd\_I}} = -\frac{\alpha_1}{2}\left[y_0^1\langle r^{-4}\rangle + y_2^1\langle r^{-6}\rangle\right] - \frac{\alpha_2}{2}\left[y_0^2\langle r^{-6}\rangle + y_2^2\langle r^{-8}\rangle\right] \qquad (16)$$

defining the four parameters y_i^j, which quantify the nonadiabatic corrections to the fine structure energies. The four parameters, y_i^j, were state dependent, i.e., they had to be calculated independently for each (n, L) state considered. Nonadiabatic correction factors of this type were used to analyze fine structure data in Ca Rydberg levels (Vaidyanathan and Shorer, 1982).

A somewhat simpler approximate treatment of nonadiabatic corrections to the long-range model was discussed by (Gallagher et al., 1982). He introduced correction factors κ_1 and κ_2 such that the fine structure energies were given by:

$$\Delta E_{\text{NonAd\_II}} = -\frac{1}{2}\kappa_1\alpha_1\langle r^{-4}\rangle - \frac{1}{2}\kappa_2\alpha_2\langle r^{-6}\rangle + \cdots. \qquad (17)$$

The correction factors, κ_1 and κ_2, were also state dependent, and had to be calculated separately for each Rydberg level. To a good approximation they could be calculated by assuming that only a single excited core state contributes to the polarizabilities, α_1 and α_2. Explicit formulas for these correction factors, using this approximation, have been given (Gallagher et al., 1982). By calculating these correction factors in the case of Ba Rydberg levels, Gallagher was able to account for the observed fine structure in h, i, and k levels with $n \sim 20$, where nonadiabatic effects are quite significant.

Neglecting the variation of the correction factors κ_i or y_i^j, with state, both approaches predict fine structure that is consistent with the form expected in the adiabatic model.

$$\Delta E = A\langle r^{-4}\rangle + B\langle r^{-6}\rangle. \qquad (18)$$

However, the interpretation of the coefficients is complicated by the nonadiabatic corrections. The coefficient A is now only approximately equal to $\alpha_1/2$. Similarly, the coefficient B is no longer $\alpha_2/2$. Because of this, many of the early experimental studies, in which experimental data was fitted to the form of Eq. (17), referred to the fitted coefficients $2A$ and $2B$ as α_1' and α_2', respectively, delaying the determination of α_1 and α_2 until after the application of nonadiabatic corrections. This distinction was relatively insignificant for α_1, where the nonadiabatic corrections seldom amounted to more than a few percent, but could be very important for α_2, where factors of two or more were often encountered.

A different approach to treating nonadiabatic effects was used by Drachman in studies of helium Rydberg fine structure (Drachman, 1982). His approach extended the model used earlier (Kleinman et al., 1968) to describe nonadiabatic effects in the long-range interactions responsible for electron scattering. The starting point is a formal expansion of the energy denominator in Eq. (11) in terms of the ratio between Rydberg and core energy differences.

$$\frac{1}{\Delta E_C + \Delta E_R} = \frac{1}{\Delta E_C} - \frac{\Delta E_R}{(\Delta E_C)^2} + \frac{(\Delta E_R)^2}{(\Delta E_C)^3} - \cdots. \tag{19}$$

We will refer to this as the "adiabatic expansion". Substituting it back into Eq. (11), the first term gives the usual adiabatic expression. The succeeding terms can be simplified, using the properties of the Rydberg radial wavefunctions and integration by parts, to express the nonadiabatic corrections to each multipole order of the adiabatic fine structure energy as expectation values of higher inverse powers of the Rydberg radial coordinate. For example, the dipole and quadrupole terms give:

$$\begin{aligned}
\Delta E_{\text{NonAd\_III}}^{\kappa=1} &= -\frac{1}{2}\alpha_1 \langle r^{-4} \rangle + 3\beta_1 \langle r^{-6} \rangle + \cdots, \\
\Delta E_{\text{NonAd\_III}}^{\kappa=2} &= -\frac{1}{2}\alpha_2 \langle r^{-6} \rangle + \frac{15}{2}\beta_2 \langle r^{-8} \rangle + \cdots
\end{aligned} \tag{20}$$

with

$$\beta_\kappa \equiv \sum_\gamma \frac{|\langle g| \sum_{i=1}^{N-1} r_i^\kappa C_0^{[\kappa]}(\hat{r}_i)|\gamma, L_C = \kappa \rangle|^2}{(E_C^{[0]}(\gamma) - E_C^{[0]}(g))^2}. \tag{21}$$

This again predicts a fine structure of the form of Eq. (18), but with the important distinction that the coefficient of the first term is precisely one half α_1, without the need for any numerical correction. Another important difference from the other approaches is that none of the coefficients in this approach ($\alpha_1, \beta_1, \alpha_2, \beta_2$) are state dependent. That is, these parameters should suffice to predict the energy of any (nL) Rydberg level, as long as L is large enough to satisfy the assumptions of the model.

The treatment of helium fine structure is particularly simple because in this case, the core wavefunctions, like the Rydberg wavefunctions, are hydrogenic and therefore well known. In this case, Drachman was able to show that the fine structure energies can be written as the expectation value of an effective potential consisting of increasing negative powers of r, and calculated the coefficients of all terms up to r^{-10} analytically. He used the Feshbach projection operator technique to systematize the calculation, but in essence, it consists of carefully applying three expansions to the calculation,

(a) The basic perturbation expansion in the perturbation V.
(b) The multipole expansion of V.
(c) The "adiabatic expansion".

The first two expansions are, of course, unremarkable. The third expansion is the critical one. It is important to note that its utility depends only on the fact that the Rydberg electron's zeroth order wavefunction is hydrogenic. Thus, the same approach should be equally applicable to more complex atoms, though in that case the core parameters could not be calculated so readily. Drachman has, in fact, applied the same approach to Rydberg states of lithium (Drachman and Bhatia, 1995), and others have applied similar approaches to studies of more complex Rydberg systems (Laughlin, 1995). One obvious limitation to this approach, however, is that the adiabatic expansion may not converge if low-lying excited states of the core are important in any multipole order. This is the case in Ba Rydberg levels, where a low-lying 5d excitation of the core produces very large nonadiabatic corrections to the quadrupole terms in $E^{[2]}$ (Snow et al., 2005).

Even when the underlying adiabatic expansion converges well, the resulting effective potential for the Rydberg system, like other expansions of long-range forces in terms of increasing negative powers of r, is still formally divergent at any finite r. In practice, even if the coefficients of all terms are known, as is true to a high degree for helium, the expectation value of successive terms in any particular (n, L) state will begin to increase for some term, and the best estimate of the Rydberg energy would be found by truncating the expectation value at the smallest term. The power of r at which this occurs increases steadily as L increases. For helium, the terms proportional to r^{-9} and r^{-10} are still decreasing for $L > 5$ (Drachman, 1993), and they contribute only about 10^{-5} of the total fine structure energy. So, in spite of the formal divergence, this "asymptotic expansion" of the long-range potential can provide very precise predictions of Rydberg energies (Dalgarno and Lewis, 1956). In the helium case, results obtained with the long-range model can be compared with independent variational calculations, discussed further in Section 3.3 (Drake, 1993). Comparison of the fine structure energies computed with these two independent approaches shows good agreement to within the expected precision of the asymptotic expansion method. At about $L = 7$, the precision of the asymptotic expansion method exceeds that of the variational method, and consequently no variational calculations have been reported for levels with $L > 7$ (Drachman, 1993).

3.1.3. Other Corrections

3.1.3.1. Core penetration and exchange In the long-range models, it is assumed at the outset that the Rydberg electron is distinguishable and nonpenetrating. For sufficiently high-L, these assumptions are surely valid, but to establish this and set limits on possible contributions from core penetration and exchange, it is important to model these contributions using other methods. Contributions to high-L fine structure due to core penetration and exchange in alkali atoms have been discussed by Patil (1986). He estimates that contributions to fine structure energies from core penetration and exchange are comparable in size and given by

$$\Delta E_{\text{pen}} \cong -\frac{\delta_{\text{pen}}}{n^3} \text{ a.u.} \tag{22}$$

where $\delta_{\text{pen}} \sim 10^{-5}$ for nf states of Na (Patil, 1986), and approximately two orders of magnitude smaller for Na ng states. For heavier alkalis, the predicted penetration effects increase approximately as $Z^{2.5}$, where Z is the nuclear charge. Drachman estimated the contributions to helium Rydberg states from core penetration, finding $\delta_{\text{pen}} \sim 10^{-6}$ for nf states, and decreasing by about a factor of 500 for each unit increase in L (Drachman, 1982). Theodosiou estimated penetration effects in alkaline-earth Rydberg atoms, using several different methods, and gives detailed predictions for $L = 3, 4, 5$, and 6 Rydberg levels of all alkaline earths (Theodosiou, 1983). For the heaviest case, Ba, He estimates $\delta_{\text{pen}} \sim 10^{-5}$ for $L = 5$ levels. A good experimental study of penetration effects is the study of nf and ng Rydberg levels in Cs (Sansonetti et al., 1981). By comparing polarization model fits of the nf and ng series, Sansonetti et al. determined the net penetration/exchange contributions to the nf energies, finding $\delta \sim 0.006$, in agreement with calculations to a few percent. The same calculation predicted smaller effects in the ng levels by a factor of about 400. While there is agreement that the effects of penetration decrease very rapidly with L, there is no unique experimental signature to distinguish the contributions to Rydberg fine structure from this cause. Consequently, there is, as yet, no clear experimental test of the existing calculations in high-L levels and therefore no clear indication of their accuracy. Depending on the size of the ion core and the desired experimental precision, penetration effects should become negligible at $L = 4, 5$, or 6 in neutral Rydberg atoms.

3.1.3.2. Second-order polarization energies The most significant portion of the fourth-order perturbation energies from V in Eq. (6) has been shown to be equivalent to the action of the polarization potential acting in second-order in the space of Rydberg levels (Drachman, 1982). For any Rydberg system with only scalar fine structure dominated by the dipole polarizability, these contributions to the Rydberg fine structure are a universal function of α_1, n, and L, and this function has been evaluated analytically (Drake and Swanson, 1991). These contributions are a very small correction to the Rydberg energies in helium, but since they are proportional to the square of α_1, they can be more significant in other systems with larger polarizabilities. For example, the calculated second-order correction to the 10I–10K interval in helium ($\alpha_1 \sim 0.28\ a_0^3$) amounts to about 36 ppm of the interval, while the calculated correction to the 20I–K interval in barium ($\alpha_1 \sim 124\ a_0^3$) is about 1.8% of the interval.

3.1.4. Spin Structure

A finer level of structure is expected in high-L Rydberg states due to the spin of the Rydberg electron and possibly the spin of the core. In the alkali atoms, the core is a $^1S_0$ state, so only the Rydberg spin contributes. Naively, one would expect a hydrogenic spin–orbit splitting given by

$$\Delta E_{\text{SpinOrbit}} = \frac{\alpha^2}{2n^3 L(L+1)}\ \text{a.u.} \tag{23}$$

The Na nf states exhibit spin–orbit splittings that are this order of magnitude, but about 5% smaller (Tran et al., 1984; Sun and MacAdam, 1994). The higher L states would be expected to be closer to the Dirac splitting, although this is not yet confirmed by measurements (Sun and MacAdam, 1994).

The helium Rydberg states and the alkali-earth Rydberg states, Ca and Ba, have, in addition to the Rydberg spin, the spin of the $^2S_{1/2}$ core electron. The spin structure of helium high-L states has been thoroughly studied, both experimentally (Hessels et al., 1987; Stevens and Lundeen, 1999) and theoretically (MacAdam and Wing, 1975; Cok and Lundeen, 1979; Drake, 1996). The observed structure changes smoothly from a conventional singlet/triplet structure in the nD states to a qualitatively different four-fold structure in nG states and all states of higher L. In the D states, the exchange energy dominates, forcing the eigenstates to be either symmetric (triplet) or antisymmetric (singlet) spin states. The spin–orbit energies are small corrections. However, as L increases, the exchange energy decreases very rapidly, two to three orders of magnitude for each change in L. By $L = 4$, the exchange energy is much smaller than the spin–orbit energies, which decrease only as L^{-2}. The resulting structure is dominated by the spin orbit interactions with the Rydberg spin and the core spin. These differ in sign and by a factor of two in magnitude since the interaction with the Rydberg spin is reduced by the Thomas precession. The two spin–orbit interactions give rise to an approximately equally spaced four-fold structure, as illustrated in Fig. 5, which is common to all states with $L \geq 4$. To a very good approximation, the spin structure in high-L states is described by the Hamiltonian

$$H_{\text{spin}} = \alpha^2 \langle r^{-3} \rangle \left[\frac{\vec{L} \cdot \vec{S}_R}{2} - \vec{L} \cdot \vec{S}_C + (\vec{S}_R \cdot \vec{S}_C) - 3(\vec{S}_R \cdot \hat{r})(\hat{r} \cdot \vec{S}_C) \right]$$
$$+ V_x \left[\frac{1}{2} + 2(\vec{S}_R \cdot \vec{S}_C) \right] \tag{24}$$

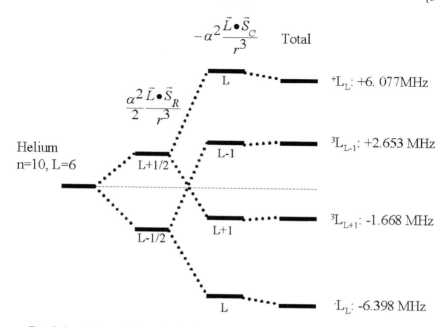

FIG. 5. Level diagram illustrating the formation of the typical four-fold spin splitting in high-L Rydberg levels of helium. The dominant interactions are the two spin–orbit interactions with the Rydberg spin, S_R, and the core spin, S_C. The first of these gives the usual Dirac spin–orbit splitting. The second is of opposite sign and twice as large because it is not reduced by the Thomas precession. Additional contributions from the spin-spin interaction are much smaller, and in high-L states the exchange interaction is negligible on this scale. The two eigenstates with $J = L$ are approximately equal mixtures of singlet and triplet spin states. The numerical energies correspond to the $n = 10$, $L = 6$ states of helium.

in which V_x is a small parameter representing exchange effects. Spin–spin and exchange interactions, the third and fourth terms in Eq. (24), are small corrections to the dominant spin–orbit structure. The eigenstates are no longer, except by accident, eigenstates of total electronic spin. Instead, the two states with $J = L$ are approximately equal mixtures of ^1L_L and ^3L_L states. This is unrelated to the assumption of distinguishability for the Rydberg electron; the same result is obtained with totally antisymmetric wavefunctions. Comparison between measured and calculated spin structure intervals in G, H, and I states of helium show good agreement with this essentially hydrogenic picture with precision of 0.1% (Hessels et al., 1987). The nearly complete spoiling of the singlet–triplet quantum number in states of high-L is sometimes obscured by the continued use of the labels ^1L_L and ^3L_L for these states. The alternate labels shown in Fig. 5 emphasize the complete mixing.

The Rydberg states of Ba and Ca, might be expected to show similar spin structure, since they also have a $^2S_{1/2}$ core with the same magnetic moment. Measurements in Ba, however, revealed a spin splitting that was several orders of magnitude larger than the hydrogenic spin–orbit splitting (Gallagher et al., 1982). More recently, measurements in Rydberg states of Mg-like Si also gave evidence of a spin structure that was inconsistent with the helium picture (Komara et al., 2003). A crude theoretical model that accounted for both the Mg-like Si observations and the Ba observations attributed the splittings to the indirect effects of nP and nD admixtures into the core wavefunction caused by the electric field of the Rydberg electron (Snow et al., 2003). The effect of these admixtures was shown to be approximately equivalent to modifying the spin–orbit terms in the spin Hamiltonian with factors b_d and b_Q

$$H_{\text{spin}} = \alpha^2 \langle r^{-3} \rangle \left[\frac{\vec{L} \cdot \vec{S}_R}{2} - (1 - b_d - b_Q) \vec{L} \cdot \vec{S}_C \right] \tag{25}$$

where the correction factors b_d and b_Q are related to the *nonadiabatic* contributions to the perturbed core wavefunction. In the case of Ba nH and nI Rydberg levels, where the nonadiabatic effects are large, the correction factors were several orders of magnitude greater than 1, in approximate agreement with measured spin splittings. In Mg-like Si, $n = 29$, the corrections were noticeable but much smaller. In helium, the correction factors are insignificant. Because these contributions decrease more rapidly with L than the normal spin–orbit contributions, the net spin structure should approach the helium-like structure at sufficiently high L.

3.1.5. Tensor Fine Structure

Rydberg states in which the core ion has $L_c \geq 1$ may display a tensor fine structure due to the core's permanent electric quadrupole moment. This type of structure has been observed in doubly excited $6p_{3/2}nl$ and $6d_{5/2}nl$ states of barium (Pruvost et al., 1991) and studied with high precision is the neon atom, where the observed structure was illustrated in Fig. 2 above (Ward et al., 1996). Those measurements were interpreted by simply extending the effective potential formulation used for S-state cores by adding terms corresponding to the permanent electric moments and tensor polarizability (Schoenfeld, 1994).

$$
\begin{aligned}
V_{\text{eff}}(r, \theta_{cN}) = &-\left(\frac{1}{2} \frac{\alpha_1}{r^4} + \frac{1}{2} \frac{(\alpha_2 - 6\beta_1)}{r^6} + \cdots \right) \\
&- \left(\frac{Q}{r^3} + \frac{\alpha_{1T}}{r^4} + \frac{C}{r^6} + \cdots \right) P_2(\cos \theta_{cN}).
\end{aligned}
\tag{26}
$$

The scalar term appears just as for S-state cores. The second-rank tensor term contains the quadrupole moment term, the tensor polarizability term, and a term proportional to r^{-6} which stands for a combination of adiabatic quadrupole–quadrupole and dipole–octupole tensor polarizabilities with nonadiabatic corrections to the dipole–dipole tensor polarizabilities. In the neon atom, the expectation value of this potential accounts for most of the features of the measured structure illustrated in Fig. 2 (Ward et al., 1996). The scalar term determines the center of gravity of the three states with common L. The coefficient of the leading term can be extracted from the measured pattern, as before, by taking advantage of the variation of hydrogenic radial expectation values with L. The second-rank tensor term produces the energy differences between states of common L. The variation of these splittings with L reveals the relative contributions of the several terms with different dependences on the Rydberg radial coordinate. In this way, once the fine structure pattern is determined experimentally across a sufficient range of L, both α_1 and Q can be determined.

Close study of the neon fine structure pattern indicates that, in addition to the scalar and tensor contributions predicted by Eq. (26), a small vector component, proportional to $\vec{J}_c \cdot \vec{L}$, is also present. Most of this can be attributed to the magnetic interactions between the Rydberg electron and the Ne^+ core. A straightforward generalization of Eq. (24), substituting the magnetic moment of Ne^+ for the magnetic moment of He^+ leads to the expected form of the dominant magnetic interactions.

$$H_{spin} = \alpha^2 \langle r^{-3} \rangle \left[\frac{\vec{L} \cdot \vec{S}_R}{2} - g_J \frac{\vec{L} \cdot \vec{J}_c}{2} \right] \tag{27}$$

where g_J is the g-factor of the $^2P_{3/2}$ ground state of Ne^+, expected to be 4/3 in pure LS coupling. An additional term, also proportional to $\vec{J}_c \cdot \vec{L}$ but proportional to r^{-6} was predicted by Zygelman, using arguments based on geometric phases (Zygelman, 1990). The presence of this term, which has since been named "vector hyperpolarizability", was confirmed by the experimental measurements of (Ward et al., 1996), and later recalculated with more conventional methods by Clark and Greene (1999). It was distinguished from the vector structure due to the second term in Eq. (27) by its different dependence on L. In the comparison, g_J was found to be consistent with the expected value of 4/3.

In comparing the measured neon fine structure with predictions based on Eqs. (25) and (26), the second-order energies were an important factor. These are the energy shifts generated when the potential of Eq. (26) is applied in second-order, mixing different Rydberg series. These were calculated, using approximate values of the coefficients in Eq. (26). The calculated energy shifts were small but not insignificant (Ward et al., 1996; Komara et al., 1999). For example, they contribute less than 1% to the intervals between 10I and 10K states in neon. These energy shifts are, of course, analogous to the shifts described in MQDT treatments, since they correspond to mixing between different Rydberg series. However, since the magnitude of the energy shifts is much too small to significantly affect the relative positions of Rydberg states in different series, they are more conveniently treated with perturbation theory.

Stimulated, in part, by the prediction of what has now become known as vector hyperpolarizability, Greene and collaborators carried out a completely independent theoretical treatment of the neon fine structure (Clark et al., 1996; Clark and Greene, 1999). Their treatment, which is discussed in more detail in Section 3.3, reached conclusions very similar to those obtained using the long-range model discussed here.

3.2. LONG-RANGE MODEL FOR H_2

3.2.1. Coulomb Interactions

The structure of high-L Rydberg states of H_2 is closely analogous to the structure of atoms such as neon whose cores possess permanent quadrupole moments. In the case of H_2, however, the H_2^+ core has a quadrupole moment oriented along its internuclear axis. In low-lying excited electronic states of H_2, the anisotropic interactions with the H_2^+ ion are strong enough that the electronic eigenstates are characterized by their projection along the internuclear axis, $\lambda \equiv \vec{L} \cdot \hat{\rho}$. For highly excited Rydberg levels, especially those with high-L, the coupling to the internuclear axis is so weak that the H_2^+ ion rotates freely within the orbit of the Rydberg electron. In this "case d" coupling limit both the rotational angular momentum of the H_2^+ ion, $\vec{R}$, and the angular momentum of the Rydberg electron, $\vec{L}$, are good quantum numbers for the eigenstates. These then couple together to form the total angular momentum, exclusive of spin, $\vec{N} = \vec{R} + \vec{L}$. As long as $R < L$, this results in a tensor fine structure consisting of $2R + 1$ fine structure levels for each value of L, with the energy differences between the different levels predominantly due to the quadrupole moment of H_2^+. Of course when the core is in an $R = 0$ rotational state it presents an isotropic appearance to the Rydberg electron, and consequently the fine structure of Rydberg levels bound to $R = 0$ cores is similar to the helium atom, showing only a scalar fine structure pattern. In general, each ro-vibrational state of the H_2^+ core, (v, R), will bind an entire series of Rydberg levels, and the fine structure of high-L Rydberg states in each series will be quite different. The dependence on R is dramatic because this determines the multiplicity of the fine structure pattern, $2R + 1$. The dependence on v is also noticeable because the core parameters, Q, α, etc. depend on the internuclear separation and so have a different average value in each vibrational level. Thus, a large variety of H_2 Rydberg levels is expected, each characterized by the quantum numbers of the core, (v, R), of the Rydberg electron, (n, L), and of their coupled angular momentum, N. Some experimental studies of these states (Sturrus et al., 1986) have denoted them in a quasi-atomic notation

$(v, R)nL_N.$

An effective potential model of the fine structure of high-L Rydberg states of H_2 has been worked out by analogy with the atomic cases (Sturrus et al., 1985). The Rydberg state is represented by a reduced Rydberg wavefunction

$$\Psi_{\upsilon RnLNm}^{\text{Ryd}} = \frac{g_{\upsilon R}(\rho)}{\rho} \sum_{M,m_1} \langle RMLm_1 | RLNm \rangle Y_{RM}(\hat{\rho}) \Psi_{nLm_1}(\vec{r}) \tag{28}$$

where $\vec{\rho}$ is the internuclear axis and $\vec{r}$ is the Rydberg electron's coordinate with respect to the center of the internuclear axis. In this reduced wavefunction, no reference is made to the coordinate of the core electron, which is assumed to be in its ground electronic state. The Rydberg fine structure energy is found by taking the expectation value of the effective potential

$$V_{\text{Pol}}^{\text{LOPM}}(\vec{\rho}, \vec{r}) = -\frac{Q(\rho)}{r^3} P_2(\cos\theta) - \frac{1}{2} \frac{\alpha_S(\rho)}{r^4} - \frac{1}{3} \frac{\alpha_T(\rho)}{r^4} P_2(\cos\theta) + \cdots \tag{29}$$

where

$$\cos\theta \equiv \hat{\rho} \cdot \hat{r},$$
$$\alpha_S(\rho) \equiv \frac{2\alpha_\perp(\rho) + \alpha_\parallel(\rho)}{3}, \tag{30}$$
$$\alpha_T(\rho) \equiv \alpha_\parallel(\rho) - \alpha_\perp(\rho)$$

and $Q(\rho)$ is the quadrupole moment of the core and $\alpha_\perp(\rho)$ and $\alpha_\parallel(\rho)$ are the dipole polarizabilities for fields perpendicular and parallel to the internuclear axis, respectively. All three of these parameters are functions of the internuclear separation, ρ. This form of the effective potential, which includes only the most significant terms, is referred to as the Lowest Order Polarization Model (LOPM). In later work, it was extended to form the Higher Order Polarization Model (HOPM), by including terms proportional to r^{-5} and r^{-6}. These include the permanent hexadecapole moment, the scalar and tensor components of the nonadiabatic dipole polarizability, the scalar, tensor, and 4th rank components of the adiabatic quadrupole polarizability, and the tensor and 4th rank components of the adiabatic dipole–octupole polarizability (Sturrus et al., 1988; Arcuni et al., 1990b). All of these coefficients have been calculated as a function of internuclear separation and averaged over the appropriate ro-vibrational wavefunctions by others (Bishop and Lam, 1987). Certain other adiabatic contributions can be estimated from calculated higher polarizabilities, such as (dipole)^4 hyperpolarizabilities, proportional to r^{-8}, and dipole–dipole–quadrupole polarizabilities, proportional to r^{-7} (Sturrus et al., 1988), but since other nonadiabatic terms which are as yet uncalculated are expected to be of similar size, these terms are omitted from the HOPM. Using calculated values of the various moments and polarizabilities, the HOPM predicts the fine structure of $(0,1)10F_N$

and $(0,1)10G_N$ states within a few percent (Sturrus et al., 1988), and also provides a guide to the expected form of the results which can be used to extract more precise values of the leading parameters from experimental data. As in the case of helium and neon, the second-order effects of the potential of Eq. (29) make small but significant contributions to the Rydberg fine structure.

3.2.2. Spin and Hyperfine Interactions

Four spins are present in a H_2 Rydberg state, two electron spins and two proton spins. These lead to a finer level of structure that has been studied experimentally in some detail. Because of the Pauli Principle, the two proton spins are restricted to be in a symmetric or antisymmetric spin state depending on whether the spatial function is antisymmetric or symmetric on interchange of the two protons. In other words, if H_2^+ is in an even rotational state, the total proton spin, I, must be zero. On the other hand, if it is in an odd rotational state, the total proton spin must be one. To date, the most carefully studied Rydberg states are those bound to the $(0,1)$ state of H_2^+. They must have $I = 1$. In the free H_2^+ ion, the total proton spin, the core electron spin, and the rotational angular momentum couple according to the H_2^+ hyperfine Hamiltonian (McEachran et al., 1978)

$$H_{\mathrm{Hyp}} = b(\vec{I} \cdot \vec{S}_c) + c(\vec{I} \cdot \hat{\rho})(\hat{\rho} \cdot \vec{S}_c) + d(\vec{S}_c \cdot \vec{R}). \qquad (31)$$

In H_2 Rydberg states, the coupling of the internuclear axis, $\hat{\rho}$, to the Rydberg electron's angular momentum $\vec{L}$ is generally stronger than either the c or d hyperfine couplings. However, the dipole hyperfine coupling b is independent of the axis orientation and is therefore unaffected by the Rydberg electron. Consequently, this term splits the core state $(0,1)$ into two hyperfine levels according to whether $\vec{F} = \vec{I} + \vec{S}_c$ is equal to $3/2$ or $1/2$. The $F = 3/2$ state is higher in energy by about 1384 MHz. For many purposes, this can be regarded as giving rise to two types of Rydberg states, depending on the core hyperfine state. Although the $(v, R, 1/2)nL_N$ and $(v, R, 3/2)nL_N$ states differ in absolute energy by 1384 MHz, electric dipole transitions between different high-L states generally do not change the hyperfine quantum number, so this hyperfine splitting is not apparent in the usual Rydberg spectra. In contrast, the hyperfine constants c and d, do contribute to the experimental spectra. The core spin, $\vec{F} = \vec{I} + \vec{S}_c$, couples to $\vec{N} = \vec{R} + \vec{L}$ to form $\vec{J}_1 = \vec{F} + \vec{N}$, the total angular momentum, exclusive of Rydberg spin. At this stage, each $(0, 1, 3/2)nL_N$ state splits into four J_1 states with splittings on the order of the hyperfine constant $c \sim 128$ MHz and $d \sim 42$ MHz. Similarly, each $(0, 1, 1/2)nL_N$ state splits into two states due to the spin rotation hyperfine constant, d. Finally, each of these six levels splits into two as the Rydberg spin is coupled in to form $\vec{J} = \vec{J}_1 + \vec{S}_R$. The interactions responsible for these smallest splittings are expected to be the usual spin–orbit interactions, as in helium or neon Rydberg levels. The spin/hyperfine structure of $(0, 1)nL_N$ Rydberg states of H_2 has been discussed further in connection with experimental studies (Sturrus et al., 1986; Fu et al., 1992).

The spin and hyperfine structure of even R levels of H_2 should be much simpler since $I = 0$. An especially simple case is that of $R = 0$ levels. There, since the rotational wavefunction of the core is isotropic, the Rydberg fine structure should be completely analogous to the helium atom structure, with no vector or tensor structure. The spin structure should also be virtually identical to the helium spin structure since $I = 0$ and $R = 0$, completely eliminating hyperfine structure and leaving only the core electron spin to interact with the Rydberg electron. This is confirmed by experimental studies (Jacobson et al., 1997).

The spin and hyperfine structure of D_2 is more complex than H_2. Since the deuterons are spin 1 bosons, they must be in a symmetric spin state when D_2^+ is in an even rotational level, meaning $I = 2, 0$. On the other hand, when D_2^+ is in an odd rotational level, the spin state must also be antisymmetric, meaning $I = 1$. D_2 Rydberg states bound to both even and odd rotational levels of D_2^+ show hyperfine structure, and each $(v, R)nL_N$ level splits into either 24 or 12 spin/hyperfine levels according to whether R is even or odd, respectively. The resulting spin and hyperfine structure of high-L Rydberg states of D_2 with $R = 0$ has been partially resolved experimentally (Jacobson et al., 1997).

3.3. COMPARISON WITH TRADITIONAL METHODS

The long-range models used to describe high-L Rydberg structure in atoms and molecules are specifically tailored to describe nonpenetrating, distinguishable Rydberg electrons, and to parameterize their fine structure energies in terms of properties of the isolated core ion. They assume a hydrogenic zeroth order solution and develop the fine structure energies as expectation values of an asymptotic potential in increasing negative powers of the Rydberg radial coordinate. Most of the theoretical work to date has specialized to the case of Rydberg atoms or ions with S-state cores, but similar models have been used to describe atomic Rydberg states with anisotropic cores and diatomic molecules. Smaller scale fine structure due to Rydberg spin and core spin(s) has been included with phenomenological models.

Confidence in these special theoretical techniques rests in part on comparison with complementary theoretical studies using more traditional techniques. The most extensive such comparison, by far, has been achieved in the helium atom. In this case, the long-range model has been derived rigorously, including all terms up to r^{-10} in the effective potential, with the coefficients occurring in the potential calculated analytically using the exact nonrelativistic wavefunctions of the He^+ core (Drachman, 1993). In addition, relativistic corrections have been evaluated with some care in the long-range picture (Drachman, 1985;

Hessels, 1992). For comparison, the same states have also been described using precise variational calculations of the nonrelativistic two-electron wavefunction, and including relativistic and radiative corrections by standard perturbative methods (Drake, 1990). The final comparison between these two theoretical techniques shows very good agreement (Drachman, 1993). For Rydberg states with $L < 7$, the variational calculation is more precise, although the long-range model agrees to within the precision estimated from its convergence. For states with $L \geq 7$ the long-range model is more precise than the variational calculation. Many details of the theoretical description have been clarified through the comparison of the two methods. Some of these are discussed further in Section 4.1. The success of the long-range model in describing helium increases confidence in its description of atoms and ions with S-state cores, although clearly there are other factors involved in cases of heavier multielectron cores, such as larger nonadiabatic corrections, which are not so conveniently checked in the absence of an independent and precise theoretical model.

The status of the long-range model in atoms with anisotropic cores is somewhat less secure. Of course the gross structure due to the core quadrupole moment is similar to that found in traditional Hartree–Fock calculations in the "pair-coupling" limit (Eriksson, 1956), but these first-order calculations were never intended to give precise descriptions of Rydberg structure. In the case of neon, there has been a completely independent calculation, from first principles, using the multichannel approach (Clark and Greene, 1999). This calculation is nonperturbative in the sense that it avoids assuming the initial Rydberg radial function is hydrogenic. Instead, the problem is treated within the broader context of adiabatic and "post-adiabatic" problems in which one coordinate is slowly varying compared with all others. In this case the slowly varying coordinate is the Rydberg electron's radial coordinate. In this picture, the problem reduces to a set of coupled differential equations for the radial functions. Successive transformations reduce the coupling between different Rydberg channels, leading to channel potentials which closely resemble the effective potential given above in Eq. (26). Comparison between this theoretical model and the measured fine structure pattern in neon, $n = 10$ (Ward et al., 1996), showed improved agreement with experiment, compared with the effective potential model, although the underlying reason for the improved agreement is not entirely clear. The fact that this independent calculation gives results similar to the long-range model is reassuring. However, the reasons for the small differences in the theoretical predictions of the multichannel and long-range models of neon have not been understood in detail, and the comparison is clouded by lack of precise independent knowledge of the core properties treated as fitting parameters in both approaches. Of course, it is also important to note that the multichannel calculation predicts contributions to the fine structure from the vector hyperpolarizability term, which would otherwise have been omitted from the long-range model.

The status of the long-range model for H_2 is even less well tested by alternate theoretical calculations. There has been a multichannel treatment of H_2 Rydberg structure (Greene and Jungen, 1985; Jungen et al., 1989). The approach is similar to the nonperturbative method used to describe neon Rydberg states (Clark and Greene, 1999) in that it does not begin with the assumption of hydrogenic radial functions for the Rydberg electron, but carries out a numerical integration in channel potentials which reflect the long-range interactions. In contrast to the neon treatment, however, it appears that the asymptotic potential has not been independently derived, but rather simply adopted *ad hoc*. Where direct comparison with data can be made, the multichannel and long-range models appear to give predictions of comparable precision. For example, for $n = 10$, G–H and H–I fine structure intervals in H_2, Jungen finds results that are very similar to those obtained using the effective potential model (Jungen et al., 1989).

An indirect test of the adequacy of long-range models of high-L Rydberg fine structure is the accuracy of core properties deduced from the measured fine structure. In cases where precise calculations of these properties have been reported, agreement with values deduced from high-L fine structure appears to be very good. For example, the core polarizability deduced for Na-like Si, Si^{3+} agrees to within the quoted calculation precision (0.15%) with relativistic many-body perturbation theory calculations (Komara et al., 2003). The value obtained for the Ne^+ quadrupole moment from the measured $n = 10$ neon fine structure (Ward et al., 1996) agrees to within 0.4(3)% with a multiconfiguration calculation (Sundholm and Olsen, 1994). Of course, in these cases, the uncertainties associated with the calculations of the properties of these multielectron core ions could exceed the uncertainties associated with the long-range model of the Rydberg fine structure. More details on the measured core properties and some typical comparisons with independent calculations are given in Section 4.2.

4. Results

Having discussed the experimental and theoretical approaches to study of high-L Rydberg states, what remains is to review the results obtained to date from these studies. What are these studies teaching us? Do interesting questions remain? For purposes of this discussion, the progress achieved in these studies seems to fall into three categories: (a) theoretical progress, (b) ion property measurements, and (c) applications.

4.1. THEORETICAL PROGRESS

Some of the experimental studies have sufficient precision that they illustrate the need for improvements in the theoretical models used to describe high-L fine structure. These continue to stimulate improved theoretical models, and therefore they are valuable, in part, for advancing the understanding of this class of Rydberg states. Some of the most significant studies in this group have been of the helium atom. As the simplest nonhydrogenic atom, helium was an early target of careful experimental studies, and an attractive subject of theoretical investigation. The high-L Rydberg states of helium are probably the simplest two-electron system in nature, and so these are especially attractive as a testing ground for fundamental atomic structure issues. This has been the underlying motivation for much of the work, both experimental and theoretical. One issue that received much attention in this regard was the question of the effects of retardation on the Rydberg structure. Obviously, as larger and larger Rydberg states are considered, at some point it becomes implausible to describe the interactions occurring between the Rydberg electron and the core ion as resulting from instantaneous Coulomb interactions, as is implicit in Eq. (6). Instead, there must be some way to account for the effects of retardation in the propagation of electromagnetic forces between the core and the Rydberg electron. A close analogy exists with the long-range force between two neutral atoms. In 1948, Casimir and Polder showed that the interaction between two neutral atoms at distances $r \gg 137a_0$ was altered by retardation, changing from the usual R^{-6} Van der Waals interaction to a R^{-7} interaction now known as the Casimir–Polder interaction (Casimir and Polder, 1948). In 1978, Kelsey and Spruch predicted that an analogous change occurs in the interaction between a charged particle and a polarizable system, such as a Rydberg core ion (Kelsey and Spruch, 1978). They predicted that at distances $r \gg 137a_0$, in addition to the adiabatic polarization potential, an additional interaction would arise given by:

$$V_{KS}(r) = \frac{11}{4\pi} \frac{\hbar}{mc} \frac{e^2 \alpha_1}{r^5} \tag{32}$$

where α_1 is the polarizability of the object. Because of the presence of $\hbar$ in the potential, this is clearly a nonclassical interaction. Spruch also showed that it could be viewed as a consequence of vacuum fluctuations of the electromagnetic field (Spruch and Kelsey, 1978), and suggested that the prediction could be tested by measurements of the fine structure of helium Rydberg states. This intriguing suggestion immediately motivated experimental and theoretical efforts towards a test, and was a primary factor responsible for the remarkable progress achieved in both theory and experiment in the intervening years. One illustration of that progress is shown in Fig. 6, which shows the measurement precision and theoretical accuracy of Rydberg G–H intervals in helium as a function of time. Between 1976, when the first calculation appeared, and 1999 when the most recent measurement was reported, the precision of measurements and calculations have both increased by approximately five orders of magnitude, from a few percent to a few tenths of a part-per-million. While measurements of this precision exist only for a few Rydberg levels, $n = 7, 9$, and 10, high precision variational calculations have now been reported for all

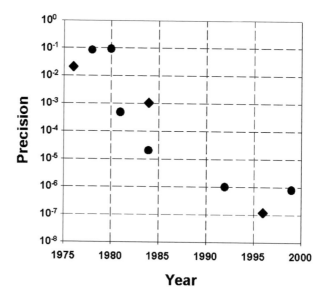

FIG. 6. Recent history of experimental and theoretical precision for G–H fine structure intervals in helium, illustrating the progress made in studying high-L Rydberg states of helium over the past three decades. The circles are experimental measurements, where the precision is taken from the quoted experimental errors. The diamonds show theoretical calculations. The precision of the 1976 calculation is inferred from its accuracy. The newer calculations estimate their precision, which is now somewhat better than experimental measurements. The illustrated measurements, all in MHz, are: 8G–8H: 931.34(44) (Cok and Lundeen, 1981); 9G–9H: 713(58) (Beyer and Kollath, 1978), 696(63) (Panock et al., 1980), 665.84935(58) (Stevens and Lundeen, 1999); 10G–10H: 490.990(10) (Palfrey and Lundeen, 1984), 491.00523(49) (Hessels et al., 1992), 491.00722(39) (Stevens and Lundeen, 1999). The theoretical calculations illustrated are all predictions of the 10G–10H interval. They are: 481 MHz (Deutsch, 1976), 491.01(52) (Drachman, 1985), 491.00823(6) (Drake, 1996).

Rydberg states of helium with $n \leq 10$ (Drake, 1996). Predictions using the effective potential model approach are available for all helium Rydberg states, without the need for detailed calculation, and for states with $L > 7$, these are at least as precise as the variational calculations (Drachman, 1993). The existing measurements are not in complete agreement with these calculations, but the remaining discrepancies are only on the order of 1 kHz. A detailed comparison between theory and experiment has been give elsewhere (Hessels, 1992; Stevens and Lundeen, 2000).

In order to reach the present degree of agreement with experiment, many subtle features have had to be incorporated into the theoretical treatment. Clarifying these often subtle effects has been greatly aided by the existence of two completely independent approaches to the calculations. One approach, the effective potential approach of Drachman, expands on the simplified discussion given in

Section 3 above, and treats the Rydberg structure in terms of the Long-Range Interactions. The other approach that has been applied very successfully is a variational approach, much more similar to the approaches generally used to treat atomic structure problems. This approach begins with a variational solution to the nonrelativistic Coulomb Hamiltonian, and includes a host of smaller corrections as perturbations on this picture (Drake, 1993). Together, these two approaches have led to the remarkable advances in calculation precision illustrated by Fig. 6. Among the subtle contributions now predicted and clearly confirmed by measurements are:

Relativistic corrections. Several different types of relativistic corrections enter into the predictions of high-L structure. Their relative importance can be illustrated by their respective contributions to the 10G–H interval (Hessels, 1992). For comparison, this interval is about 491 MHz and has been measured to a precision of 0.36 kHz (Stevens and Lundeen, 1999).

(a) p^4 correction to the nonrelativistic zeroth order kinetic energy. This term is given by the usual solution to the Dirac equation and is by far the largest relativistic correction, increasing the 10G–H interval by 7077.4 kHz;
(b) relativistic correction to the He$^+$ ion's dipole polarizability. This decreases the 10G–H interval by 108.0 kHz;
(c) p^4 correction to the polarization energies. This increases the 10G–H interval by 7.1 kHz;
(d) relativistic correction to the He$^+$ ion's quadrupole polarizability, α_2, and to the first nonadiabatic correction to the dipole polarizability, β_1. Together, these contribute 1.4 kHz to the 10G–H interval.

Retardation corrections. The influence of retardation on the Rydberg structure has now been calculated in several ways. The physics underlying the contributions of retardation to the fine structure now appears to be well understood, even though it has not yet been possible to test the original Kelsey and Spruch suggestion by precise measurements in Rydberg states confined to $r > 137a_0$ ($L > 16$). For the 10G–H interval, inclusion of retardation reduces the interval by 42.2 kHz. An additional small correction to the retardation effect, denoted V'', which amounts to 0.7 kHz for the 10G–H interval, has not yet been confirmed by experiment (Stevens and Lundeen, 2000).

Reduced mass corrections. Terms of order m_e/m_α and $(m_e/m_\alpha)^2$ have been carefully considered and included in the calculations of helium fine structure. An interesting treatment (Drachman, 1988) includes these terms by reformulating the Rydberg problem in Jacobi coordinates, thus greatly simplifying the evaluation of these terms by making the distinguishability of the high-L Rydberg electron explicit. These terms contribute 134 kHz to the 10G–H interval (Drachman, 1985).

Lamb shift corrections. Naively, one would expect that Lamb shift corrections would be negligible for nonpenetrating Rydberg levels, and indeed this is true for the Rydberg electron itself. However, the He^+ $1^2S_{1/2}$ core's energy contains a large contribution from the 1S Lamb shift ($\sim$128,000 MHz). When the core electron is polarized by the electric field of the Rydberg electron, its wavefunction becomes a mixture of the 1S with nP states, where the radiative shifts are much smaller. The more polarized the core, the less its effective Lamb shift. Thus, the modified radiative shifts in the core contribute indirectly a radiative correction to the energy of nonpenetrating Rydberg levels. This contributes $+12.9$ kHz to the 10G–H interval.

All these effects have been confirmed by experimental measurements in high-L Rydberg states of helium. They are also important in achieving agreement with measurements in lithium high-L Rydberg levels (Bhatia and Drachman, 1997). Experiments in other Rydberg systems are as yet only sensitive to the largest relativistic correction (p^4).

One area with much potential for future study is the study of He-like and Li-like Rydberg ions. Relativistic and QED effects are already significant for existing experimental studies of neutral helium and lithium Rydberg states, and the theoretical methods needed to predict the structure of isoelectronic Rydberg ions are already well developed (Bhatia and Drachman, 1999). Relativistic effects would, of course, be expected to increase in importance for these systems. However, to date, there is very little experimental data regarding these systems. It has been suggested that experimental studies of these ions could be carried out by stimulated recombination at Storage Ring facilities (Babb et al., 1992). Some studies of Be-like high-Z Rydberg ions have already been carried out in this way, but to date these studies have not given high precision measurements of the fine structure patterns (Schussler et al., 1995; Wolf et al., 2000). A potential alternative could be the RESIS/microwave method, which has recently achieved high precision studies in two Rydberg ions (Komara et al., 2003; Komara et al., 2005).

4.2. Ion Property Measurements

When high-L fine structure patterns are measured well within the range of accepted theoretical methods, the primary products of the measurements are determinations of properties of the core ions, such as dipole polarizabilities and quadrupole moments. Sometimes the measured properties are directly useful in applications. Other times, their value rests in the test they provide of calculations predicting the properties of the core ions from first principles. Since long-range forces are important in many other applications, and are always controlled by similar properties, increased confidence in the calculation of those properties can be very valuable.

4.2.1. Dipole Polarizability

The dipole polarizabilities of neutral atoms and molecules have been widely studied (Miller and Bederson, 1988). For an isotropic system, the polarizability is characterized by a single number

$$\alpha \equiv \frac{\vec{d}}{\vec{E}} \qquad (33)$$

where $\vec{d}$ is the electric dipole moment induced by the application of an external field $\vec{E}$. The most precise measured and calculated values of the polarizability of neutral atoms are updated and summarized elsewhere (Miller, 2003). The most precise measured values ($<0.1\%$) are obtained for atomic vapors, where they can be deduced from measurements of refractive index (Goebel et al., 1996). For many other neutral atoms and molecules, the best measurements of α are obtained using thermal beam deflection measurements, and are precise to about 2%. Recent advances in laser manipulation of atoms have led to new experimental techniques which have now made possible measurements of $\sim0.1\%$ precision in favorable cases (Amini and Gould, 2003; Ekstrom et al., 1995). For atoms, there is a close theoretical connection between the measured polarizabilities and other measurable properties of these neutral systems, such as resonance lifetimes and van der Waals coefficients, since all of them are largely determined by the dipole matrix element between the ground and first excited state. This connection has been exploited by Derevianko and Porsev to take advantage of the most precise measurements in atomic Cs (Derevianko and Porsev, 2002) to test the accuracy of relativistic atomic structure calculations for the Cs system, critical to measurements of Parity Nonconservation. Experimental measurements of α continue to be one of the best quantitative tests of atomic structure calculations in systems of interest for Parity Nonconservation studies.

Measurements of high-L Rydberg fine structure can provide precise measurements of dipole polarizabilities for a wide range of positive ions. Deducing the polarizabilities depends on knowledge of a pattern of fine structure energies, so that contributions from higher-order terms can be reliably separated from the leading term. Table I lists the ion polarizabilities that have been reported, based on observations of high-L Rydberg fine structure patterns. The precision of these measurements, in many cases, is at least equal to the precision of the best methods available for measurement of neutral atom polarizabilities. In the cases where these results have been compared with calculations of comparable precision, generally good agreement is found. For example, using relativistic Many Body Perturbation Theory for the alkali-like ions, Si^{3+} and Ba^+, Safronova and Derevianko report $\alpha_1 = 7.418(10)$ for Si^{3+} (Safronova, 2003) and $\alpha_1 = 124$ for Ba^+ (Derevianko, 2003). The most precise experimental numbers are for the ground states of H_2^+ and D_2^+. A very careful comparison with theory for these numbers

Table I

Ion polarizabilities which have been reported, based on observations of high-L Rydberg fine structure.

Ion	$\alpha(a_0^3)$	Method	Reference
Li^+	0.188(2)	Microwave/optical	(Cooke et al., 1977)
C^+	5.48(2)	RESIS/optical	(Ward, 1994)
N^+	3.77(11)	RESIS/optical	(Jacobson et al., 1996)
O^+	2.576(3)	RESIS/microwave	(Ward, 1994)
Ne^+	1.3028(13)	RESIS/microwave	(Ward et al., 1996)
S^+	10.6	RESIS/optical	(Deck et al., 1993)
Na^+	0.994(3)	Microwave/optical	(Gray et al., 1988)
Mg^+	33.0(5)	Solar Spectrum	(Chang and Noyes, 1983)
Mg^+	33.8(4)	Microwave/optical	(Lyons and Gallagher, 1998)
Si^+	19.0	Solar Spectrum	(Chang, 1984)
Ca^+	87(2)	Microwave/optical	(Vaidyanathan et al., 1982)
Cs^+	15.54(3)	Microwave/optical	(Safinya et al., 1980)
Ba^+	125.5(1.0)	Microwave/optical	(Gallagher et al., 1982)
Ba^+	124.30(16)	RESIS/optical	(Snow et al., 2005)
Si^{2+}	11.666(4)	RESIS/microwave	(Komara et al., 2005)
Si^{3+}	7.404(11)	RESIS/microwave	(Komara et al., 2003)
O^{5+}	1.05	Stim. Recomb.	(Schussler et al., 1995)
$H_2^+ (0,0)$	3.16796(15)	RESIS/microwave	(Jacobson et al., 2000)
$H_2^+ (0,1)$	3.1770(34)	RESIS/microwave	(Sturrus et al., 1991)
$D_2^+ (0,0)$	3.07187(54)	RESIS/microwave	(Jacobson et al., 2000)

reveals small but significant discrepancies of about 0.023(5)% (Jacobson et al., 2000). Differences of this order could possibly be due to uncalculated relativistic and radiative corrections such as those studied in the helium atom.

4.2.2. Quadrupole Moments

The fine structure patterns of high-L Rydberg electrons bound to nonisotropic core ions are sensitive to the quadrupole moment of the core ion. Table II lists the positive ion quadrupole moments that have been reported, based on observations of high-L fine structure patterns. There are very few comparable measurements of neutral atom quadrupole moments. One example is the 6% measurement of the quadrupole moment of the ground state of Al (Angel et al., 1967). A pattern similar to the dipole polarizabilities is found when these measurements are compared with theory. For example, Sundholm calculates the quadrupole moment of Ne^+ and finds agreement to within 0.41(25)% with the RESIS measurement (Sundholm and Olsen, 1994). The H_2^+ quadrupole moment appears to be in reasonable agreement with theory, but only if relativistic corrections are included in the calculation (Bishop, 1989).

Table II

Ion quadrupole moments that have been reported based on study of fine structure patterns in high-*L* Rydberg states. We note that (Chang, 1984) use a different definition of the atomic quadrupole moment than used here and in (Angel et al., 1967).

Ion	$Q(ea_0^2)$	Method	Reference
$Si^+(^2P_{3/2})$	1.33	Solar Spectrum	(Chang, 1984)
$C^+(^2P_{3/2})$	0.475(2)	RESIS/optical	(Ward, 1994)
$N^+(^3P_2)$	−0.364(6)	RESIS/optical	(Jacobson et al., 1996)
$Ne^+(^2P_{3/2})$	−0.20403(5)	RESIS/microwave	(Ward et al., 1996)
$H_2^+(0,1)$	1.64295(30)	RESIS/microwave	(Sturrus et al., 1991)

4.2.3. Other Ion Properties

In addition to the polarizability and quadrupole moment, high-*L* fine structure measurements are in principle sensitive to many higher-order long-range interactions, and could be used to determine quadrupole polarizabilities, hexadecapole moments, magnetic moments, etc. In practice, this would require very precise measurements of fine structure in a wide range of high-*L* states, and there are few examples where this has been reported. Determination of quadrupole polarizability is complicated by the significant nonadiabatic corrections to the dipole polarization energy which have the same r^{-6} signature. Nevertheless, quadrupole polarizabilities have been reported for the Ba^+ ion in two studies (Gallagher et al., 1982; Snow et al., 2005). In high-*L* Rydberg states with anisotropic cores, the dipole polarizability has both scalar and tensor components, and the tensor component can be inferred from the *L*-dependence of the tensor fine structure pattern. Such measurements have been reported both for neon (Ward et al., 1996) and for H_2^+ (Sturrus et al., 1991). The magnetic moment of the Ne^+ ground state has also been deduced from the observed spin structure of high-*L* states of neon (Ward et al., 1996). Fine structure studies in H_2 have been used to determine a rotational energy splitting in the H_2^+ ion and the hyperfine constants of the $H_2^+(0,1)$. In the case of the rotational splitting, the close coincidence between the zeroth order energies of the $n = 27$ states bound to $H_2^+(0,1)$ and the $n = 16$ states bound to $H_2^+(0,3)$ was exploited to measure the $(0,1)$ to $(0,3)$ rotational energy splitting. This was done by observing microwave transitions between high-*L* Rydberg states bound to the different cores (Arcuni et al., 1990b). The result,

$$E(0,3) - E(0,1) = 288.85900(8) \text{ cm}^{-1} \tag{34}$$

is found to be in excellent agreement with careful calculations (Moss, 1990).

In the case of the hyperfine constants, measurements of the resolved hyperfine components of microwave transitions between different $n = 27$ Rydberg states of H_2 bound to the $(0,1)$ state of H_2^+ were used to determine the core hyperfine constants (Fu et al., 1992). The results were:

$$b + \frac{c}{3} = 922.940(20) \text{ MHz},$$
$$c = 128.259(26) \text{ MHz}, \tag{35}$$
$$d = 42.348(29) \text{ MHz}.$$

These are in reasonable agreement with the best calculations, but again it seems possible that the calculations are incomplete (Babb and Dalgarno, 1992).

These several examples illustrate the potential of high-L Rydberg studies to reveal a wider range of ion properties than simply the leading coefficients of the long-range interactions.

4.3. APPLICATIONS

The importance of high-L Rydberg levels in atomic and molecular processes, such as dielectronic recombination (Jacobs et al., 1976) and Zero Electron Kinetic Energy (ZEKE) spectroscopy (Chupka, 1992) has become clear in many studies. The detailed spectroscopy of the high-L levels does not figure prominently in these applications, which depend instead on their high statistical weights and long radiative lifetimes. For example, dielectronic recombination occurs initially into Rydberg states of relatively low-L, but in the presence of small electric fields, the initial population can be mixed into the large manifold of high-L states, stabilizing the recombination (Jacobs et al., 1976). Similarly, in ZEKE spectroscopy of molecules, a pulsed laser excitation is followed by delayed field ionization, and collisional or Stark mixing from the initially excited low-L level into the manifold of long-lived high-L levels can be a very important factor in forming the ZEKE signal. The primary impact of the high-L fine structure pattern in these applications is in determining the L-mixing probability in a given environment. The processes, collisional and Stark L-mixing, which bring the high-L states into play in these applications also make spectroscopic observations of high-L fine structures difficult in many laboratory environments. These factors, however, may not apply in other natural environments, and this has led to increased interest in observing spectrally resolved high-L Rydberg emission in these other circumstances.

The outstanding example of this is undoubtedly the observation and study of weak emission lines (WEL) in solar and stellar spectra. This effort began with the observation and identification of high-L emission lines from Mg, Al, and Si in the solar spectrum near 12 microns (Breault and Noyes, 1983; Chang and Noyes, 1983; Chang, 1984). These lines typically appear as a narrow emission line superposed upon a broader absorption line. The precise mechanisms of their formation are still being studied (Chang et al., 1991; Sigut and Lester,

1996), but it is clear that Rydberg states of high-*L* are naturally inclined to produce these lines. There are good reasons to believe that future high resolution searches may reveal such emission lines from a wide range of atoms and ions, not only on the sun but also in other stellar spectra (Sigut and Lester, 1996). Recent reports cataloging emission features on one such star, 3 Cen A, may be a first step in that direction (Wahlgren and Hubrig, 2004). The study of the 12 micron emission lines has already been very fruitful on the sun. Because of the low Doppler sensitivity of these lines and their narrow widths, they are prime candidates for studies of the solar magnetic field (Jennings et al., 2002). The high-*L* emission lines are also very sensitive to the electric environment in the stellar atmosphere, and can also be used for other unique diagnostic studies. They are now widely studied (Deming et al., 1998 and references therein).

Another area of potential importance is the search for high-*L* emission lines from H_2 in astrophysical environments. As understanding of the spectroscopy of these states has improved, a number of researchers have suggested searches for such lines (Babb and Chang, 1992; Stickland and Cotterell, 1996). Although there appear to be a number of astrophysical environments where such emission lines might be formed, there have not yet been any reported observations. Nevertheless, this is another area of continued interest where improved understanding of high-*L* Rydberg spectroscopy plays a role.

5. Summary and Outlook

By any measure, recent decades have witnessed tremendous advances in understanding the fine structure of high-*L* Rydberg states. New experimental techniques have revealed fine structure patterns that had always been hidden from view by limited experimental resolution. Theoretical progress has made it possible to interpret these patterns confidently in terms of a limited number of properties of the positive ion cores binding the Rydberg electron. The special case of helium Rydberg states has been at the forefront of this progress, both in experiment and theory. The new experimental techniques have now been used to study many other high-*L* atomic Rydberg fine structure patterns, and molecular and ionic high-*L* Rydberg systems are increasingly being studied with similar techniques. As a result of this progress, properties of positive ions that control their long-range interactions, such as quadrupole moments and dipole polarizabilities, are beginning to be measurable with more ease and precision than is possible for similar properties of neutral atoms.

Many challenges and opportunities still exist. In the case of helium and lithium, and their isoelectronic ions, theoretical predictions of high-*L* fine structure appear to be more precise than existing measurements. This historically unusual situation poses a special challenge to experimenters to find innovative methods to explore these systems, especially in the more highly relativistic ions where virtually no experimental measurements now exist. Another challenge to experiment is to extend these studies to heavier atoms and ions, where the theoretical calculation of measurable core properties becomes increasingly difficult because of relativistic and many-body effects. For example, Rydberg atoms or ions built upon Fr-like ion cores would be an attractive subject of study, since atomic theory for alkali-like ions is relatively advanced and is critical to interpretation of measurements of Parity Nonconservation in atoms. Other challenges include the experimental study of a wider range of high-L Rydberg molecules. To date, only H_2 and D_2 have been studied with high resolution, but the experimental techniques used there should be applicable to a much wider range of molecules.

Challenges to theory also abound. Except for the helium-like and lithium-like atoms and ions, the precision of experimental measurements generally exceeds theory. Within the context of long-range models, the task of theory is to express the high-L fine structure patterns in terms of well-defined core parameters, possibly leaving the actual calculation of these parameters as a separate exercise. The case of neon Rydberg states illustrates the importance of this task, with the introduction of a significant new term in the long-range interactions. However, it seems unlikely that the present calculations in neon, important though they are, can be the most efficient approach since they fail to incorporate the nearly exact zeroth order solution for the Rydberg wavefunction. In the case of H_2 Rydberg levels, the need for a more complete *a-priori* calculation is even greater. There is also a clear need for incorporation of spin effects into theoretical models. At present, with the exception of helium, theoretical treatments assume only Coulomb forces, and experimental observations of spin effects have been treated phenomenologically by experimenters. This is especially dramatic in the case of Ba Rydberg levels where extremely large spin splittings have been observed and only a very crude model "explains" the effects. Since any such theoretical efforts would be, of necessity, well outside the scope of traditional theoretical work, continued progress will likely continue to depend on the stimulation of precise experimental measurements.

It is remarkable that more than a century after the "discovery" of Rydberg states, this very large class of atomic and molecular excited states is so little explored. Exploring them, and understanding the relatively simple physics that controls their behavior promises to provide interesting challenges for some time yet.

6. Acknowledgements

This work was supported by the Chemical Sciences, Geosciences, and Biosciences Division of the Office of Basic Energy Sciences, Office of Science, U.S. Department of Energy. The author gratefully acknowledges the efforts of students who have contributed to his own education in the subject, including David Cok, Stephen Palfrey, W. Gregg Sturrus, Eric Hessels, Francis Deck, Joe Fu, Nelson Claytor, R. Frank Ward, Jr., Phil Jacobson, Dan Fisher, Bob Komara, and Erica Snow, as well as post-docs Phil Arcuni, Jerry Stevens, Charles Fehrenbach, and Alina Gearba, and faculty colleague Brett DePaola. Theoretical colleagues who have generously shared their insights and calculations include Richard Drachman, Gordon Drake, Walter Johnson, Steve Blundell, Chris Greene, Andrei Derevianko, Marianna Safronova, and Samuel Cohen.

7. References

Amini, J.M., Gould, H. (2003). High precision measurement of the static dipole polarizability of cesium. *Phys. Rev. Lett.* **91**, 153001.

Angel, J.R.P., Sandars, P.G.H., Woodgate, G.K. (1967). Direct measurements of an atomic quadrupole moment. *J. Chem. Phys.* **47**, 1552.

Arcuni, P.W., Fu, Z.W., Lundeen, S.R. (1990a). Energy difference between the ($v = 0$, $R = 1$) and ($v = 0$, $R = 3$) states of H_2^+, measured with interseries microwave spectroscopy of H_2 Rydberg states. *Phys. Rev. A* **42**, 6950.

Arcuni, P.W., Hessels, E.A., Lundeen, S.R. (1990b). Series mixing in high-*L* Rydberg states of H_2: An experimental test of polarization model predictions. *Phys. Rev. A* **41**, 3648.

Aymar, M., Greene, C.H., Luc-Koenig, E. (1996). Multichannel Rydberg spectroscopy of complex atoms. *Rev. Mod. Phys.* **68**, 1015.

Babb, J.F., Chang, E.S. (1992). The Rydberg electronic transitions of the hydrogen molecule. *At. Data Nucl. Data Tables* **50**, 137.

Babb, J.F., Dalgarno, A. (1992). Spin coupling constants and hyperfine transition frequencies for the hydrogen molecular ion. *Phys. Rev. A* **46**, R5317.

Babb, J.F., Habs, D., Spruch, L., Wolf, A. (1992). Retardation (Casimir) energy shifts for Rydberg helium-like low-*Z* ions. *Z. Phys. D* **23**, 197.

Basterrechea, F.J., Davies, P.B., Smith, D.M., Stickland, R.J. (1994). Diode laser spectroscopy of the 7I–6H and 7H–6I transitions in H_2. *Molecular Phys.* **81**, 1435.

Beyer, H.J., Kollath, K.J. (1978). Measurement of intervals between [1,3]D and high-*L* states of He for $n = 7$ to 10 by electric-field-induced anticrossings. *J. Phys. B. At. Mol. Phys.* **11**, 979.

Bhatia, A.K., Drachman, R.J. (1997). Relativistic, retardation, and radiative corrections in Rydberg states of lithium. *Phys. Rev. A* **55**, 1842.

Bhatia, A.K., Drachman, R.J. (1999). Energy levels of triply ionized carbon (CIV): Polarization model. *Phys. Rev. A* **60**, 2848.

Bishop, D.M. (1989). Comment on microwave spectroscopy of high-*L* H_2 Rydberg states: the (0,1) 10 G, H, I, and K states. *Phys. Rev. Lett.* **62**, 3008.

Bishop, D.M., Lam, B. (1987). Vibrational polarizabilities for H_2^+, H_2, and N_2. *Chem. Phys. Lett.* **134**, 283.

Bockasten, K. (1974). Mean values of powers of the radius for hydrogenic electron orbits. *Phys. Rev. A* **9**, 1087.

Brandenberger, J.R., Lundeen, S.R., Pipkin, F.M. (1976). Multiphoton radiofrequency resonances in He_4^+. *Phys. Rev. A* **14**, 341.

Breault, J., Noyes, R. (1983). Solar emission lines near 12 microns. *Astrophys. J.* **269**, L61.

Casimir, H.B.G., Polder, D. (1948). The influence of retardation on the London–van der Waals forces. *Phys. Rev.* **73**, 360.

Chang, E.S. (1984). Nonpenetrating Rydberg states of silicon from solar data. *J. Phys. B. At. Mol. Phys.* **17**, L11.

Chang, E.S., Noyes, R.W. (1983). Identification of the solar emission lines near 12 microns. *Astrophys. J.* **275** (1), L11.

Chang, E.S., Sakai, H. (1982). Properties of ions from spectroscopic data. *J. Phys. B. At. Mol. Phys.* **15**, L649.

Chang, E.S., Avrett, E.H., Mauas, P.J., Noyes, R.W., Loeser, R. (1991). Formation of the infrared emission lines of MgI in the solar atmosphere. *Astrophys. J.* **379**, L79.

Chupka, W.A. (1992). Factors affecting lifetimes and resolution of Rydberg states observed in zero-kinetic-energy spectroscopy. *J. Chem. Phys.* **96**, 4520.

Clark, W., Greene, C.H. (1999). Adventures of a Rydberg electron in an anisotropic world. *Rev. Mod. Phys.* **71**, 821.

Clark, W., Greene, C.H., Miecznik, G. (1996). Anisotropic interaction potential between a Rydberg electron and an open-shell ion. *Phys. Rev. A* **53**, 2248.

Cok, D.R., Lundeen, S.R. (1979). Magnetic and electric fine structure in helium Rydberg states. *Phys. Rev. A* **19**, 1830; 24, 3283.

Cok, D.R., Lundeen, S.R. (1981). Atomic beam measurements of helium F–G, F–H, and F–I intervals. *Phys. Rev. A* **23**, 2488.

Cooke, W.E., Gallagher, T.F., Hill, R.M., Edelstein, S.A. (1977). Resonance measurements of $d-f$ and $d-g$ intervals in lithium Rydberg states. *Phys. Rev. A* **16**, 1141.

Dalgarno, A., Lewis, J.T. (1956). The representation of long range forces by series expansion I: The divergence of the series, II: The complete perturbation calculation of long range forces. *Proc. Phys. Soc. London Sect. A* **69**, 57.

Deck, F.J., Hessels, E.A., Lundeen, S.R. (1993). Population of high-L sulfur Rydberg levels by ion–Rydberg-atom charge transfer. *Phys. Rev. A* **48**, 4400.

Deming, D., Jennings, D.E., McCabe, G., Moran, T., Lowenstein, R. (1998). Limb observations of the 12.32-micron MgI emission line during the 1994 annular eclipse. *Solar Phys.* **182**, 283.

Derevianko, A., Porsev, S.G. (2002). Determination of lifetimes of 6P$_J$ levels and ground state polarizability of Cs from the van der Waals coefficient C$_6$. *Phys. Rev. A* **65**, 053403.

Deutsch, C. (1976). Rydberg states of HeI using the polarization model. *Phys. Rev. A* **13**, 2311.

Drachman, R.J. (1982). Rydberg states of helium: An optical potential analysis. *Phys. Rev. A* **26**, 1228.

Drachman, R.J. (1985). Rydberg states of helium: Relativistic and second-order corrections. *Phys. Rev. A* **31**, 1253.

Drachman, R.J. (1988). Rydberg states of helium: A new recoil term. *Phys. Rev. A* **37**, 979.

Drachman, R.J. (1993). Rydberg states of helium: Some further small corrections. *Phys. Rev. A* **47**, 694.

Drachman, R.J., Bhatia, A.K. (1995). Rydberg levels of lithium. *Phys. Rev. A* **51**, 2926.

Drake, G.W.F. (1990). Variational eigenvalues for the Rydberg states of helium: Comparison with experiment and with asymptotic expansions. *Phys. Rev. Lett.* **65**, 2769.

Drake, G.W.F. (1993). High-precision calculations for the Rydberg states of helium. In: Levin, F.S., Micha, D.A. (Eds.), "Long-Range Casimir Forces: Theory and Recent Experiments on Atomic Systems", Plenum Press, New York, pp. 107–217.

Drake, G.W.F. (1996). Precision calculations for helium. In: Drake, G.W.F. (Ed.), "Atomic, Molecular, and Optical Physics Handbook", AIP Press, Woodbury, NY, pp. 154–171.

Drake, G.W.F., Swainson, R.A. (1991). Quantum defects and the $1/N$ dependence of Rydberg energies – 2nd-order polarization effects. *Phys. Rev. A* **44**, 5448.

Dyubko, S.F., Efremov, V.A., Gerasimov, V.G., MacAdam, K.B. (2003). Microwave spectroscopy of Al I atoms in Rydberg states: D and G terms. *J. Phys. B. At. Mol. Opt. Phys.* **36**, 4827.

Edlen, B. (1964). Atomic Spectra. "Encyclopedia of Physics", vol. XXVII. Springer, Berlin.

Eissa, H., Opik, U. (1967). The polarization of a closed-shell core of an atomic system by an outer electron I. A correction to the adiabatic approximation. *Proc. Phys. Soc. London* **92**, 556.

Ekstrom, C.R., Schmiedmayer, J., Chapman, M.S., Hammond, T.D., Pritchard, D.E. (1995). Measurement of the electric polarizability of sodium with an atom interferometer. *Phys. Rev. A* **51**, 3883.

Eriksson, K.B.S. (1956). Coupling of electrons with high orbital angular momentum, illustrated by $2pnf$ and $2png$ in NII. *Phys. Rev.* **102**, 102.

Farley, J.W., MacAdam, K.B., Wing, W.H. (1979). Fine structure of Rydberg states. IV. Completely resolved fine structure in D, F, and G states of ^{4}He. *Phys. Rev. A* **20**, 1754.

Fisher, D.S., Lundeen, S.R., Fehrenbach, C.W., DePaola, B.D. (2001). Energy transfer in ion–Rydberg-atom charge exchange. *Phys. Rev. A* **63**, 052712.

Freeman, R.R., Kleppner, D. (1976). Core polarization and quantum defects in high-angular momentum states of alkali atoms. *Phys. Rev. A* **14**, 1614.

Fu, Z.W., Hessels, E.A., Lundeen, S.R. (1992). Determination of the hyperfine constants of H_2^+ ($v = 0$, $R = 1$) by microwave spectroscopy of high-*L* $n = 27$ Rydberg states of H_2. *Phys. Rev. A* **46**, R5313.

Gallagher, T.F. (1994). "Rydberg Atoms". Cambridge Univ. Press, Cambridge.

Gallagher, T.F., Hill, R.M., Edelstein, S.A. (1976). Resonance measurements of d–f–g–h splittings in highly excited sates of sodium. *Phys. Rev. A* **14**, 744.

Gallagher, T.F., Kachru, R., Tran, N.H. (1982). Radio frequency resonance measurements of the Ba $6sng$–$6snh$–$6sni$–$6snk$ intervals: An investigation of the non-adiabatic effects in core polarization. *Phys. Rev. A* **26**, 2611.

Goebel, D., Hohm, U., Maroulis, G. (1996). Theoretical and experimental determination of the polarizabilities of the zinc $^1S_0$ state. *Phys. Rev. A* **54**, 1973.

Gray, L.G., Sun, X., MacAdam, K.B. (1988). Resonance measurements of d–f–g–h intervals in Rydberg states of sodium and a redetermination of the core polarizabilities. *Phys. Rev. A* **38**, 4985.

Greene, C.H., Jungen, Ch. (1985). Molecular applications of quantum defect theory. *Adv. At. Mol. Phys.* **21**, 51.

Herzberg, G., Jungen, Ch. (1982). High orbital angular momentum states in H_2 and D_2. *J. Chem. Phys.* **77** (12), 5876.

Hessels, E.A. (1992). Higher-order relativistic corrections to the polarization energies of helium Rydberg states. *Phys. Rev. A* **46**, 5389.

Hessels, E.A., Sturrus, W.G., Lundeen, S.R., Cok, D.R. (1987). Measurement of the magnetic fine structure of the 10G and 10H states of helium. *Phys. Rev. A* **35**, 4489.

Hessels, E.A., Arcuni, P.W., Deck, F.J., Lundeen, S.R. (1992). Microwave spectroscopy of high-*L*, $n = 10$ Rydberg states of helium. *Phys. Rev. A* **46**, 2622.

Jacobs, V.L., Davis, J., Kepple, P.C. (1976). Enhancement of dielectronic recombination by plasma electric microfields. *Phys. Rev. Lett.* **37**, 1390.

Jacobson, P.L., Labelle, R.D., Sturrus, W.G., Ward Jr., R.F., Lundeen, S.R. (1996). Optical spectroscopy of high-*L* $n = 10$ Rydberg states of nitrogen. *Phys. Rev. A* **54**, 314.

Jacobson, P.L., Fisher, D.S., Fehrenbach, C.W., Sturrus, W.G., Lundeen, S.R. (1997). Determination of the dipole polarizabilities of H_2^+ (0, 0) and D_2^+ (0, 0) by microwave spectroscopy of high-*L* Rydberg states of H_2 and D_2. *Phys. Rev. A* **56**, R4361; 57, 4065 (E).

Jacobson, P.L., Komara, R.A., Sturrus, W.G., Lundeen, S.R. (2000). Microwave spectroscopy of helium-like Rydberg states of H_2 and D_2: Determination of the dipole polarizabilities of H_2^+ and D_2^+ ground states. *Phys. Rev. A* **62**, 102509.

Jennings, D.E., Deming, D., McCabe, G., Sada, P.V., Moran, T. (2002). Solar magnetic field studies using the 12 micron emission lines. IV. Observations of a Delta region solar flare. *Astrophys. J.* **568**, 1043.

Jones, R.R., Gallagher, T.F. (1988). Autoionization of high-*L* Ba $6p_{1/2}nl$ states. *Phys. Rev. A* **38**, 2846.

Jungen, Ch., Dabrowski, I., Herzberg, G., Kendall, D.J.W. (1989). High orbital angular momentum states in H_2 and D_2. II. The $6h$–$5g$ and $6g$–$5f$ transitions. *J. Chem. Phys.* **91**, 3926.

Kelsey, E.J., Spruch, L. (1978). Retardation effects on high Rydberg states: A retarded R^{-5} polarization potential. *Phys. Rev. A* **18**, 15.

Kleinman, C.J., Hahn, Y., Spruch, L. (1968). Dominant nonadiabatic contribution to the long-range electron–atom interaction. *Phys. Rev.* **165**, 53.

Komara, R.A., Sturrus, W.G., Pollack, D.H., Cochran, W.R. (1999). Dalgarno–Lewis method for second-order energies of Rydberg states of neon. *Phys. Rev. A* **59**, 251.

Komara, R.A., Gearba, M.A., Lundeen, S.R., Fehrenbach, C.W. (2003). Determination of the polarizability of Na-like silicon by study of the high-L Rydberg states of Si^{2+}. *Phys. Rev. A* **67**, 062502.

Komara, R.A., Gearba, M.A., Fehrenbach, C.W., Lundeen, S.R. (2005). Ion properties from high-L Rydberg fine structure: Dipole polarizability of Si^{2+}. *J. Phys. B At. Mol. Opt. Phys.* **38**, S87.

Laughlin, C. (1995). An asymptotic potential method for high-L Rydberg levels of calcium. *J. Phys. B. At. Mol. Opt. Phys.* **28**, 2787.

Lemoine, B., Petitprez, D., Destombes, J.L., Chang, E.S. (1990). High precision infrared diode laser spectrum of Mg I. *J. Phys. B. At. Mol. Opt. Phys.* **23**, 2217S.

Lyons, B.J., Gallagher, T.F. (1998). Mg 3snf–3sng–3sni intervals and the Mg^+ dipole polarizability. *Phys. Rev. A* **57**, 2426.

MacAdam, K.B., Wing, W.H. (1975). Fine structure of Rydberg states: $n = 6$ and 7D and F states of 4He. *Phys. Rev. A* **12**, 1464.

MacAdam, K.B., Gray, L.G., Rolfes, R.G. (1990). Projectile n distributions following charge transfer of Ar^+ and Na^+ in a Na Rydberg target. *Phys. Rev. A* **42**, 5269.

Martin, W.C., Fuhr, J.R., Kelleher, D.E., Musgrove, A., Podobedova, L., Reader, J., Saloman, E.B., Sansonetti, C.J., Wiese, W.L., Mohr, P.J., Olsen, K. (1999). NIST Atomic Spectra Database (Version 2.0), (online). Available: http://physics.nist.gov/asd (January 11, 2005), National Institute of Standards and Technology, Gaithersburg, MD.

Mayer, J.E., Goeppert Mayer, M. (1933). The polarizabilities of ions from spectra. *Phys. Rev.* **43**, 605.

McEachran, R.P., Veenstra, C.J., Cohen, M. (1978). Hyperfine structure in the hydrogen molecular ion. *Chem. Phys. Lett.* **59**, 275.

Miller, T.M. (2003). Atomic and molecular polarizabilities. In: "CRC Handbook of Chemistry and Physics", CRC Press, Boca Raton, FL, pp. 10–163.

Miller, T.M., Bederson, B. (1988). Electric dipole polarizability measurements. *Adv. At. Mol. Phys.* **25**, 37.

Moss, R.E. (1990). Calculations on the (0,1) and (0,3) vibration–rotation levels of H_2^+. *Chem. Phys. Lett.* **172**, 458.

Palfrey, S.L., Lundeen, S.R. (1984). Measurement of high-angular-momentum fine structure in helium: An experimental test of long-range electromagnetic forces. *Phys. Rev. Lett.* **53**, 1141.

Panock, R., Rosenbluth, M., Lax, B., Miller, T.A. (1980). Laser magnetic resonance spectroscopy of normally forbidden transitions: Electrostatic fine structure of the $n = 9$, $L = 1$–8 4He singlet states. *Phys. Rev. A* **22**, 1050.

Pascale, J., Olson, R.E., Reinhold, C.O. (1990). State-selective capture in collisions between ions and ground and excited-state alkali-metal atoms. *Phys. Rev. A* **42**, 5305.

Patil, S.H. (1986). Atomic potentials, polarizabilities, and nonadiabatic corrections in high-angular-momentum Rydberg states. *Phys. Rev. A* **33**, 90.

Pruvost, L., Camus, P., Lecompte, J.-M., Mahon, C.R., Pillet, P. (1991). High angular momentum $6pnl$ and $6dnl$ doubly excited Rydberg states of barium. *J. Phys. B. At. Mol. Opt. Phys.* **24**, 4723.

Risberg, P. (1955). The spectrum of singly-ionized magnesium, MgII. *Arkiv for Fysik* **9**, 483.

Ruff, G.A., Safinya, K.A., Gallagher, T.F. (1980). Measurements of the $n = 15$–17 f–g intervals in Cs. *Phys. Rev. A* **22**, 183.

Rydberg, J.R. (1890). On the structure of the line-spectra of the chemical elements. *Philosoph. Mag. (London) Ser. 5* **29**, 331.

Safinya, K.A., Gallagher, T.F., Sandner, W. (1980). Resonance measurements of f–h and f–i intervals in cesium using selective and delayed field ionization. *Phys. Rev. A* **22**, 2672.

Sansonetti, C.J., Andrew, K.L., Verges, J. (1981). Polarization, penetration, and exchange effects in the hydrogen-like nf and ng terms of cesium. *J. Opt. Soc. Am.* **71**, 423.

Schoenfeld, W.G. (1994). Analysis of high-L Rydberg levels of silicon from the solar spectrum: A test of the extended polarization model. PhD Thesis, University of Massachusetts, Amherst, MA.

Schussler, T., Schramm, U., Ruter, T., Broude, C., Grieser, M., Habs, D., Schwalm, D.E., Wolf, A. (1995). Laser-stimulated recombination spectroscopy for the study of long-range interactions in highly charged Rydberg ions. *Phys. Rev. Lett.* **75**, 802.

Sigut, T.A.A., Lester, J.B. (1996). Infrared emission lines in early type stars. I. MgII. *Astrophys. J.* **461**, 972.

Snow, E.L., Komara, R.A., Gearba, M.A., Lundeen, S.R. (2003). Indirect spin–orbit interaction in high-L Rydberg states with $^2S_{1/2}$ cores. *Phys. Rev. A* **68**, 022510.

Snow, E.L., Gearba, M.A., Komara, R.A., Lundeen, S.R. (2005). Determination of the dipole and quadrupole polarizabilities of Ba$^+$ by measurement of the fine structure of high-L $n = 9$ and 10 Rydberg states of barium. *Phys. Rev. A* (submitted for publication).

Spruch, L., Kelsey, E.J. (1978). Vacuum fluctuation and retardation effects on long-range potentials. *Phys. Rev. A* **18**, 845.

Stebbings, R.F., Dunning, F.B. (1983). "Rydberg States of Atoms and Molecules". Cambridge Univ. Press, Cambridge.

Stevens, G.D., Lundeen, S.R. (1999). Measurements of G–H, H–I fine structure intervals in $n = 7, 9$, and 10 helium Rydberg states. *Phys. Rev. A* **60**, 4379.

Stevens, G.D., Lundeen, S.R. (2000). Experimental studies of helium Rydberg fine structure. In: "Comments on Atomic and Molecular Physics, Comments on Modern Physics", vol. 1, Part D, pp. 207–219.

Stickland, R.J., Cotterell, B.J. (1996). Infrared inter-Rydberg emission spectra as a possible probe of molecular hydrogen in astrophysical environments. *Chem. Phys. Lett.* **251**, 287.

Storry, C.H., Rothery, N.E., Hessels, E.A. (1997a). Measurement of the $n = 9$ F-to-G intervals in atomic lithium. *Phys. Rev. A* **55**, 128.

Storry, C.H., Rothery, N.E., Hessels, E.E. (1997b). Separated oscillatory field measurement of the $n = 10$ $^+F_3$–$^+G_4$ interval in helium: A 200 part per billion measurement. *Phys. Rev. A* **55**, 967.

Sturrus, W.G., Sobol, P.E., Lundeen, S.R. (1985). Observation of high-angular-momentum Rydberg states of H$_2$ in a fast beam. *Phys. Rev. Lett.* **54**, 792.

Sturrus, W.G., Hessels, E.A., Lundeen, S.R. (1986). High-resolution microwave spectroscopy of the 10G–10H Rydberg transition in H$_2$. *Phys. Rev. Lett.* **57**, 1863.

Sturrus, W.G., Hessels, E.A., Arcuni, P.W., Lundeen, S.R. (1988). Laser spectroscopy of ($v = 0$, $R = 1$)10F and ($v = 0$, $R = 1$)10G states of H$_2$: A test of the polarization model. *Phys. Rev. A* **38**, 135.

Sturrus, W.G., Hessels, E.A., Arcuni, P.W., Lundeen, S.R. (1991). Microwave spectroscopy of high-L H$_2$ Rydberg states ($v = 0$, $R = 1$) $n = 10$ G, H, I, and K. *Phys. Rev. A* **44**, 3032.

Sun, X., MacAdam, K.B. (1994). Microwave measurements of d–f–g–h intervals and d and f fine structure of sodium Rydberg states. *Phys. Rev. A* **49**, 2453.

Sundholm, D., Olsen, J. (1994). Finite-element multiconfiguration Hartree–Fock calculations of the atomic quadrupole moments of C$^+$(^{2}P) and Ne$^+$(^{2}P). *Phys. Rev. A* **49**, 3453.

Theodosiou, C.E. (1983). Evaluation of penetration effects in high-L Rydberg states. *Phys. Rev. A* **28**, 3098.

Tran, N.H., van Linden van den Heuvell, H.B., Kachru, R., Gallagher, T.F. (1984). Radiofrequency resonance measurements of Na d–f intervals. *Phys. Rev. A* **30**, 2097.

Vaidyanathan, A.G., Shorer, P. (1982). Dynamical contributions to the quantum defects of calcium. *Phys. Rev. A* **25**, 3108.

Vaidyanathan, A.G., Spencer, W.P., Rubbmark, J.R., Kuiper, H., Fabre, C., Kleppner, D., Ducas, T.W. (1982). Experimental study of nonadiabatic core interactions in Rydberg states of calcium. *Phys. Rev. A* **26**, 3346.

Van Deurzen, C.H.H. (1977). Analysis of 4-ionized vanadium (VV). *J. Opt. Soc. Am.* **67**, 476.

Wahlgren, G.M., Hubrig, S. (2004). Emission lines in the optical spectrum of 3 Cen A. *Astronomy Astrophys.* **418**, 1073.

Ward Jr., R.F. (1994). Spectroscopy of high-L $n = 10$ Rydberg states of carbon, oxygen, and neon. PhD Thesis, Univ. of Notre Dame.

Ward Jr., R.F., Sturrus, W.G., Lundeen, S.R. (1996). Microwave spectroscopy of high-L Rydberg states of neon. *Phys. Rev. A* **53**, 113.

White, H.E. (1934). "Introduction to Atomic Spectra". McGraw-Hill, New York, p. 8.

Wolf, A., Uhlenberg, G., Schramm, U., Schussler, T., Livingston, A.E., Gwinner, G., Saathoff, G., Schwalm, D. (2000). Spectroscopy using stimulated electron–ion recombination. *Hyp. Interact.* **127**, 203.

Zygelman, B. (1990). Non-Abelian geometric phase and long-range atomic forces. *Phys. Rev. Lett.* **64**, 256.

A STORAGE RING FOR NEUTRAL MOLECULES

FLORIS M.H. CROMPVOETS[1], *HENDRICK L. BETHLEM*[1,2] *and*
GERARD MEIJER[2]

[1] *FOM—Institute for Plasma Physics Rijnhuizen, P.O. Box 1207, NL-3430 BE Nieuwegein,*
The Netherlands

[2] *Fritz-Haber-Institut der Max-Planck-Gesellschaft, Faradayweg 4-6, D-14195 Berlin, Germany*

Abstract

Electrostatic hexapoles are common tools in physical chemistry for performing
state-selection and focusing of molecules. A hexapole can be viewed as a 2D-trap

ISSN 1049-250X
DOI 10.1016/S1049-250X(05)52005-6

for polar molecules in low-field seeking states. Therefore, by bending such a hexapole into a torus a storage ring for neutral molecules can be formed. We here review our experiments on such a prototype storage ring. The ring is loaded by a decelerated ammonia beam. The forward velocity as well as the longitudinal and transverse velocity spread of this beam can be adjusted prior to injection into the ring. Using a phase-space description it is detailed how loading of a molecular packet into the storage ring can be performed optimally. Various techniques are used to investigate the motion of the molecules in the ring. The stored molecules can be observed for up to 50 round trips, at which point the packet fills up the entire ring. A design is presented for a sectional storage ring in which the longitudinal spreading of the packet inside the ring can be controlled.

1. Introduction

The art of experimental physics is to control the conditions under which experiments take place. Over the last decades tremendous progress has been made in control over the internal and external degrees of freedom of particles, ranging from ions (Dehmelt, 1990; Paul, 1990) to neutrons (Paul, 1990) and atoms (Chu, 1998; Cohen-Tannoudji, 1998; Phillips, 1998), and finally to molecules. In the experiments described in this paper, a packet of ammonia molecules in a single rovibrational state, spatially oriented, is decelerated and focused using time-varying electric fields and is subsequently stored in a hexapole storage ring. The stored molecules are observed after having made up to 50 round trips in the ring, corresponding to a flight path of 40 meters. The absolute velocity as well as the width of the velocity distribution of a molecular packet that is injected into the ring is under computer control. The level of control over neutral molecules demonstrated in this experiment holds unique possibilities for various experiments studying molecular properties and interactions. An overview of applications of trapped cold molecules in the fields of high-precision measurements and collision studies is given by Bethlem and Meijer (2003). For many of these applications confining molecules in a ring rather than in a more conventional trap has great advantages. In a ring, bunches of cold molecules can be made to interact repeatedly, at well-defined times and at distinct locations, with electromagnetic fields and/or other molecules. A sectional version of the storage ring allows one to design specific sections of the ring for specific tasks; sections can be optimized for instance for detection or for performing evaporative cooling. This great flexibility might be crucial to obtain the sensitivity required for performing various collision studies, for in-beam lifetime measurements and to obtain quantum degeneracy in a molecular gas. In a ring multiple packets of molecules can be stored simultaneously;

counter-propagating bunches can be stored and made to collide. The sensitivity obtained in such experiments increases linearly with the number of times that the bunches of molecules collide. In a ring, this can lead to enhancements of several orders of magnitude. The possibility to store multiple packets simultaneously in the ring, will make it possible to study molecules over extended times without making concessions to the repetition frequency at which the experiments runs; e.g., molecular packets can be stored for seconds while injection and detection takes place at 10 Hz. If evaporative cooling, or any other cooling mechanism, proves to be viable, molecular packets can be cooled in the ring and coupled out at any convenient moment in time.

Our paper is organized as follows. In Section 2 the interaction between molecules and electric fields is described and it is detailed how this interaction can be used to manipulate the trajectories of polar molecules. In particular, the Stark effect in deuterated ammonia and hexapole focusing of a beam of deuterated ammonia molecules are discussed in detail. There is a large similarity between the dynamics of polar molecules in inhomogeneous electric fields and that of ions in electric fields. Concepts that are developed to describe charged particle accelerators and storage rings can, therefore, with a suitable translation, be applied to decelerators and storage rings for polar molecules. The concepts of phase space and phase-space matching are introduced in Section 3. Section 4 starts with a theoretical description of the motion of molecules in a storage ring. The design of the storage ring for neutral molecules is presented, including its theoretical possibilities and limitations. The molecular beam machine—the injection beamline for the ring—is explained together with the detection scheme for the ammonia molecules that is used in the experiments. Finally, the experimental results obtained with a prototype storage ring are shown and discussed. In Section 5 an improved injection beamline for the storage ring is described. A so-called 'buncher' is inserted in the beamline to be able to adjust the longitudinal phase-space distribution of the molecular beam. Furthermore, the beamline is upgraded with additional hexapoles to keep the molecular beam transversely together. It is experimentally demonstrated that it is possible with this buncher to longitudinally focus and cool a molecular beam. In Section 6 we describe how this improved injection beamline can be used to inject molecules into the storage ring. The hexapoles and buncher make it possible to load more molecules into the ring at lower translational (longitudinal) temperatures resulting in higher densities of the stored molecules. In Section 7 the design of a sectional storage ring is presented. In such a storage ring a package of molecules can be kept together as buncher sections are part of the ring. These buncher sections can compensate for the tangential spreading out of the molecular packet in the ring. Calculations are performed to investigate the stability of a sectional storage ring and to determine the right parameters for operation of the ring.

2. Manipulating Polar Molecules

Neutral molecules are said to be polar and hence have an electric dipole moment when the (time-averaged) centers of positive charge density and negative charge density do not coincide (Debye, 1929). Classically, the dipole of a molecule can be viewed as one positive point charge $+q$ and one negative point charge $-q$ separated by a distance s (Fig. 1). The electric dipole moment $\vec{\mu}$ is defined as $\vec{\mu} = q\vec{s}$, having a magnitude of $q|\vec{s}|$ and pointing from the negative charge towards the positive charge. In the homogeneous electric field of Fig. 1 there is no net force on the dipole as the forces on the positive and negative charges cancel exactly. Only a torque $\vec{N} = \vec{\mu} \times \vec{E}$ will act on the dipole and the molecule will perform a librational motion along the electric field lines (Loesch and Remscheid, 1990; Friedrich and Herschbach, 1991). In an inhomogeneous electric field the forces on both sides of the dipole no longer cancel exactly. Depending on whether the orientation of the dipole is parallel or antiparallel to the field, the force will be towards higher or lower magnitudes of the electric field, respectively.

In quantum mechanics the picture is rather different. The quantized levels of a polar molecule are mixed due to the interaction of the dipole moment of the molecule with the electric field, and consequently, these levels will shift in energy. This effect was first observed in the hydrogen spectrum by Stark (1914) and is thereafter called the Stark effect. Not only the energy of the molecule is quantized but also its orientation with respect to an external electric field. This space quantization has first been demonstrated in the famous Stern–Gerlach experiment (Stern, 1921). In case of a symmetric top molecule, the orientation depends on the total angular momentum $\vec{J}$, of the quantum state that the molecule is in, its projection M on the space fixed axis defined by the electric field direction, and its projection K on the molecular symmetry axis.

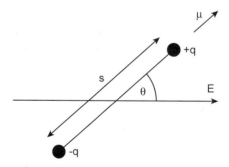

FIG. 1. Electric dipole $q\vec{s}$ in an homogeneous electric field $\vec{E}$. The dipole moment points from the negative charge towards the positive charge and makes an angle θ with the electric field.

2.1. THE STARK EFFECT IN DEUTERATED AMMONIA

For the experiments described in this paper an isotopomer of ammonia, ND_3, is used as a test molecule. Ammonia is the textbook example of a molecule that can undergo inversion, when the three hydrogen atoms tunnel simultaneously from one side of the nitrogen atom to the other. The energy barrier associated with this tunneling motion is shown in Fig. 2.

The horizontal lines indicate the levels of the ν_2 umbrella vibration of ammonia where the two lowest levels correspond to $\nu_2 = 0$ and $\nu_2 = 1$ for the case of a single potential well (barrier absent). Due to the finite barrier height, the levels are split into a symmetric level and an antisymmetric level and the energy difference between these two levels, the inversion splitting W_{inv}, is 0.053 cm^{-1} for ND_3 (van Veldhoven et al., 2002). The inversion splitting for normal ammonia, NH_3, is 0.79 cm^{-1} (Kukolich, 1967). The magnitude of the inversion splitting is important because it affects the linearity of the Stark effect, as will be discussed at length later. In the experiments described in this paper a beam of deuterated ammonia is formed in a supersonic expansion. Because of the adiabatic cooling in the expansion, only the lowest rotational levels of the vibronic ground state are populated and about 60% of all ND_3 molecules reside in the $|J, K\rangle = |1, 1\rangle$ level which is the ground state level of para-ammonia.[1]

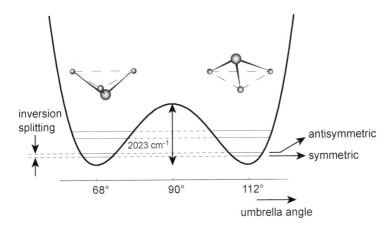

FIG. 2. Potential energy of the ν_2 (umbrella) mode of ammonia. The energy splitting between the two lowest levels is 0.053 cm^{-1} for deuterated ammonia.

---

[1] Deuterated ammonia, ND_3, is a boson and the Pauli exclusion principle demands that the total wave function including the nuclear spin has to be symmetric if two deuterium atoms are interchanged. Levels with nuclear spin wave functions of E symmetry are denoted para-ammonia while levels with A_1 or A_2 symmetry are denoted ortho-ammonia (van Veldhoven et al., 2002).

The Stark shifted energy levels are found by calculating the eigenvalues of the Stark Hamiltonian $H_{\text{Stark}} = -\vec{\mu} \cdot \vec{E}$. As the Stark shift is small compared to the vibrational and electronic energies in a molecule, only the rotational part of the molecular wave function has to be taken into account. For symmetric tops, such as ammonia, a full basis set of rotational wave functions is given by the rigid rotor wave functions $|J, K, M\rangle$ (Zare, 1987). The Stark operator has odd parity, so it is convenient to use a basis of rotational eigenfunctions with a definite parity by creating linear combinations of the rigid rotor wave functions $|J, K, M\rangle$:

$$|J, K, M, \pm\rangle = \frac{1}{\sqrt{2}}\left(|J, K, M\rangle \pm |J, -K, M\rangle\right). \tag{1}$$

The only nonzero matrix elements of H_{Stark} are then given by

$$\langle J, K, M \pm |H_{\text{Stark}}|J, K, M\mp\rangle = -\mu|\vec{E}|\frac{|MK|}{J(J+1)}, \tag{2}$$

$$\langle J, K, M \pm |H_{\text{Stark}}|J+1, K, M\pm\rangle$$
$$= -\frac{\mu|\vec{E}|}{J+1}\sqrt{\frac{((J+1)^2 - K^2)((J+1)^2 - M^2)}{(2J+1)(2J+3)}}, \tag{3}$$

with $|\vec{E}|$ the absolute electric field strength and μ the permanent electric dipole moment. At low electric field strengths and for molecules with a large separation of rotational energy levels, the first term, Eq. (2), dominates. In first order perturbation theory the Stark shifted levels of ammonia are found at energies given by

$$W_{\text{Stark}} = \pm\sqrt{\left(\frac{W_{\text{inv}}}{2}\right)^2 + \left(-\mu|\vec{E}|\frac{MK}{J(J+1)}\right)^2}, \tag{4}$$

where the electric dipole moment of ND_3 is $\mu = 1.497$ D (Debye units) (Gandhi and Bernstein, 1987) and the zero energy is taken at the center of the inversion levels. It is clear from this expression that at low electric fields, the inversion splitting gives rise to a quadratic dependence of the Stark shift on the applied electric field. The choice of ND_3 over NH_3 is made because the magnitude of the inversion splitting is significantly reduced upon deuteration. This leads to a more linear Stark effect in the range of electric fields used in the experiments. In addition, the kinetic energy that can be extracted from an ammonia molecule per electric field section in the Stark decelerator is larger for ND_3 than for NH_3 due to this decreased inversion splitting, as will be discussed in Section 2.3.

At higher electric field strengths and/or for heavier molecules with a closer spacing of rotational levels, different J-levels will start to mix and off-diagonal terms in the Stark matrix, Eq. (3), which couple different J-levels, have to be

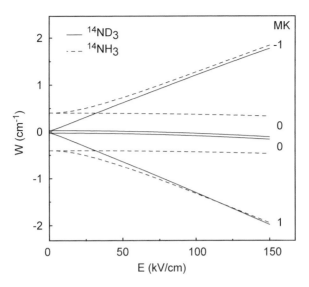

FIG. 3. Stark shift of ND$_3$ and NH$_3$ in the $|J, K\rangle = |1, 1\rangle$ level shown for electric fields up to 150 kV/cm. (Figure reproduced from (Bethlem et al., 2002a) with permission.) © 2002 The American Physical Society.

taken into account. Figure 3 shows the Stark shift of ND$_3$ for the $|J, K\rangle = |1, 1\rangle$ level. Coupling up to $J = 7$ is taken into account. It is clear from this figure that for electric fields strengths above 10 kV/cm and below 100 kV/cm the Stark shift is almost linear. For completeness, the Stark shift of NH$_3$ is plotted as well (dashed lines).

In this calculation, the hyperfine splitting due to the coupling of the total angular momentum $\vec{J}$ with the nuclear spins of the nitrogen and the deuterium (hydrogen) nuclei is neglected as it is much smaller than the Stark shift for the electric fields used in the experiments. Nevertheless, due to this coupling there are many hyperfine levels present. For instance, the $|J, K\rangle = |1, 1\rangle$ low-field seeking state of ND$_3$ consists of 48 separate hyperfine components in an electric field (van Veldhoven et al., 2002).

Molecules with a positive Stark shift, i.e., molecules whose Stark energy increases with increasing electric field strength as indicated by the positive sign in Eq. (4), will move towards lower electric fields in order to lower their potential energy. These molecules are therefore referred to as low-field seekers. Molecules with a negative Stark shift, i.e., molecules whose Stark energy decreases with increasing electric field strength, will move to higher electric fields in order to lower their potential energy. These molecules are therefore referred to as high-field seekers.

2.2. FOCUSING A BEAM OF POLAR MOLECULES

The Stark energy that a molecule acquires in an electric field is the potential energy of that molecule in the electric field. Therefore, the force on the molecule at a point $\vec{r}$ is given by the gradient of the Stark energy in that point:

$$\vec{F}(\vec{r}) = -\vec{\nabla} W_{\text{Stark}}(\vec{r}) \tag{5}$$

$$= \underbrace{-\left(\frac{\partial W_{\text{Stark}}}{\partial E}\right)}_{\mu_{\text{eff}}} \vec{\nabla} |\vec{E}(\vec{r})|. \tag{6}$$

Here, $-\partial W_{\text{Stark}}/\partial E$ can be regarded as the effective electric dipole moment μ_{eff} which includes the orientation of the dipole moment vector relative to the electric field vector. If we compare this equation to the equation for the force that a charged particle undergoes in an electric field, we see a large similarity. It is seen that the effective dipole moment of a polar molecule plays a role analogous to that of the charge of an ion, and that the gradient of the electric field strength takes over the role of the electric field when polar molecules, rather than charged particles, are manipulated. Therefore, polar molecules can be manipulated in a similar way as that ions can be manipulated, when one uses inhomogeneous electric fields acting on the effective dipole moment rather than homogeneous electric fields acting on the charge.

In contrast to the force on a charged particle in an electric field, the force on a polar molecule in an inhomogeneous electric field depends on the quantum state of the molecule. Inhomogeneous electric fields can therefore be used to select molecules in specific states. This has been used, for instance, to perform scattering experiments with state-selected beams and to perform very sensitive microwave spectroscopy. Inhomogeneous deflection fields act as a filter, rejecting molecules in unwanted states, but this does not necessarily lead to enhancement of the signal due to molecules in the selected state. An important improvement, therefore, was the use of magnetic and electrostatic lenses, focusing molecules in a selected state. Electrostatic focusing was developed independently by Gordon et al. (1954) and Bennewitz et al. (1955).

Electric multipole fields and, in particular, the hexapole field are convenient means to focus molecular beams of low-field seekers. An electrostatic multipole focuser consists of a number of rods placed equidistantly on the outside of a circle. The rods are alternately at ground potential and at high voltage (voltage difference V), creating a cylindrically symmetric electric field whose magnitude is zero at the centerline on the symmetry axis. The multipole focuser is placed such that the molecular beam axis and the symmetry axis of the multipole focuser coincide. Molecules in low-field seeking states will then experience a force towards the beam axis; molecules in high-field-seeking states, on the other hand, are defocused.

The electric potential of a multipole configuration obeys the Laplace equation in absence of any free charges. It can be derived by recognizing that, because the multipole potential does not depend on the coordinate along the molecular beam axis z, any complex analytical function $f(\zeta) = \phi(x, y) + i\psi(x, y)$, $\zeta = x + iy$ can be used to describe the equipotential lines. Generally, a $2n$-pole configuration can be derived from the complex function

$$f(\zeta) = A\zeta^n, \tag{7}$$

with n a positive integer and $A = V/2R^n$, a constant that depends on the voltage difference V between two adjacent electrodes of the multipole and on the inner radius R of the multipole. The real part of $f(\zeta)$, $\phi(x, y)$, defines the equipotential lines of normal multipoles whereas the imaginary part, $\psi(x, y)$, defines the equipotential lines of skew multipoles. The skew multipoles are rotated around the symmetry axis over an angle $2\pi/4n$ with respect to the normal multipoles. As the electrode surfaces of a physical multipole are in fact equipotential surfaces, the functions $\phi(x, y)$ and $\psi(x, y)$ give the physical shape of the electrodes, which is hyperbolic in nature.

The equipotential lines for an ideal hexapole are given for $n = 3$:

$$f(\zeta) = \underbrace{A(x^3 - 3xy^2)}_{\phi(x, y)} + i\,\underbrace{A(3x^2y - y^3)}_{\psi(x, y)}. \tag{8}$$

The electric fields of the normal hexapole and skew hexapole can be calculated by taking the gradient of the potentials $\phi(x, y)$ and $\psi(x, y)$, respectively. The magnitude of the electric field $|\vec{E}|$ is in both cases the same and is given by

$$|\vec{E}| = 3A(x^2 + y^2) = \frac{3V}{2R^3}r^2. \tag{9}$$

From Eqs. (2) and (9) it is clear that an hexapole is the ideal focuser for molecules having only a first order Stark effect. As the electric field and hence the Stark energy increases quadratically in r, the force on the molecules will be harmonic and molecules coming from a point are focused again into a point. As the electric field in a $2n$-pole focuser depends on r like $|\vec{E}| \propto r^{n-1}$, molecules with a quadratic Stark effect are better served with a quadrupole focuser.

For deuterated ammonia the harmonic force is slightly perturbed due to the zero field inversion splitting. The force exerted by the hexapole electric field on an ammonia molecule is calculated by inserting Eq. (9) into Eq. (4) and then applying Eq. (6):

$$\vec{F} = -\vec{\nabla}\left[\pm\sqrt{\left(\frac{W_{inv}}{2}\right)^2 + \left(\frac{1}{2}kr^2\right)^2}\right] \tag{10}$$

$$= \mp \frac{k^2 r^3 \vec{r}}{\sqrt{W_{\text{inv}}^2 + k^2 r^4}} \tag{11}$$

$$= \mp \frac{kr\vec{r}}{\sqrt{1+(\frac{W_{\text{inv}}}{kr^2})^2}}. \tag{12}$$

The upper sign corresponds here to low-field seeking states and the lower sign corresponds to high-field seeking states. In these equations k is the harmonic force constant given by

$$k = \mu \frac{3V}{R^3} \frac{|MK|}{J(J+1)}. \tag{13}$$

It is obvious from Eq. (12) that when the magnitude of the inversion splitting is zero, the molecule experiences a perfect harmonic force with an angular frequency given by $\omega = \sqrt{k/m}$. The inversion splitting leads to a reduction of the force near the hexapole axis compared to the situation without inversion splitting. Molecules flying closest to the hexapole axis are affected the most by this and are focused less than molecules flying further off-axis.

For an ideal hexapole the hyperbolic surfaces of the electrodes would nearly touch at large radii, leading to electrical discharges when a high voltage difference is applied between adjacent electrodes. To prevent this, and for ease of manufacturing, cylindrical electrodes are commonly used.

The hexapoles that are used in the experiments described in this paper are composed of six cylindrical rods equidistantly placed on the outside of a circle with radius R. In most experiments a voltage difference of 10 kV is applied to adjacent hexapole rods. The rods have a radius of $R/2$, following Reuss (1988). The electric field of the approximated hexapole geometry and the ideal hexapole field are shown in Figs. 4(a) and 4(b), respectively. The field in the approximated hexapole geometry is calculated using a finite element program (Dahl, 1995). The ideal field is calculated using Eq. (9). The contour lines of the electric field in our hexapole geometry show a hexagonal perturbation close to the electrodes. This deviation from the cylindrical symmetry has been experimentally observed in two-dimensional imaging experiments (Jongma et al., 1995).

A better approximation of the ideal hexapole field can be obtained for a rod radius of $0.565R$, as recommended by Anderson (1997). The contour lines of the electric field in the hexapole geometry as suggested by Anderson are more circular close to the electrodes, as shown in Fig. 4(c). This becomes clearer if one looks at the difference between the ideal field and the approximated fields. In Fig. 5 this difference is shown as a function of radial position r along the dashed lines indicated in the hexapole cross sections. It is seen from the figure that for $r > 0.5R$ the hexapole geometry as proposed by Reuss deviates more from the ideal field than the geometry proposed by Anderson. For a radial position r close to

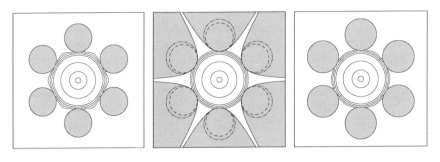

(a) Rod radius: $R/2$ (b) Ideal hexapole   (c) Rod radius: $0.565R$

FIG. 4. Lines of equal electric field strength in three different hexapole configurations. In all geometries the inner radius R is 3 mm and the potential on the electrodes is alternately at $+10$ kV and -10 kV. The lines of equal electric field strength are from the inside out: 1, 10, 40, 80, 90, 100, and 110 kV/cm. (a) Approximated hexapole with a radius of the rods equal to $R/2$. (b) Ideal hexapole with hyperbolic electrodes. (c) Approximated hexapole with a radius of the rods equal to $0.565R$. The contours of the rods in (a), (c) are shown in the ideal hexapole geometry (b) as well.

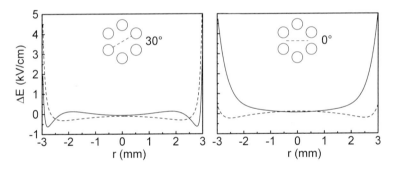

FIG. 5. Electric field difference between the ideal hexapole field and the fields of the approximated geometries as proposed by Reuss (1988) (solid line) and by Anderson (1997) (dashed line) are plotted as a function of the radial position r. Same hexapole settings are used as in Fig. 4. The differences are shown for two different cuts through the hexapole: $0°$ and $30°$.

the hexapole inner radius R, both approximate geometries deviate strongly from the ideal hexapole geometry. For future experiments it might be better to use the hexapole geometry as proposed by Anderson because it approximates the ideal hexapole field better. One has to take into account, however, that due to the larger radius of the rods the smallest distance between two adjacent electrodes is $0.435R$ compared to $0.5R$ for the geometry as proposed by Reuss. Hence, for the same voltage settings discharges are likelier to occur in the geometry as proposed by Anderson. If one wants to use a rod radius of $R/2$ then the effects due to the not cylindrically symmetrical fields close to the hexapole rods can be avoided by

Table I
Parameters used for trajectory calculations through the
hexapole.

$\lvert J, K\rangle$	$\lvert 1, 1\rangle$
μ	1.497 D
W_{inv}	0.053 cm^{-1}
velocity v_z	91.8 m/s
inner radius hexapole R	3 mm
voltage difference V	10 kV
mass, m	20 amu

making the hexapole radius R sufficiently large and the voltages on the electrodes sufficiently high.

To visualize the operation of the hexapole the equations of motion for molecules traveling through the hexapole have been solved and their trajectories have been calculated. Edge effects at the hexapole entrance and exit are neglected in the calculations. This is allowed as the hexapoles that are used are long compared to the length of the molecular packet and as they are rapidly switched on and off when the molecular packet is completely inside. In this case all the molecules experience a force during an equal time and 'chromatic aberration' of the hexapole lens is avoided. The trajectories of a few ND$_3$ molecules passing through an ideal hexapole oriented along the z-axis are numerically calculated and shown in Fig. 6. The molecules are assumed to originate from a single point. The values of the molecular and hexapole parameters that have been used for these calculations are given in Table I and are taken from the experiment. The force constant of the (ideal) hexapole is $k = 0.14$ cm^{-1}/mm^2 and the corresponding rotation frequency in phase-space is $\omega/2\pi = 1450$ Hz.

In the left figure the inversion splitting is set equal to zero and the molecular beam that is originating from a single point is focused into a single point again. The right figure demonstrates the effect of nonlinearities in the equations of motion resulting from a nonzero value of the inversion splitting. The trajectories in this figure show that molecules with inversion splitting moving close to the hexapole axis are less well focused than molecules that have no inversion splitting. The inversion splitting blurs the focus of the molecular beam: the sharp point like focus changes into a ring-shaped focus (cylindrically symmetric around the hexapole axis). This has been experimentally observed in two-dimensional imaging experiments (Jongma et al., 1995). The effect is similar to the spherical aberration caused by optical lenses that leads to caustics.

The hexapole described above focuses molecules in low-field-seeking states. Focusing of molecules in high-field-seeking states is more difficult. Maxwell's equations do not allow for a maximum of an electric field in free space (Wing, 1984; Ketterle and Pritchard, 1992a), and, therefore, molecules in high-field-

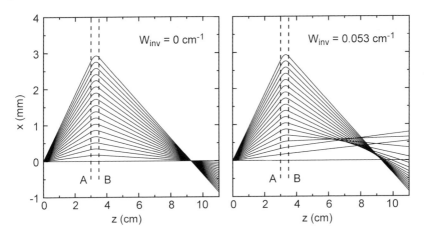

FIG. 6. Trajectories of ND_3 molecules passing through an hexapole with z along the molecular beam axis and x perpendicular to the beam axis. Curves are calculated for molecules without (left) and with (right) inversion splitting. It is assumed that all molecules start in one point and move freely towards the hexapole which is assumed to be ideal. The hexapole is switched on when the molecules are at position A and switched off when the molecules are position B. The field is on for a duration of 54 μs. A voltage difference of 10 kV is put between adjacent rods ($R = 3$ mm). The molecular packet is completely inside the hexapole when the field is applied. After the hexapole is switched off, the molecules continue in free flight.

seeking states have the tendency to crash into the electrodes, where the electric fields are the highest. A number of schemes have been demonstrated to overcome this difficulty, most notably by using a charged wire (Helmer et al., 1960; Chien et al., 1975; Loesch and Scheel, 2000) and 'alternate gradient' focusing (Auerbach et al., 1966; Kakati and Lainé, 1967).

2.3. DECELERATING AND TRAPPING OF POLAR MOLECULES

Molecules can be deflected and focused using fields that are inhomogeneous perpendicular to the molecular beam direction. If the electric field is inhomogeneous in the direction of the molecular beam, the longitudinal velocity of the beam will be altered. When molecules in a low-field seeking state enter a region of high electric field (see Fig. 7), they will gain Stark energy. This gain in Stark energy is compensated by a loss in kinetic energy. If the electric field is greatly reduced before the molecules have left this region, the molecules do not regain the lost kinetic energy and continue their motion at their instantaneous (reduced) velocities. The maximum amount of kinetic energy that can be taken out of a molecule depends on its difference in Stark energy in the high and low electric field region. For ND_3 molecules in the low-field seeking $|1, 1\rangle$ state the Stark shift in an

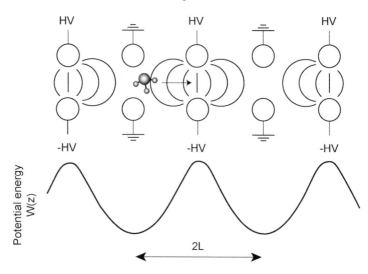

FIG. 7. Array of electric field stages. Each stage consists of two parallel cylindrical rods. The potential energy of a ND_3 molecule in the $|1, 1\rangle$ low-field seeking state is shown below as a function of the position z along the molecular beam axis.

electric field of 100 kV/cm is 1.2 cm^{-1}. The Stark shift of NH_3 molecules in the same state and electric field is only 0.9 cm^{-1}. The kinetic energy of a molecule in a supersonic beam is typically on the order of 100 cm^{-1}. So in order to make a significant change in kinetic energy, this deceleration process needs to be repeated by letting the molecules pass through an array of inhomogeneous electric field stages that are switched synchronously with the arrival time of the decelerating molecular packet. This process is referred to as Stark deceleration and the array of electric field stages is accordingly referred to as a Stark decelerator. This concept was already considered in the 1950s. However, attempts to demonstrate deceleration at MIT (Golub, 1967) or acceleration at the University of Chicago (Bromberg, 1972) remained unsuccessful. This idea was picked up 25 years later by our group. We demonstrated deceleration of metastable CO molecules in low-field-seeking states using an array of 63 switched electric fields (Bethlem et al., 1999, 2000b) and deceleration of metastable CO molecules in high-field-seeking using an array of 12 alternate gradient lenses (Bethlem et al., 2002b). Stark deceleration has now been applied to molecular beams of ND_3 and NH_3 (Bethlem et al., 2002a), OH (Bochinski et al., 2003, 2004; van de Meerakker et al., 2005), NH (van de Meerakker, 2006) and YbF (Tarbutt et al., 2004) as well as to Cs atoms (Maddi et al., 1999).

Molecules in low-field-seeking states are attracted to minima of the electric field and, when sufficiently slow, cannot escape from these minima. An electrostatic trap consisting of a ring electrode with two end caps was proposed by Wing (1980). The electric field is zero at the center of the trap and increases with distance from the center. This trap has been used to confine ammonia molecules by Bethlem et al. (2000a, 2002a) and OH radicals by van de Meerakker et al. (2005). Molecules in high-field-seeking states can be confined using a maximum of the electric field, however, as discussed earlier a maximum of the electric field is not allowed by the Maxwell equations in free space. By using a rotating saddle point of the electric field an AC trap can be devised similar to ion traps (Paul, 1990). Recently, trapping of deuterated ammonia molecules in high-field seeking states in an AC electric trap has been experimentally demonstrated (van Veldhoven et al., 2005). Rather than trapping molecules in a static point, molecules can also be confined in a ring structure, which is the topic of this review.

2.4. A STORAGE RING FOR POLAR MOLECULES

The first storage ring for charged particles was envisaged by Lawrence and Edlefson (1930) and demonstrated one year later (Lawrence and Livingston, 1931). In this so-called cyclotron a magnetic field delivers the centripetal force to keep the ions in circular orbits. In straight flight it is not possible to focus ions in both directions. In a cyclotron, however, the centrifugal force adds an extra term which stabilizes the motion of the ions. In a synchrotron the circular magnet is divided in two or more sections and, by applying a rf-voltage to these sections, the ions are accelerated and bunched in the gap (Veksler, 1944; McMillan, 1945). In such a 'weak' focusing storage ring the forces along the radial and the vertical direction are coupled; increasing the radial focusing force decreases the focusing force in the vertical direction, and vice versa. This posed a severe limitation. A major breakthrough was, therefore, the concept of 'strong'' or 'alternate gradient' focusing by Courant, Livingston and Snyder (Courant et al., 1952; Courant and Snyder, 1958). Given a certain magnetic field, the radius that is required to keep the ions confined in the ring depends on the energy of the ions. As the energy steadily increased, so did the size of the storage rings, culminating in the 27 km circumference of the LEP storage ring near Geneva. For an overview of the development of storage rings for charged particles see, for instance, (Livingston and Blewett, 1962; Humphries Jr., 1990; Conte and MacKay, 1991; Wilson, 2001).

Weak and strong focusing storage rings have their direct counterpart for neutral polar molecules. Polar molecules in high-field-seeking states can be trapped around a charged wire (Sekatskiĭ, 1995; Sekatskiĭ and Schmiedmayer, 1996; Jongma et al., 1997) or in electric alternate gradient rings (Auerbach et al., 1966;

Nishimura et al., 2004).[2] Since the force that can be exerted on neutral particles is typically 8–10 orders of magnitude smaller, the depth of these traps is rather small, on the order of 10 mK. In contrast to charged particles, neutral particles also have states which are attracted to minima rather than maxima of the field. A very simple storage ring can be devised by bending an electric (or magnetic)[3] multipole focuser into a circle and connecting the exit to its entrance. The trap depth for such a ring can be up to 1 K (Katz, 1997). We have recently demonstrated the first storage ring for neutral polar molecules (Crompvoets et al., 2001, 2004). In this paper we give a detailed account of the experiments that we have performed in this ring to date.

3. Manipulating Polar Molecules in Phase Space

In describing the dynamics of molecules in the ring it is very useful to look at the trajectories of the molecules in position–momentum space, known as phase-space. The Hamiltonian for a particle is given by $H(\vec{q}, \vec{p}, t)$, where $\vec{q}$ and $\vec{p}$ are the spatial coordinates and the canonical conjugated momentum coordinates, respectively, and t is the time. The Hamiltonian provides a complete description of the motion of a particle in phase-space and the equations of motion can generally be written as

$$\dot{\vec{q}} = \vec{\nabla}_p H, \tag{14}$$

$$\dot{\vec{p}} = -\vec{\nabla}_q H + \vec{Q}. \tag{15}$$

Here, $\vec{Q}$ is a force that is not derivable from a potential, such as frictional forces that depend on the velocity of the particle. $\vec{\nabla}_q$ and $\vec{\nabla}_p$ are the gradient operators for the three spatial coordinates and for the three momentum coordinates, respectively. For a large group of particles, a phase-space density function $\rho(\vec{q}, \vec{p}, t)$ is defined for the number of particles per unit volume in six dimensional phase-space. The time evolution of the phase space density describes how the motion of a group of particles evolves in time. It can be shown (Conte and MacKay, 1991) that when the non-Hamiltonian forces $\vec{Q}$ do not depend on the velocity, i.e., in the absence of dissipative forces that depend on the velocity and that change the

---

[2] Paramagnetic atoms in high-field-seeking states can be trapped around a current-carrying wire (Ketterle and Pritchard, 1992a; Schmiedmayer, 1995) or in a magnetic alternate gradient ring (Thompson et al., 1989).

[3] A magnetic hexapole ring was used to confine neutrons and used to determine their natural beta decay lifetime (Kügler et al., 1978, 1985). More recently cold rubidium atoms were held in a miniaturized magnetostatic storage ring (Sauer et al., 2001).

energy in the system, phase-space density remains constant;

$$\frac{d\rho}{dt} = 0. \tag{16}$$

This is the Liouville theorem (Liouville, 1838). This theorem implies that once you have a beam of particles, for instance a molecular beam, there is no way to increase or decrease the phase-space density of particles in the beam with conservative forces, i.e., forces that only depend on the position of the molecules. Two important properties of phase-space can be inferred from Liouville's theorem (Lichtenberg, 1969). First, different trajectories in phase-space do not intersect at any given instant of time. This is evident from the fact that in a Hamiltonian system the initial conditions and time uniquely determine the subsequent motion. Hence, if two trajectories crossed at a certain time, they would have the same position and momentum at that time and their subsequent motion would be identical. Second, a boundary C_1 in phase-space, which bounds a group of particles at time t_1 will transform into a boundary C_2 at time t_2, which bounds the same group of particles. This follows directly from the first property. As a consequence, the motion of a group of particles in phase-space can be calculated by tracking the boundary of that group in phase-space. In phase-space representation, the group of particles behaves as an incompressible fluid: the shape of the six dimensional volume can change but the volume itself cannot. This has major consequences for the molecular beam experiments described in this paper. The phase-space density is defined during the supersonic expansion and cannot be increased with forces generated by (time-varying) electric fields as these forces are not velocity dependent (Ketterle and Pritchard, 1992b). The momentum spread, which is a measure for the translational temperature of the molecules in the beam, can be lowered ('cooling') but only at the expense of the number density.

Liouville's theorem does not hold when dissipative forces act on the system of particles. In this case lowering the momentum spread, and hence the temperature, leads to real cooling as it is accompanied by an increase in phase-space density. This is for instance the case in laser cooling of atoms (Chu, 1998; Cohen-Tannoudji, 1998; Phillips, 1998), where the force on the atoms is velocity dependent due to the Doppler effect, and in evaporative cooling (Ketterle and van Druten, 1996), where the collisions between the particles and the subsequent escape of the fastest atoms from the trap give the required velocity dependent force. Both of these real cooling methods can increase the phase-space density considerably; evaporative cooling is the final cooling step needed for the creation of atomic Bose–Einstein condensates (Cornell and Wieman, 2002; Ketterle, 2002).

An interesting alternative method for cooling, a method that does not rely on a velocity dependent force, is stochastic cooling (van der Meer, 1972). Stochastic cooling is used to increase the brightness of charged particle beams in storage rings (van der Meer, 1985). Phase-space density is increased without violating Liouville's theorem by realizing that Liouville's theorem only holds for an infinite number of particles, i.e., a continuous distribution. For a finite number of particles the density is not continuous in phase-space and one has a discrete distribution of point-like particles surrounded by empty space. In this case phase-space may be distorted such that individual particles are displaced towards the center of the distribution and the empty space is squeezed outwards. The momenta of the particles are reduced (hence cooling) and the particle density is increased, leading to an increase in phase-space density near the center of the original volume. It must be noted that this is only valid in the classical regime where the individual particle wave packets can be ignored compared to the separation between the particles in phase-space. In experimental implementations of stochastic cooling, the fluctuations in the average positions and momenta of particles are measured at a certain time. A feedback signal is then sent to a correction element that damps these fluctuations. Stochastic cooling has been applied very successfully in charged particle storage rings, where it enabled the discovery of the W-boson and Z-boson, communicators of the weak interaction. In 1984 S. van der Meer was awarded the Nobel prize for his invention and implementation of stochastic cooling. Recently, proposals to apply stochastic cooling to atoms using laser detection schemes for feedback control have been put forward (Raizen et al., 1998; Ivanov et al., 2003). Similar schemes might be applicable to molecules as well.

It should be noted that in the description of the phase-space density function ρ and Liouville's theorem thus far, only the external degrees of freedom of a particle are taken into account. However, molecules also have many internal degrees of freedom. It is possible in principle to increase the external phase-space density, and thereby to lower the translational temperature, by decreasing the internal phase-space density. One example of translational cooling by heating internal degrees of freedom is adiabatic demagnetization (Reif, 1985) in which energy is extracted from a sample by means of a change in magnetic field strength. Another way to increase the phase-space density is based on the fact that the potential energy of a neutral particle in a trap depends on the quantum state of that particle. A proposal has been put forward to accumulate NH molecules in a magnetic trap (van de Meerakker et al., 2001). This can be accomplished by optically pumping the NH molecules from a nontrappable quantum state to a trappable quantum state via a unidirectional path using spontaneous emission from an intermediate state. Molecules already stored in the magnetic trap are not lost in this way and the phase-space density can be increased.

3.1. Phase-Space Matching

In the experiments presented in this paper, we want to load as many molecules as possible, at the highest possible density, and at the lowest possible tempera-

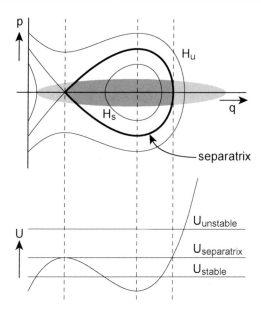

FIG. 8. The upper part of the figure shows the equipotential lines H in phase-space for an arbitrary one-dimensional potential well (shown below). The thick equipotential line is the separatrix: the boundary describing the acceptance of the potential well. The shaded ellipse represents the emittance of the incoming particles. The darker shaded area indicates the part of the emittance that is accepted by the potential well. The horizontal lines in the potential energy curve are the energy levels corresponding to (un)stable trajectories and trajectories on the separatrix in the phase-space plot.

ture from a supersonic expansion into the storage ring, i.e., we want to get the highest possible phase-space density in the storage ring. The methods employed to load the ring use time-varying electric fields. As phase-space density cannot be increased with time-varying electric fields, it is important to start with a source with a high initial phase-space density.

We start with a pulsed molecular beam created in a supersonic expansion. The initial molecular packet occupies a certain volume in phase-space. This initial phase-space volume is called the emittance of the molecular beam. According to Liouville's theorem, the emittance is constant (in absence of dissipative forces) but the shape of the emittance or the boundary of the phase-space volume may change.

The injection beamline for the storage ring consists of one or more sections that exert forces on the molecules. These forces arise from the interaction of the molecules with the electric field and as these forces are conservative they can be associated with potential wells. The equipotential lines of the potential well indicate the shape of the acceptance. The trajectories of the molecules that happen to be within the acceptance are closed orbits in phase-space and the molecules

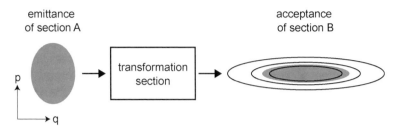

FIG. 9. Schematic representation of phase-space matching. The gray shaded emittance of section *A* is mapped onto the acceptance, or the equipotential lines, of section *B* using an interposed transformation section. Note that the volume in phase-space, i.e., the shaded area in this two-dimensional plot remains constant under this transformation.

perform an oscillatory motion in real space as well as in momentum space. Upon entering a section whose acceptance is smaller than the emittance of the molecular beam, the molecules outside the acceptance volume will be cut off and are lost from the beamline. Figure 8 shows this cutoff of an arbitrary emittance by the acceptance of an arbitrary one-dimensional potential well.

Not only the size but also the shape of the emittance at the exit of one beamline section must be mapped onto the shape of the acceptance of the next beamline section. This process is called phase-space matching (Lichtenberg, 1969). The principle of phase-space matching is shown schematically in Fig. 9. An interposed transformation section is necessary to transform the shape of the emittance at the exit of section *A* such that it fits the acceptance of section *B*. Phase-space matching is necessary to keep the phase-space density constant, especially in view of nonlinearities in potential wells. If the emittance is not properly matched onto the acceptance, filamentation of the phase-space distribution occurs and the effective phase-space volume becomes bigger. This is shown in Fig. 10(a). At $t = t_0$ particles with an elliptical phase-space distribution, the emittance, are coupled into a nonlinear potential well. It is immediately clear from this figure that the emittance does not match the acceptance of the potential well: both the shape of the emittance and its position in phase-space are wrong. The particles reside in the potential well until $t = t_1$ ($t_1 > t_0$). Due to the phase-space mismatch and the nonlinearity of the potential well, filamentation occurs. This leads to an increase of the *effective* phase-space volume, i.e., the envelope that encloses all (trapped) particles in phase-space becomes larger. The effective phase-space volume is indicated in the figure with the dashed contour line. Compared to the original phase-space volume, the effective phase-space volume is almost twice as large. This increase in effective phase-space volume does not occur when the emittance is properly matched onto the acceptance, as shown in Fig. 10(b). Here, the shape and position in phase-space of the elliptical phase-space distribution are altered with a transformation section, as schematically shown in Fig. 9, to match

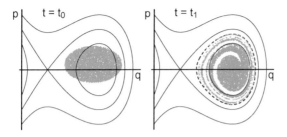

(a) Imperfect phase-space matching

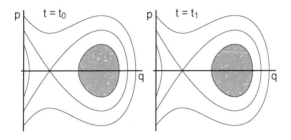

(b) Perfect phase-space matching

FIG. 10. The loading of a molecular packet into a potential well with imperfect (a) and perfect (b) phase-space matching.

the acceptance of the potential well at $t = t_0$. In this process the volume, i.e., the area of the distribution remains constant. Again, the particles reside in the potential until $t = t_1$. In this case, no filamentation occurs and the shape and volume of the phase-space distribution is unchanged: the phase-space density remains constant.

3.2. PHASE-SPACE TRANSFORMATIONS

There are two important and general transformations in phase-space: translation and rotation. Translation corresponds to linear free flight of a particle in real space, while rotation corresponds to the situation when a linear force is applied to a particle. As both transformations are linear in nature, the transformation of the position and velocity coordinates in phase-space can be conveniently expressed in matrix notation.

In free flight there are no forces acting on the particles and therefore the velocity spread is constant. For free flight during a time t with velocity v_z along the z-axis

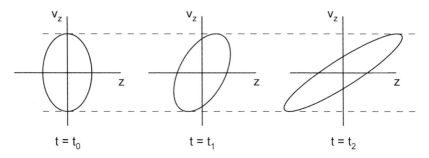

FIG. 11. Phase-space representation of free flight motion. Ellipses represent phase-space distributions in the z direction at $t = t_0$, $t = t_1$, and $t = t_2$ during free flight. The position spread grows but
the velocity spread remains unchanged as does the area of the ellipse.

the transformation of coordinates is given by

$$\begin{pmatrix} z_f \\ v_{z_f} \end{pmatrix} = \begin{pmatrix} 1 & t \\ 0 & 1 \end{pmatrix} \begin{pmatrix} z_i \\ v_{z_i} \end{pmatrix}, \tag{17}$$

where the subscripts i and f indicate the initial and final coordinates, respectively.
A phase-space representation of free flight of a package of molecules is shown
schematically in Fig. 11 for motion along the z-axis.

At $t = t_0$, the boundary of the distribution of molecules in phase-space is
indicated by an ellipse. Later, at $t = t_1$ the position spread is wider due to the
velocity spread and the ellipse is tilted and stretched. At $t = t_2$, the ellipse is tilted
and stretched even more. The area ('volume') of the ellipse remains constant in
this process.

If a group of particles resides in an harmonic potential well then a linear force
acts on the particles. The position of a particle in phase-space starts to rotate as
long as the force is present. This leads to a change in both the positions and the
momenta of the particles. For particles in an harmonic potential well the transformation of coordinates is given by

$$\begin{pmatrix} z_f \\ v_{z_f} \end{pmatrix} = \begin{pmatrix} \cos \omega t & \frac{1}{\omega} \sin \omega t \\ -\omega \sin \omega t & \cos \omega t \end{pmatrix} \begin{pmatrix} z_i \\ v_{z_i} \end{pmatrix}, \tag{18}$$

where the subscripts i and f denote the initial and final coordinates, respectively.
In this case the molecules rotate at a constant angular frequency ω and the rotation
is uniform or isochronous for all molecules in the potential well and the shape of
the phase-space distribution remains constant. Figure 12 schematically shows this
rotation of the distribution of molecules in phase-space.

Phase-space rotations are conveniently used to minimize the velocity spread
and hence to create a velocity focus, as shown in the figure at $t = t_1$. As the phase-
space volume remains constant, the position spread increases in this process.

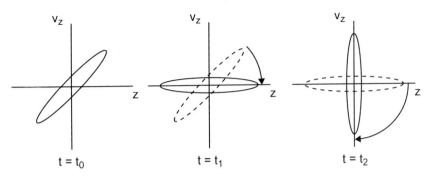

FIG. 12. Rotation in phase-space. Ellipses represent the phase-space distribution along one direction (z) at $t = t_0$, $t = t_1$, and $t = t_2$ during rotation in an harmonic potential well. The distribution rotates clockwise. The initial orientation of the distribution is shown at $t = t_0$. At $t = t_1$ the distribution is rotated such that the velocity spread is minimized. At $t = t_2$ the distribution is rotated further, such that the position spread is minimal.

Reducing the velocity spread is in fact reducing the translational temperature of the package of molecules but as the phase-space density is not increased, this is not 'real' cooling, opposed to the laser cooling and evaporative cooling schemes mentioned earlier. If the phase-space distribution is rotated 90° further then the position spread is minimized leading to a spatial focus. The velocity spread is thereby increased.

In practice, it is rather difficult to make perfectly linear transformations due to mechanical limitations and/or due to the nonlinear nature of the force acting on the particles. This will make the matching from one section to the next less than perfect, ultimately leading to a loss in phase-space density. In an anharmonic potential well, molecules near the edge of the well have a different oscillation frequency from molecules near the bottom of the potential well. Therefore, the distribution does not rotate uniformly in phase space but becomes distorted. In an harmonic potential well on the other hand, all molecules have the same oscillation frequency and the package of molecules rotates uniformly in phase-space.

In the hexapole focusing of deuterated ammonia discussed in Section 2.2, the nonlinearity introduced by the inversion splitting leads to distortion which hinders phase-space matching as shown in Fig. 13. In this figure, numerically calculated transverse phase-space distributions are shown at the four different positions along the beam axis (z-axis) shown in Fig. 6. From top to bottom: at $z = 0$, at the moment when the hexapole is switched on (position A), at the moment when the hexapole is switched off (position B), and near the focus of the molecular beam (at $z \approx 0.093$ m). Calculations are shown for molecules *without* (left column) and *with* (right column) inversion splitting. Due to the inversion splitting, the force near the hexapole axis is weaker and hence the rotation frequency in phase-space is lower. This leads to a spiraling of the phase-space distribution:

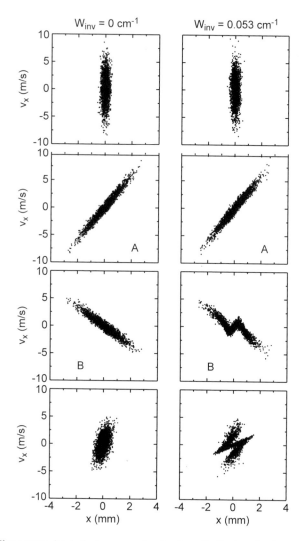

FIG. 13. Phase-space plots in the transverse direction for a distribution of molecules flying through an ideal hexapole. The left column shows the phase-space plots without inversion splitting and the right column shows them with inversion splitting. The upper row shows the initial distribution at $z = 0$. Full widths at half maximum are $\Delta x = 0.4$ mm and $\Delta v_x = 6$ m/s. The second row shows the distribution after free flight to the hexapole (position A in Fig. 6). The third row shows the distribution in phase-space after passing through the hexapole (position B in Fig. 6). Finally, the fourth row shows the distribution near the focus of the hexapole.

molecules further off-axis rotate faster than molecules close to the hexapole axis. As the calculation was done in one direction (x) the spiraling effect is rather pronounced. The effect is less pronounced when the calculation is performed in three dimensions as most molecules in the distribution then reside off-axis.

4. A Prototype Storage Ring for Neutral Molecules

In this section the design and first performance testing results of a storage ring for neutral polar molecules are described. The principle of the storage ring is based on the linear, electrostatic hexapole lens that transversely confines a beam of low-field seeking molecules. By bending such a linear hexapole around into a torus, a storage ring can be created (Katz, 1997). The electric field geometry of the storage ring that is actually used for the experiments, deviates from the commonly used cylindrically symmetric hexapole fields that have been discussed in Section 2.2. Instead, we have used a dipole configuration because this simplifies the detection of the ammonia molecules. An analytical description of the potential well and the motion of the molecules is very difficult in this configuration, however, and the trajectories are calculated numerically. In the experiment, decelerated bunches of deuterated ammonia molecules are injected into the storage ring and are detected after up to six times around the ring. Each bunch contains about 10^6 molecules in a single quantum state and has a translational temperature of 10 mK.

4.1. MOTION OF MOLECULES IN A HEXAPOLE RING

To understand the motion of the molecules in the storage ring it is useful to look first at a ring with an electric field that has an hexapole geometry as this can be described analytically by Eq. (12). Later, we extend this analysis for the geometry that is actually used with numerical calculations.

4.1.1. Equilibrium Orbit

As already mentioned, a storage ring can be created by bending an hexapole focuser around and by connecting its exit to its entrance. The electric field of the hexapole ring transversely confines low-field seeking molecules. In the forward, tangential direction they fly freely. The centripetal force required to keep the molecules in a circular orbit arises from the Stark interaction of the molecules in the inhomogeneous electric field that is created by putting high voltages on alternating rods of the hexapole. In Fig. 14 a cross section through the hexapole storage ring is shown.

The position of a molecule with respect to the center of the hexapole geometry is given by its radial coordinate r' and its vertical coordinate y' (the z coordinate

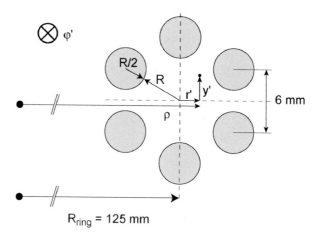

FIG. 14. Cross-sectional view of the hexapole storage ring.

is used for the direction along the molecular beam axis throughout this paper). The tangential coordinate ϕ' points into the plane of the paper. The position of a molecule with respect to the center of the ring is given by the radial coordinate ρ, the tangential coordinate ϕ, and the vertical coordinate y. The relation between r' and ρ is $\rho = R_{ring} + r'$, where R_{ring} is the distance from the center of the ring to the center of the hexapole. The coordinates with a prime indicate the system rotating with a angular frequency Ω at a distance R_{ring} around the origin. The coordinates without a prime indicate the inertial system fixed at the origin. It should be noted that $\rho \parallel r'$, $\phi = \phi'$, and $y = y'$.

When the hexapole inner radius, R, is small compared to the radius of the ring, R_{ring} (from the center of the ring to the center of the hexapole geometry), the electric field will be only slightly distorted. In the upper left of Fig. 15 the calculated contours of equal electric field strength are shown for a linear hexapole. An hexapole geometry is taken with an inner radius $R = 4$ mm and with a radius of the cylindrical rods equal to $R/2$. The voltage difference between adjacent electrodes is taken to be 10 kV in these calculations. The electric field is numerically calculated using Simion (Dahl, 1995). When this hexapole is bent into a large torus with $R_{ring} = 125$ mm, the electric field lines are hardly distorted (upper right). By reducing the torus radius to $R_{ring} = 20$ mm (lower left) or to the smallest radius possible $R_{ring} = 5.2$ mm (lower right), the electric field lines deviate more and more from the cylindrically symmetric electric field of the linear hexapole.

In the case of an hexapole electric field the force on the molecule is given by Eq. (12) and points radially inwards for a molecule in a low-field seeking state. The radial and vertical equilibrium position of the molecules can be calculated by

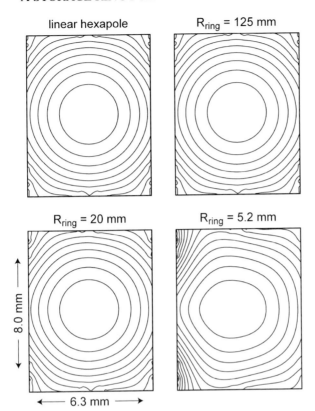

FIG. 15. Contour lines indicating the electric hexapole field for a linear hexapole (upper left), for a storage ring with a radius $R_{ring} = 125$ mm (upper right), with a radius $R_{ring} = 20$ mm (lower left), and with the smallest possible radius $R_{ring} = 5.2$ mm (lower right). For the latter radius, the inner ring electrodes are spheres. The inner radius $R = 4$ mm and the voltage difference between adjacent electrodes is 10 kV. The field of view that is shown has a width of 6.3 mm and a height of 8.0 mm. The center of the ring is to the left in the storage ring plots. In each graph the inner most electric field line is 5 kV/cm. Other field lines are shown at 5 kV/cm intervals.

balancing the forces in either one of these directions. In the vertical direction, the gravitational force is balanced by the vertical component of the hexapole force

$$mg = -\frac{ky'}{\sqrt{1 + \left(\frac{W_{iny}}{kr'^2}\right)^2}}, \tag{19}$$

where m is the mass of the molecule and g is the gravitational acceleration. In the horizontal plane, the centrifugal force is balanced by the radial component of the hexapole force

$$\frac{m v_\phi^2}{R_{\text{ring}} + r'} = \frac{k r'}{\sqrt{1 + \left(\frac{W_{\text{inv}}}{k r'^2}\right)^2}}, \tag{20}$$

where v_ϕ is the tangential or longitudinal velocity of the molecule in the ring. Solving Eqs. (19) and (20) simultaneously for r' and y' gives the radial equilibrium position, r_0', and the vertical equilibrium position, y_0', respectively. In the ideal (hypothetical) case, when the inversion splitting is zero and the hexapole force is truly linear, the equilibrium radius r_0' is found to be

$$r_0' = \frac{R_{\text{ring}}}{2} \left[\sqrt{1 + \left(\frac{2 v_\phi}{R_{\text{ring}}\omega}\right)^2} - 1 \right] \approx \frac{v_\phi^2}{R_{\text{ring}}\omega^2}, \tag{21}$$

where $\omega = \sqrt{k/m}$ and k is given by Eq. (13). For a ND$_3$ molecule in the low-field seeking $|J, K\rangle = |1, 1\rangle$ state with a tangential velocity $v_\phi = 91.8$ m/s one finds in this ideal case an equilibrium radius $r_0' = 1.90$ mm (or $\rho_0 = 126.90$ mm) and a vertical equilibrium position $y_0' = -0.28$ μm. When the nonzero inversion splitting of deuterated ammonia is accounted for, the equilibrium radius and the vertical equilibrium position are numerically calculated to be $r_0' = 1.95$ mm and $y_0' = -0.29$ μm, respectively. Consequently, the linear approximation is good enough to determine these equilibrium orbits. The effect of gravity on the motion of the molecules in the storage ring is seen to be small and will be neglected from now on. It is clear that for a given maximum attainable electric field strength, the radius of the storage ring, R_{ring}, is proportional to the square of the velocity v_ϕ. In other words, the maximum attainable electric field strength and the size of the ring set an upper limit on the tangential velocity of molecules that will be accepted. The availability of beams of slow neutral molecules, with a tunable absolute velocity (Bethlem et al., 1999; Gupta and Herschbach, 1999), is therefore an important asset in the performance testing of a storage ring for molecules.

An important consequence of the circular motion in the storage ring is that molecules with different longitudinal velocities v_ϕ have closed orbits at different equilibrium radii. If a molecule with velocity $(v_\phi)_0$ orbits at its equilibrium radius ρ_0 then a molecule with velocity $v_\phi = (v_\phi)_0 + \Delta v_\phi$ orbits at radius $\rho = \rho_0 + \Delta\rho$. Consequently, faster molecules orbiting the ring have a longer flight path while slower molecules have a shorter flight path. This difference in flight paths between faster and slower molecules counteracts the longitudinal spreading out of the molecular packet, and one could wonder whether this effect could be used to completely cancel this longitudinal spreading. For this, the difference in round trip time ΔT between the fastest and slowest molecule has to be zero. The ring is then said to be isochronous and all molecules arrive at the same time at the detection position. The round trip time of a molecule is given by $T = 2\pi\rho/v_\phi$.

The relative change in round trip time is then given by

$$\frac{\Delta T}{T_0} = \frac{\Delta \rho}{\rho_0} - \frac{\Delta v_\phi}{(v_\phi)_0} = \left[\frac{1}{\rho_0} \left(\frac{\partial \rho}{\partial v_\phi} \right) - \frac{1}{(v_\phi)_0} \right] \Delta v_\phi, \tag{22}$$

where the subscript 0 represents the molecule on the equilibrium orbit. In case of isochronous operation of the ring, the term between the square brackets of Eq. (22) has to be zero. Realizing that $(\partial \rho / \partial v_\phi) = (\partial r' / \partial v_\phi)$ and using Eq. (21) one finds for the linear case where the inversion splitting is zero

$$v_\phi = \frac{\omega^2 \rho_0 R_{ring}}{2 \sqrt{(v_\phi)_0^2 - \omega^2 \rho_0^2}}. \tag{23}$$

This solution is real only when $(v_\phi)_0 > \omega \rho_0$. When we introduce the angular vector $\vec{\Omega} = \Omega \vec{y}$ with which the molecules orbit in the ring, $(v_\phi)_0$ can be written as $\Omega \rho_0$. Equation (22) therefore implies that isochronous operation of the storage ring is only possible when $\omega < \Omega$. On the other hand the required identity in Eq. (20) can be rewritten in the linear case as $\Omega^2(R_{ring} + r') = \omega^2 r'$ which is only possible when $\omega \geq \Omega$. It can thus be concluded that an hexapole torus storage ring as such is not isochronous (unless $R_{ring} = 0!$); for isochronous operation of a storage ring the longitudinal momentum distribution in the ring needs to be manipulated actively.

4.1.2. Betatron Oscillations

When the molecules are stored inside the storage ring, they will oscillate around the equilibrium orbit. These oscillations are called betatron oscillations after the oscillatory motion of charged particles first studied in betatron storage rings. In the harmonic potential well, the oscillations in the vertical and radial (horizontal) direction are not coupled, but, due to conservation of angular momentum $\vec{L}$, the radial motion of the molecules is coupled to the forward, tangential motion; as $|\vec{L}| = m\rho v_\phi$, a change in ρ has to be accompanied by a change in v_ϕ (Goldstein et al., 2002). It is readily shown that the amplitude of the oscillation in the tangential velocity, Δv_ϕ, is proportional to the amplitude of the velocity in the radial direction with a proportionality constant of Ω / ω.

The acceptance of the potential well of the hexapole ring depends on the tangential velocity of the molecules. In the vertical direction the potential well is given by

$$W(y') = \frac{1}{2} \frac{k y'^2}{\sqrt{1 + \left(\frac{W_{inv}}{kr^2} \right)^2}}. \tag{24}$$

In the horizontal direction the potential well is the sum of the Stark energy in the electric field of the hexapole and a pseudo-potential energy W_{centri} associated with the centrifugal force:

$$W_{\text{centri}} = -\int \frac{m v_\phi^2}{R + r'} dr'. \tag{25}$$

When the constant of integration is set by our definition that $W_{\text{centri}} = 0$ for $r' = 0$, the potential well in the radial direction is given by

$$W(r') = \frac{1}{2} \frac{k r'^2}{\sqrt{1 + (\frac{W_{\text{inv}}}{k r'^2})^2}} - m v_\phi^2 \ln \left| 1 + \frac{r'}{R} \right|. \tag{26}$$

When the tangential velocity increases, the radial equilibrium orbit moves outwards and for a fixed electric field strength the potential well depth decreases. This is shown in Fig. 16. Here, the dashed line is the centrifugal potential energy which is zero at $r' = 0$. The solid line is a plot of the radial potential well at $y = 0$, as given by Eq. (26). The plots are shown for increasing tangential velocities. At $v_\phi = 0$ m/s the radial potential well is deepest while for velocities larger than approximately 135 m/s the potential well is nonexistent as the equilibrium orbit lies beyond the radius of the hexapole ($r_0 > 4$ mm).

Of course, as the tangential velocity increases and the radial potential well becomes less deep, the vertical potential well also becomes less deep. In principle, the acceptance runs up to a radius of 5.2 mm, right in between the ring electrodes where the electric field is maximum. The electrodes act like a mask, however, and cut away a part of the phase-space distribution. To assign a number to the acceptance, we will only take into account a radius of 4 mm, which is the radius of the circle enclosed by the ring electrodes. The rationale for this is that nonlinearities in the potential can lead to a coupling between the radial and vertical motion and a molecule originally accepted by the ring at a radius larger than 4 mm (in between the electrodes) will eventually crash onto one of the electrodes and can be assumed lost. So, the acceptance is limited by the positions of the ring electrodes.

The longitudinal acceptance is limited by the curvature of the ring and the radial position at which the molecules are coupled into the storage ring. For instance, molecules entering the storage ring with a tangential velocity of about 90 m/s at their radial equilibrium position of $r' = 1.9$ mm are accepted only over a length of 36 mm. The accepted velocity range runs from 0 m/s to about 135 m/s, at which point the molecules are no longer able to follow the curvature of the ring. The transverse acceptance depends on the tangential velocity, since for higher tangential velocities the transverse potential well decreases. For a tangential velocity of 90 m/s, the radial acceptance is calculated to be $[\Delta r \times \Delta v_r] = [4.2 \text{ mm} \times 12.9 \text{ m/s}]$ and the vertical acceptance is calculated to be $[\Delta y \times \Delta v_y] = [7.0 \text{ mm} \times 20.8 \text{ m/s}]$.

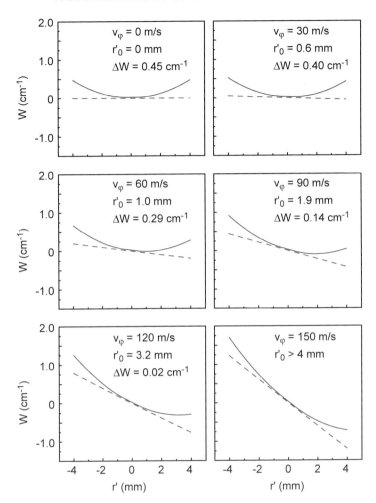

FIG. 16. Radial potential energy well (solid lines) for ND_3 in a hexapole storage ring for different tangential velocities. The potential well is taken at $y = 0$. The centrifugal potential energy is indicated with dashed lines.

4.2. MOTION OF MOLECULES IN THE DIPOLE RING

The actual electric field geometry in our hexapole torus storage ring deviates from the cylindrically symmetric hexapole fields. The reason for this is that apart from trapping the molecules inside the storage ring, we also need to be able to detect them. As the detection of ammonia is performed via a resonance enhanced ionization scheme we need to be able to extract the laser-produced ions from the ring,

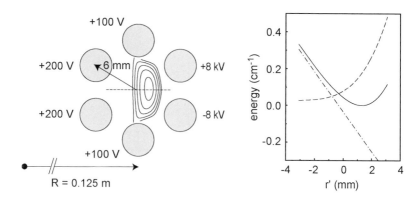

FIG. 17. Left: cross section of the hexapole storage ring in the dipole configuration. Contour lines of equal potential energy for ND$_3$ molecules in the $|J, K\rangle = |1, 1\rangle$ state with positive Stark-shift and with a tangential velocity of 89 m/s are shown at 0.02 cm^{-1} intervals. Right: the dashed curve shows the Stark energy as a function of the displacement from the center of the hexapole (r') in the plane of the hexapole ring (along the dashed axis as shown in the left part of the figure). The solid curve is obtained by adding the potential energy as a result of the centrifugal force for an ammonia molecule with a tangential velocity of 89 m/s (dot-dashed line) to the Stark energy (dashed line). The minimum of the resulting potential well is located 1.3 mm offset from the geometrical center of the hexapole. (Reprinted by permission from Nature (Crompvoets et al., 2001).)
© 2001 MacMillan Publishers Ltd.

which will be discussed in Section 4.3. A concomitant advantage of the electric field geometry used, is that transitions from trapped to untrapped states which can occur at zero electric field, so-called Majorana transitions, are avoided.

Figure 17 illustrates the voltages applied to the different rods of the ring and the equipotential lines in the well experienced by deuterated ammonia molecules in the low-field seeking $|1, 1\rangle$ state travelling with a tangential velocity of 89 m/s. The potential well is calculated in a similar fashion, as outlined above, for an ideal hexapole field. It is the sum of the Stark energy and the potential energy due to the centrifugal force. In this case an analytical expression for the electric field is not available and the electric field has been numerically calculated using Simion (Dahl, 1995). Although the shape of the potential well differs from the hexapole case, the motion of the molecules in this 'dipole' ring is in principle the same as for the hexapole ring: angular momentum is a conserved quantity and the molecules perform betatron oscillations around the equilibrium orbit. The difference is the introduction of additional nonlinearity in the equations of motion as well as an asymmetry in the shape of the potential well. The trajectories of the molecules in the ring are numerically calculated by correctly incorporating the Stark energy in the equations of motion. From Fig. 17 it is clear that the potential well is anharmonic (equipotential lines are not ellipses) and asymmetric. Therefore the betatron oscillations of the molecules have different frequencies in the vertical and

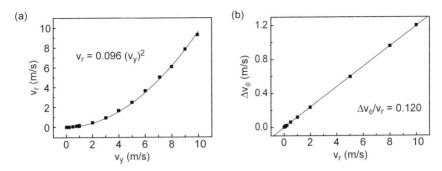

FIG. 18. (a) Coupling in the dipole ring between v_y and v_r. The plot shows the velocity amplitude in the radial direction as a function of the (initial) velocity amplitude in the vertical direction. (b) Coupling in the dipole ring between v_r and Δv_ϕ. The plot shows the velocity amplitude in the tangential direction as a function of the (initial) velocity amplitude in the radial direction.

radial direction. The frequencies also depend on the position of the molecules in the potential well. The oscillation frequencies at the equilibrium orbit in the vertical and horizontal (radial) direction for the potential well shown in Fig. 17 are 390 Hz and 970 Hz, respectively. This corresponds to 3.4 oscillations per round trip in the vertical direction and to 8.4 oscillations per round trip in the radial (and therefore also in the tangential) direction. These oscillation frequencies are found by numerical integration of the path that a molecule traverses through the ring. To stay in the harmonic region, the amplitude of the oscillations is chosen to be small in this simulation.

Larger amplitudes lead to strong coupling between the radial and vertical motion. On the left-hand side of Fig. 18 the result of a numerical calculation of the coupling between v_y and v_r for different initial values of v_y is shown. It appears that this coupling is quadratic in the dipole ring. The coupling between the radial and tangential motion is shown on the right-hand side of Fig. 18. The coupling constant, i.e., the slope of the graph, is numerically found to be 0.120 in good agreement with the expected value of $\Omega/\omega = 0.118$.

As for the hexapole storage ring the potential well of the dipole storage ring depends on the tangential velocity. Figure 19 shows the radial potential well as a function of the tangential velocity v_ϕ. In contrast to the potential well of the hexapole ring, the potential well of the dipole ring vanishes both for high and low tangential velocities. At high velocities the molecules cannot make the turn; the centrifugal force is larger than the force exerted by the electric field. At low velocities the potential well opens up at the inside of the storage ring because the Stark energy is a monotonically rising function of r'. The well is deepest around a tangential velocity of approximately 90 m/s. In contrast to the hexapole storage ring, the longitudinal velocity acceptance of the dipole ring has a lower

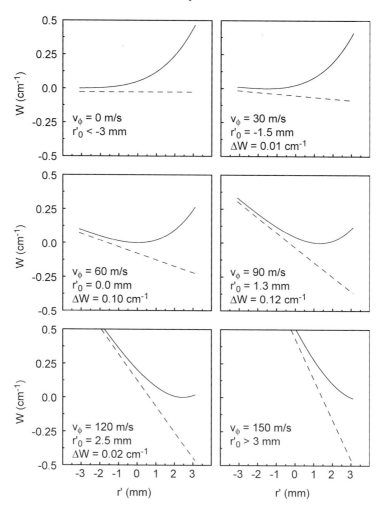

FIG. 19. Radial potential well ($y' = 0$) of the dipole storage ring as a function of the tangential velocity. The solid lines are the sum of the Stark energy and the potential energy due to the centrifugal energy (dashed lines).

bound of approximately 20 m/s; the upper bound is approximately 130 m/s. The longitudinal position acceptance is 44 mm. The transverse acceptance in the radial direction for an equilibrium radius of 1.3 mm (90 m/s) is found from numerical calculation to be $[\Delta r \times \Delta v_r] = [6.2 \text{ mm} \times 19.1 \text{ m/s}]$. If the cutoff due to the rods of the hexapole is taken into account, the effective position acceptance in the radial direction is limited to $\Delta r = 5.4$ mm. The acceptance in the vertical direction is found to be $[\Delta y \times \Delta v_y] = [7.5 \text{ mm} \times 11.5 \text{ m/s}]$.

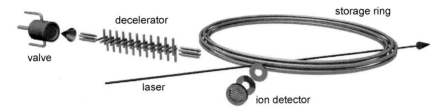

FIG. 20. From left to right: a gas pulse of less than 1% ND$_3$ in Xe exits a cooled ($T = 200$ K) pulsed valve with a mean velocity of 285 m/s. After passage through the skimmer the beam is collimated with a pulsed hexapole and sub-sets of these molecules are slowed down to velocities in the 76 m/s to 110 m/s range upon passage through the Stark decelerator. The bunches of slow molecules are tangentially injected into the electrostatic storage ring (12.5 cm radius; 4 mm diameter rods). Molecules in the detection region of the storage ring are ionized using a pulsed laser, after which they are extracted, and detected with an ion detector. The flight path of the molecules from the pulsed valve to the detection region in the storage ring is about 65 cm.

4.3. EXPERIMENTAL SET-UP

The experimental set-up shown schematically in Fig. 20 consists of a compact molecular beam machine that tangentially injects molecules into a 25 cm diameter hexapole torus storage ring. A beam of deuterated ammonia is formed by expanding a mixture of less than 1% ^{14}ND$_3$ seeded in Xenon through a pulsed valve into vacuum. The valve is operated at a temperature of 200 K, resulting in an average velocity of the molecular beam of approximately 285 m/s. The relative velocity spread in the beam is typically 20%, corresponding to a longitudinal temperature of 1.5 K. Due to adiabatic cooling in the expansion, only the lowest rotational levels in the vibrational and electronic ground state are populated in the beam. Roughly 60% of all the ND$_3$ molecules in the beam reside in the $|J, K\rangle = |1, 1\rangle$ level, the ground-state level for para-ammonia molecules. For the experiments reported here, only ammonia molecules in the low field seeking levels of the upper component of this inversion doublet (1/3 of the ground-state molecules) are used.

The ND$_3$ molecules pass through a 1.0-mm-diameter skimmer into a second, differentially pumped vacuum chamber and then move into a short hexapole with radius $R = 3$ mm that is pulsed to 10 kV and acts as a positive lens for the selected ND$_3$ molecules. Alternatively, one could leave the hexapole on for a much longer period, or even leave it on constantly at a lower voltage. Pulsing the hexapole at high voltage ensures that the force acting on the molecules is more linear because they are in the linear regime of the Stark shift and that all molecules are equally strongly focused. Another advantage is that it is possible to vary both the position and the strength of the hexapole lens by adjusting the timings when the hexapole is switched on and off. In addition, this approach bypasses the need for separate power supplies for each hexapole. The molecules are focused into a 35-cm-long

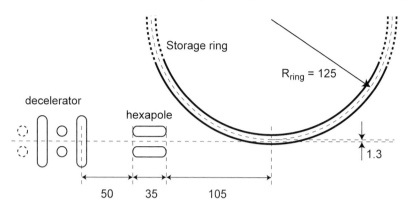

FIG. 21. Distances (in mm) between the Stark decelerator, the hexapole and the ring are shown schematically.

Stark decelerator containing an array of 64 equidistant electric-field stages oper-ated at $+10/-10$ kV. When the ammonia molecules in low field seeking levels enter a region of high electric field (up to 90 kV/cm), they will gain Stark energy. This gain in Stark energy ('potential' energy) is compensated by a loss in kinetic energy. If the electric field is greatly reduced before the molecules have left this region, they will not regain the lost kinetic energy. This process is repeated by letting the molecules pass through the array of electric field stages, which are switched synchronously with the arrival of the packet of decelerating molecules. The process in the Stark decelerator, the equivalent of a charged particle linear accelerator, can be viewed as slicing a bunch of molecules with both a narrow spatial distribution and a narrow velocity distribution (determined by the settings of the decelerator) out of the original beam, and decelerating these to arbitrarily low absolute velocities. In this process the phase-space density remains constant, and one can thus translate the high phase-space densities from the moving frame of the molecular beam to the laboratory frame. A detailed description of the de-celeration of ammonia using time-varying electric fields has been given elsewhere (Bethlem et al., 2002a).

A second pulsed hexapole, connected to a separate power supply, with a length of 35 mm and radius $R = 3$ mm is positioned approximately 50 mm downstream from the exit of the decelerator as shown schematically in Fig. 21. This hexapole focuses the decelerated molecules into the storage ring. The hexapole is switched on and off such that the signal of the molecules in the ring is maximized. The total distance from the exit of the decelerator to the ring is approximately 190 mm. The ammonia molecules enter the ring tangentially, passing through the 2-mm-wide 'slit' between two hexapole rods. Hence, only part of the transverse phase-space distribution will be coupled into the ring when the beam is too wide. The injection

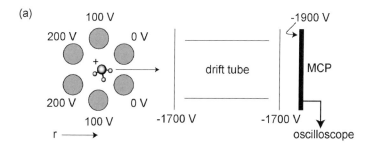

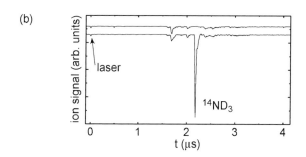

FIG. 22. (a) Time-of-flight mass spectrometer. The electric field created by the low voltages on the ring electrodes accelerates the positive ND_3-ions out of the ring. The ions are further accelerated towards a drift tube which is at -1700 V. The ions are detected with a multichannel plate detector, which is typically at -1900 V. The ion signal is recorded on an oscilloscope. (b) Typical time-of-flight mass spectrum taken with (lower trace) and without (upper trace) ammonia beam.

of the molecular beam into the ring is optimized by aligning the ring vertically and horizontally by means of mechanical actuators. As shown in Fig. 21 the ring is given an horizontal offset such that the molecular beam axis coincides with the radial equilibrium position of $r'_0 = 1.3$ mm for a tangential velocity of 89 m/s, which is actually used in the experiments.

The density of ND_3 molecules inside the storage ring is probed using the focused radiation of a pulsed laser at 317 nm, which ionizes in a $2 + 1$ resonance enhanced multiphoton process (single color) the ND_3 molecules that are in the selected quantum state (Ashfold et al., 1988; Bethlem et al., 2002a). Just before firing the laser, the high voltages on the outer two ring electrodes are switched off. Then the residual electric field created by the low voltages (100 V and 200 V) on the other electrodes accelerates the ions radially outwards as shown in Fig. 22. The storage ring itself serves then as the extraction region of a linear time-of-flight mass spectrometer. The ND_3-ions are detected with a multichannel plate detector which is at -1900 V and the ion signal is used as a measure for the density of neutral ammonia molecules in the ring.

4.4. RESULTS AND DISCUSSION

Fig. 23 shows measurements of the relative density of neutral ammonia molecules in the detection region of the storage ring, as a function of the time after the molecules left the pulsed valve. Without any voltage applied to the Stark decelerator (that is, without deceleration), the ammonia density in the storage ring peaks around 2.4 ms after the molecules left the valve. This is the time it takes molecules travelling with an average velocity of around 285 m/s to cover the approximately 65-cm distance between the pulsed valve and the storage ring. The velocity spread is approximately 20% (full width at half-maximum, FWHM), implying a translational temperature of around 1.5 K in the moving frame. With the Stark decelerator switched on and configured to decelerate molecules from 275 m/s to a final velocity of 100 m/s, a series of peaks occur at later times. These different bunches, whose velocity is indicated in Fig. 23, originate from the spatially extended molecular gas pulse at the entrance of the decelerator. The

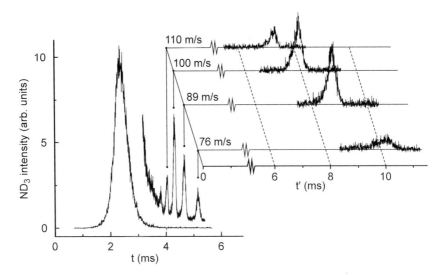

FIG. 23. Time-of-flight profiles. Density of ammonia molecules inside the storage ring as a function of time after opening of the pulsed valve (t), without (single large peak around 2.4 ms) and with the Stark-decelerator on (series of peaks at later arrival times). Note that the intensity of the decelerated bunches is on the same absolute scale as the original beam; the focusing effects in the decelerator typically enhance the signal on the beam axis by a factor of 5 compared to the situation that the decelerator is not switched on at all. The inset shows the density of ammonia molecules inside the storage ring measured as a function of the time after switching on the storage ring (t'). Bunches of decelerated molecules with four different center velocities are injected in the ring, with settings as shown in Fig. 17, and their time-of-flight distributions after the first round trip are recorded. (Reprinted by permission from Nature (Crompvoets et al., 2001).)

© 2001 MacMillan Publishers Ltd.

molecules that exit the decelerator with a velocity of 110 m/s were already in the second electric-field section of the decelerator when the decelerator was switched on, and so missed one deceleration cycle. The molecules that exit the decelerator with velocities of 89 m/s and 76 m/s are molecules that entered the decelerator later, with an initial velocity that was lower than the average beam velocity. The individual bunches are still clearly separated in the probe region, which is located 19 cm away from the exit of the decelerator. The velocity spread in the decelerated bunches of molecules is only 4–5 m/s (FWHM), corresponding to a translational (longitudinal) temperature of around 10 mK. The signal we obtain corresponds to about a hundred ND_3-ions being generated per laser pulse. As ionization of the ND_3 molecules takes place only in the focus of the laser beam, the actual detection volume in our system is about 10^{-4} cm^3. Each bunch contains a minimum of about 10^6 state-selected ND_3 molecules. When a selected bunch of decelerated ammonia molecules has entered the storage ring, high voltages are abruptly applied to the outer two hexapole rods as indicated in Fig. 17. The separation between the different bunches at the entrance region of the storage ring is sufficiently large that only one selected bunch of decelerated molecules is captured. The molecules are detected after they have been stored in the ring for a certain time.

The inset to Fig. 23 shows the density of neutral ammonia in the detection region of the storage ring as a function of time after switching on the ring. After separately injecting each of the four selected bunches of molecules, the corresponding time-of-flight distributions after one round trip are measured. The measurements show that ammonia molecules in the velocity interval from 76 m/s to 110 m/s are captured in the ring with the settings as applied. Molecules that move significantly faster do not experience a force large enough to keep them in a circular orbit, whereas molecules that move significantly slower will be deflected too much by the repulsive outer wall of the ring, and either hit the inner hexapole rods or escape from the storage ring on the inside otherwise. With the settings that have been used, the injection efficiency is optimum for velocities around 89 m/s, where the potential well in the radial direction is deepest and hence more molecules can be captured. Figure 24 shows the peak intensity of the ion signal for the first round trip ($t' = 8.76$ ms) as a function of the positive and negative high voltages on the outer ring electrodes. The dashed line indicates the background level, which corresponds to the signal intensity at 0 kV. The point in the graph on the left is taken when the laser is blocked. The signal drops rapidly when the voltages decrease from ±8 kV to ±6 kV. Lowering the trapping voltages results in an increase of the equilibrium radius and in a shallower potential well, in which the molecules are less tightly confined. As a consequence, less molecules are captured at a lower density.

Upon making successive round trips the bunch of molecules gradually spreads out in the tangential direction as a result of the residual velocity spread. The ex-

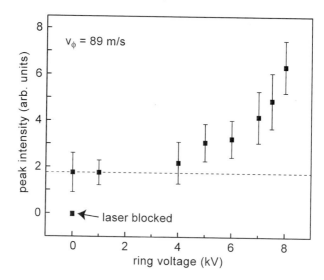

FIG. 24. Peak signal of the first round trip in the ring as a function of the applied voltages ($\pm$) to the outer ring electrodes.

perimental data in the upper trace of Fig. 25 demonstrate that a bunch of ammonia molecules with a velocity of 89 m/s can still be identified after it has made six round trips in the storage ring. The observed gradual broadening and decrease in peak intensity after each additional round trip is largely explained by the above-mentioned spreading out of the package along the ring, as substantiated by the results from trajectory simulations, shown in the lower trace of Fig. 25. Loss of molecules from the storage ring by collisions with background gas could explain the observed faster decay of signal at early times after the gas pulse, when the background pressure is still rather high.

5. Longitudinal Focusing and Cooling of a Molecular Beam

In the storage ring described in the previous section, molecules are confined in the transverse directions, however, they are essentially free along the tangential direction. As a consequence the density of the stored molecules drops rather quickly resulting in the observation of six round trips only. In this section we describe a method to reduce the tangential velocity spread by means of a so-called buncher (Crompvoets et al., 2002). In the buncher, a beam of polar molecules is exposed to an harmonic potential in the forward direction, i.e., along the molecular beam axis. This results in a uniform rotation of the longitudinal phase-space distribu-

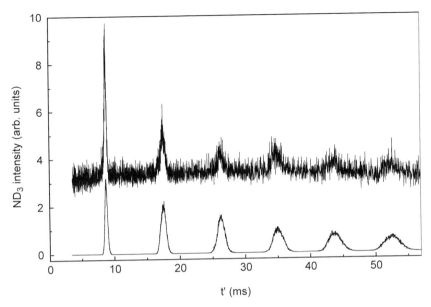

FIG. 25. Multiple round trips. The measurements shown in the upper trace show that bunches containing approximately a million ammonia molecules make up to six round trips in the storage ring. The packet of stored molecules is seen to gradually broaden and to decrease in peak intensity, mainly due to the tangential velocity spread (dispersion). In the lower trace the results of Monte Carlo simulations, calculating trajectories of molecules through the whole experimental set-up, are shown. From the simulations a translational temperature of 10 mK is deduced for the stored bunch of molecules. (Reprinted by permission from Nature (Crompvoets et al., 2001).)
© 2001 MacMillan Publishers Ltd.

tion of the ensemble of molecules. By switching the buncher on and off at the appropriate times, it can be used either to produce a narrow spatial distribution at a certain position downstream from the buncher or to produce a narrow velocity distribution. A reduction in the width of the spatial distribution is accompanied by an increase in the width of the velocity distribution, and vice versa, as the phase-space density remains constant in this process. In the field of charged particle beams, where bunching-elements are routinely used, these operations are commonly referred to as re-bunching and bunch-rotation, respectively (Humphries Jr., 1990). Proposals have been put forward to use time-varying magnetic fields for longitudinal focusing of atoms (Chu et al., 1986; Ammann and Christensen, 1997) and neutrons (Summhammer et al., 1986), both in real space and in velocity space. For atoms, where bunch-rotation is often referred to as 'δ-kick cooling', these operations have recently been experimentally demonstrated (Maréchal et al., 1999; Myrskog et al., 2000).

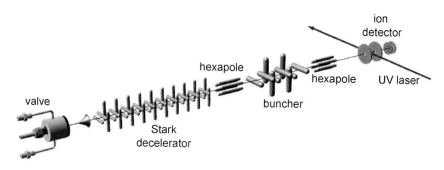

FIG. 26. Scheme of the experimental set-up. A pulsed beam of ammonia molecules is decelerated and passes through a buncher. The arrival time distribution of the packet of molecules at the laser interaction zone is recorded using a UV-laser based ionization detection scheme. (Figure reproduced from (Crompvoets et al., 2002) with permission.)
© 2002 The American Physical Society.

5.1. PRINCIPLE AND DESIGN OF THE BUNCHER

The molecular beam machine that is used for the longitudinal focusing and cooling experiments is schematically depicted in Fig. 26. A beam of deuterated ammonia is formed by expanding a ND_3/Xe mixture into vacuum. The molecular beam has a velocity of approximately 285 m/s and a longitudinal temperature of about 1.5 K. The molecular beam passes through a skimmer and enters the Stark decelerator, mounted in a second vacuum chamber.

In the experiments reported here, the Stark decelerator is operated at a phase angle of 70° creating a 1 mm long bunch of ammonia molecules with an average forward velocity of 91.8 m/s and with a longitudinal velocity spread of about 6.5 m/s at the exit of the decelerator. The calculated longitudinal phase-space distribution of this initially prepared molecular beam is shown in Fig. 27 at $t = 0$ ms. This is the distribution relative to the position in phase-space of the synchronous molecule, at the moment that the decelerator is switched off. At this moment, the synchronous molecule is located on the molecular beam axis (z-axis) 0.6 mm upstream from the center of the last electric field stage of the decelerator, and is moving with a velocity of $v_z = 91.8$ m/s along the z-axis. The entrance of the buncher is located some 15 cm downstream from the exit of the decelerator. A 5 cm long hexapole is installed between the decelerator and the buncher. The hexapole is switched on for a few tens of μs once the packet of ammonia molecules is completely inside it; the effective length of the hexapole is thus only a few mm. The hexapole maps the transverse phase-space distribution of the molecules in low field seeking states from the decelerator onto the buncher. Of course, the forward phase-space distribution remains unchanged. In going from the exit of the decelerator to the buncher the packet of ammonia molecules spreads out

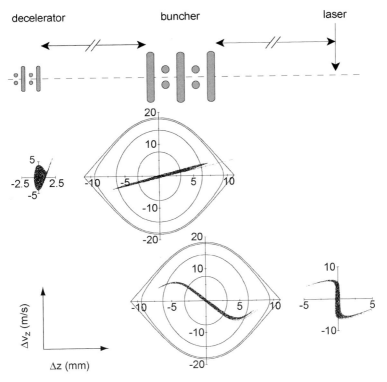

FIG. 27. Expanded view of the end of the decelerator, the buncher and the detection region. The calculated longitudinal phase-space distribution of the ammonia molecules is given at the exit of the Stark decelerator ($t = 0$ ms), at the entrance ($t = 1.743$ ms) and exit ($t = 2.100$ ms) of the buncher and in the laser detection region ($t = 2.652$ ms), relative to the position in phase-space of the synchronous molecule. Hexapoles are not shown in the figure. (Figure reproduced from (Crompvoets et al., 2002) with permission.)
© 2002 The American Physical Society.

along the molecular beam axis. This results in the elongated and tilted distribution in longitudinal phase-space as shown in Fig. 27 for $t = 1.743$ ms, the time that the buncher is switched on.

The buncher consists of an array of five electric field stages, with a center-to-center distance along the molecular beam axis of 11 mm. Each stage consists of two parallel cylindrical metal rods with a diameter of 6 mm, centered at a distance of 10 mm. One of the rods is connected to a positive and the other to a negative switchable high-voltage power supply. Alternating stages are connected to each other. As the electric field close to the electrodes is higher than that on the molecular beam axis, molecules in low-field seeking states will experience a force focusing them towards this axis. This focusing occurs only in the plane perpendicular to the electrodes. By alternately orienting the electric field stages horizontally and vertically, molecules are focused in either transverse direction.

When the synchronous molecule is exactly in between the first and second electric field stage of the buncher, the high voltage on the odd numbered electric field stages is switched on. The synchronous molecule is now on the downward slope of a potential well, and will be accelerated. When the synchronous molecule is 11 mm further, on the upward slope of the potential well, it is decelerated again to its original velocity. At this time the high voltage on the odd numbered stages is switched off. Molecules that are originally slightly ahead of the synchronous molecule, i.e., molecules that are originally faster, will spend more time on the upward slope than on the downward slope of the potential well, and are therefore decelerated relative to the synchronous molecule. Molecules that are originally slightly behind the synchronous molecule, i.e., molecules that are originally slower, will spend less time on the upward slope than on the downward slope of the potential well, and are therefore accelerated relative to the synchronous molecule. This process can be repeated, by switching on the high voltage on the even numbered stages when the odd numbered stages are switched off. Evidently, this can be done for a maximum of three times in our present buncher. The electric fields in the buncher are designed such, that the (series of) potential well(s) experienced by the ammonia molecules along the molecular beam axis is harmonic over an interval of approximately 1 cm around the minimum of the well. It is noted that the operation principle of the buncher is equivalent to that of the Stark decelerator at a phase angle of $0°$ (Bethlem et al., 2000b). The electrode geometry is, in fact, the same as the one of the decelerator except that it is scaled up by a factor two. This scaling-up increases the acceptance in position space by a factor of two but in momentum space the acceptance is reduced by a factor of two for the same voltages on the electrodes. The rotation frequency of the distribution in phase-space consequently changes as well. At a phase angle of $0°$, the rotation frequency is now 600 Hz.

5.2. Longitudinal Focusing of a Molecular Beam

During the time that the buncher is on, the longitudinal phase-space distribution is rotated in the (z, v_z)-plane. The calculated distribution at $t = 2.100$ ms, i.e., at the time that the buncher is switched off, is shown in Fig. 27 for the case that a series of three potential wells is used in the buncher with voltages of $+9.5$ kV and -9.5 kV applied to the electrodes. In this figure, the contours of equal energy for the ammonia molecules in the potential well are shown, relative to the position in phase-space of the synchronous molecule. It is seen that the rotation in phase-space is uniform, i.e., that the potential is harmonic, near the center and that the rotation is slower further outward, reaching zero at the separatrix. For optimum performance, the experiment has to be designed such that, for the most part, the harmonic part of the acceptance of the buncher is used. At the time that the buncher is switched off, slow molecules are ahead while fast molecules are lagging behind, leading to a longitudinal spatial focus further downstream. In the particular situation depicted in Fig. 27, the ammonia beam has a longitudinal focus with a width of about 0.5 mm, some 4.5 cm behind the buncher.

In the experiments the ammonia beam is probed either at a distance of 4.5 cm or at a distance of 33 cm downstream from the buncher. In the latter case, a second pulsed hexapole is used to transversely focus the ammonia beam exiting the buncher into the laser detection region. Using a $(2 + 1)$-resonance enhanced multi photon ionization scheme with a pulsed laser around 317 nm, $^{14}ND_3$ molecules in the upper component of the $|J, K\rangle = |1, 1\rangle$ inversion doublet are selectively ionized. Mass-selective detection of the parent ions is performed in a short linear time-of-flight set-up. The ion signal is proportional to the density of neutral ammonia molecules in the laser interaction region.

In Fig. 28 the measured time-of-flight (TOF) distributions for ammonia molecules over the 52.8 cm distance from the exit of the decelerator to the laser detection region (33 cm behind the buncher) are shown for various voltages on the buncher. The vertical scale is the same for all measurements, and the measurements are given an offset for clarity. With the buncher switched on for 239 μs, i.e., when two bunching stages are used, a longitudinal spatial focus is obtained with a voltage of $+5.75$ kV and -5.75 kV on the buncher. From the observed width of the TOF distribution a longitudinal spatial focus of about 2.0 mm is deduced; the magnification of the packet exiting the decelerator is about a factor 2, which is what is expected with the thin-lens approximation for this geometry. With higher voltages, the beam is over-focused, resulting in a broader TOF distribution. Approximately 30% of the molecules are within the full width half maximum of the TOF distribution when the buncher is operated at 5.75 kV. This is clear from the time-integrated signal of the bunched (5.75 kV) signal as shown in the upper graph of Fig. 28. In the inset, the measured TOF distributions recorded 4.5 cm behind the buncher (24.3 cm from the exit of the decelerator) and using three bunching stages are shown. A considerably tighter longitudinal spatial focus is now obtained with voltages of $+9.5$ kV and -9.5 kV on the buncher. The observed 6.0 μs width of the TOF distribution corresponds to a longitudinal spatial focus of approximately 0.5 mm; the calculated phase-space distributions for this particular situation are those shown already in Fig. 27. The observed TOF distributions are quantitatively reproduced in trajectory calculations, shown underneath the experimental data.

5.3. Longitudinal Cooling of a Molecular Beam

It is evident from the description of the operation of the buncher that it is also possible to rotate the phase-space distribution of the bunch such that an elongated

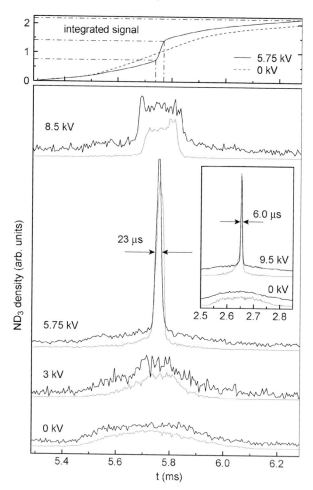

FIG. 28. Lower graph: density of ammonia molecules 33 cm behind the buncher as a function of time (at $t = 0$ ms the synchronous molecule is at the exit of the decelerator) for different voltages on the buncher, using two bunching stages. In the inset similar measurements recorded at a 4.5 cm distance behind the buncher, using three bunching stages, are shown. The results of trajectory calculations through the entire set-up are shown underneath the experimental data. Upper graph: integrated signals of bunched (5.75 kV) and unbunched (0 kV) molecular package. (Figure reproduced from (Crompvoets et al., 2002) with permission.)
© 2002 The American Physical Society.

horizontal distribution is formed, i.e., that a focus is created in velocity space. The optimum settings for longitudinal cooling of the molecular beam can be deduced from the experimentally determined settings for spatial focusing.

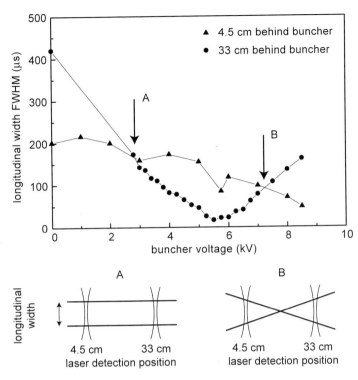

FIG. 29. Longitudinal width (FWHM) of the molecular packet as a function of the applied buncher voltage at 33 cm (circles) and 4.5 cm (triangles) behind the buncher. Arrow A indicates the voltage crossing at which the beam is in principle longitudinally collimated. Arrow B indicates the voltage crossing when there is a focus in between the two measurement positions. The situations at arrow A and arrow B are shown schematically in the lower part of the figure.

In Fig. 29 the widths (FWHM) of the measured TOF distributions are shown as a function of the applied buncher voltage. The data indicated with circles are measured 33 cm behind the buncher. The data indicated with triangles are measured 4.5 cm behind the buncher. The longitudinal spatial focus measured at 33 cm behind the buncher can be observed clearly in the corresponding set of measurements. The two sets of measurement points cross twice. At the voltages at which these crossings occur, the longitudinal width of the molecular packet is the same. For the higher voltage crossing, the package is spatially focused halfway between the two measurement positions at 4.5 cm and 33 cm behind the buncher. For the lower voltage crossing, the longitudinal velocity of the molecular beam is focused and the molecular packet hardly spreads out in the forward direction; the width is practically the same at both positions. The experimentally determined voltage settings where the two sets of measurements cross for the first time enable one to

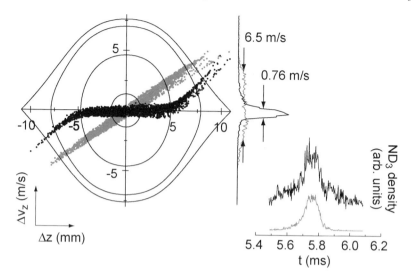

FIG. 30. Calculated longitudinal phase-space distribution at the moment that the buncher is switched on ($t = 1.743$ ms) and off ($t = 1.982$) with voltages of +3.8 kV and −3.8 kV. The projections of the longitudinal velocity distributions on the v_z-axis are given for both cases. On the right-hand side, the measured (upper trace) and calculated (lower trace) TOF distributions 33 cm behind the buncher are shown. (Figure reproduced from (Crompvoets et al., 2002) with permission.) © 2002 The American Physical Society.

find the settings for the production of a molecular packet with the lowest possible longitudinal velocity spread, i.e., with the lowest possible longitudinal temperature.

In Fig. 30 the calculated longitudinal phase-space distribution for longitudinal cooling is shown at the moment when the buncher is switched on and off. In this case two bunching stages are used, and voltages of +3.8 kV and −3.8 kV are applied. These voltages are somewhat higher than those found from the measurements presented in Fig. 29. This is due to the difficulty in determining the longitudinal widths of the molecular packet close to the buncher. Near the buncher the TOF distribution is a convolution of the bunched part of the molecular packet with the unbunched part of the packet. This unbunched part still has a large density directly behind the buncher and hence contributes much to the measured longitudinal widths. This leads to an overestimate of the width of the bunched part of the molecular packet.

The projection of the phase-space distribution onto the vertical axis after bunch-rotation gives a longitudinal velocity distribution with a full width at half maximum of 0.76 m/s. This corresponds to a record-low longitudinal temperature of our molecular beam of 250 μK. The measured and calculated TOF distributions 33 cm behind the buncher are shown in the figure as well.

6. Dynamics of Molecules in the Storage Ring

In the previous section longitudinal focusing of a beam of polar molecules was demonstrated. In combination with electrostatic multipole lenses, full six-dimensional phase-space matching of one element onto another is possible. In this section the buncher is used to decrease the longitudinal (tangential) velocity spread of the molecular packet prior to injecting them in the ring. The density of the molecular packet in the ring therefore drops less quickly than in the injection scheme without a buncher. Furthermore, by properly matching the transverse emittance of the Stark decelerator onto the transverse acceptance of the ring with hexapole lenses, molecules are coupled into the ring more efficiently, resulting in more observable round trips than in previous experiments, as reported in Section 4. Together, the buncher and hexapoles enable us to control the full six-dimensional phase-space distribution of the molecular packet entering the ring: with the buncher we can control the longitudinal phase-space distribution while with the hexapole we can independently control the transverse phase-space distribution.

6.1. EXPERIMENTAL SETUP AND ALTERNATIVE BUNCHING SCHEME

The setup, consisting of the injection beamline and the storage ring, is depicted schematically in Fig. 31. A beam of deuterated ammonia is formed by expanding a ND_3/Xe mixture into vacuum. After the beam has passed through a skimmer into a second vacuum chamber the ammonia molecules in low-field seeking states of the upper level of the inversion doublet are transversely focused with a pulsed electrostatic hexapole lens into the Stark-decelerator. The ammonia molecules are decelerated to 91.8 m/s for the experiments described here.

At the exit of the decelerator the calculated longitudinal phase-space distribution is $[\Delta z \times \Delta v_z] = [1.1 \text{ mm} \times 5.9 \text{ m/s}]$ and the transverse phase-space distributions are $[\Delta x \times \Delta v_x] = [\Delta y \times \Delta v_y] = [1.0 \text{ mm} \times 5.0 \text{ m/s}]$, where z lies along the molecular beam axis. Here the position spread and velocity spread are the full widths at half maximum (FWHM) of a fitted Gaussian distribution. The transverse phase-space distribution is mapped onto the acceptance of the ring using a telescope of two electrostatic hexapole lenses. The acceptance of the ring is in fact set by the vertical (y) and radial (r) eigenfrequencies of the potential well which are numerically found as $f_y = 390$ Hz and $f_r = 910$ Hz near the minimum of the well for the voltage settings used. The relation between the frequency f_s,

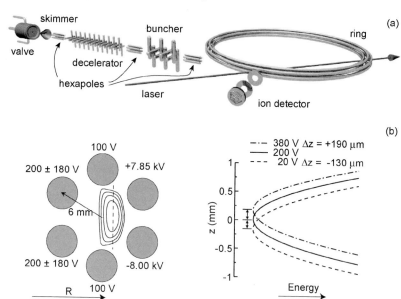

FIG. 31. (a) Scheme of the experimental set-up. The injection beamline consists of a pulsed valve, a Stark decelerator, hexapoles, and a buncher. (b) The molecules are trapped in a 2-dimensional potential well. The center of the ring is to the left. Equipotential lines are shown at 0.02 cm^{-1} intervals. Molecules are detected in the ring using a UV-laser based ionization detection scheme. On the right the vertical position of the potential well along the equilibrium radius is shown for three values of the (AC modulation) voltage on the inner ring electrodes. (Figure reproduced from (Crompvoets et al., 2004) with permission.)

the accepted position spread Δs, and the accepted velocity spread Δv_s is given by $\Delta v_s = 2\pi f_s \Delta s$. The potential well shown in Fig. 31 contains both the Stark energy and the quasi-potential centrifugal energy for molecules moving in circular orbits with a tangential velocity of 91.8 m/s. It is evident from the equipotential lines in Fig. 31 that the potential well is asymmetric in the dipolar configuration that we have used. To load the ring the hexapoles rotate the transverse phase-space distribution of the molecules uniformly such that it matches the vertical acceptance of the ring, since in this direction the potential well is the shallowest. For this, the hexapoles are switched on for only a few tens of microseconds, corresponding to an effective length of a few millimeters for molecules moving with a velocity of 91.8 m/s. As explained in Section 5, the buncher creates an harmonic potential well that rotates the longitudinal phase-space distribution until the longitudinal velocity spread has been minimized. The buncher and the hexapoles are offset by a few hundred volts, creating a sufficiently large residual electric field to prevent Majorana transitions. In addition, possible transitions between low-field

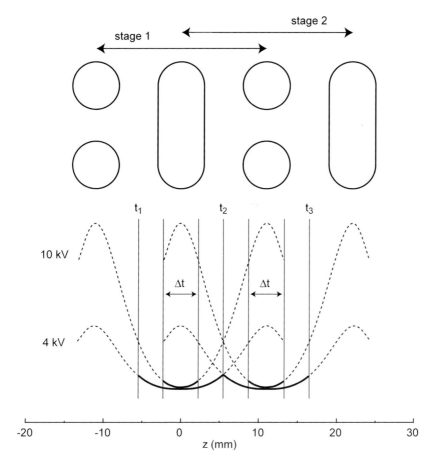

FIG. 32. Comparison of the longitudinal potential wells and their duration for the two bunching methods. The potential wells (thick lines) and the timings for longitudinal cooling are shown for the synchronous molecule. The position of the buncher electrodes is schematically shown above the potential wells. In the bunching method demonstrated in Section 5, stage 1 of the buncher is switched on at t_1. At t_2 stage 1 is switched off and stage 2 is switched on until t_3. In the alternative bunching method an extra time delay is added/subtracted to the timings, shortening the length of the potential well symmetrically around its minimum.

seeking and non low-field seeking hyperfine levels that can be induced by rapid switching of the high voltages, are avoided by this offset field as well.

Here, we describe and demonstrate an alternative bunching scheme. In the bunching method detailed in the previous section, the length of the longitudinal potential well was kept constant while the amount of rotation of the molecules in longitudinal phase-space was controlled by adjusting the force constant. The latter was accomplished by varying the voltages applied to the buncher electrodes. The advantage of this method is that the time intervals between switching can be kept constant; these are simply the length of the buncher stage divided by the velocity of the synchronous molecule. It is often easier to adjust a voltage than to adjust several timings.

In the experiment described here, the hexapoles before and behind the buncher and the buncher itself are electrically connected to the same positive voltage power supply. In this case only one voltage setting can be used and hence the timings for the buncher and hexapoles have to be adjusted in order to obtain good longitudinal and transverse focusing, respectively. The hexapoles operate optimally at high voltages as has been discussed in Section 2. The voltages applied to the buncher that are necessary for longitudinal cooling, though, are only about ± 4 kV. Hence, in this case the voltage difference between the rods of the hexapoles is only 4 kV and the nonlinearity in the focusing properties of the hexapoles due to the inversion splitting is more pronounced. By operating the hexapoles, and therefore also the buncher, at fixed high voltages of ± 10 kV, the electric fields are sufficiently high such that the nonlinearity due to the inversion splitting can be neglected. This means, however, that in the buncher the longitudinal phase-space distribution always rotates with a fixed angular frequency. By shortening the longitudinal potential well of the buncher symmetrically around its minimum by adjusting the switching times, i.e., by adding and subtracting time-delays symmetrically around the original timings, one can alter the angle over which the longitudinal phase-space distribution of the molecules rotates. The potential well is then switched on for a time duration Δt as shown in Fig. 32. In this way it is also possible to longitudinally cool a molecular packet. Figure 33 shows the distributions of molecules in phase-space at different positions along the beamline in the case of longitudinal cooling. It should be noted that by shortening the potential well in the buncher, the molecules will move freely in between the potential wells. As free flight is a linear transformation, this will not lead to an increase in effective phase-space volume.

The bunching scheme, in which the buncher is operated at constant voltage and variable duration, is compared with the bunching scheme, in which the buncher is operated at constant duration and variable voltage. We refer to these bunching schemes from now on as $f(t)$-bunching and $f(V)$-bunching, respectively. First we discuss rebunching; the longitudinal spatial focusing of a molecular beam. Figure 34 shows the recorded ND_3 density at the detection position in the storage ring as a function of the time after the synchronous molecule has exited the decelerator. On the left, the time-of-flight distribution is shown when the buncher is operated with the $f(t)$-bunching scheme. The voltages on the buncher electrodes are fixed to ± 10 kV and the time during which the buncher is on is set to $\Delta t = 88.1$ μs. With these settings a longitudinal spatial focus is created at the detection position in the storage ring. On the right-hand side of Fig. 34 the

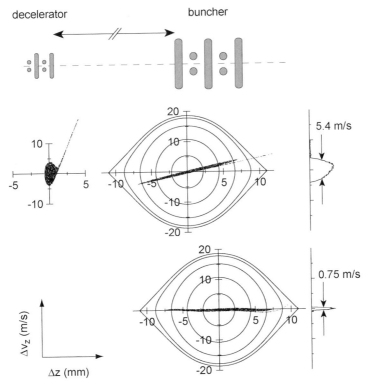

FIG. 33. Calculated longitudinal phase-space distributions of a packet of ND_3 molecules are given at the exit ($t = 0$ ms) of the Stark decelerator, at the entrance ($t = 1.778$ ms), and at the exit ($t = 1.946$ ms) of the buncher, relative to the position of the synchronous molecule. The potential well in each buncher stage is switched on for a duration $\Delta t = 49.1$ μs. Two buncher stages are used to rotate the packet of molecules in phase-space.

time-of-flight distribution is shown when the buncher is operated with the $f(V)$-bunching scheme. In this case the molecular beam is longitudinally focused at the detection position in the ring when voltages of ± 7.5 kV are put on the buncher electrodes. It is immediately clear that there is little difference between the two bunching schemes. The longitudinal widths (FWHM) of the molecular package obtained from a Gauss fit are practically the same: $\Delta t_{f(t)} = 35.8 \pm 0.6$ μs vs. $\Delta t_{f(V)} = 38.5 \pm 0.7$ μs. The peak position and the peak signal intensity are also practically the same for both cases. The $f(t)$-bunching scheme apparently works equally well as the $f(V)$-bunching scheme.

   The main use of the buncher between the decelerator and the storage ring is that it can be used for bunch rotation, i.e., for longitudinal cooling of the molecular beam prior to injection in the storage ring. To be able to accurately determine the

f(t)-bunching   f(V)-bunching

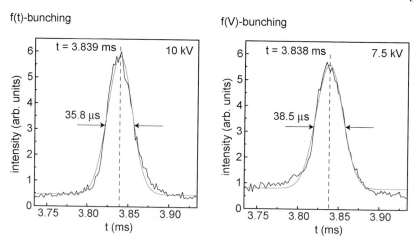

FIG. 34. Rebunching: longitudinal spatial focusing of a molecular beam. Time-of-flight distributions obtained with the $f(t)$-bunching scheme (left) and $f(V)$-bunching scheme (right). The measured ND_3 parent ion signal is plotted as a function of time, where t is the moment when the synchronous molecule exits the decelerator. Peak positions as well as FWHM of the distributions are indicated.

width of the velocity distribution as produced by the buncher, a measurement of the time-of-flight profile needs to be done in the far field, as far away from the exit of the buncher as possible. Only in that case can it be assumed that the observed width of the time-of-flight distribution is dominated by the velocity spread of the molecular packet and that the initial spatial spread of the packet can be neglected. Therefore, it is best to perform this measurement after many round trips in the storage ring. In the next section such a measurement, using the $f(t)$-bunching scheme, is presented.

Due to the limited opening angle of the entrance slit of the ring, a part of the molecular packet is cut off by the outer pair of electrodes. The numerically calculated phase-space distribution of the molecular packet entering the ring is given by $[\Delta r \times \Delta v_r] = [2.3 \text{ mm} \times 2.5 \text{ m/s}]$, $[\Delta y \times \Delta v_y] = [1.4 \text{ mm} \times 1.4 \text{ m/s}]$, $[\Delta z \times \Delta v_z] = [9.5 \text{ mm} \times 0.64 \text{ m/s}]$ for the $f(t)$-bunching scheme.

After the molecules have entered the 25-cm diameter electrostatic storage ring, the two outer ring electrodes are rapidly (<100 ns) switched to high voltages. This creates an electric field that delivers the centripetal force on the molecules. Just prior to detection the high voltages on these electrodes are switched off again. Then the ammonia molecules are ionized in a $2+1$ resonance enhanced multiphoton ionization (REMPI) process, using 317 nm radiation. The UV-light is focused and only molecules in the approximately 3 mm long, 100 μm diameter beam waist are detected. The ions are repelled from the storage ring by the residual electric field created by the permanently present 200 V and 100 V on the inner and middle ring electrodes, respectively, and are then detected with a linear time-of-flight mass spectrometer.

6.2. LONGITUDINAL TEMPERATURE OF MOLECULES IN THE RING

In Fig. 35 the ND_3 parent ion signal is shown as a function of the storage time in the ring; the origin of the time axis is the time when the high voltages on the ring are switched on. Each data point represents the parent ion signal intensity averaged over 20 laser shots, i.e., averaged over 20 subsequent deceleration, loading and detection cycles. The experiment normally runs at a 10 Hz repetition rate. However, for the measurements that exceed 100 ms storage times in the ring the repetition rate is reduced; the laser system still runs at 10 Hz, but the effective repetition rate is reduced using a mechanical shutter.

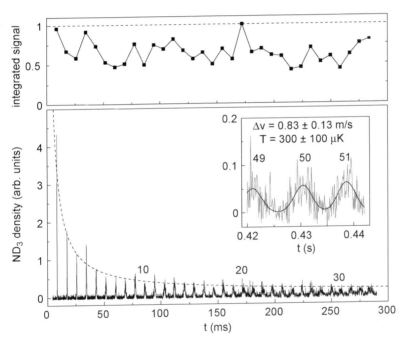

FIG. 35. In the lower graph the ammonia density is shown at the loading position in the ring as a function of storage time. The dashed line is a $1/t$ plot, used to guide the eye. In the inset, a measurement of the ammonia density after 49–51 round trips is shown, together with a multipeak Gaussian fit, from which a longitudinal temperature of 300 ± 100 μK is deduced for the molecular packet in the ring. The upper graph shows the time-integrated signal after each round trip. (Figure reproduced from (Crompvoets et al., 2004) with permission.)

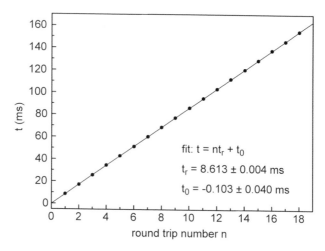

FIG. 36. The center positions of the first 18 peaks are plotted as a function of the round trip number n. A linear fit (solid line) yields a value for the round trip time of $t_r = 8.613 \pm 0.004$ ms. The offset $t_0 = -0.103 \pm 0.040$ ms is caused by the delay between the arrival time of the decelerated packet in the detection region of the storage ring and the moment when the ring is switched on.

The peaks in the signal indicate the passage of a molecular packet through the laser focus. In Fig. 36 the center positions, determined by Gaussian fits, of the first 18 peaks are plotted as a function of the round trip number. From a linear fit the round trip time can be derived, which is $t_r = 8.613 \pm 0.004$ ms. For a tangential velocity of 91.8 m/s and a corresponding equilibrium radius of $r_0' = 1.3$ mm the round trip time is theoretically expected to be 8.645 ms. The discrepancy with the experimental value can be attributed to the uncertainty in the tangential velocity and in the radial position at which the molecular packet is coupled into the ring. The offset $t_0 = -0.103 \pm 0.004$ ms is caused by the fact that the ring is switched on ≈ 0.1 ms later than the arrival time of the decelerated packet in the detection region of the storage ring. This delay is experimentally determined during the optimization of the round trip peak signals in the ring and probably indicates a less than perfect phase-space matching of the molecular beam emittance onto the acceptance of the ring.

It is seen that the peak intensity after each round trip gradually decreases while the width of the peaks increases. This is a result of the residual tangential velocity spread of the molecular packet in the ring. Molecules with a tangential velocity of 91.8 m/s traverse the ring at the equilibrium orbit, which is a distance $r = 1.3$ mm away from the center of the hexapole geometry. Faster molecules orbit the ring at a larger radius and have a longer flight path; slower molecules orbit at a smaller radius and have a shorter flight path. This difference in flight paths between faster and slower molecules counteracts the longitudinal spreading out of the molecular packet. As the relative change in orbit radius is small, however, this effect is very limited in the present case.[4] Therefore the tangential spreading out of the packet in the ring can be approximated very well by the longitudinal spreading out of a packet in linear free flight. The time-spread of the signal at the nth round trip is given by $(1/v)\sqrt{(\Delta z_0)^2 + (n\Delta v_0 t_r)^2}$, where Δz_0 is the initial longitudinal position spread, v the average velocity, and Δv_0 the longitudinal velocity spread. Without further losses from the ring, the peak density is inversely proportional to this width. When the initial position spread can be neglected, this implies an expected $1/n$ behavior for the peak intensities, as indicated with the dashed line in Fig. 35.

It is seen that the peak heights actually do not decrease monotonically. This is explained by the two-dimensional oscillatory motion of the molecular packet in the ring in combination with the position sensitive detection scheme. The oscillatory motion can be particularly prominent when there is a slight transverse phase-space mismatch when the molecules are injected into the ring. In the upper graph of Fig. 35 the time-integrated signal for each round-trip is shown. This signal clearly shows the relatively large intensity fluctuations just discussed, but it is also seen that, on average, the signal remains practically constant. Loss due to collisions with background gas can be neglected as the background pressure in the ring chamber is 10^{-9} mbar, corresponding to an anticipated $1/e$ decay time of several seconds. Furthermore, in the electric field geometry used in this experiment, the molecules are never in zero electric field so zero-field crossing (Majorana) transitions are avoided.

The inset of Fig. 35 shows the signal after 49, 50, and 51 round trips. The molecules have then been stored in the ring for about 0.43 s, corresponding to a total flight path of more than 40 m. It is seen from the data that the molecular packet has stretched out so far that it fills about half the ring, i.e., its FWHM length has become about 40 cm. A longitudinal temperature can be deduced from the relative temporal width of the signal from the 50th round trip. After such a long flight distance, the contribution of the initial position spread to the width can be neglected, and the relative temporal width is in a good approximation equal to $\Delta v/v$. From the multipeak Gaussian fit of the data (dark solid line), in combination with the known 91.8 m/s tangential velocity of the molecules, the absolute longitudinal velocity spread is found as 0.83 ± 0.13 m/s. This corresponds to a longitudinal temperature of 300 ± 100 μK, in close agreement with the numerically calculated value at the entrance of the ring. It is noted that this value for the longitudinal velocity spread is obtained with the $f(t)$-bunching scheme and that it is almost

---

[4] Isochronous operation of the ring, where the longitudinal spreading out is completely cancelled, is fundamentally impossible as discussed earlier (see Section 4.1.1).

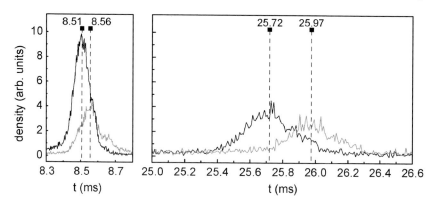

FIG. 37. Time-of-flight distributions of molecules in the ring for the first (left) and the third (right) round trip. The distributions in black are obtained with ±8 kV on the outer ring electrodes while the distributions in gray are obtained with ±6 kV on the outer ring electrodes. The vertical dashed lines indicate the centers of the peaks as obtained by fitting the peaks to Gaussian distributions. Horizontal scale and vertical scale are the same for both graphs.

identical to the value of 0.76 m/s found in Section 5 with the $f(V)$-bunching scheme.

As shown in Section 4 the equilibrium radius and hence the orbit time depends strongly on the electric field that delivers the confinement force. If the voltages on the ring electrodes are lowered, then the transverse trapping potential well widens and the equilibrium radius, as well as the orbit time of the molecules, increases for a given fixed forward velocity. In Fig. 37 the recorded ND_3 parent ion signal is shown as a function of the storage time in the ring for the first and the third round trip. Each round trip is measured at two different voltage settings: ±6 kV and ±8 kV on the outer ring electrodes. Interestingly, while the peak density on the third round trip for the 8 kV setting is about half as large as the peak density at the first round trip, the peak density at the third round trip for the 6 kV setting is almost as large as the peak density at the first round trip. This is the result of the position sensitive detection method and a slight mismatch of the emittance of the molecular beam onto the acceptance of the storage ring, as will be discussed in Section 6.3.

It is seen from Fig. 37 that with the 8 kV setting the package of molecules arrives earlier than with the 6 kV setting. There is a difference of about 1% in round trip time; the average round trip times are 8.61 ms and 8.71 ms for the 8 kV and 6 kV setting, respectively. As the molecules have the same tangential velocity of 91.8 m/s for both settings, this difference is due to a difference in the equilibrium radii: $r'_{8\,kV} = 0.7$ mm and $r'_{6\,kV} = 2.2$ mm. This corresponds roughly to the calculated values of 1.3 mm and 1.9 mm.

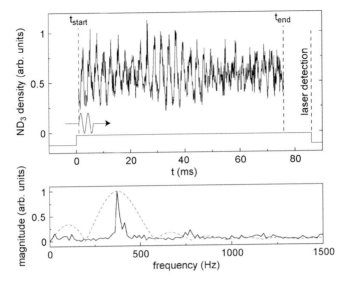

FIG. 38. The upper graph shows the density of ammonia molecules in the ring after 10 round trips as a function of the start time of the AC voltage modulation. The modulation signal is a two-period sine-wave with a period of $t = 2.56$ ms. The starting point of this modulation signal is scanned stepwise from t_{start} to t_{end}. A Fourier transformation of the data is shown in the lower graph (solid curve) together with the frequency spectrum of the two-period sine-wave (dashed curve). (Figure reproduced from (Crompvoets et al., 2004) with permission.)
© 2004 The American Physical Society.

6.3. BETATRON OSCILLATIONS IN THE DIPOLE RING

The collective, transverse motion of the molecules in the ring can be excited by applying an AC voltage on top of the 200 V DC voltage on the inner ring electrodes. This leads to an oscillatory motion of the potential well, which, for the electric field configuration used, is almost exclusively in the vertical direction, as shown in Fig. 31. In the experiment, the vertical motion of the molecular packet is driven by a two-period sine-wave with an amplitude of 180 V and with a period of $t = 2.56$ ms. This modulation is started at a variable time t_{start} after the molecules have entered the ring, which is at $t = 0$. The block pulse indicates the time that the molecules are confined in the ring. In the upper graph of Fig. 38 the density of ammonia molecules in the ring after ten round trips, i.e., after the molecules have been in the ring for 86 ms, is shown as a function of the time when the AC modulation starts (relative to the switching on of the ring). Each data point represents the averaged signal over 30 laser shots and is proportional to the ammonia density. The ammonia density in the laser detection area after these 10 round trips shows a strong modulation, the equivalent of a vertical betatron oscillation in charged particle storage rings. In our experimental set-up the tightly

focused detection laser crosses the ring in the horizontal plane, and vertical oscillations are therefore more readily detected than radial oscillations. Although
the AC modulation voltage shakes the potential well roughly only 160 μm up and
down from the vertical equilibrium position, the resulting driven vertical oscillation of the molecules is calculated to have an amplitude of about 1 mm. It is seen
from the measurements that even when the AC modulation is induced during the
first round trip, the molecular packet coherently oscillates in the vertical direction after 10 round trips. From these measurements, the eigen-frequency for the
vertical oscillation in the ring can be accurately determined.

In the lower graph of Fig. 38 a Fourier transform of the data is shown. The main
peak in the frequency spectrum is found at 373 Hz, with a shoulder at 413 Hz, and
the first overtone of this peak is visible as well. Also shown in this figure is the
frequency spectrum of the two-period sine-wave (dashed), which is set to drive the
fundamental vertical oscillation. The experimentally determined value of 373 Hz
for this oscillation frequency is close to the calculated value of $f_y = 390$ Hz.
If detection were to be performed exactly at the center of the vertical motion,
the molecular packet would pass the laser focus twice per oscillation and only the
overtone would be observed. If the laser detection region would be at an extremum
of the vertical motion, only the fundamental frequency would appear. The actual
relative intensities of these two frequency components in the spectrum therefore
depends on the details of the alignment of the laser relative to the ring as well as of
the coupling of the injection beamline to the ring. The envelope of the modulation
of the ammonia density appears as a typical beating pattern, indicative of coupled
motion at closely spaced oscillation frequencies. In the Fourier transform this
shows up as the 413 Hz shoulder of the main 373 Hz resonance. It turns out
that, rather than being due to typical beating, the intensity of the envelope of
the modulation correlates with the intensity of the corresponding time-integrated
signal shown in the upper graph of Fig. 35; the vertical oscillation of the package
of molecules shows the largest modulation depth in the detection region when this
oscillation is initiated at a time when the package has the best overlap with the
detection region.

Numerical calculations were performed trying to simulate the oscillatory signal
of Fig. 38. It turned out to be practically impossible to fit the measured data.
The main reason for this is that the detection of the molecules is very position
sensitive. In the experiment it is practically impossible to determine the exact
detection position and hence it is difficult to find the right parameter settings for
the simulations.

6.4. BETATRON OSCILLATIONS IN THE HEXAPOLE RING

In the storage ring experiments described in the previous section the potential well
is rather nonlinear and asymmetric. Nonlinearity leads to coupling between the

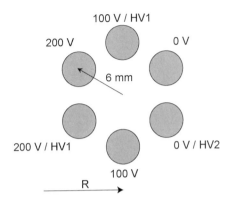

100 V / HV1

200 V   0 V

6 mm

200 V / HV1   0 V / HV2

100 V

R

FIG. 39. Cross section through hexapole storage ring. Voltages on the ring electrodes are shown for the hexapole configuration. For the electrodes that are switched between two voltages both voltages are indicated.

radial and vertical motion of the molecules stored in the ring. Asymmetry leads to different betatron oscillation frequencies for the radial and vertical direction, which together with the nonlinear coupling leads to complex oscillatory behavior of the molecules in the ring.

The linearity and symmetry of the potential well of the storage ring can be improved by approximating the ideal hexapole electric field. An additional advantage is that the potential well does not open up at the inside of the ring, as is the case for the potential well of the dipole ring. This makes the storage ring suitable for molecules with low tangential velocities which would otherwise escape at the inside of the ring. Furthermore, an hexapole field with the same potential well strength as the field in the dipole ring can be obtained at lower voltages. A disadvantage of the hexapole field geometry is the existence of a zero electric field in the ring where the molecules can undergo Majorana transitions to untrapped states, after which they would be lost from the ring. However, as the equilibrium radius of the molecules is about 2 mm away from this zero field and as the amplitude of the betatron oscillation is not large, it is unlikely that these Majorana transitions will occur.

In this section the effect of the trapping voltages on the dynamics of the molecules inside the hexapole ring is investigated. Altering the voltages leads to a change in betatron oscillation frequency. The voltages on the ring electrodes for the hexapole configuration are shown in Fig. 39. The high voltages $HV1$ and $HV2$ are used for trapping of the molecules and are switched off for the ion extraction. The hexapole field is best approximated when $HV2 \approx HV1 + 360$ V, as is found from calculations using Simion (Dahl, 1995).

The same injection beamline and detection scheme is used as in the dipole ring experiments. Only the timings of the hexapoles in front of and behind the buncher

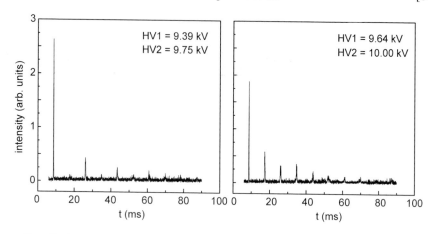

FIG. 40. Ammonia density in the hexapole storage ring as a function of the storage time t, shown for two different high voltage settings: left graph $(HV1, HV2) = (9.39, 9.75)$ kV and right graph $(HV1, HV2) = (9.64, 10.00)$ kV. Horizontal and vertical scales are the same for both graphs.

are adjusted to match the emittance of the molecular beam onto the slightly different acceptance of the hexapole ring. To detect the molecules, the voltages on the ring electrodes are switched to 0 V, 100 V, or 200 V as indicated in Fig. 39. This creates an electric field that pushes the laser produced ions out of the ring towards a time-of-flight mass spectrometer as discussed in Section 4.

The effect of the trapping voltages on the motion of molecules in the ring is quite strong. Figure 40 shows the ND_3 density as a function of the storage time in the ring for two different voltage pair settings $(HV1, HV2)$. The left graph shows the signal for $(HV1, HV2) = (9.39, 9.75)$ kV and the right graph shows the signal for $(HV1, HV2) = (9.64, 10.00)$ kV. Compared to the measurements performed with the dipole ring, the signal drops much faster over time. Now, the tenth round trip is hardly visible. As the buncher settings are the same in both cases, this is most likely explained by a mismatch of the transverse emittance of the molecular beam onto the transverse acceptance of the hexapole ring. Attempts to improve the number of observable round trips by better spatial alignment of the storage ring onto the molecular beam and by optimizing hexapole and buncher timing settings proved to be unsuccessful. In particular, the exact position of the ring with respect to the laser beam was difficult to control, but also the combined transverse focusing effect of the hexapoles and buncher differed from the theoretically expected focusing effect. Comparing the two graphs in Fig. 40 it appears that the transverse motion is very sensitive to changes in the trapping voltages and hence the potential well strength. In the left graph of this figure the even round trips are missing in the time-of-flight profile whereas in the right graph all the peaks are present. Due to the improved linearity of the hexapole ring over the dipole ring the package of molecules remains more localized in phase-space (less filamentation). As the detection of the molecules is very position sensitive and the injection of the molecules into the ring might not be perfect, it is possible that, for some round trips, the molecules miss the laser focus and are not detected.

Due to this sensitive dependence on the trapping voltages on the ring electrodes, the motion of the molecules in the ring can be investigated as a function of these voltages. The wavelengths of the betatron oscillations that the molecules make in the hexapole ring are the same in the vertical and radial direction because the potential well of the hexapole ring is cylindrically symmetric. The betatron wavelength depends on the hexapole force constant k given by Eq. (13) in the following way:

$$\lambda = 2\pi v_\phi \sqrt{\frac{m}{k}} \propto \sqrt{\frac{1}{V}}. \tag{27}$$

Here, λ is the wavelength of the betatron oscillation, v_ϕ is the tangential velocity, m is the mass of the molecule and V is the voltage difference between the adjacent electrodes of the hexapole storage ring. The inversion splitting is neglected here but near the hexapole axis, where the electric field is relatively small, the force constant k is effectively reduced and, as a consequence, the wavelength is effectively increased. It is clear from this equation that it is possible to scan the wavelength of the betatron oscillations by altering the voltage on the hexapole ring.

Figure 41 shows the ammonia density at the third round trip ($t = 25.93$ ms) as a function of the ring voltage $HV2$. $HV1$ is scanned together with $HV2$ such that the difference between the two remains constant. The measurement is performed for three different vertical laser detection positions. Instead of translating the laser focus vertically, it was experimentally easier and more accurate to translate the ring vertically. The ring is translated vertically from $\Delta y = +0.5$ mm to $\Delta y = -0.5$ mm. It is clear that the oscillations depend on the detection position as the peak positions shift and the peak shapes change. At all three detection positions the signal shows an oscillatory behavior. This can be attributed to the mismatch of the emittance of the molecular beam onto the acceptance of the ring and to the position sensitive detection scheme that is employed. For certain voltage settings the density of ammonia molecules in the laser focus has a maximum. The laser focus then overlaps optimally with the path of the molecular packet. For other voltage settings the density is almost zero. In this case the overlap of the laser focus with the path of the molecules is minimal. The top axis of Fig. 41 indicates the betatron wavelength, which is calculated using Eq. (27) assuming no inversion splitting. The wavelengths λ_{peak} corresponding to the peak positions are determined from a fit. At the third round trip the synchronous molecule has travelled a distance of 2.38 m. Dividing this distance by the determined wavelengths gives the number of betatron oscillations n that have occurred up to the

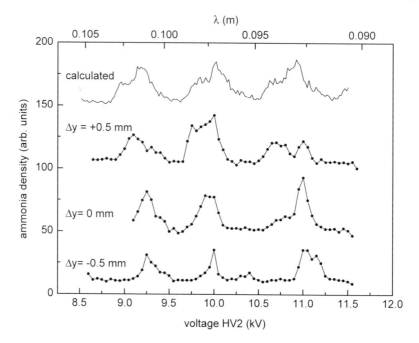

FIG. 41. Ammonia density at third round trip ($t = 25.93$ ms) as a function of the voltage on the ring electrodes ($HV2 = HV1 + 360$ V) for three different heights of the laser beam: $\Delta y = -0.5$ mm, $\Delta y = 0$ mm, and $\Delta y = +0.5$ mm. Upper curve is a numerical simulation of the $\Delta y = +0.5$ mm measurement. The curves are given an offset for clarity. The top axis indicates the corresponding wavelength scale.

Table II
Trapping voltages $HV2$, the corresponding wavelengths λ_{peak}, and the number of betatron oscillations n for the three peaks of the $\Delta y = +0.5$ mm signal shown in Fig. 41.

	$HV2$ (kV)	λ_{peak} (m)	n
peak 1	9.14	0.102	23.3
peak 2	9.90	0.0977	24.3
peak 3	10.8	0.0934	25.5

third round trip for the given voltage settings. The relevant data are listed in Table II. The wavelengths correspond to oscillation frequencies of about 940 Hz, in good agreement with the calculations.

The voltage scan is simulated numerically in order to analyze the signal. The input phase-space distribution is taken from the calculated values in Section 6.1. From the calculations it is again found that the oscillating signal is rather sensitive to the vertical laser detection position and to the alignment of the ring onto the molecular beam. A numerical simulation is shown in Fig. 41 and corresponds to a vertical shift in detection position of $\Delta y = +0.5$ mm. A qualitative fit of the experimental data is obtained when the vertical position of the ring relative to the incoming molecular beam is set to $\Delta y = -0.5$ mm. The peak positions agree rather well. The asymmetric shapes result from the asymmetry between the radial and vertical phase-space distributions; calculations with radially and vertically symmetric phase-space distributions show that the peaks are smoother and more symmetric.

7. Design of a Sectional Storage Ring

In the storage ring discussed so far, the molecular packet is transversely confined but it is still free to spread tangentially. As a consequence, the density of molecules stored in the ring decreases with $1/t$, where t is the storage time. In this section a simple and compact design for a sectional storage ring is presented in which the molecules are also confined tangentially. A more elaborate design can be found in (Nishimura et al., 2003).

In the present storage ring the transverse potential well is the same at every position along the ring and the orbits of the molecules in the ring are stable as long as the kinetic energy of the molecules does not exceed the depth of the potential well. In a sectional storage ring, the strength of the potential well, and hence the confinement force, changes when the molecules move from one section to the next. These changes may lead to parametric amplification of the betatron motion and, eventually, to beam loss. The simplest conceivable storage ring is a ring consisting out of two half hexapole rings with a small gap in between to allow for bunching and detection, as schematically depicted in Fig. 42. In this section the conditions for stable operation of this prototype sectional storage ring are discussed.

7.1. TRANSVERSE STABILITY IN A SECTIONAL STORAGE RING

7.1.1. A Linear Array of Hexapoles

A qualitative understanding of the stability in the sectional ring can be obtained by neglecting the centrifugal force, i.e., by considering an array of straight hexapoles. This simple model contains essentially all the important features of a sectional ring. In first order approximation, when the force is assumed to be perfectly linear (no inversion splitting), the transverse motion of the molecules in phase-space can be conveniently described with the matrix method (Conte and MacKay, 1991). The matrix method is used extensively in the design of charged particle storage

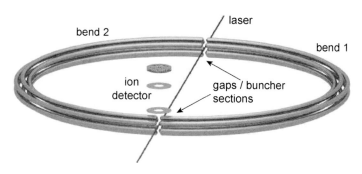

FIG. 42. Sectional storage ring consisting of two semicircle hexapoles. The detection laser is shot right through the gaps, ionizing the ammonia molecules that are present in the laser focus. The parent ions are then extracted from the ring and accelerated towards an ion detector.

rings and gives a first order estimate of the stability of a storage ring. The transformation of the position and velocity coordinates of the molecules by each section of the ring is given by a single matrix. In order to calculate the transformation matrix of the entire storage ring it is useful to divide the ring in unit cells. Each unit cell is a block of one or more ring sections and it can occur many times in the ring. The simple sectional storage ring presented here consists of two cells, where each cell consists of one hexapole bend section and a free flight section.

The transformation matrix M_{ring} for one round trip through the sectional ring presented here is a multiplication of two cell matrices M_{cell}. The matrix M_{cell} is again a multiplication of an hexapole focusing matrix and a free flight matrix as given in Eqs. (18) and (17), respectively. For stable operation of the ring the motion of the molecules in the ring has to remain bound after n propagations through the matrix M_{cell}, with $n \to \infty$. This requires that $-2 \leq \mathrm{Tr}(M_{cell}) \leq 2$, where $\mathrm{Tr}(M_{cell})$ is the trace of the matrix M_{cell} (Conte and MacKay, 1991). In this case, the stability criterion is found to be

$$-2 \leq 2\cos\left(\frac{\omega L_{hex}}{v}\right) - \frac{\omega L_{gap}}{v}\sin\left(\omega L_{hex}/v\right) \leq 2, \qquad (28)$$

where ω is the oscillation frequency in the field of the hexapoles, v is the velocity of the molecule, L_{hex} is the length of the hexapoles, and L_{gap} is the length of free flight path.

When the inversion splitting is taken into account, the matrix formalism can no longer be used. A numerical calculation has to be performed to investigate the transverse stability of the ring. The trajectories of 100 ND_3 molecules in the $|J, K\rangle = |1, 1\rangle$ state are numerically calculated while passing through the 25 cm diameter ring with a tangential velocity of 91.8 m/s. The initial transverse phase-space distributions are taken to be $[\Delta r \times \Delta v_r] = [\Delta y \times \Delta v_y] = [1\ \mathrm{mm} \times 5\ \mathrm{m/s}]$, with r and y in the horizontal (radial) and vertical direction, respectively. Here, the widths are at FWHM of Gaussian distributions. The number of molecules that is still in the ring after 100 round trips is counted. The ratio of this number to the number of molecules injected into the ring can be regarded as the transmission efficiency of the ring for the parameters used. Figure 43(a) shows a plot of the transmission efficiency as a function of k_{hex} and L_{gap}. The horizontal axis is linear in the voltage difference between the hexapole electrodes. The maximum value of $k_{hex} = 0.06$ cm^{-1}/mm^2 corresponds to a voltage of 10.18 kV for an hexapole with an inner radius of $R = 4$ mm. The dark bands in the diagram indicate the regions of stability. At larger values of k_{hex} the transmission efficiency decreases strongly when the free flight path length L_{gap} is increased. Figure 43(b) shows the same calculation but now for the hypothetical case when the molecules have no inversion splitting. In this case the hexapole force is perfectly linear. Due to the improved linearity the bands are more pronounced and extend over a wider range of free flight path lengths. The thick white horizontal lines in this figure indicate the regions of stability according to Eq. (28) obtained with the matrix method.

As observed, for some values of k_{hex} hardly any molecule remains in the ring, even when L_{gap} approaches zero. The occurrence of these so-called stop bands is a result of half-integer resonances in the ring. These can be understood by examining the phase-space diagram shown in Fig. 44. Assume a molecule orbiting the ring with a round trip frequency Ω and a betatron oscillation frequency ω_{beta}. Figure 44 shows the motion of such a molecule in phase-space (inner circle). The momentum axis is divided by ω_{beta} to make the trajectories in phase-space perfect circles. Inversion splitting is neglected. At a certain orbital position in the ring the molecule exits an hexapole bend section and enters a free flight section. The position of the molecule in phase space shifts horizontally during free flight as indicated by arrow 1 in Fig. 44. Then the molecule enters the next hexapole bend section. The trajectory of the molecule describes now a larger circle in phase-space: its betatron amplitude has grown. During its flight through the hexapole the betatron phase of the molecule increments by an amount $\Delta\alpha$. This difference, $\Delta\alpha$, between the betatron phase at the exit and the betatron phase at the entrance of a cell is called the phase advance of that cell, like in charged particle accelerators. The half-integer resonances arise when the phase advance is a multiple of π. Every time the molecule has rotated over an angle of π in phase space and it enters a free flight section, it moves to a circle with a larger radius in phase space (arrow 2 in Fig. 44). In this way the betatron amplitude grows quickly out of bounds and eventually the molecule escapes from the ring or crashes into the electrodes.

Parametric amplification of the amplitude of the betatron motion can be avoided by choosing a phase advance different from (a multiple of) π. When, for instance, a phase advance of $2\pi/5$ is taken, as indicated by arrow 1' in Fig. 44, the effect of the second gap will partly cancel the effect of the first gap. In this case the

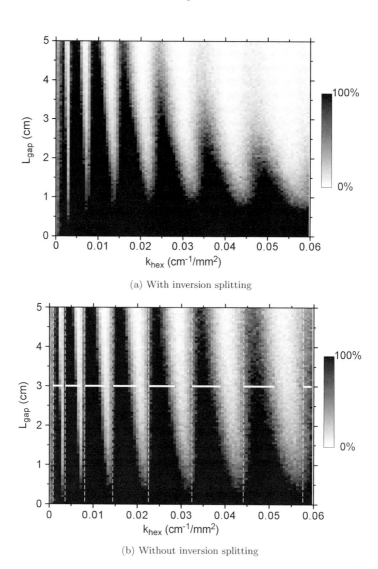

(a) With inversion splitting

(b) Without inversion splitting

FIG. 43. (a) Stability diagram of a 25 cm diameter sectional storage ring consisting of two bend sections and two free flight sections, when the centrifugal force is neglected. At each coordinate (k_{hex}, L_{gap}) the transmission efficiency is calculated by sending 100 ammonia molecules through the ring and determining how many molecules are still in the ring after 100 round trips. (b) Same diagram but now for the hypothetical situation when the ammonia molecules have no inversion splitting. The thick white horizontal lines indicate the regions of stability according to Eq. (28). The vertical dashed white lines indicate the position of the stop bands according to Eq. (31).

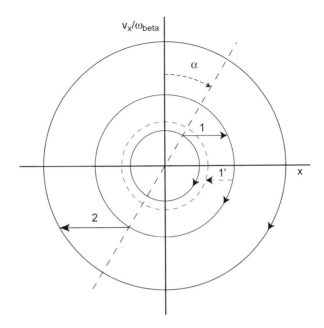

FIG. 44. Phase-space plot in one transverse direction showing an half-integer betatron resonance in the sectional storage ring consisting of hexapole bend sections and free flight sections. The betatron amplitude of a molecule (radius of the circles) grows without bound when the phase advance is a multiple of π. The solid arrows (1, 2) indicate the free flight paths of the molecule, which occur every time the betatron phase has increased by π. The dashed arrow ($1'$) indicates the free flight path when the phase advance is $2\pi/5$.

betatron amplitude sometimes grows and sometimes shrinks, averaging out in the end.

The position k_{hex} of the stop bands can easily be calculated as the phase advance $\Delta\alpha$ is given by

$$\Delta\alpha = \omega_{\text{beta}} \frac{L_{\text{hex}} + L_{\text{gap}}}{v}. \tag{29}$$

As the stop bands occur when the phase advance is a multiple of π, one finds for the betatron angular frequency

$$\omega_{\text{beta}} = n \frac{\pi v}{L_{\text{hex}} + L_{\text{gap}}}, \tag{30}$$

with n an integer. By substituting k_{hex} for ω_{beta} using the relation $\omega_{\text{beta}} = \sqrt{k_{\text{hex}}/m}$, one finds for the force constant

$$k_{\text{hex}} = n^2 \left(\frac{\pi v}{L_{\text{hex}} + L_{\text{gap}}} \right)^2 m, \tag{31}$$

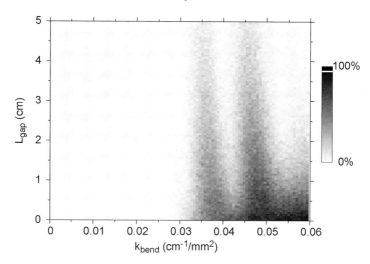

where m is the mass of the molecule. This explains the observed dependence of the position of the stop bands on k_{hex}, as seen in Fig. 43. The stability plots are calculated for ND_3 with $v = 91.8$ m/s and $L_{\text{hex}} + L_{\text{gap}} = 39.3$ cm. In that case the proportionality constant between k_{hex} and n^2 is 9.0×10^{-4} cm^{-1}/mm^2. The stop bands as found from Eq. (31) are indicated in Fig. 43(b) with vertical dashed white lines.

7.1.2. Bend Hexapoles

Until now we have completely neglected the centrifugal force. The centrifugal force shifts the equilibrium orbit away from the center, thereby decreasing the depth of the transverse well (see Fig. 16). For a given tangential velocity, the transverse well will vanish all together below a certain value of the force constant, k_{bend}. Figure 45 shows again the transmission efficiency as a function of k_{bend} and L_{gap} now including the centrifugal force and incorporating the inversion splitting of ammonia. For $k_{\text{bend}} < 0.03$ cm^{-1}/mm^2 no molecules are detected after 100 round trips because the force constant is too weak to keep the molecules inside the ring.

To improve the transmission efficiency, the voltages on the ring are increased shortly before and after the gap. We will refer to this operation as 'kicking'. In Fig. 46 the transmission is calculated when the force constant is doubled over a certain length, L_{kick}, while L_{gap} is kept at 10 mm. The diagram shows the transmission efficiency as a function of the force constant k_{bend} and the length L_{kick}.

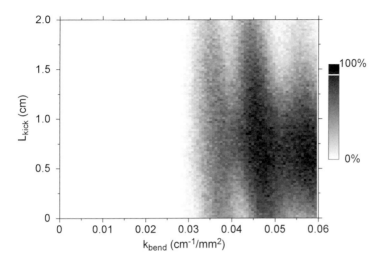

FIG. 46. Stability diagram of the sectional storage ring with hexapole kickers. The free flight path length is kept constant at $L_{\text{gap}} = 10$ mm.

It is seen, that with 'kicking' the transmission is improved considerably. When $L_{\text{gap}} + 2L_{\text{kick}}$ is much shorter than the distance molecules travel before making one betatron oscillation, $2\pi v / \omega_{\text{beta}}$, we expect a maximum in the transmission when the average of the force constant over $L_{\text{gap}} + 2L_{\text{kick}}$ is equal to k_{bend}, such that the molecules experience approximately the same force constant over the whole ring. We thus expect a maximum at $L_{\text{kick}} = 5$ mm. The actual maximum is seen to be shifted to a somewhat longer length.

7.2. LONGITUDINALLY FOCUSING IN A SECTIONAL RING

In the previous section it was shown that, at the appropriate voltage settings, gaps can be introduced in the storage ring without a dramatic drop in transmission. Moreover, by switching the ring at higher voltages before and after the gap, one can reduce the loss in molecules to a minimum. In this section it will be shown that, if the voltages are switched appropriately, the molecular packet can be re-bunched. Let us first determine the operation point of the ring. From Figs. 45 and 46 it follows that for the current design the optimal force constant, k_{bend}, is 0.047 cm^{-1}/mm^2. This corresponds to a voltage difference of approximately 8 kV between adjacent hexapole rods as shown in Fig. 47(a). For the kicker we apply double this voltage difference as shown in Fig. 47(b). For completeness, the voltages which are used for detection are shown in Fig. 47(d).

To achieve bunching inside the ring, molecules that are faster than the synchronous molecule have to lose kinetic energy while slower molecules have to gain

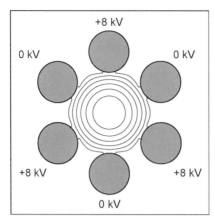

(a) Confinement configuration

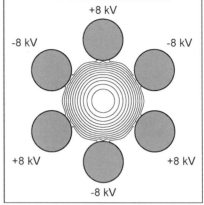

(b) Hexapole kicker configuration

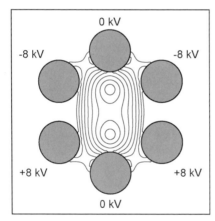

(c) Buncher configuration

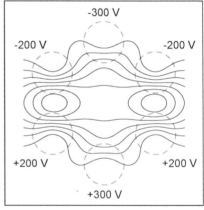

(d) Detection configuration

FIG. 47. Cross section through the hexapole bend section showing voltage settings and electric field contour lines for confinement configuration (a), hexapole kicker configuration (b), buncher configuration (c) and detection configuration (d). The latter shows the electric field in the middle of a gap. The hexapole radius is $R = 4$ mm. Contour lines are shown at 5 kV/cm intervals for (a), (b), (c) and 0.1 kV/cm intervals for (d). Contour lines with highest electric field value: 30 kV/cm (a), 60 kV/cm (b), 45 kV/cm (c) and 0.8 kV/cm (d).

kinetic energy. This is achieved by switching the voltages on section 2 to the configuration shown in Fig. 47(c). The voltages on section 1 are switched off. This creates, in the longitudinal direction, a smoothly rising potential energy curve W_1

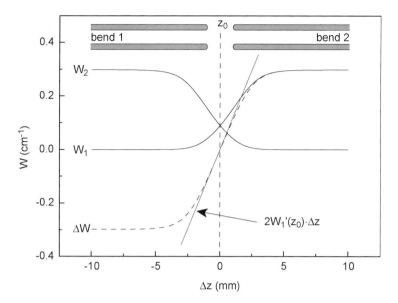

FIG. 48. Potential energy curves W_1 and W_2 are shown as a function of the position z of a molecule traversing the ring at r_0. The center of the gap is $z = 0$. Also shown is the potential energy difference ΔW. The slope $W_1'(z_0)$ is determined by a linear fit of ΔW.

for ammonia molecules in the low-field seeking $|J, K\rangle = |1, 1\rangle$ state as shown in Fig. 48. When the synchronous molecule is exactly at the center of the gap at z_0, bend section 1 is switched to the buncher configuration and bend section 2 is switched off. This creates the mirror image W_2 of the potential energy curve W_1, with respect to the center of the gap z_0. The difference between W_1 and W_2 leads to a restoring force in the longitudinal direction.

The potential energy of the synchronous molecule remains unchanged in this process, $W_1(z_0) = W_2(z_0)$, but the energy of a molecule that passes a distance Δz in front of the synchronous molecule changes by an amount $\Delta W(\Delta z) = W_1(z_0 + \Delta z) - W_2(z_0 + \Delta z)$, which is the dashed curve shown in Fig. 48. It is clear that a molecule that passes in front of the synchronous molecule loses kinetic energy while a molecule that lags behind the synchronous molecule gains kinetic energy. Because of the mirror symmetry we have $W_2(z_0 + \Delta z) = W_1(z_0 - \Delta z)$ and the potential energy difference becomes

$$\Delta W = W_1(z + \Delta z) - W_1(z - \Delta z). \tag{32}$$

Expanding this in a Taylor series around z_0 yields

$$\Delta W = 2W_1'(z_0)\Delta z + \frac{1}{3}W_1'''(z_0)\Delta z^3 + \cdots, \tag{33}$$

 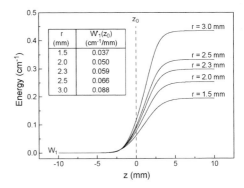

r (mm)	$W_1'(z_0)$ (cm⁻¹/mm)
1.5	0.037
2.0	0.050
2.3	0.059
2.5	0.066
3.0	0.088

FIG. 49. Left: contour lines showing the transverse electric field in the middle of the gap at z_0. The dashed lines indicate the contours of the hexapole electrodes. Right: radial dependence of the buncher potential W_1. The potential energy curves are shown for $r = 1.5, 2.0, 2.3$ (equilibrium orbit), 2.5 and 3.0 mm. The inset lists $W_1'(z_0)$ for the different radii.

where the n primes denote the nth derivative of W_1 with respect to z. The average force $\langle F \rangle$ over one section of the ring (one gap and one bend) with length L is for small values of Δz approximately given by

$$\langle F \rangle = -\frac{2W_1'(z_0)\Delta z}{L}. \tag{34}$$

This is a linear restoring force with an angular frequency given by

$$\omega_{\text{bunch}} = \sqrt{\frac{2W_1'(z_0)}{mL}}. \tag{35}$$

The value for $W_1'(z_0) = 0.059$ cm⁻¹/mm in case of ND₃ and is found from a linear fit of ΔW through z_0. This results in a frequency $\omega_{\text{bunch}}/2\pi = 67$ Hz for a section with length $L = 0.4$ m, amounting to one oscillation in longitudinal phase-space per 1.76 round trips. As the velocity spread, Δv_z, and the position spread, Δz, are related by $\Delta v_z = \omega_{\text{bunch}}\Delta z$, one finds for $\Delta z = 4$ mm, the distance over which the buncher is still linear, a velocity spread of $\Delta v_z = 0.85$ m/s. The longitudinal emittance of the decelerator is about $[\Delta z \times \Delta v_z] = [2 \text{ mm} \times 8 \text{ m/s}]$ for the currently used deceleration settings ($\phi_s = 70°$, $v_{\text{sync}} = 91.8$ m/s). One needs to map the emittance onto the longitudinal acceptance of the ring for which the buncher in the injection beamline is used.

Figure 49 depicts the electric field of the buncher at $z = z_0$ in the transverse direction. It is clear that the field is not homogeneous in the radial direction. Hence, the potential energy curve W_1 depends on the radial position as shown on the right-hand side of Fig. 49. The effect of this radial dependence is that the molecules experience a small force towards the center of the storage ring as the electric field increases radially outwards. The radial dependence couples the longitudinal motion to the transverse motion. However, the transverse betatron oscillation frequency of about 1 kHz is much larger than the longitudinal oscillation frequency of 67 Hz and, as a consequence, the coupling between the two directions is expected to be weak.

8. Conclusions and Outlook

The manipulation of polar molecules using time-varying inhomogeneous electric fields has many similarities with the manipulation of charged particles using time-varying electric fields. Previously, we have demonstrated this by showing that concepts that are used throughout in charged particle accelerators, such as phase-stability, bunch rotation, phase-space matching and alternate gradient focusing can be applied to polar molecules as well. Here, we explore the possibilities of a storage ring for polar molecules. Confining molecules in storage rings rather than in traps, as is common in atomic physics, seems to offer many advantages. We have demonstrated a small prototype storage ring created by bending a hexapole focuser into a torus. In this ring the molecules are confined in the transverse directions but are essentially free in the tangential direction. Therefore, an injected molecular packet will gradually spread out until it fills the entire ring. We have shown that by cutting the ring in two halves, one obtains a geometry that can be used to rebunch the packet. This design is currently being tested at the Fritz Haber Institute in Berlin. Although useful by itself, we consider this ring as a stepping stone to more sophisticated rings which will store tens of packets simultaneously. We believe that such a ring offers unique possibilities for various cold molecules experiments. One of the applications of such a ring would be to study collisions between counter-propagating bunches, i.e., to measure collision cross-sections as a function of beam energy and to record resonances in the collision complex; a μeV molecular physics experiment as the ultimate copy of a high energy nuclear physics experiment.

9. Acknowledgements

This paper reviews five years of work performed in Nijmegen, Nieuwegein and Berlin. We gratefully acknowledge the contributions of many colleagues to the experiments described in this paper, in particular, Giel Berden, David Carty, Cynthia Heiner, Rienk Jongma, Jochen Küpper, Bas van de Meerakker, Allard Mosk, André van Roij, Paul Smeets and Jacqueline van Veldhoven. This work is part of the research programme of the Stichting voor Fundamenteel Onderzoek der Materie (FOM), which is financially supported by the Nederlandse Organisatie voor Wetenschappelijk Onderzoek (NWO).

10. References

Ammann, H., Christensen, N. (1997). Delta kick cooling: A new method for cooling atoms. *Phys. Rev. Lett.* **78**, 2088–2091.

Anderson, R.W. (1997). Tracks of symmetric top molecules in hexapole electric fields. *J. Phys. Chem. A* **101**, 7664–7673.

Ashfold, M.N.R., Dixon, R.N., Little, N., Stickland, R.J., Western, C.M. (1988). The $\tilde{B}^1 E''$ state of ammonia: Sub-Doppler spectroscopy at vacuum ultraviolet energies. *J. Chem. Phys.* **89**, 1754–1761.

Auerbach, D., Bromberg, E.E.A., Wharton, L. (1966). Alternate-gradient focusing of molecular beams. *J. Chem. Phys.* **45**, 2160–2166.

Bennewitz, H.G., Paul, W., Schlier, Ch. (1955). Fokussierung polarer Moleküle. *Z. Phys.* **141**, 6–15.

Bethlem, H.L., Meijer, G. (2003). Production and application of translationally cold molecules. *Internat. Rev. Phys. Chem.* **22**, 73–128.

Bethlem, H.L., Berden, G., Meijer, G. (1999). Decelerating neutral dipolar molecules. *Phys. Rev. Lett.* **83**, 1558–1561.

Bethlem, H.L, Berden, G., Crompvoets, F.M.H., Jongma, R.T., van Roij, A.J.A., Meijer, G. (2000a). Electrostatic trapping of ammonia molecules. *Nature (London)* **406**, 491–494.

Bethlem, H.L., Berden, G., van Roij, A.J.A., Crompvoets, F.M.H., Meijer, G. (2000b). Trapping neutral molecules in a traveling potential well. *Phys. Rev. Lett.* **84**, 5744–5747.

Bethlem, H.L., Crompvoets, F.M.H., Jongma, R.T., van de Meerakker, S.Y.T., Meijer, G. (2002a). Deceleration and trapping of ammonia using time-varying electric fields. *Phys. Rev. A* **65**, 053416(1–20).

Bethlem, H.L., van Roij, A.J.A., Jongma, R.T., Meijer, G. (2002b). Alternate gradient focusing and deceleration of a molecular beam. *Phys. Rev. Lett.* **88**, 133003(1–4).

Bochinski, J.R., Hudson, E.R., Lewandowski, H.J., Meijer, G., Ye, J. (2003). Phase space manipulation of cold free radical OH molecules. *Phys. Rev. Lett.* **91**, 243001(1–4).

Bochinski, J.R., Hudson, E.R., Lewandowski, H.J., Ye, J. (2004). Phase space manipulation of cold free radical OH molecules. *Phys. Rev. A* **70**, 043410(1–18).

Bromberg, E.E.A., Acceleration and alternate-gradient focusing of neutral polar diatomic molecules. Ph.D. Thesis, University of Chicago.

Chien, K.-R., Foreman, P.B., Castleton, K.H., Kukolich, S.G. (1975). Relaxation cross section measurements on NH_3 and lower state focussing in a beam maser. *Chem. Phys.* **7**, 161–163.

Chu, S. (1998). Nobel lecture: The manipulation of neutral particles. *Rev. Mod. Phys.* **70**, 685–706.

Chu, S., Bjorkholm, J.E., Ashkin, A., Gordon, J.P., Hollberg, L.W. (1986). Proposal for optically cooling atoms to temperatures of the order of 10^{-6} K. *Opt. Lett.* **11**, 73–75.

Cohen-Tannoudji, C.N. (1998). Nobel lecture: Manipulating atoms with photons. *Rev. Mod. Phys.* **70**, 707–719.

Conte, M., MacKay, W.W. (1991). "An Introduction to the Physics of Particle Accelerators". World Scientific, Singapore.

Cornell, E.A., Wieman, C.E. (2002). Nobel lecture: Bose–Einstein condensation in a dilute gas, the first 70 years and some recent experiments. *Rev. Mod. Phys.* **74**, 875–893.

Courant, E.D., Snyder, H.S. (1958). Theory of the alternating-gradient synchrotron. *Ann. Phys.* **3**, 1–48.

Courant, E.D., Livingstone, M.S., Snyder, H.S. (1952). The strong-focusing synchrotron—A new high energy accelerator. *Phys. Rev.* **88**, 1190–1196.

Crompvoets, F.M.H., Bethlem, H.L., Jongma, R.T., Meijer, G. (2001). A prototype storage ring for neutral molecules. *Nature (London)* **411**, 174–176.

Crompvoets, F.M.H., Jongma, R.T., Bethlem, H.L., van Roij, A.J.A., Meijer, G. (2002). Longitudinal focusing and cooling of a molecular beam. *Phys. Rev. Lett.* **89**, 093004(1–4).

Crompvoets, F.M.H., Bethlem, H.L., Küpper, J., van Roij, A.J.A., Meijer, G. (2004). Dynamics of neutral molecules stored in a ring. *Phys. Rev. A* **69**, 063406(1–5).

Dahl, D.A. (1995). Simion 3D Version 6.0, Idaho National Engineering Laboratory, Idaho Falls.

Debye, P. (1929). "Polar Molecules". Dover, New York.

Dehmelt, H. (1990). Nobel lecture: Experiments with an isolated subatomic particle at rest. *Rev. Mod. Phys.* **62**, 525–530.

Friedrich, B., Herschbach, D.R. (1991). On the possibility of orienting rotationally cooled polar molecules in an electric field. *Z. Phys. D* **18**, 153–161.

Gandhi, S.R., Bernstein, R.B. (1987). Focusing and state selection of NH_3 and OCS by the electrostatic hexapole via first- and second-order Stark effects. *J. Chem. Phys.* **87**, 6457–6467.

Goldstein, H., Poole, C., Safko, J. (2002). "Classical Mechanics", 3rd Edn. Addison-Wesley, San Francisco.

Golub, R., On decelerating molecules, Ph.D. Thesis, Massachusetts Institute of Technology.

Gordon, J.P., Zeiger, H.J., Townes, C.H. (1954). Molecular microwave oscillator and new hyperfine structure in the microwave spectrum of NH_3. *Phys. Rev.* **95**, 282–284.

Gupta, M., Herschbach, D. (1999). A mechanical means to produce intense beams of slow molecules. *J. Phys. Chem. A* **103**, 10670–10673.

Helmer, J.C., Jacobus, F.B., Sturrock, P.A. (1960). Focusing molecular beams of NH_3. *J. Appl. Phys.* **31**, 458–463.

Humphries Jr., S. (1990). "Principles of Charged Particle Acceleration". Wiley, New York. Digital Edn.

Ivanov, D., Wallentowitz, S., Walmsley, I.A. (2003). Quantum limits of stochastic cooling of a bosonic gas. *Phys. Rev. A* **67**, 061401(1–4).

Jongma, R.T., Rasing, T., Meijer, G. (1995). Two-dimensional imaging of metastable CO molecules. *J. Chem. Phys.* **102**, 1925–1933.

Jongma, R.T., von Helden, G., Berden, G., Meijer, G. (1997). Confining CO molecules in stable orbits. *Chem. Phys. Lett.* **270**, 304–308.

Kakati, D., Lainé, D.C. (1967). Alternate-gradient focusing of a molecular beam of ammonia. *Phys. Lett. A* **24**, 676.

Katz, D.P. (1997). A storage ring for polar molecules. *J. Chem. Phys.* **107**, 8491–8501.

Ketterle, W. (2002). Nobel lecture: When atoms behave as waves: Bose–Einstein condensation and the atom laser. *Rev. Mod. Phys.* **74**, 1131–1151.

Ketterle, W., Pritchard, D.E. (1992a). Trapping and focusing ground state atoms with static fields. *Appl. Phys. B* **54**, 403–406.

Ketterle, W., Pritchard, D.E. (1992b). Atom cooling by time-dependent potentials. *Phys. Rev. A* **46**, 4051–4054.

Ketterle, W., van Druten, N.J. (1996). Evaporative cooling of trapped atoms. *Adv. Atom. Mol. Opt. Phys.* **37**, 181–236.

Kukolich, S.G. (1967). Measurement of ammonia hyperfine structure with a two-cavity maser. *Phys. Rev.* **156**, 83–92.

Kügler, K.-J., Paul, W., Trinks, U. (1978). Magnetic storage ring for neutrons. *Phys. Lett. B* **72**, 422–424.

Kügler, K.-J., Moritz, K., Paul, W., Trinks, U. (1985). NESTOR—A magnetic storage ring for slow neutrons. *Nucl. Instrum. Methods A* **228**, 240–258.

Lawrence, E.O., Edlefson, N.F. (1930). On the production of high speed protons. *Science* **72**, 376–377.

Lawrence, E.O., Livingston, M.S. (1931). The production of high speed protons without the use of high voltages. *Phys. Rev.* **38**, 834.

Lichtenberg, A.J. (1969). "Phase Space Dynamics of Particles". *Wiley Series in Plasma Physics*. Wiley, New York.

Liouville, J. (1838). Sur la théorie de la variation des constantes arbitraires. *J. Math. Pures Appl.* **3**, 342.

Livingston, M.S., Blewett, J.P. (1962). "Particle Accelerators". *International Series in Pure and Applied Physics*. McGraw-Hill, New York.

Loesch, H.J., Remscheid, A. (1990). Brute force in molecular reaction dynamics: A novel technique for measuring steric effects. *J. Chem. Phys.* **93**, 4779–4790.

Loesch, H.J., Scheel, B. (2000). Molecules on Kepler orbits: An experimental study. *Phys. Rev. Lett.* **85**, 2709–2712.

Maddi, J.A., Dinneen, T.P., Gould, H. (1999). Slowing and cooling molecules and neutral atoms by time-varying electric-field gradients. *Phys. Rev. A* **60**, 3882–3891.

Maréchal, E., Guibal, S., Bossennec, J.-L., Barbé, R., Keller, J.-C., Gorceix, O. (1999). Longitudinal focusing of an atomic cloud using pulsed magnetic forces. *Phys. Rev. A* **59**, 4636–4640.

McMillan, E.M. (1945). The synchrotron—A proposed high energy particle accelerator. *Phys. Rev.* **68**, 143–144.

Myrskog, S.H., Fox, J.K., Moon, H.S., Kim, J.B., Steinberg, A.M. (2000). Modified "δ-kick cooling" using magnetic field gradients. *Phys. Rev. A* **61**, 053412(1–6).

Nishimura, H., Lambertson, G., Kalnins, J.G., Gould, H. (2003). Feasibility of a synchrotron storage ring for neutral polar molecules. *Rev. Sci. Instrum.* **74**, 3271–3278.

Nishimura, H., Lambertson, G., Kalnins, J.G., Gould, H. (2004). Feasibility of a storage ring for polar molecules in strong-field-seeking states. *European J. Phys. D* **31**, 359–364.

Paul, W. (1990). Nobel lecture: Electromagnetic traps for charged and neutral particles. *Rev. Mod. Phys.* **62**, 531–540.

Phillips, W.D. (1998). Nobel lecture: Laser cooling and trapping of neutral atoms. *Rev. Mod. Phys.* **70**, 721–741.

Raizen, M.G., Koga, J., Sundaram, B., Kishimoto, Y., Takuma, H., Tajima, T. (1998). Stochastic cooling of atoms using lasers. *Phys. Rev. A* **58**, 4757–4760.

Reif, F. (1985). "Fundamentals of Statistical and Thermal Physics", Internat. Edn. McGraw-Hill, Singapore.

Reuss, J. (1988). State-selection by nonoptical methods. In: Scoles, G. (Ed.), In: *Atomic and molecular beam methods*, vol. 1. Oxford Univ. Press, New York, pp. 276–292.

Sauer, J.A., Barrett, M.D., Chapman, M.S. (2001). Storage ring for neutral atoms. *Phys. Rev. Lett.* **87**, 270401(1–4).

Schmiedmayer, J. (1995). A wire trap for neutral atoms. *Appl. Phys. B* **60**, 169–179.

Sekatskiĭ, S.K. (1995). Electrostatic traps for polar molecules. *JETP Lett.* **62**, 916–920.

Sekatskiĭ, S.K., Schmiedmayer, J. (1996). Trapping polar molecules with a charged wire. *Europhys. Lett.* **36**, 407–412.

Stark, J. (1914). Beobachtungen über der Effect des elektrischen Feldes auf Spektrallinien, I. Querefect. *Ann. Phys.* **43**, 965–982.

Stern, O. (1921). Ein Weg zur experimentellen Prüfung der Richtungsquantelung im Magnetfeld. *Z. Phys.* **7**, 249–253.

Summhammer, J., Niel, L., Rauch, H. (1986). Focusing of pulsed neutrons by traveling magnetic potentials. *Z. Phys. B* **62**, 269–278.

Tarbutt, M.R., Bethlem, H.L., Hudson, J.J., Ryabov, V.L., Ryzhov, V.A., Sauer, B.E., Meijer, G., Hinds, E.A. (2004). Slowing heavy, ground-state molecules using an alternating gradient decelerator. *Phys. Rev. Lett.* **92**, 173002(1–4).

Thompson, D., Lovelace, R.V.E., Lee, D.M. (1989). Storage rings for spin-polarized hydrogen. *J. Opt. Soc. Amer. B* **6**, 2227–2234.

van de Meerakker, S.Y.T. (2006). Deceleration and electrostatic trapping of OH radicals. Ph.D. Thesis, Radboud University, Nijmegen.

van de Meerakker, S.Y.T., Jongma, R.T., Bethlem, H.L., Meijer, G. (2001). Accumulating NH radicals in a magnetic trap. *Phys. Rev. A* **64**, 041401(1–4).

van de Meerakker, S.Y.T., Smeets, P.H.M., Vanhaecke, N., Jongma, R.T., Meijer, G. (2005). Deceleration and electrostatic trapping of OH radicals. *Phys. Rev. Lett.* **94**, 023004(1–4).

van der Meer, S. (1972). Stochastic damping of betatron oscillations in the ISR. CERN Internal Report CERN/ISR-PO/72-31, CERN, Geneva.

van der Meer, S. (1985). Stochastic cooling and accumulation of antiprotons. *Rev. Mod. Phys.* **57**, 689–697.

van Veldhoven, J., Jongma, R.T., Sartakov, B., Bongers, W.A., Meijer, G. (2002). Hyperfine structure of ND_3. *Phys. Rev. A* **66**, 032501(1–9).

van Veldhoven, J., Bethlem, H.L., Meijer, G. (2005). AC electric trap for ground-state molecules. *Phys. Rev. Lett.* **94**, 083001(1–4).

Veksler, V.I. (1944). A new method of acceleration of relativistic particles. *Doklady Akad. Nauk SSSR* **43**, 346–351.

Wilson, E.J.N. (2001). "An Introduction to Particle Accelerators". Oxford Univ. Press, Oxford.

Wing, W.H. (1980). Electrostatic trapping of neutral atomic particles. *Phys. Rev. Lett.* **45**, 631–634.

Wing, W.H. (1984). On neutral particle trapping in quasistatic electromagnetic fields. *Prog. Quant. Electr.* **8**, 181–199.

Zare, R.N. (1987). "Angular Momentum". Wiley, New York.

ADVANCES IN ATOMIC, MOLECULAR AND OPTICAL PHYSICS, VOL. 52

NONADIABATIC ALIGNMENT BY INTENSE PULSES. CONCEPTS, THEORY, AND DIRECTIONS

TAMAR SEIDEMAN[*] and EDWARD HAMILTON

Department of Chemistry, Northwestern University, 2145 Sheridan Road, Evanston, IL 60208-3113, USA

Abstract

We review the theory of intense laser alignment of molecules and present a survey of the many recent developments in this rapidly evolving field. Starting with a qualitative discussion that emphasizes the physical mechanism responsible for laser alignment, we proceed with a detailed exposition of the underlying theory, focusing on aspects that have not been presented in the past. Application of the theory in several pedagogical illustrations is then followed by a review of the recent experimental and theoretical advances in the field. We conclude with a discussion of new directions, future opportunities, and areas where we expect intense laser align-

[*] e-mail: seideman@chem.northwestern.edu

ISSN 1049-250X
DOI 10.1016/S1049-250X(05)52006-8

ment to play a role in the future. Throughout we emphasize the recent evolution of the method of nonadiabatic alignment from isolated diatomic molecules to complex media.

1. Preliminaries

Alignment by short intense laser pulses—the application of coherent light to produce rotationally-broad, spatially well-defined wavepackets—is emerging as a fascinating problem in fundamental research with a broad range of potential applications. Interestingly interdisciplinary, the problem of short-pulse-induced alignment interfaces with problems and methods in stereochemistry [1,2], intense laser physics,[1] time–domain spectroscopy [3], and the theory of coherent states [4]. Potential applications range from study and manipulation of chemical reactions [5–8] and elucidation of molecular structure [3,9,10], through generation of ultra-short light pulses [11] and of high-order harmonics of light [12–17], to fundamental studies in coherence and dissipation [18] and new routes to quantum information processing [19].

The present review has no ambition to cover all of these topics. We have attempted to provide a balance between a basic tutorial and an overview of the current status of the field, emphasizing the former aspect, which is covered to a much lesser extent in the existing literature than the latter. We thus expect several sections to be useful for readers who are familiar with the problem, while others serve readers who are new to the field and seek an introductory discussion. Given the richness of the subject matter and the large number of important contributions made by groups in a diverse range of disciplines in recent years, the task of reviewing the relevant literature is particularly challenging, if possible. Here we

---

[1] Alignment in high intensity multielectron dissociative ionization (MEDI) was addressed in early studies by several groups, see, for instance, D.T. Strickland, Y. Beaudoiin, P. Dietrich and P.B. Corkum (1992). *Phys. Rev. Lett.* **68**, 2755; D. Normand, L.A. Lompré and C. Cornaggia (1992). *J. Phys. B* **25**, L497; V.R. Bhardwaj, D. Mathur and F.A. Rajgara (1998). *Phys. Rev. Lett.* **80**, 3220; Ch. Ellert and P.B. Corkum (1999). *Phys. Rev. A* **59**, R3170; L.J. Frasinski, J. Plumridge, J.H. Posthumus, K. Codling, P.F. Taday, E.J. Divall and A.J. Langley (2001). *Phys. Rev. Lett.* **86**, 2541. The physical origin of alignment in such experiments is the same as that at moderate intensities (below ionization thresholds) but their interpretation is more complicated [121] due to the angular dependence of the phenomenon of enhanced ionization—strong enhancement of the ionization rate of aligned molecular ions in a narrow range of internuclear distances. Importantly, by contrast to alignment in moderately intense fields, alignment in MEDI experiments can take place only for a combination of pulse duration and molecular mass that allows the molecule to align *during* the pulse, as the molecule is destroyed before the end of the interaction.

take advantage of the availability of a review of the theoretical and experimental aspects of strong field alignment that appeared in press in 2003 [20], to limit our attention to material that is excluded from that paper. We focus on review of the (substantial amount of) work that was published subsequent to submission of the earlier review, on the underlying theory, which we chose to omit from [20], and on discussion of new directions and future opportunities that we did not yet envision at the time of its writing. Our review of the literature is thus regretfully incomplete, although we have made an effort to include the relevant literature that the reader will not find in [20]. It is our hope that the article will provide our audience with easy access to the literature of interest along with appetite to explore more of it than what we have been able to cover here.

Before proceeding to outline the structure of the present article, we refer the reader to several review articles on related topics that space considerations preclude from inclusion in what follows. The problem of adiabatic alignment (alignment by continuous wave laser fields, or, equivalently, by pulses long compared to the system time-scales), is reviewed in [20,21] (see also [22]). This problem is formally and numerically (though not experimentally) equivalent to the extensively studied problem of alignment of nonpolar molecules in a static (direct current, DC) electric field [23–30], recently discussed in the context of field-modified spectroscopy, e.g., in [31,32]. Intense laser alignment in pump-probe experiments and its characterization by time-resolved photoelectron imaging are included in the review article [33]. The related field of rotational coherence spectroscopy (RCS) has been thoroughly reviewed in the early literature [34,35]. A brief discussion of strong field effects in RCS, as well as a comprehensive list of references, are included in the recent review article [3]. An early but certainly not outdated review of coherent states is provided in [36]. Finally, the more general but closely related problem of wavepacket revivals is reviewed in [37].

Since we expect different sections of this review to interest different readers, we have attempted to structure the article such that each section could be read independently of the others. In the next section we discuss the qualitative physics responsible for intense laser alignment and place the underlying concepts in a broader perspective, making ties with related problems and methods in different subdisciplines of physics and chemistry. Section 3 discusses the theory of short, intense pulse alignment for a general polyatomic molecule. In Section 4 we apply the theory of Section 3, first to model systems that serve to illustrate the general concepts in a molecule-independent way, and next to realistic molecules that allow direct comparison with experiments. We deviate from the standard practice in review articles by basing Section 4 and part of Section 3 on formal and numerical results obtained in our laboratory during the past year that were not published elsewhere.

Section 5 provides a review of recent accomplishments in the area of intense laser alignment. We focus on work that appeared subsequent to submission of [20] and attempt to include a balance of theoretical, numerical and experimental advances. These include theoretical explorations of the correspondence between classical and quantum mechanical descriptions of post-pulse alignment [38–41], experimental advances in the production and characterization of orientation [42–46], and studies of rotational wavepacket engineering and manipulation with shaped pulses [47]. Section 6 is devoted to new directions, future opportunities, and areas where we expect that intense laser alignment will play a role in the future. These include control of charge transfer reactions in solution, field-guided molecular assembly as a method of device fabrication, structural studies of bio-molecules, the combination of molecular alignment with molecular optics as a route to material nano-processing, opportunities in quantum computing and the application of rotational wavepackets to probe dissipative media. We conclude Section 6 with a brief summary. An appendix details the derivation of a result of more general interest that is utilized Section 3.

2. Basic Concepts

In this section we provide a qualitative picture of coherent laser alignment, making contact with, and drawing analogies to, familiar problems in other subdisciplines of physics and chemistry, where relevant. We start by discussing the mechanism underlying laser alignment within a quantum mechanical framework (Section 2.1) and noting the role played by the field and system parameters (Section 2.2). To that end we focus on the simplest case scenario of a linear, isolated rigid rotor subject to a linearly polarized field. While this simple model serves here primarily to bring across the general concepts, it corresponds to the vast majority of realistic systems considered to date in experiments and numerical calculations. In Section 2.3 we note, via discussion of the corresponding classical dynamics, that short-pulse-induced alignment of nonlinear rotors is expected to introduce new and interesting physics at the fundamental level. In Section 2.4 we extend alignment from one to three dimensions through choice of the laser polarization. The last two subsections discuss briefly the possibility of orienting molecules with intense pulses, and the extension of alignment from isolated molecules to dissipative media.

2.1. ROTATIONAL EXCITATION AND COHERENT ALIGNMENT

Consider a rigid, linear molecule subject to a linearly polarized laser field whose frequency is tuned near resonance with a vibronic transition. In the weak field limit, if the system has been initially prepared in a rotational level J_0, electric dipole selection rules allow J_0 and $J_0 \pm 1$ in the excited state. The interference between these levels gives rise to a mildly aligned excited state population, depending in sense on the type of the dipole transition. At nonperturbative intensities, the system undergoes Rabi-type oscillations between the two resonant rotational manifolds, exchanging another unit of angular momentum with the field on each transition. Consequently, a rotationally-broad wavepacket is produced in both states. The associated alignment is considerably sharper than the weak-field distribution, as it arises from the interference of many levels [48,49]. The degree of rotational excitation is determined by either the pulse duration or the balance between the laser intensity and the detuning from resonance. In the limit of a short, intense pulse, $\tau_{\text{pulse}}^2 < (B_e\Omega_R)^{-1}$, the number of levels that can be excited is roughly the number of cycles the system has undergone between the two states, $J_{\max} \sim \tau_{\text{pulse}}/\Omega_R^{-1}$, where τ_{pulse} is the pulse duration, B_e is the rotational constant,[2] Ω_R is the Rabi coupling[3] and Ω_R^{-1} is the corresponding period. In the case of a long, or lower intensity pulse, $\tau_{\text{pulse}}^2 > (B_e\Omega_R)^{-1}$, the degree of rotational excitation $J_{\max}$ is determined by the accumulated detuning from resonance, i.e., by the requirement that the Rabi coupling be sufficient to access rotational levels that are further detuned from resonance as the excitation proceeds, $J_{\max} \sim \sqrt{\Omega_R/2B_e}$. A quantitative discussion is given in [50].

A rather similar coherent rotational excitation process takes place at nonresonant frequencies, well below electronic transition frequencies. In this case a rotationally-broad superposition state is produced via sequential Raman-type ($|\Delta J| = 0, 2$) transitions, the system remaining in the ground vibronic state [50]. The qualitative criteria determining the degree of rotational excitation remain as discussed in connection with the near-resonance case. In the case of near-resonance excitation, however, the Rabi coupling is proportional to the laser electric field, whereas in the case of nonresonant excitation it is proportional to the intensity, i.e., to the square of the field, due to the two-photon nature of the cycles. Consequently, much higher intensities are required to achieve a given degree of alignment, as expected for a nonresonant process. On the other hand, much higher intensities can be exerted at nonresonant frequencies, as undesired competing processes scale similarly with detuning from resonance. It is interesting to note that the dynamics of rotational excitation and the wavepacket properties are essentially identical in the near and nonresonance excitation schemes. In fact it is possible to formally transform the equations of motion corresponding to one case into those corresponding to the other [50]. In reality the near- and the nonresonant mechanisms may act simultaneously, with the former typically corresponding to a vibrational resonance [51]. The degree of spatial localization and the time-evolution of the alignment are essentially independent of the frequency

---

[2] See Section 3.2.
[3] The precise form of the Rabi coupling (the matrix element of the field matter interaction) depends on the experiment in mind and is identified for specific scenarios in Section 3.

regime; they are largely controlled by the temporal characteristics of the laser pulse, as illustrated in the following subsection.

2.2. TIME EVOLUTION

The simplest case scenario is that of alignment in a continuous wave (CW) field, proposed in [52,53], where dynamical considerations play no role. In practice, intense field experiments are generally carried out in pulsed mode, but, provided that the pulse is long with respect to the rotational period, $\tau_{rot} = \pi\hbar/B_e$, each eigenstate of the field-free Hamiltonian is guaranteed to evolve adiabatically into the corresponding state of the complete Hamiltonian during the turn-on, returning to the original (isotropic) field-free eigenstate upon turn-off [50]. Formally, the problem is thus reduced to alignment in a CW field of the peak intensity. The latter problem is formally equivalent to alignment of nonpolar molecules in a strong DC field, since the oscillations of the laser field at the light frequency can be eliminated at both near- and nonresonance frequencies [48,50] (see Section 3 and Appendix A). Thus, in the long pulse limit, $\tau_{pulse} \gg \tau_{rot}$, laser alignment reduces to an intensively studied problem that is readily understood in classical terms, namely the problem of field-induced pendular states [23–30]. In this limit, the sole requirement for alignment is that the Rabi coupling be large as compared to the rotational energy of the molecules at a given rotational temperature. Intense laser alignment in the CW limit (termed adiabatic alignment) grew out of research on alignment and orientation in DC fields, and is very similar to DC alignment conceptually and numerically. It shares the advantage of DC field alignment of offering an analytical solution in the linear, rigid rotor case [52]. It shares, however, also the main drawback of DC field alignment, namely, the alignment is lost once the laser pulse has been turned off. For applications one desires field-free aligned molecules.

Short-pulse-induced alignment (termed nonadiabatic, or dynamical alignment) was introduced [48] at the same time as the adiabatic counterpart and is similar in application but qualitatively different in concept. This field of research grew out of research on wavepacket dynamics and shares several of the well-studied features of vibrational and electronic wavepackets while exhibiting several unique properties (*vide infra*). A short laser pulse, $\tau_{pulse} < \tau_{rot}$, leaves the system in a coherent superposition of rotational levels that is aligned upon turn-off, dephases at a rate proportional to the square of the wavepacket width in J-space, and subsequently revives and dephases periodically in time. As long as coherence is maintained, the alignment is reconstructed at predetermined times and survives for a controllable period. As with other discrete state wavepackets of stable motions, all observables obtained from the wavepacket are periodic in the full revival time, given, in the case of rotational wavepackets, by the rotational period τ_{rot}. Interestingly, under rather general conditions, the alignment is significantly enhanced after turn-off of the laser pulse [49]. The origin of the phenomenon of enhanced field-free alignment is unraveled in [49] by means of an analytical model of the revival structure of rotational wavepackets. The first experiment to realize the prediction of field-free, post pulse alignment was reported in 2001 [54]. Several of the more recent experimental realizations [55–59] are discussed in Section 5.

The ultrashort pulse limit of enhanced alignment, where the interaction is an impulse as compared to rotations, is again readily understood in classical terms [49]. In this limit an intense pulse imparts a "kick" to the molecule that rapidly transfers a large amount of angular momentum to the system and gives rise to alignment that sets in only after the turn-off. We note that gas-phase femto-chemistry experiments are often carried out under conditions close to the impulse limit. High intensities are difficult to avoid in ultrafast pump-probe spectroscopy experiments. For relatively heavy systems the pump pulse is thus instantaneous on the rotational time scale, $\tau_{pulse} \ll \tau_{rot}$, but sufficiently long to coherently excite rotations, $\tau_{pulse} \gg \Omega_R^{-1}$ (recall that Ω_R is proportional to the field amplitude in the near-resonance case). The molecule is rotationally frozen during the interaction but the sudden "kick" is encoded in the wavepacket rotational composition and gives rise to alignment after the pulse turn-off. Thus, in both the long, $\tau_{pulse} \gg \tau_{rot}$, and the ultrashort, $\tau_{pulse} \ll \tau_{rot}$, pulse limits intense laser alignment is readily understood in classical terms, at least on a qualitative level. Both limits, moreover, allow for an analytical solution [50,52], hence offering useful insight. The intermediate pulse-length case is not amenable to analytical formulation but offers control over the alignment dynamics through choice of the field parameters.

A third mode of alignment is introduced in [60] in the context of simultaneous alignment and focusing of molecular beams, potentially an approach to nanoscale surface processing [61]. A route to that end is provided by the combination of long turn-on with rapid turn-off of the laser pulse. During the slow turn-on, the isotropic free rotor eigenstate evolves into an eigenstate of the complete Hamiltonian—an aligned many-J superposition—as in the long-pulse limit. Upon rapid turn-off the wavepacket components are phased together to make the full Hamiltonian eigenstate but now evolve subject to the field-free Hamiltonian. The alignment dephases at a rate determined by the rotational level content of the wavepacket (hence the intensity, the pulse duration and the rotational constant), and subsequently revives. At integer multiples of the rotational period, $t = t_0 + n\tau_{rot}$, t_0 being the pulse turn-off, the fully interacting eigenstate of the complete Hamiltonian attained at the pulse peak is precisely reconstructed and the alignment characterizing the adiabatic limit is transiently available under field-free conditions. Experimental realization of field-free alignment via the combination of slow turn-on with rapid turn-off was recently reported [62]. Numerical illustrations of the alignment resulting from a pulse combining slow turn-on with rapid turn-off are provided in Section 4. A discussion of potential applications of this mode of alignment is given in Section 6.

Our discussion so far has been limited to linear systems, which have been the focus of the vast majority of theoretical and experimental contributions to the field so far. We show in the next section that the case of nonlinear, in particular asymmetric, rotors offers new and interesting physics.

2.3. Role of the Molecular Symmetry. From Diatomic Molecules to Complex Systems

While the potential practical interest in the nonadiabatic alignment dynamics of complex polyatomic molecules is evident (see Section 6), the fundamental interest in nonlinear structures, in particular asymmetric tops, is readily appreciated by consideration of the rotation of rigid bodies within classical mechanics.

To that end we begin by introducing in Fig. 1 two sets of Cartesian axes systems; a space-fixed reference frame (x, y, z), later to be defined by the field polarization and propagation direction, and a body-fixed reference frame, (X, Y, Z),

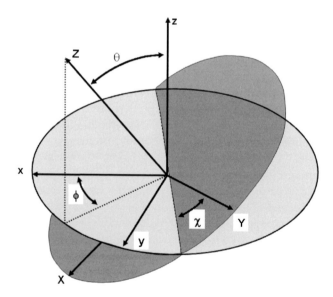

FIG. 1. Definition of the body-fixed (X, Y, Z) and space-fixed (x, y, z) coordinate systems in terms of the Euler angles that characterize their relative rotation.

typically defined via the molecular inertia tensor and (when relevant) symmetry.[4] Also introduced in Fig. 1 are the Euler angles of rotation of the body- with respect to the space-fixed frame. We use the convention of Hertzberg [63] and Bohr [64] in choosing the Euler angles [65,66], and adopt the common notation, wherein the polar Euler angle, $\cos^{-1}(\widehat{Z} \cdot \hat{z})$, is denoted θ and the azimuthal angles of rotation about the space- and body-fixed z-axes are denoted ϕ and χ, respectively. The collective variable $\widehat{R} = (\theta, \phi, \chi)$ is used whenever no ambiguity may arise.

Solution of Euler's equations of motion for the time-evolution of θ, ϕ and χ is particularly simple for a freely rotating symmetric top [67]. In free space a unique axis in space is not naturally defined and it is conventional to take the space-fixed z-axis in the direction of the (constant) angular momentum vector of the top. It is then readily shown that the angle θ between the top axis and the direction of J is constant, $\dot{\theta} = 0$, whereas the two azimuthal angles execute stable periodic motion as functions of time, $\dot{\phi} = J/I_{aa}$, $\dot{\chi} = J \cos\theta (I_{cc}^{-1} - I_{aa}^{-1})$. Here I is the inertia tensor with elements $I_{kk} = \sum_i m_i(q_i^2 - q_{ik}^2)$, $I_{kk'} = I_{k'k} = -\sum_i m_i q_{ik} q_{ik'}$, where $\mathbf{q}_i$ is the position vector of particle i with mass m_i in the body-fixed frame and $k, k' = X, Y, Z$. Thus, the top axis undergoes regular precession about the angular momentum vector $\mathbf{J}$, describing a circular cone with apex half-angle θ, while rotating uniformly about its own axis. The same result can be derived through angular momentum conservation arguments [67].

This attractive simplicity is lost in the asymmetric top case. Rotation about the c- and a-axes (corresponding to the largest and smallest inertia moments, respectively) remains stable in the sense that small deviation of the top from its path produces motion close to the unperturbed one. Rotation about the b-axis, however, is no longer regular, hence a small deviation suffices to give rise to motion that diverges strongly from the original path of the top. (The instability is readily visualized when one attempts to spin a book about its three axes.) It follows that θ and χ remain periodic functions of time but ϕ, the azimuthal Euler angle corresponding to rotation about the space-fixed z-axis, does not. Rather, the motion of $\phi(t)$ is a combination of two periodic functions with incommensurable periods. It is due to this incommensurability that the asymmetric top does not return to its original position at any later time.

One of the attractive features of wavepackets is that they provide a route to the classical limit; by contrast to an eigenstate, a wavepacket approaches in a well-defined limit (the limit of a sufficiently broad superposition in the quantum number space) the motion of a classical particle whose position is well defined at all times. This feature is particularly interesting in the context of rotational wavepackets, because the nature of their excitation and the small level spacings of

---

[4] Part of the literature uses capital letters to denote the body-fixed Cartesian coordinates and small letters to denote the space-fixed coordinates but the reverse notation is also common.

rotational spectra allow the population of extremely broad wavepackets under realistic experimental conditions. The ability of rotational wavepackets to approach the classical motion of a single linear rotor is evident from Fig. 4 below and was illustrated experimentally and numerically before (see Section 5), but for linear rotors this motion does not reveal nontrivial physics. The foregoing discussion suggests that the asymmetric top case would be more interesting. We return to this opportunity in Sections 3 and 4.

2.4. ROLE OF THE FIELD POLARIZATION. THREE-DIMENSIONAL ALIGNMENT

A linearly-polarized laser field defines one axis in space and can therefore induce 1-dimensional orientational order; it introduces an effective potential well as a function of a single angle—the polar angle between the polarization vector and the most polarizable molecular axis—thus confining the motion in this angular variable while leaving the motion in the two azimuthal angles free. Put alternatively (but equivalently), a linearly polarized field excites a broad superposition of angular momentum levels J while conserving the projection of $\mathbf{J}$ onto the space-fixed z-axis (the magnetic quantum number M up to a factor $\hbar$). The projection of $\mathbf{J}$ onto the body-fixed z-axis (the helicity quantum number K up to $\hbar$) is either rigorously conserved (as, e.g., in a field tuned near resonance with a parallel transition) or changes by only one or two quanta (as, e.g., in a field tuned near resonance with a perpendicular transition or a nonresonant field interacting with an asymmetric top molecule). Such wavepackets can be sharply defined in θ-space but are isotropic in ϕ-space and at most mildly defined with respect to χ. Similarly, a circularly polarized intense pulse excites a broad superposition of J-levels but cannot produce a broad wavepacket in either M- or K-space; the molecule is confined to a plane but within this plane it is free to rotate.

The examples of Section 4, along with several of the potential applications proposed in Section 6, suggest that, although linearly polarized laser fields induce rich dynamics, particularly so for asymmetric top molecules, it is the case of asymmetric tops for which the usefulness of linear polarization as a control tool becomes limited, requiring a more general approach to eliminating molecular rotations and inducing orientational order. Going beyond one-dimensional order becomes more pertinent as one considers increasingly complex systems. The discussion of Section 2.3 suggests that it would be also of fundamental interest to produce a wavepacket that is spatially localized in all three angular variables and investigate in how far can this localization be prolonged into the field free domain.

Elliptically-polarized fields suggest themselves as a simple, intuitive route to three-dimensional orientational control. In particular, they introduce the possibility of tailoring the field polarization to the molecular polarizability tensor by varying the eccentricity from the linear (best adapted to linear, or prolate symmetric tops) to the circular (best adapted for planar symmetric tops) polarization limit. The theoretical and experimental techniques to spatially localize rotational wavepackets of asymmetric top molecules in the three angular variables were developed already in 2000 but were applied only in the adiabatic domain [68]. This work is reviewed in Sections II.B and III.C of [20] and hence excluded from the present review. The underlying theory is not discussed in either the original brief report [68] or the review [20] and is thus outlined in Section 3 below. A detailed analysis is given in a forthcoming publication [69], which provides also numerical illustrations of nonadiabatic, three-dimensional alignment.

2.5. MOLECULAR ORIENTATION

A purely alternating current (AC) electromagnetic field polarized along the (say) space-fixed z-axis preserves the symmetry of the field-free Hamiltonian with respect to $z \rightarrow -z$ (by contrast to a DC electric field that defines a direction, as well as an axis, in space); it aligns, but cannot orient molecules. The motivation for augmenting laser alignment to produce orientation comes from the field of stereochemistry, where orientation techniques have proven to make a valuable tool for elucidating reaction mechanisms [1,2]. Several different methods have been proposed to that end [70–81], two of which have been also tested in the laboratory [42–46].

Within the method of [70], an intense laser field is combined with a relatively weak DC electric field, with the former serving to produce a broad superposition of J-levels via the nonlinear polarizability interaction and the latter providing a means of breaking the $z \rightarrow -z$ symmetry. This approach is applied in [42] to orient HXeI, and in [43] and [44] to orient OCS molecules. An alternative approach, advanced by several groups, takes advantage of the natural asymmetry of half-cycle pulses to orient molecules. Substantial experience has been gained in recent years in optimizing the ensuing orientation and manipulating its temporal characteristics by different control schemes [75,76,78,80]. A third method, which likewise received significant attention [46,72,74,77], exploits the possibility of breaking the spatial $z \rightarrow -z$ symmetry through coherent interference, e.g., by two-color phase-locked laser excitation or by combining the fundamental frequency with its second harmonic. (The second harmonic is often taken to be resonant with a vibrational transition but a nonresonant variant of the same procedure has also been suggested [81].) The most recent developments in the area of molecular orientation are highlighted in Section 5. The exciting task of three-dimensional orientation through the combination of elliptical polarization with one of these routes to symmetry breaking remains to be investigated.

2.6. ALIGNMENT IN DISSIPATIVE MEDIA

The foregoing discussion has been restricted to the limit of isolated molecules, corresponding to a molecular beam experiment, where collisions do not take place on the time-scale of relevance and coherence is fully maintained. The motivations for extending alignment to dissipative media, where collisions give rise to decoherence[5] and population relaxation on relevant time-scales are multifold.

First, one can show [18] that the unique coherence properties of rotationally-broad wavepackets provide a sensitive probe of the dissipative properties of the medium. In particular, it is found that the experimental observables of alignment disentangle decoherence from population relaxation effects, providing independent measures of the relaxation and decoherence dynamics that go beyond rate measurements. Second, we expect laser alignment to become a versatile tool in chemistry, once the effects of dissipative media on alignment are properly understood, see Section 6 for several of the potential applications that may be envisioned. A third motivation comes from recent experiments on rotational coherence spectroscopy in a dense gas environment [3]. So far interpreted within a weak field approach, experimental work in this area has recently provided evidence for strong field effects [82], calling for a nonperturbative theory.

Reference [18] explores the evolution of nonadiabatic alignment in dissipative media within a quantum mechanical density matrix approach, illustrating both the sensitivity of rotationally broad wavepackets to the dissipative properties of the medium and the possibility of inhibiting rotational relaxation, so as to prolong the alignment lifetime, by choice of the field parameters. A classical study of alignment in a liquid is provided in [83]. The application of intense laser alignment in solutions to control charge transfer reactions is illustrated in [8]. Experimentally, nonadiabatic, intense pulse alignment in the dense gas environment has been probed in several studies, although not in all cases reported as such. A particularly quantitative study is provided in [56], which compares the alignment measured in rotationally-cold molecular beam environments with that obtained in the dense gas medium. Details of recent advances in this research are given in Section 5.

---

[5] We follow the conventions of the gas phase wavepacket dynamics literature, where "dephasing" is used for change of the relative phases of the wavepacket components due to anharmonicity of the molecular spectrum, whereas "decoherence" is used for loss of phase. The former takes place in the isolated molecule limit whereas the latter requires collisions or photon emission. It is important to note that this usage differs from that standard in the condensed phase literature, where loss of phase information due to collisions is referred to as "dephasing" while the term "quantum decoherence" is generally reserved for loss of phase information due to decay of the overlap between bath wavefunctions reacting to different system states.

3. Theory

In this section we review the theory of intense laser alignment. We start in Section 3.1 with a general formulation, applicable to arbitrary system and field characteristics, and note the form of the interaction in the limits of near- and non-resonant frequencies.[6] In Section 3.2 we focus on the case of linearly polarized, nonresonant fields, which is currently the center of growing experimental and theoretical activity, and examine the form of the field-matter interaction for different classes of molecules. In Section 3.3 we provide prescriptions for quantum dynamical calculation of the alignment characteristics and note the instances in which each prescription is most efficient. Technical details are deferred to an appendix. We attempt to address readers of widely varying familiarity with the subject of matter interaction with intense light and hence include as many footnotes comments and clarifications that would be trivial to some, but new to other, potential readers. For pedagogical purposes, we depart from our standard usage of atomic units and explicitly include the $\hbar$, such that the units of all variables are readily identified.

3.1. GENERAL FORMULATION

We consider an isolated molecular system subject to a nonperturbative radiation pulse, treating the material system as a quantal, and the electromagnetic field as a classical entity,

$$\epsilon(t) = \frac{1}{2}\left[\varepsilon(t)e^{i\omega t} + \text{c.c.}\right]. \tag{1}$$

In Eq. (1), $\varepsilon(t) = \hat{\varepsilon}\varepsilon(t)$, $\hat{\varepsilon}$ is a unit vector in the field polarization direction, $\varepsilon(t)$ is an envelope function, and ω is the central frequency. Within the electric dipole approximation the field-matter interaction is given as,

$$V = -\boldsymbol{\mu} \cdot \epsilon(t), \tag{2}$$

where $\boldsymbol{\mu}$ is the electric dipole operator, $\boldsymbol{\mu} = e\sum_j \mathbf{r}_j$, e is the electron charge, and $\mathbf{r}_j$ denotes collectively the coordinates of electron j. The wavepacket is formally expanded in a complete set of rovibronic eigenstates $|\xi\mathbf{vn}\rangle$ as,

$$\left|\Psi(t)\right\rangle = \sum_{\xi,\mathbf{v},\mathbf{n}} C^{\xi\mathbf{vn}}|\xi\mathbf{vn}\rangle, \tag{3}$$

---

[6] We follow the common, although perhaps ambiguous usage, where nonresonant is intended to imply far-off-resonance.

where ξ is an electronic index, $\mathbf{v}$ denotes collectively the vibrational quantum numbers, and $\mathbf{n}$ is a collective rotational index.[7]

Two limiting cases are of particular formal and experimental interest. In case the field frequency is tuned near an electronic resonance, the two electronic levels approximation is generally valid[8] and the interaction Hamiltonian reduces to,

$$V_{\xi\xi'} = -\boldsymbol{\mu}_{\xi,\xi'} \cdot \boldsymbol{\epsilon}(t), \tag{4}$$

where $\boldsymbol{\mu}_{\xi,\xi'} = \langle\xi|\boldsymbol{\mu}|\xi'\rangle$ is the matrix element of the dipole operator in the electronic subspace. Most common is the case of excitation from the ground state, $\xi = 0$. In case the frequency is far-detuned from vibronic transition frequencies, a specific excited state that dominates the interaction cannot be singled out, real population resides solely in the initial vibronic state, and the effect of all excited vibronic states on the dynamics in the initial state need be considered. As shown in Appendix A, the field-matter interaction can be cast in this situation in the form of an approximate induced Hamiltonian [84,85],

$$H_{\text{ind}} = -\frac{1}{4}\sum_{\rho\rho'}\varepsilon_\rho\alpha_{\rho\rho'}\varepsilon_{\rho'}^* = \sum_\rho \varepsilon_\rho\mu_\rho^{\text{ind}}, \quad \mu_\rho^{\text{ind}} = \sum_{\rho'}\alpha_{\rho\rho'}\varepsilon_{\rho'}^*, \tag{5}$$

where $\rho = x, y, z$ are the space-fixed Cartesian coordinates (Section 2.3), ε_ρ are the Cartesian components of the field, α is the polarizability tensor and $\boldsymbol{\mu}^{\text{ind}}$ defines an induced dipole operator. In terms of the body-fixed Cartesian coordinates, $k = X, Y, Z$, the polarizability tensor takes the form,

$$\alpha_{\rho\rho'} = \sum_{kk'}\langle\rho|k\rangle\alpha_{kk'}\langle k'|\rho'\rangle, \tag{6}$$

where $\langle k|\rho\rangle$ are elements of the transformation matrix between the space-fixed and body-fixed frames.

---

[7] An equivalent expansion to Eq. (3), often found in the literature, is

$$|\Psi(t)\rangle = \sum_{\xi,\mathbf{v},\mathbf{n}} F^{\xi,\mathbf{v},\mathbf{n}}(t)|\xi, \mathbf{v}, \mathbf{n}\rangle e^{-iE^{\xi,\mathbf{v},\mathbf{n}}t/\hbar},$$

with the expansion coefficients related through

$$C^{\xi,\mathbf{v},\mathbf{n}} = F^{\xi,\mathbf{v},\mathbf{n}}(t)e^{-iE^{\xi,\mathbf{v},\mathbf{n}}t/\hbar},$$

see Appendix A. The choice between the two forms is a matter of efficiency and convenience and is application-dependent.

[8] The 2-electronic levels approximation holds if the pulse duration and frequency are such that the laser band width is small as compared to the electronic levels spacing. This approximation clearly fails when the central frequency is tuned to the regime of high Rydberg levels, but typically holds in near-resonance excitation into low energy excited states.

The formal and numerical aspects of alignment at near-resonance frequencies are addressed in a recent review about the related topic of time-resolved photoelectron imaging [33]. In the following subsections we focus on the case of alignment at nonresonant frequencies. As noted in Section 2.1, and is clarified through the derivation of Appendix A, in practice both near- and nonresonant alignment may contribute to the signal [51], the former corresponding, in general, to a vibrational resonance.

3.2. NONRESONANT, NONADIABATIC ALIGNMENT

An explicit form of the induced Hamiltonian of Eq. (5) in terms of the spatial coordinates is obtained by inserting Eq. (6) in Eq. (5) and expressing the transformation elements $\langle k|\rho \rangle$ in terms of the Euler angles of rotation, $\widehat{R} = (\theta, \phi, \chi)$, see Section 2.3 and Fig. 1. Table I of [86] provides expressions for the $\langle k|\rho \rangle$ in terms of $\widehat{R}$. The resultant general form of the interaction Hamiltonian is analyzed in a forthcoming publication [69], where we illustrate theoretically and numerically the possibility of inducing field-free 3D alignment by means of a short, elliptically polarized pulse. A numerical implementation of the general formalism appeared in an early publication [68], where, however, the underlying theory is not detailed. As discussed in Section 6, we expect the general case of elliptical polarization to play an important role in future research on intense laser alignment, as one progresses to polyatomic molecules. Here, however, we limit attention to the case of linearly polarized fields, which has been the center of theoretical and experimental attention so far.

For linearly-polarized fields the double sum in Eq. (5) collapses to a single term and the induced interaction takes the form,

$$
\begin{aligned}
H_{\text{ind}} &= -\frac{\varepsilon^2(t)}{4} \left[\alpha^{ZX} \cos^2 \theta + \alpha^{YX} \sin^2 \theta \sin^2 \chi \right] \\
&= -\frac{\varepsilon^2(t)}{4} \left\{ \frac{\alpha^{ZX} + \alpha^{ZY}}{3} D_{00}^2(\widehat{R}) - \frac{\alpha^{YX}}{\sqrt{6}} \left[D_{02}^2(\widehat{R}) + D_{0-2}^2(\widehat{R}) \right] \right\}, \quad (7)
\end{aligned}
$$

where we followed the standard convention of defining the space-fixed z-axis as the field polarization direction, $\hat{\varepsilon}_\rho = \hat{\varepsilon}_{\rho'} = \hat{z}$, and introduced generalized polarizability anisotropies as $\alpha^{ZX} = \alpha_{ZZ} - \alpha_{XX}$, $\alpha^{YX} = \alpha_{YY} - \alpha_{XX}$, α_{kk} being the body-fixed components of the polarizability tensor. The D_{qs}^2 in the second equality are rotation matrices (often termed Wigner matrices) and we use the notation of Zare [66].[9] Whereas the first equality of Eq. (7) is useful for visualization

---

[9] The book [66] is fairly recent and much of the literature uses earlier notational conventions. We adopt here the notation of [66], as it has become very popular in recent years, but note that our early articles on the alignment problem followed Edmonds' notation [65]. The latter differs only slightly from Zare's and uses the same definition of the Euler angles.

purposes, the second will provide below an easy access to the form of the matrix elements of H_{ind} and the nature of the transitions it induces. The two forms of Eq. (7) omit (different) terms that are independent of the angles and simply shift the potential, without effecting the alignment dynamics. The dependence of the interaction on θ and χ—the polar Euler angle and the angle of rotation about the body-fixed z-axis—gives rise to the population of a broad wavepacket in J- and K-spaces, J being the matter angular momentum and K the helicity (the eigenvalue of the projection of $\mathbf{J}$ onto the body-fixed z-axis up to a factor $\hbar$). The independence of Eq. (7) on ϕ—the angle of rotation about the space-fixed z-axis—leads to conservation of the magnetic quantum number M (the eigenvalue of the space-fixed z-projection of $\mathbf{J}$ up to $\hbar$).

In the case of a symmetric top molecule, $\alpha_{XX} = \alpha_{YY}$, and Eq. (7) simplifies as,

$$H_{ind} = -\frac{1}{4}\varepsilon^2(t)\Delta\alpha\cos^2\theta, \quad \Delta\alpha = \alpha_\| - \alpha_\perp, \tag{8}$$

where $\alpha_\|$ and $\alpha_\perp$ are the components of the polarizability tensor parallel and perpendicular to the molecular axis, respectively, $\alpha_\| = \alpha_{ZZ}, \alpha_\perp = \alpha_{XX} = \alpha_{YY}$. Due to the cylindrical symmetry of the system, the field-matter interaction is independent of χ. As a consequence, the projection of the angular momentum vector onto the top axis, the operator J_Z, remains well-defined and the associated quantum number, K, remains a conserved quantum number in the presence of the field. Thus, a broad superposition of total angular momentum eigenstates is coherently excited during the pulse, while the projection of the angular momentum vector onto both the space- and the body-fixed z-axes is unaltered. We remark that the interaction Hamiltonian takes the same form, Eq. (8), in the cases of linear and symmetric top molecules. Nonetheless, the nonradiative Hamiltonian (and hence also the alignment dynamics) differ qualitatively, as will become evident below.

With the field matter interaction given by Eqs. (5) and (6) (or by its special limits (7) or (8)), the complete Hamiltonian takes the form,

$$H = H_{mol} + H_{ind}, \tag{9}$$

in the strictly nonresonant case, where H_{mol} is the field-free molecular Hamiltonian. Both the nonradiative and the radiative terms in Eq. (9) are functions of the vibrational coordinates $\mathbf{Q}$ and the rotational coordinates $\widehat{R}$, where the dependence of the field-free Hamiltonian on the molecular vibrations is due to centrifugal and Coriolis interactions whereas the $\mathbf{Q}$-dependence of the interaction term derives from the dependence of the transition dipole elements and hence the polarizability tensor on the vibrational coordinates. In bound state problems, where the system does not explore regions of vibrational space remote from the equilibrium configuration, it is common to neglect the $\mathbf{Q}$-dependence of the interaction. Although cases where this approximation fails are known, its popularity is understandable, given the numerical labor involved in computing the polarizability tensor for a polyatomic system. The dependence of H_{mol} on the vibrational coordinates can be likewise neglected provided that attention is restricted to short time scales and/or relatively low temperatures, where the observation time is short with respect to inverse the rotation–vibration coupling strength.

Within the rigid rotor approximation the **Q**-dependence of both terms in Eq. (9) is neglected and H_{mol} reduces to the rotational Hamiltonian,

$$H_{mol} \approx H_{rot} = \frac{J_X^2}{2I_{XX}^e} + \frac{J_Y^2}{2I_{YY}^e} + \frac{J_Z^2}{2I_{ZZ}^e}. \tag{10}$$

In Eq. (10), the operators J_k, $k = X, Y, Z$, are the Cartesian components of the total material angular momentum vector, I_{kk}^e are the corresponding principal moments of inertia (Section 2.3), computed at the equilibrium configuration, and it is assumed that the body-fixed coordinate system has been chosen such that cross terms of the inertia tensor are eliminated.[10] By convention, the three principal moments of inertia are denoted I_{aa}^e, I_{bb}^e, I_{cc}^e with $I_{aa}^e \leq I_{bb}^e \leq I_{cc}^e$. Thus, for the case of linear molecules $I_{aa}^e = 0$, $I_{bb}^e = I_{cc}^e$ whereas for prolate and oblate symmetric top molecules the principal moments of inertia satisfy,

$$I_{aa}^e < I_{bb}^e = I_{cc}^e \quad \text{(prolate symmetric top)},$$

and

$$I_{aa}^e = I_{bb}^e < I_{cc}^e \quad \text{(oblate symmetric top)},$$

respectively. The rotational Hamiltonian is commonly expressed in terms of the rotational constants $A_e = 1/2I_{aa}^e$, $B_e = 1/2I_{bb}^e$, and $C_e = 1/2I_{cc}^e$ as,

$$H_{rot} = \frac{J_a^2}{2I_{aa}^e} + \frac{J_b^2}{2I_{bb}^e} + \frac{J_c^2}{2I_{cc}^e} = A_e J_a^2 + B_e J_b^2 + C_e J_c^2. \tag{11}$$

(Rotational constants are given in part of the literature in units of inverse time, where

$$A_e = \frac{\hbar}{4\pi I_{aa}^e}, \qquad B_e = \frac{\hbar}{4\pi I_{bb}^e}, \qquad C_e = \frac{\hbar}{4\pi I_{cc}^e},$$

or in units of inverse length, where $A_e = \hbar/(4\pi I_{aa}^e c)$, etc. We adopt here the more common usage, where energy units are employed, $A_e = \hbar^2/(2I_{aa}^e)$, etc.)[11] In the

---

[10] Since the inertia tensor is Hermitian, there exists a transformation of the coordinates for which the inertia tensor is diagonal [66,122].

[11] Often the equilibrium rotational constants are denoted A, B, and C with the subscript e implied. We prefer the more orthodox notation of [123], where the subscript is retained as a reminder that attention is confined to the rigid rotor limit.

case of prolate symmetric top rotors it is conventional to choose the body-fixed z-axis as the a-axis, whereby

$$\hbar^2 H_{\text{rot}} = C_e \mathbf{J}^2 + (A_e - C_e) J_Z^2 \quad \text{(prolate symmetric top)}, \tag{12}$$

with eigenvalues

$$E^{JK} = C_e J(J+1) + (A_e - C_e) K^2, \tag{13}$$

whereas in the oblate case choice of the c-axis to define the body fixed z-axis is conventional, whereby

$$\hbar^2 H_{\text{rot}} = A_e \mathbf{J}^2 + (C_e - A_e) J_Z^2 \quad \text{(oblate symmetric top)}, \tag{14}$$

with eigenvalues

$$E^{JK} = A_e J(J+1) + (C_e - A_e) K^2. \tag{15}$$

Finally, in the limit of linear rigid rotors, Eq. (12) reduces to $\hbar^2 H_{\text{rot}} = B_e \mathbf{J}^2$, $E^J = B_e J(J+1)$, and the complete Hamiltonian takes the form,

$$H = \hbar^{-2} B_e \mathbf{J}^2 - \frac{1}{4} \varepsilon^2(t) \Delta\alpha \cos^2\theta. \tag{16}$$

Equation (16) is familiar; it has been the starting point in a substantial body of theoretical studies of the alignment dynamics of diatomic molecules during the past decade.

3.3. NUMERICAL IMPLEMENTATION

With the form of the Hamiltonian for the molecular symmetry in question identified, we proceed to solve the time-dependent Schrödinger equation subject to the nonperturbative interaction. We focus on the nonresonant interaction case and restrict attention to the range of validity of the rigid rotor approximation. The complementary case of near-resonance alignment and a full (nonrigid) Hamiltonian is reviewed in [33] in the context of time-resolved photoelectron imaging. The wavepacket is thus re-expanded (see Appendix A) in a suitable rotational basis as,

$$\left| \Psi_{\xi \mathbf{v} \mathbf{n}_i}(t) \right\rangle = \sum_{\mathbf{n}} C_{\xi \mathbf{v} \mathbf{n}_i}^{\mathbf{n}}(t) |\mathbf{n}\rangle, \tag{17}$$

where ξ and $\mathbf{v}$ are the electronic and vibrational indices (which are conserved) and $\mathbf{n}_i$ is the set of rotational quantum numbers of the parent state, which define the initial conditions. The dependence of the expansion coefficients on the initial state indices $\{\xi, \mathbf{v}, \mathbf{n}_i\}$ is indicated explicitly in Eq. (17) but omitted for clarity of notation in what follows, $C^{\mathbf{n}}(t) \equiv C_{\xi \mathbf{v} \mathbf{n}_i}^{\mathbf{n}}(t)$. To simplify the presentation we omit the conserved quantum numbers ξ and $\mathbf{v}$ also from the left-hand side, $\Psi_{\mathbf{n}_i} \equiv \Psi_{\xi \mathbf{v} \mathbf{n}_i}$.

The choice of a rotational basis set, $\{|\mathbf{n}\rangle\}$, is largely a matter of convenience and numerical efficiency. In the case of nonadiabatic alignment, one is often interested in exploring the long-time evolution subject to the molecular Hamiltonian, where the system exhibits field-free, post-pulse alignment and rotational revivals. A numerically advantageous basis in that case is the set of orthonormal eigenstates of H_{rot}, as with this choice the propagation subsequent to the turn-off is analytical. In the general case of an asymmetric top molecule one finds,

$$\left|\Psi_{J_i \tau_{\text{at}}, i M_i}(t)\right\rangle = \sum_{J \tau_{\text{at}}} C^{J \tau_{\text{at}} M_i}(t) |J \tau_{\text{at}} M_i\rangle, \tag{18}$$

where we label the eigenstates of H_{rot} by the matter angular momentum quantum number J, the asymmetric top quantum number,[12] $\tau_{\text{at}} = -J, -J+1, \ldots J$, and the magnetic quantum number M, $\mathbf{n} = \{J, \tau_{\text{at}}, M\}$. The right-hand side of Eq. (18) notes that M is conserved in a linearly polarized field, $M = M_i$. A general analytical form of the corresponding eigenfunctions, $\langle \widehat{R} | \mathbf{n} \rangle = \langle \widehat{R} | J \tau_{\text{at}} M \rangle$, is not available, and hence for numerical purposes it is conventional to expand the $|J \tau_{\text{at}} M\rangle$ in a suitable symmetric top basis, $\{|JKM\rangle\}$,

$$|J \tau_{\text{at}} M\rangle = \sum_K a^K_{J \tau_{\text{at}}} |JKM\rangle, \tag{19}$$

where the expansion coefficients $a^K_{J \tau_{\text{at}}}$ and the eigenvalues $E^{\mathbf{n}} = E^{J \tau_{\text{at}}}$ are obtained by diagonalization of the rotational Hamiltonian H_{rot}, expressed in the symmetric top representation [66]. In Eq. (19) $\langle \widehat{R} | JKM \rangle$ are normalized symmetric top eigenfunctions,

$$\langle \widehat{R} | JKM \rangle = \sqrt{\frac{2J+1}{8\pi^2}} D^{J*}_{MK}(\widehat{R}), \tag{20}$$

where D^J_{MK} are the rotation matrices introduced in Section 3.2. The matrix elements of H_{rot} in the symmetric top basis are readily evaluated analytically, e.g., by expressing the J_k in Eq. (10) in terms of raising and lowering operators [65,66]. The diagonalization procedure is numerically straightforward since conservation of the total (material) angular momentum in the field-free system restricts the matrix representation of H_{rot} to a single J manifold.

In cases where the field is present throughout most of the period of relevance, expansion of $\Psi(t)$ in a symmetric top basis offers an advantage over that in an asymmetric basis, as numerical propagation is inevitable. Often it is advantageous to employ a hybrid approach, where the symmetric top basis is used during the laser pulse and the asymmetric top basis is used during the field-free evolution.

---

[12] The asymmetric top quantum number is conventionally denoted τ; we use here the subscripts "at" to avoid confusion with the pulse duration.

The unitary transformation between the symmetric and asymmetric top representation (given as the matrix of eigenvectors of $H_{\rm rot}$ and consisting of the $a^K_{J\tau_{\rm at}}$), is then applied sufficiently long subsequent to the pulse for the system to be considered field-free.

In the case of a symmetric top molecule, the eigenstates of $H_{\rm rot}$ are the normalized Wigner matrices of Eq. (20), $|\mathbf{n}\rangle = |JKM\rangle$, with $E^{\mathbf{n}} = E^{JK}$ of Eqs. (13), (15), and Eq. (17) takes the form,

$$\left|\Psi_{J_i K_i M_i}(t)\right\rangle = \sum_{JK} C^{JKM_i}(t)|JKM_i\rangle. \tag{21}$$

As mentioned above, the dependence of the wavepacket on the set of initial conditions $\{J_i K_i M_i\}$ (indicated explicitly on the left-hand side of Eq. (21)) is implicit but not explicitly noted in the $C^{JKM_i}(t)$ on the right-hand side. Finally, in the case of linear molecules, the normalized rotation matrices reduce to spherical harmonics, $\langle \widehat{R}|JK = 0M\rangle = \sqrt{(2J+1)/(4\pi)}D^{J^*}_{M0}(\widehat{R}) = \langle \widehat{R}|JM\rangle = Y_{JM}(\theta,\phi)$, and the eigenenergies become functions of a single quantum number J, $E^{\mathbf{n}} = E^J = B_e J(J+1)$. An interesting exception is the case of a linear molecule that carries electronic angular momentum. Here the eigenstates of $H_{\rm rot}$ are given as rotation matrices, $|J\lambda M\rangle$ (λ being the projection of the electronic angular momentum onto the molecular axis), as for the symmetric top Hamiltonian, but $|\lambda|$ is upper limited by the electronic, rather than by the total material angular momentum.

Substituting Eq. (17) in the time-dependent Schrödinger equation,

$$i\hbar \frac{\partial \Psi(t)}{\partial t} = H(t)\Psi(t),$$

and using the orthonormality of the expansion basis states, one derives a set of coupled differential equations for the expansion coefficients,

$$
\begin{aligned}
i\hbar \dot{C}^{\mathbf{n}}(t) &= \sum_{\mathbf{n'}} \langle \mathbf{n}|H(t)|\mathbf{n'}\rangle C^{\mathbf{n'}}(t) \\
&= \sum_{\mathbf{n'}} \{\langle \mathbf{n}|H_{\rm rot}|\mathbf{n'}\rangle + \langle \mathbf{n}|H_{\rm ind}(t)|\mathbf{n'}\rangle\} C^{\mathbf{n'}}(t),
\end{aligned}
\tag{22}
$$

where the choice of partitioning of H into H_0 and an interaction term is system and application dependent, as discussed above. Both the nonradiative and the radiative matrix elements in Eq. (22) are analytically soluble in the rotational basis. In the general case of an asymmetric top system, choosing the asymmetric top eigenstates as an expansion basis, $H_0 = H_{\rm rot}$, we have,

$$
\begin{aligned}
\langle \mathbf{n}|H(t)|\mathbf{n'}\rangle &= E^{J\tau_{\rm at}}\delta_{JJ'}\delta_{\tau_{\rm at},\tau'_{\rm at}} + \langle J\tau_{\rm at}M|H_{\rm ind}(t)|J'\tau'_{\rm at}M\rangle \\
&= E^{J\tau_{\rm at}}\delta_{JJ'}\delta_{\tau_{\rm at},\tau'_{\rm at}} - \frac{\varepsilon^2(t)}{4}\sum_{KK'} a^K_{J\tau_{\rm at}} a^{K'}_{J'\tau'_{\rm at}}
\end{aligned}
$$

$$\times \left\{ \frac{\alpha^{ZX} + \alpha^{ZY}}{3} \langle JKM|D_{00}^2|J'KM\rangle \delta_{KK'} \right.$$
$$\left. - \frac{\alpha^{YX}}{\sqrt{6}} \langle JKM|[D_{02}^2 + D_{0-2}^2]|J'K'M\rangle \right\} \tag{23}$$

where $\alpha^{kk'} = \alpha_{kk} - \alpha_{k'k'}$ are the generalized polarizability anisotropies introduced in Eq. (7). With the choice of an appropriate symmetric top basis for the expansion, H_0 is given by Eqs. (12), (14), and

$$\langle \mathbf{n}|H(t)|\mathbf{n}'\rangle = \langle JKM|H_{\mathrm{rot}}|JK'M\rangle \delta_{JJ'}$$
$$- \frac{\varepsilon^2(t)}{4} \left\{ \frac{\alpha^{ZX} + \alpha^{ZY}}{3} \langle JKM|D_{00}^2|J'KM\rangle \delta_{KK'} \right.$$
$$\left. - \frac{\alpha^{YX}}{\sqrt{6}} \langle JKM|[D_{02}^2 + D_{0-2}^2]|J'K'M\rangle \right\}. \tag{24}$$

In the case of a symmetric top molecule, Eq. (8), both Eqs. (23) and (24) reduce to

$$\langle \mathbf{n}|H(t)|\mathbf{n}'\rangle = \delta_{KK'} \left\{ E^{JK}\delta_{JJ'} - \frac{2}{3}\frac{\varepsilon^2(t)\Delta\alpha}{4}\langle JKM|D_{00}^2|J'KM\rangle \right\}. \tag{25}$$

The matrix elements of the field matter interaction in Eqs. (23)–(25) are given as superpositions of integrals of the form,

$$\langle jkm|D_{qs}^2|j'k'm'\rangle$$
$$= (-1)^{k'+m'}\sqrt{(2j+1)(2j'+1)} \begin{pmatrix} j & 2 & j' \\ m & q & -m' \end{pmatrix} \begin{pmatrix} j & 2 & j' \\ k & s & -k' \end{pmatrix}, \tag{26}$$

where the selection rules $|J-2| \leq J' \leq J+2$, $K' = K$, $K \pm 2$, follow from the properties of the 3-j symbols, which embody the underlying angular momentum algebra.

With the form of the wavepacket determined via solution of the set of coupled differential Eqs. (22), all observables of interest can be computed nonperturbatively as a function of time. (A more economical solution can be used in the case of fully adiabatic alignment, $\tau_{\mathrm{pulse}} \gg \tau_{\mathrm{rot}}$, where each eigenstate of the field-free Hamiltonian is guaranteed to evolve adiabatically into the corresponding state of the complete Hamiltonian as the pulse turns on, returning adiabatically to the free rotor eigenstate upon turn-off. In this situation the dynamics is eliminated and the problem is formally equivalent to alignment of nonpolar molecules in a DC electric field. Numerically it is then economical to diagonalize the complete Hamiltonian at the peak of the field, and thus compute all observables in terms of the fully interacting eigenstates of H.) Solution of the time-dependent Schrödinger equation with an initial condition $\mathbf{n} = \mathbf{n}_i$ produces observables

$O_{\mathbf{n}_i}(t)$ corresponding to a pure state, that is, to an experiment where the parent state has been prepared in an eigenstate of the field-free Hamiltonian. More common are experiments where the parent state is a thermal average – a mixed state defined by a rotational temperature T_{rot}, corresponding to Boltzmann averaged observables $O_T(t)$.[13] The latter observables are obtained from the pure state analogues as,

$$O_T(t) = \sum_{\mathbf{n}_i} w_{\mathbf{n}_i}(T_{\mathrm{rot}}) O_{\mathbf{n}_i}(t), \tag{27}$$

where $w_{\mathbf{n}_i}(T_{\mathrm{rot}})$ are normalized weight functions, consisting of the Boltzmann factor $Q_{\mathrm{rot}}^{-1} \exp(-E^{\mathbf{n}_i}/kT_{\mathrm{rot}})$ (Q_{rot} being the rotational partitioning function and k Bolzmann's constant) and, when relevant, a spin statistics weight.

The observable that has been most commonly used to quantify the degree and the time-evolution of the alignment is the expectation value of $\langle \cos^2 \theta \rangle$ in the wavepacket,

$$
\begin{aligned}
\left\langle \cos^2 \theta \right\rangle_{\mathbf{n}_i}(t) &= \left\langle \Psi_{\mathbf{n}_i}(t) \middle| \cos^2 \theta \middle| \Psi_{\mathbf{n}_i}(t) \right\rangle \\
&= \sum_{\mathbf{n}\mathbf{n}'} C^{\mathbf{n}*}(t) C^{\mathbf{n}'}(t) \langle \mathbf{n} | \cos^2 \theta | \mathbf{n}' \rangle
\end{aligned} \tag{28}
$$

with the corresponding thermally averaged observable $\langle \cos^2 \theta \rangle_T(t)$ determined through Eq. (27). Complementing this average in coordinate space is the expectation values of $\mathbf{J}^2$ in the wavepacket,

$$\left\langle J^2 \right\rangle_{\mathbf{n}_i}(t) = \left\langle \Psi_{\mathbf{n}_i}(t) \middle| \mathbf{J}^2 \middle| \Psi_{\mathbf{n}_i}(t) \right\rangle = \sum_{\mathbf{n}} \left| C^{\mathbf{n}}(t) \right|^2 J(J+1), \tag{29}$$

which quantifies the extent of rotational excitation.

As was noted in early theoretical studies [48,49] and recently demonstrated experimentally [87], substantially more information regarding the alignment dynamics is available from the complete probability distribution than what the expectation value of $\cos^2 \theta$ (equivalently, the second moment of the distribution) captures. In Section 4 we characterize the alignment both via $\langle \cos^2 \theta \rangle_T(t)$, which provides a quantitative and transferable measure that is readily compared with the literature, and via the more general time-evolving probability density. A set of wavepacket attributes that fully characterize the alignment evolution, i.e., is equivalent in information content to the θ-dependence of the probability density, but is quantitative and transferable, is provided by the set of all moments of the

---

[13] Isolated molecule experiments are typically carried out in a molecular beam environment, where the rotational and translational temperatures are not equilibrated. For the alignment problem it is the rotational temperature that is relevant.

alignment, $P_\kappa = \langle \Psi(t) | P_\kappa(\cos\theta) | \Psi(t) \rangle$, where $P_\kappa(\cos\theta)$ are Legendre polynomials. Application of these moments to elucidate wavepacket dynamics, and their mapping onto experimental observables are discussed in [88]. Experimentally, however, only the lowest moment has been measured so far.

Defined in analogy to Eqs. (28), (29) and relevant in the case of asymmetric top molecules are the expectation values of $\cos^2\chi$ and J_Z^2 in the wavepacket,

$$
\begin{aligned}
\langle \cos^2\chi \rangle_{\mathbf{n}_i}(t) &= \langle \Psi_{\mathbf{n}_i}(t) | \cos^2\chi | \Psi_{\mathbf{n}_i}(t) \rangle \\
&= \sum_{\mathbf{nn'}} C^{\mathbf{n}*}(t) C^{\mathbf{n'}}(t) \langle \mathbf{n} | \cos^2\chi | \mathbf{n'} \rangle,
\end{aligned} \tag{30}
$$

$$
\begin{aligned}
\langle J_Z^2 \rangle_{\mathbf{n}_i}(t) &= \langle \Psi_{\mathbf{n}_i}(t) | J_Z^2 | \Psi_{\mathbf{n}_i}(t) \rangle \\
&= \sum_{\mathbf{n,n'}} C^{\mathbf{n}*}(t) C^{\mathbf{n'}}(t) \langle \mathbf{n} | J_Z^2 | \mathbf{n'} \rangle,
\end{aligned} \tag{31}
$$

whereas $\langle \cos^2\theta \rangle$ provides an average measure of the degree of localization in θ, $\langle \cos^2\chi \rangle$ measures the degree of localization in χ and is sensitive to both the radiative interaction and the asymmetric top coupling. Likewise, $\langle J_Z^2 \rangle$ quantifies the orientation of the angular momentum vector in the body-fixed frame.

4. Case Studies

In this section we apply the theory reviewed in Section 3 to illustrate several of the concepts underlying nonadiabatic alignment, as summarized in Section 2. We begin by illustrating a number of general properties in a system-independent fashion, and proceed with numerical results for a specific polyatomic molecule for which experimental data is available.

In order to explore general aspects of nonadiabatic alignment, it is convenient to introduce molecule- and field-independent, interaction, time, and temperature variables. We proceed by re-writing Eq. (9) with the general forms (7) and (10) for the radiative and nonradiative Hamiltonians as,

$$
\overline{H} = \sum_{k=X,Y,Z} \overline{B}_K \bar{J}_k^2 + \overline{\Omega}_{\mathrm{R}} \big[\bar{\alpha}^{ZX} \cos^2\theta + \bar{\alpha}^{YX} \sin^2\theta \sin^2\chi \big], \tag{32}
$$

where we define dimensionless Hamiltonian and angular momentum operators as,

$$
\overline{H} = \frac{\bar{I} H}{\hbar^2}, \quad \bar{I} = \frac{1}{2}\big(I_{XX}^{\mathrm{e}} + I_{YY}^{\mathrm{e}} \big),
$$

$$
\bar{J}_k = \frac{J_K}{\hbar}, \quad k = X, Y, Z, \tag{33}
$$

respectively, and introduce dimensionless rotational constants, inertia components, and interaction parameters as,

$$\bar{B}_k = \frac{\bar{I}}{2I_{kk}^e}, \qquad \bar{\alpha}_k = \frac{\alpha_k}{\bar{\alpha}},$$ (34)

and

$$\bar{\Omega}_R = \frac{\bar{I}\bar{\alpha}\varepsilon^2}{4\hbar^2},$$ (35)

respectively. In Eqs. (32)–(35) $\bar{\alpha}$ is the trace of the polarizability tensor (the average polarizability), $\bar{\alpha} = (\alpha_{XX} + \alpha_{YY} + \alpha_{ZZ})/3$ and the generalized anisotropies $\bar{\alpha}^{kk'} = \bar{\alpha}_{kk} - \bar{\alpha}_{k'k'}$ are dimensionless versions of the analogous parameters introduced in Eq. (7). In terms of the Hamiltonian (32) the time dependent Schrödinger equation is transformed as,

$$i\frac{\partial}{\partial \bar{t}}\Psi(\bar{t}) = \overline{H}\Psi(\bar{t}),$$ (36)

where $\bar{t}$ defines a dimensionless time variable,

$$\bar{t} = \frac{\hbar}{\bar{I}}t.$$ (37)

Finally, we introduce a reduced temperature variable $\overline{T} = \bar{I}kT/\hbar^2$. The transformation of Eqs. (32)–(37) is structured such that the time variable will take a physical significance in the limit of a linear molecule, $\bar{t} = (2\pi/\tau_{\text{rot}})t$, with transparent analogs in the prolate and oblate symmetric top cases (*vide infra*).

In Figs. 2 and 3 we summarize the time-scale considerations in nonadiabatic alignment discussed in Section 2.2 through the example of alignment by a slowly turning on, rapidly switching off, nonresonant pulse. This choice allows us to illustrate in a single example both effects that are common to adiabatic and nonadiabatic alignment and ones that are unique to the nonadiabatic case.

Figure 2 varies the pulse turn-on from the fully adiabatic to the sudden limit, using a pulse envelope of the form,

$$\varepsilon(t) = \begin{cases} \varepsilon_0 e^{-(t/\tau_{\text{on}})^2}, & t \leq 0, \\ \varepsilon_0 e^{-(t/\tau_{\text{off}})^2}, & t \geq 0, \end{cases}$$ (38)

for a linear rotor at zero temperature with (dimensionless) interaction strength $\bar{\Omega}_R = 100$ (see Eq. (35)). As the pulse turn-on approaches the natural system time-scale τ_{rot} from above, the alignment gradually deviates from the adiabatic process limit, and a rich and complex structure of quantum oscillations emerges after the field has turned off. Whereas a turn-on of ca. 2π, $\tau_{\text{on}} \gtrsim 5\tau_{\text{rot}}$, adiabatically switches the $J_i = 0$, $M_i = 0$ free rotor state into the lowest eigenstate of the Hamiltonian (16) (panel (a)), a less gradual turn-on, $\tau_{\text{on}} \sim 2.5\tau_{\text{rot}}$, is seen to

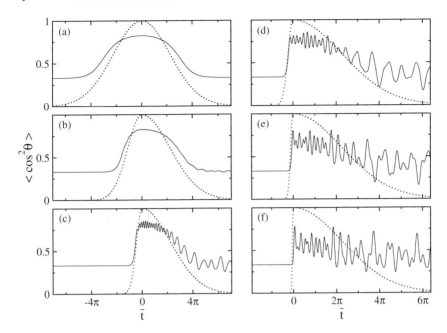

FIG. 2. Variation of the revival spectrum as a function of the pulse turn-on for constant Rabi coupling and pulse turn-off, $\bar{\Omega}_R = 10^2$, $\tau_{off} = 5$ (see Eqs. (35), (37) for definition of the dimensionless unit system used). (a) $\tau_{on} = 5$, (b) $\tau_{on} = 2.5$, (c) $\tau_{on} = 1$, (d) $\tau_{on} = 0.5$, (e) $\tau_{on} = 0.25$, (f) $\tau_{on} = 0.1$. The dotted curves show the pulse shape envelope normalized to a peak intensity of unity.

populate a minor component of the first excited eigenstate of (16), giving rise to a two-level beat pattern between the $J = 0$ and the $J = 2$ free rotor states after the adiabatic turn-off (panel (b)). High order rotational coherences are marked when the turn-on falls significantly below τ_{rot}, panels (c) and (d). For sufficiently short turn-on, the pulse duration drops below the value required for the rotational excitation that the field intensity allows, J_{max} becomes time-limited (see Section 2.2), and the alignment degrades (panels (e) and (f)).

Figure 3 complements the discussion by varying the pulse turn-off from comparable to short with respect to τ_{rot}. As τ_{off} falls below the system time-scale (panel (b)), deviation from the adiabatic limit is observed as population of two rotational levels that beat at the inverse of their energy spacing, $2\pi/(E_2 - E_0)$. For short turn-off with respect to τ_{rot} (panel (d)), the signature of higher order rotational coherences is observed as marked deviation from the two-level beat pattern. In the instantaneous turn-off limit, where τ_{rot}/τ_{off} approaches and exceeds an order of magnitude (panel (f)), the alignment attained upon turn-off is precisely reproduced at multiples of the rotational period and is invariant to further changes of τ_{off}.

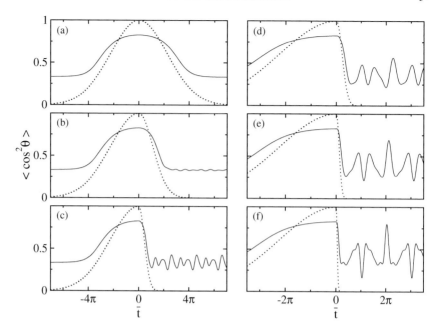

FIG. 3. Variation of the revival spectrum as a function of the pulse turn-off for constant Rabi coupling and pulse turn-on, $\overline{\Omega}_R = 10^2$, $\tau_{on} = 5$ (see Eqs. (35), (37) for definition of the dimensionless unit system used). (a) $\tau_{off} = 5$, (b) $\tau_{off} = 2.5$, (c) $\tau_{off} = 1$, (d) $\tau_{off} = 0.5$, (e) $\tau_{off} = 0.25$, (f) $\tau_{off} = 0.1$. The dotted curves show the pulse shape envelope normalized to a peak intensity of unity.

A similar behavior is observed in the case of a symmetric (e.g., Gaussian) pulse (not shown). As the pulse duration drops below ca. $\tau_{rot}/2$, quantum oscillations begin to appear in the alignment, surviving after the pulse turn-off. As the system passes from the adiabatic to the short-pulse limit, the recurrence features assume a simple form, with alignment revivals manifesting at regular intervals. Once the pulse becomes too short to excite an adequately broad range of J-states, the alignment features shrink and eventually vanish. Elsewhere it is shown analytically that in the short pulse, high intensity case, $\tau_{pulse}^2 < (B_e\Omega_R)^{-1}$, where the degree of rotational excitation is time-limited, $J_{max} \sim \tau_{pulse}/\Omega_R^{-1}$, the alignment depends solely on J_{max}, i.e., is invariant to modifications in the pulse characteristics that keep constant the product of the pulse duration by the Rabi coupling [49].

Interestingly, rotational temperature plays a major role in nonadiabatic alignment. The rich quantum mechanical structure of Figs. 2 and 3 exhibits the revival structure of a pure state, corresponding to an experiment in which the molecule is initially prepared in a single rotational eigenstate $\{J_i, M_i\}$ (J_i being the initial material angular momentum and M_i its projection onto a space-fixed z-axis—

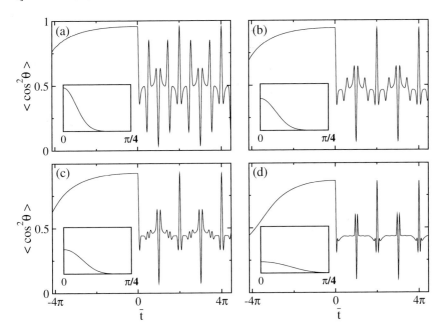

FIG. 4. Variation of the revival spectrum as a function of the rotational temperature for Rabi coupling $\overline{\Omega}_R = 10^3$, adiabatic turn-on ($\tau_{on} = 5$) and rapid turn-off ($\tau_{off} = 0.005$). (a) $\overline{T} = 0$, (b) $\overline{T} = 1$, (c) $\overline{T} = 2$, (d) $\overline{T} = 10$. The insets show the Boltzmann-averaged probability distribution associated with the wavepacket at the corresponding temperatures, calculated at the peak of the pulse ($t = 0$).

the magnetic quantum number that is conserved in linearly polarized fields). More common are experiments where the parent state is a thermal average, corresponding to a mixed state initial condition, specified by a temperature (see Eq. (27)). One effect of the finite rotational temperature is trivial: with increasing temperature, higher rotational levels are initially populated, corresponding to larger detuning from resonance (see Section 2.1). Since with adiabatic turn-on the degree of rotational excitation is controlled by the balance between the Rabi coupling and the J-dependent detuning, we find that $\langle J \rangle (t = 0) - \langle J \rangle (t \rightarrow -\infty)$ decreases with increasing temperature. (By contrast, with short pulse excitation, where the extent of rotational excitation is time-limited, $J_{max} \sim \Omega_R \tau$, $\langle J \rangle (t = 0) - \langle J \rangle (t \rightarrow -\infty)$ is temperature-independent.) Corresponding to the decreasing rotational excitation with increasing temperature is a broadening of the wavepacket probability density at $t = 0$ and a drop of $\langle \cos^2 \theta \rangle (t = 0)$. Both effects are illustrated in Fig. 4, where we contrast the $\overline{T} = 0$ alignment dynamics (panel (a)) with the finite temperature case. The insets illustrate the broadening of $|\Psi(t = 0)|^2$ with temperature while the main panels show the corresponding drop in $\langle \cos^2 \theta \rangle (t = 0)$.

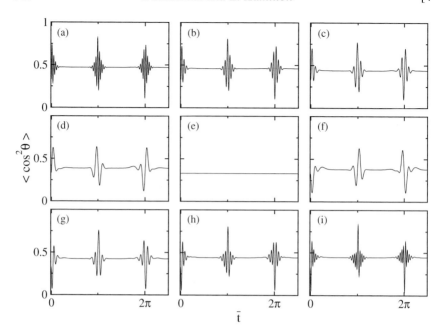

FIG. 5. Variation of the revival spectrum of a symmetric top rotor as a function of the ratio of the two distinct inertia moments R_B with the parameters varied in such a way as to hold the trace of the moment-of-inertia tensor constant. The polarizability tensor components are varied proportionately with the rotational constants. (a) $R_B = 0.5$, (b) $R_B = 0.625$, (c) $R_B = 0.75$, (d) $R_B = 0.875$, (e) $R_B = 1$ (spherical symmetry), (f) $R_B = 1.14$, (g) $R_B = 1.33$, (h) $R_B = 1.6$, (i) $R_B = 2$.

Less trivial and more dramatic is the effect of temperature on the long time, field-free evolution. The incoherent sum over the thermally populated initial states is seen to wash out much of the quantum mechanical revival structure of panel (a), leaving only the recurrences that are independent of the initial condition. The rich structure of the pure state case is thus reduced to a simple, intuitive pattern that can be readily understood in classical terms; in the limit of a sufficiently broad distribution in the quantum number space, the rotational wavepacket approaches the motion of a classical linear rotor that returns to its original position at integer multiples of the rotational period, $\bar{t} = 2\pi n$.

Having illustrated the role played by the field parameters and the rotational temperature (Sections 2.1, 2.2) through the simplest case example of a linear molecule subject to linearly polarized field in Figs. 2–4, we proceed in Figs. 5 and 6 to examine numerically the role of the molecular symmetry (Section 2.3). Figure 5 investigates systematically the case of a symmetric top rotor (see also Eqs. (13), (15)) by varying the ratio of the inertia moments parallel and perpendicular to the symmetry axis, denoted R_B, from the prolate (panel (a)) through

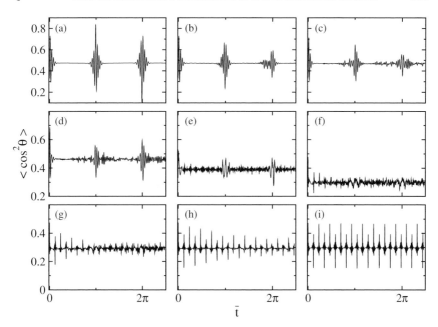

FIG. 6. Variation of the revival spectrum as a function of the asymmetry parameter $\kappa = (2B_e - A_e - C_e)/(A_e - C_e)$, with the parameters varied in such a way as to hold the trace of the moment-of-inertia tensor constant. The polarizability tensor components are varied proportionately with the rotational constants. (a) $\kappa = -1.0$ (prolate limit), (b) $\kappa = -0.965$ (iodobenzene case), (c) $\kappa = -0.95$, (d) $\kappa = -0.9$, (e) $\kappa = 0.0$, (f) $\kappa = 0.9$, (g) $\kappa = 0.95$, (h) $\kappa = 0.965$, (i) $\kappa = 1.0$ (oblate limit). Note that the scale of the abscissa changes for each of the three rows of panels.

the spherical (panel (e)) to the oblate (panel (i)) limit. The polarizability tensor components are varied proportionately with the rotational constants so as to keep fixed the relation of the inertia and polarizability tensors. Panel (a), corresponding to the extreme prolate case, $R_B = I_{aa}/I_{bb} = 1/2$, is readily understood by reference to the familiar revival structure of linear molecules. At integer multiples of the revival time that corresponds to rotation of the body-fixed z-axis, $\bar{t} = 2\pi n$, the initial alignment is precisely reconstructed, whereas at half revivals, $\bar{t} = 2\pi(n + 1/2)$, the distribution is rotated by $\pi/2$. The new feature as compared to the linear case is the fine structure of the revival structure, arising from the second term in Eq. (13) and reflecting the availability of different orientations of the angular momentum vector with respect to the body-fixed frame. As the ratio of the two distinct inertia moment R_B approaches unity, the polarizability anisotropy decreases, the interaction strength falls, and the rotational periods about the two axes become comparable. Correspondingly, the fine structure is simplified (see Eq. (13)) and $\langle \cos^2 \theta \rangle$ decreases. In the spherical limit the polarizability tensor is isotropic, the interaction vanishes, and so do the rotational excitation and the alignment. The baseline of the revival pattern approaches the linear molecule value of $\frac{1}{2}$ in the limit of small R_B and falls to the isotropic value of $\frac{1}{3}$ as $R_B \rightarrow 1$. As R_B is further increased, rotational excitation is restored, the structure of the revival spectrum reappears and the baseline approaches $\frac{1}{2}$. The extreme oblate case, $R_B = I_{cc}/I_{bb} = 2$, corresponding to the rotations of a disk, is illustrated in panel (i).

Along with the discussion following Eq. (8), Fig. 5 illustrates the differences in rotational dynamics between the linear and symmetric top rotors but also their formal similarity and common motive. In particular, both symmetries exhibit regular periodic dynamics. The discussion of Section 2.3 suggests that the rotational dynamics of asymmetric top molecules would differ in concept. In particular, stable periodic patterns are not expected. Figure 6 provides a systematic study of the rotational revivals of asymmetric top rotors. The asymmetry is quantified using the asymmetry parameter $\kappa = (2B_e - A_e - C_e)/(A_e - C_e)$ and we vary the inertia tensor from the symmetric prolate ($\kappa = -1$) to the symmetric oblate ($\kappa = 1$) through the strongly asymmetric ($\kappa = 0$) limit, keeping the trace of the inertia tensor constant. As in Fig. 5, the polarizability tensor components are varied proportionately with the rotational constants. Panel (b) of Fig. 6 corresponds to the parameters of iodobenzene, a near-prolate symmetric top molecule with $\kappa = -0.965$. The other panels depict model systems, constructed so as to investigate the complete $-1 \rightarrow 1$ asymmetry range while making reference to an actual molecule. The revival pattern is seen to be extremely sensitive to deviations from perfect symmetry, the periodic structure being significantly distorted with minor changes in κ (panels (b), (h)). It is important to note, however, that a clear revival pattern appears throughout the κ range $-1 \leq \kappa \leq 1$, including the strongly asymmetric case $\kappa \sim 0$, see panel (e). We remark that the revival structure depends not only on the anisotropy of the inertia tensor but also on that of the polarizability tensor. Hence Fig. 6 represents one, but not all classes of asymmetric top molecules.

Before concluding this section we expand briefly on the alignment dynamics of iodobenzene, a condensed view of which is presented in Fig. 6(b). This molecule is at present a unique example of an asymmetric top whose nonadiabatic alignment was studied intensively, both theoretically and experimentally. Figure 7 shows a contour plot of the probability density of the wavepacket, $P(\theta, t) = \sum_{\mathbf{v}_i} w_{\mathbf{v}_i}(T_{\text{rot}}) \int d\chi d\phi |\Psi_{\mathbf{v}_i}(\phi, \theta, \chi; t)|^2$, as a function of the polar Euler angle θ and time, focussing on the early alignment dynamics, prior to rephasing. Also shown in Fig. 7 is the expectation value of $\cos^2 \theta$ vs time, computed for the same set of parameters. It is evident that the probability density provides more complete information with regard to the evolution and quality of the alignment than the common $\langle \cos^2 \theta \rangle$ criterion. In particular, $P(\theta, t)$ illustrates the correlation between temporal and spatial confinement that underlies the post-

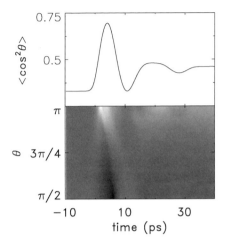

FIG. 7. Contour plot of the thermally averaged probability distribution for iodobenzene as a function of time and the polar Euler angle. The pulse envelope is a Gaussian of 3.82 ps duration and 6×10^{11} W/cm^2 peak intensity (corresponding to a reduced interaction parameter $\bar{\Omega}_R = 2445$) and the rotational temperature is 400 mK (corresponding to a reduced temperature parameter $\bar{T} = 4.2$).

pulse dynamics and is concealed in 1D measures. We note [87] that the maximum value of the probability distribution in Fig. 7 occurs earlier in time than the maximum in the expectation value of $\langle \cos^2 \theta \rangle$ corresponding to the same parameters, an effect that is also observed experimentally [87].

At present, experiments are capable of providing the entire wavepacket probability distribution, as well as directly measuring $\langle \cos^2 \theta \rangle$, while our numerical tools are capable of addressing complex molecules of experimental and practical interest on a quantum mechanical level. It is nevertheless important to stress that quantitative comparison of computed and experimental alignment results has not been reported as yet and, for several reasons, would be challenging. While proper account of temperature is essential, in most experiments the temperature is not known with precision. The accuracy to which the intensity can be experimentally determined is likewise limited. Further, in many experiments the probe samples much or all of the focal volume of the alignment laser. Consequently the intensity is not uniform and it is necessary to average calculations performed at different intensities with proper weight functions in order to contrast numerical and experimental results, a procedure that washes out several of the features that are observed in single intensity calculations [59]. To be reliably performed, the focal averaging requires experimental determination of not only the peak, but also the spatial distribution of the intensity. Finally, whereas rotational constants are available to good precision for a large variety of systems, polarizability tensors are typically known to much lesser accuracy.

5. Recent Developments

The past 8 years have witnessed an explosion of interest in nonadiabatic alignment by short, intense laser fields, spurring a wide range of applications. For work completed before 2002, [20] presents a thorough survey of both the adiabatic and the nonadiabatic alignment modes, and a shorter review of the same material is also available [22]. This section aims to provide a broad survey of work completed since the submission of [20]. Although we have attempted to include all the relevant literature emanating from this period, we recognize that the breadth of related topics prohibits an exhaustive account of this rapidly emerging field.

Several specialized topics in the field of nonadiabatic alignment have been particularly active over the last two years. Granucci and Persico [89] consider the case of resonant laser alignment in detail, and demonstrate the utility of a simple analytical theory, beginning from the short pulse approximation, for finding the delay time and intensity of the maximum alignment. For a two-dimensional model system, Volkova and coworkers examine the competition between rotational alignment and the electronic and vibrational excitations that could give rise to ionization or dissociation [90]. Such theoretical models may be applied in reverse to infer physical parameters of exotic molecular systems that would otherwise be difficult to measure directly, such as estimated polarizability anisotropies in very weakly bound rare-gas dimers [91]. Although most experimental and theoretical studies to date have focused on linear molecules, the work of Péronne et al. [57] has shown that post-pulse alignment is also possible with asymmetric top molecules, heralding the prospect of research on systems with more complicated temporal dynamics (see Figs. 6 and 7 in Section 4).

The problem of optimizing and controlling the degree and duration of post pulse alignment and orientation has been actively pursued by several groups. Renard and coworkers [47] report an experimental demonstration of the controlled alignment of N_2 molecules with a pulse shaping technique. Several theoretical studies suggest the application of sequences of two or more pulses to enhance the alignment (orientation) [75,76,78,80,92,93], and a number of experiments explore the extent to which the enhancement can be observed in the laboratory [94,95]. Very recent work by Ortigoso [93] suggests that, with proper spacing, a periodic sequence of pulses can sustain alignment for arbitrarily long times.

Rotational revivals continue to make an active research topic, owing to both the interesting physics involved and the variety of potential applications. (For a general review of the theory of wavepacket revivals, which includes a discussion of rotational revivals, the reader is referred to the work of Robinett [37].) Since the first measurement of rotational revival features in I_2 [54], several other linear molecules have been investigated as media for observing rotational revival structures. These include N_2 and O_2, which were probed by Coulomb explosion followed by ion imaging [55], and CO_2, which was analyzed through a nonintrusive weak field polarization spectroscopy technique [56,58]. More complicated revival structures occur in the time evolution of asymmetric top molecules, which produce multiple series of overlapping revival sequences that correspond to different rotational periodicities controlled by the molecular rotational constants [59] (Section 2.3). Gelin et al. [9] consider the effect of centrifugal distortions in nonrigid molecules, finding first-order corrections to the popular rigid-rotor approximation and illustrating its range of validity (see Section 3.2).

The fundamentally quantum nature of nonadiabatic alignment has been emphasized by Tikhonova and Molodenskii [38], who estimated the tunneling time between the two possible orientations of heteronuclear pendular states. Interestingly, these authors find that a classical ensemble with a uniform distribution of angular momenta would disperse without the possibility of revivals. An investigation by Child [39], however, illustrates that the quantum recurrences can be described in terms of a thermally averaged ensemble of classically evolving quantized rotors, at least during the post-pulse, field-free evolution. Several other theoretical studies have used the rotational revivals of aligned or oriented rigid rotors as a model to explore the intriguing correspondence between classical and quantum dynamics. Leibscher et al. use a semiclassical catastrophe theory to predict the possibility of generating unlimited angular squeezing using trains of multiple short laser pulses [40]. Arango et al. compare the phase space topology of the quantum and the classical monodromy for the case of diatomic rotors in combined DC and AC fields with a treatment of the onset of chaotic motion as the fields deviate from collinearity [41]. In a related study, Adelsward describes the contributions of spontaneous Raman transitions to the rotational decoherence of an ensemble of quantum rotors [96].

The laser-induced orientation of polar molecules, introduced in Section 2.5, remains an exciting topic for theoretical and experimental research. Recent experiments demonstrate the possibility of controlling molecular orientation using either a combination of DC and AC fields [42–45], within the approach pioneered in [70,97], or a two-color optical field [46], modelled, e.g., in [72,77]. The latter method is particularly amenable to optimization by means of evolutionary algorithms [77]. Quite recently the literature has contained a flurry of articles exploring the application of half-cycle pulses for the optimization and control of orientation [75,76,78–80]. Variations of this basic concept include the application of half-cycle pulses at fractional revival times to convert alignment into orientation [80], and use of vibrational wavepacket dynamics that, in the presence of a weak DC field, simulate the effect of a train of half-cycle pulses [98].

Another avenue now opening for research involves the alignment of molecules that interact with one another or are otherwise coupled to an environment in the solid or liquid phase. Shima and Nakayama [99] suggest that the dipole coupling of two adjacent quantum rotors may give rise to a notable enhancement of the alignment of each of the partners with respect to the isolated molecule case. The extension of alignment to dense media is shown in [18] to provide a sensitive probe of the dissipative properties of the medium. Alignment in a crystalline cage is shown in [100,101] to introduce interesting effects; both static effects, arising from the symmetry of the combined crystal and laser fields, and dynamical effects, resulting from the possibility of light triggering directed collisions in the confined environment. Similarly interesting are the implications of alignment of molecules in liquids and solutions [8,83].

Considerable research has focused in the past two years on a variety of applications of post-pulse, field-free alignment in optics and spectroscopy. One promising class of applications to receive recent attention is the generation of high-order harmonics of light by aligned molecules. Experimental work by de Nalda et al. [14] elucidates the dependence of high-order harmonics generation (HHG) from aligned CO_2 on the angle between the molecular axis and the polarization vector, and interprets the results in terms of the orbital symmetry of the highest occupied molecular orbital. Kaku et al. show that alignment-dependent HHG can be used to sensitively measure rotational revival features in N_2. Itatani et al. [17] find substantial enhancement of the yield of high harmonics by nonadiabatic alignment and stress the importance of the ability of the method to generate field-free alignment. Another interesting avenue of research within the same framework is use of the highly symmetric environment that may be generated through molecular alignment to engineer exotic HG patterns. A case under current theoretical investigation is benzene molecules aligned to the polarization plan of a circularly polarized field [16]. In the field of nonlinear optics, Bartels et al., extending earlier work on the use of rotational revivals to compress optical pulses [11], demonstrate the application of the birefringence created by an ensemble of aligned molecules to phase-match nonlinear frequency conversion [102].

While this brief overview illustrates that much has been already accomplished in the field of nonadiabatic alignment by short, intense pulses, we suggest in the next section that a variety of new opportunities for discovery remain available.

6. Conclusions and Outlook

Our goal in the previous sections has been to review the topic of nonadiabatic alignment by moderately intense laser pulses—a young area of research at the rich interface among molecular physics, chemical dynamics, spectroscopy and nonlinear optics. We did not attempt an exhaustive survey of all aspects of this problem but rather a complementary review to a recently published colloquium [20]. To that end we focused on the underlying concepts, the theory and numerical methods, and the very recent progress. Before concluding we propose several possible avenues for future research and note areas where we expect new applications.

The recent generalization of post-pulse alignment from isolated linear systems to polyatomic molecules of arbitrary symmetry, and to molecules interacting with dense environments, invites applications in solution chemistry, biology and, possibly, engineering. A forthcoming publication extends alignment by elliptically polarized fields from control of only the overall rotations to the more general task of manipulating also the torsion angles of polyatomic molecules. One of the applications under consideration is a new approach to manipulation of charge transfer reactions in solution, with a view to light-controlled molecular switches [8]. Another ongoing study in our laboratory utilizes the concepts of alignment and three-dimensional alignment to guide molecular assembly in the fabrication of molecular electronic and photonic devices. We are motivated by the realization that the electric, magnetic, and optical properties of molecular devices are crucially dependent on the orientational order of the molecular moiety. An inviting area of application of intense laser alignment in the domain of large molecules is structural determination via spectroscopies such as nuclear magnetic resonance (NMR), where means of alignment have long been sought [103]. Likewise intriguing is a recent proposal [10] to use diffraction imaging from a beam of laser-aligned proteins as a means of resolving the secondary structure of the proteins. In the context of coherent control, rotational wavepackets have been proposed as a route to separation of racemic mixtures of chiral species into pure enantiomers [104] and to quantum information processing [19].

Closely related to the problem of laser alignment is molecular optics—the use of moderately intense laser pulses to focus, steer and disperse molecular beams [60,61,84,105–111], and to trap [21], accelerate and angularly accelerate [112–115] molecules. In much the same way that molecular alignment utilizes the angular gradients of polarized laser fields to manipulate the rotational motions, molecular optics utilizes the spatial gradient of laser beams to manipulate the center-of-mass motion. The combination of alignment with molecular optics is an interesting avenue for future research, due to both the intriguing fundamental questions and the potential technological applications it offers. In particular, one may envision advances such as nanoscale processing of surfaces [60,61,116], molecular separation techniques [105,111], and molecular waveguides based on the combination of spatial and orientational control.

Other areas of research where one might expect short-pulse-induced alignment and orientation to become a valuable tool in future years include stereochemistry, where alignment via other approaches has already helped in elucidating reaction pathways and mechanisms [1,2], and quantum control via polarization shaping techniques [117–120], where self-learning algorithms are expected to provide new insights into the role of molecular alignment in molecular dynamics. We thus expect the nonadiabatic alignment of molecular systems to remain a rich area of research in physics, chemistry, and possibly beyond.

7. Acknowledgements

The authors thank Prof. H. Stapelfeldt for many interesting conversations and for an exciting collaboration. T.S. gratefully acknowledges support of the Department of Energy (under grant DAAD19-03-R0017) and awards of the Guggenheim Foundation, the Alexander-von-Humboldt Foundation and NATO for research leading to this review article.

8. Appendix

Appendix A. Derivation of Equation (5)

In order to expose the different time-scales underlying the evolution of the wavepacket (3), it is useful to reexpress the expansion as,

$$\Psi_{\xi_i \mathbf{v}_i \mathbf{n}_i}(t) = \sum_{\mathbf{n}} F^{\xi_i \mathbf{v}_i \mathbf{n}}(t)|\xi_i \mathbf{v}_i \mathbf{n}\rangle e^{-iE^{\xi_i \mathbf{v}_i \mathbf{n}}t/\hbar} + \sum_{\xi \mathbf{v} \mathbf{n}} F^{\xi \mathbf{v} \mathbf{n}}(t)|\xi \mathbf{v} \mathbf{n}\rangle e^{-iE^{\xi \mathbf{v} \mathbf{n}}t/\hbar},$$

$$(A.1)$$

where we separate out the rapidly oscillating part of the amplitudes $C^{\xi \mathbf{v} \mathbf{n}}(t)$ of Eq. (3) through the transformation $F^{\xi \mathbf{v} \mathbf{n}}(t) = C^{\xi \mathbf{v} \mathbf{n}}(t)e^{iE^{\xi \mathbf{v} \mathbf{n}}t/\hbar}$ and partition the sum into the rotational manifold of the initial vibronic state $\{\xi_i \mathbf{v}_i\}$ and the remaining rotational manifolds. The $|\xi \mathbf{v} \mathbf{n}\rangle$ are defined through $H_M|\xi \mathbf{v} \mathbf{n}\rangle = E^{\xi \mathbf{v} \mathbf{n}}|\xi \mathbf{v} \mathbf{n}\rangle$, H_M being the complete field-free Hamiltonian, and $H = H_M + V(t)$, where $V(t)$ is the electric dipole interaction of Eq. (2).

Substituting Eq. (A.1) into the time-dependent Schrödinger equation and using the orthonormality of the expansion basis states, one derives a set of coupled differential equations for the expansion coefficients,

$$i\hbar \dot{F}^{\xi_i \mathbf{v}_i \mathbf{n}}(t) = \sum_{\mathbf{n}'} \langle \xi_i \mathbf{v}_i \mathbf{n}|V(t)|\xi_i \mathbf{v}_i \mathbf{n}'\rangle F^{\xi_i \mathbf{v}_i \mathbf{n}'}(t)e^{i(E^{\xi_i \mathbf{v}_i \mathbf{n}} - E^{\xi_i \mathbf{v}_i \mathbf{n}'})t/\hbar}$$

$$+ \sum_{\xi' \mathbf{v}' \mathbf{n}'} \langle \xi_i \mathbf{v}_i \mathbf{n}|V(t)|\xi' \mathbf{v}' \mathbf{n}'\rangle F^{\xi' \mathbf{v}' \mathbf{n}'}(t)e^{i(E^{\xi_i \mathbf{v}_i \mathbf{n}} - E^{\xi' \mathbf{v}' \mathbf{n}'})t/\hbar},$$

$$(A.2)$$

$$i\hbar \dot{F}^{\xi \mathbf{v} \mathbf{n}}(t) = \sum_{\mathbf{n}'} \langle \xi \mathbf{v} \mathbf{n}|V(t)|\xi_i \mathbf{v}_i \mathbf{n}'\rangle F^{\xi_i \mathbf{v}_i \mathbf{n}'}(t)e^{i(E^{\xi \mathbf{v} \mathbf{n}} - E^{\xi_i \mathbf{v}_i \mathbf{n}'})t/\hbar}$$

$$+ \sum_{\xi' \mathbf{v}' \mathbf{n}'} \langle \xi \mathbf{v} \mathbf{n}|V(t)|\xi' \mathbf{v}' \mathbf{n}'\rangle F^{\xi' \mathbf{v}' \mathbf{n}'}(t)e^{i(E^{\xi \mathbf{v} \mathbf{n}} - E^{\xi' \mathbf{v}' \mathbf{n}'})t/\hbar}. \qquad (A.3)$$

We assume that $|V(t)| \ll E^{\xi_i \mathbf{v}_i \mathbf{n}} - E^{\xi \mathbf{v} \mathbf{n}'}$ and $\hbar\omega \ll E^{\xi_i \mathbf{v}_i \mathbf{n}} - E^{\xi \mathbf{v} \mathbf{n}'}$ for all $\xi \neq \xi_i$, while $\hbar\omega \gg E^{\xi_i \mathbf{v}_i \mathbf{n}} - E^{\xi_i \mathbf{v}_i \mathbf{n}'}$. Hence, unless there are one- or multiphoton resonances at ω, the interaction cannot mix different vibronic manifolds.

Accordingly we neglect the second line of Eq. (A.2). In the first line we use the fact that the products $\varepsilon(t)F^{\xi_i v_i n'}(t)$ vary slowly with time as compared to the oscillatory term, to obtain a closed approximation for the $F^{\xi vn}(t)$ $\xi \neq \xi_i$ amplitudes,

$$
\begin{aligned}
F^{\xi vn}(t) &= \frac{i}{2\hbar} \sum_{n'} F^{\xi_i v_i n'} \Bigg\{ \langle \xi vn | \boldsymbol{\varepsilon} \cdot \boldsymbol{\mu} | \xi_i v_i n' \rangle \int^t dt' e^{i(E^{\xi vn} - E^{\xi_i v_i n'})t'/\hbar + \omega t'} \\
&\quad + \langle \xi vn | \boldsymbol{\varepsilon}^* \cdot \boldsymbol{\mu} | \xi_i v_i n' \rangle \int^t dt' e^{i(E^{\xi vn} - E^{\xi_i v_i n'})t'/\hbar - \omega t'} \Bigg\} \\
&= \frac{1}{2} \sum_{n'} F^{\xi_i v_i n'} \Bigg\{ \frac{\langle \xi vn | \boldsymbol{\varepsilon} \cdot \boldsymbol{\mu} | \xi_i v_i n' \rangle}{E^{\xi vn} - E^{\xi_i v_i n'} + \hbar\omega} [e^{i(E^{\xi vn} - E^{\xi_i v_i n'})t/\hbar + \omega t} - 1] \\
&\quad + \frac{\langle \xi vn | \boldsymbol{\varepsilon}^* \cdot \boldsymbol{\mu} | \xi_i v_i n' \rangle}{E^{\xi vn} - E^{\xi_i v_i n'} - \hbar\omega} [e^{i(E^{\xi vn} - E^{\xi_i v_i n'})t/\hbar - \omega t} - 1] \Bigg\}.
\end{aligned}
\tag{A.4}
$$

Inserting Eq. (A.4) in Eq. (A.2) and neglecting the rapidly oscillating terms one finds,

$$
\begin{aligned}
i\hbar \dot{F}^{\xi_i v_i n}(t) &= \frac{1}{2} \sum_{\xi' v' n'} \langle \xi_i v_i n | V(t) | \xi' v' n' \rangle e^{i(E^{\xi_i v_i n} - E^{\xi' v' n'})t/\hbar} \sum_{n''} F^{\xi_i v_i n''} \\
&\quad \times \Bigg\{ \frac{\langle \xi' v' n' | \boldsymbol{\varepsilon} \cdot \boldsymbol{\mu} | \xi_i v_i n'' \rangle}{E^{\xi' v' n'} - E^{\xi_i v_i n''} + \hbar\omega} [e^{i(E^{\xi' v' n'} - E^{\xi_i v_i n''})t/\hbar + \omega t} - 1] \\
&\quad + \frac{\langle \xi' v' n' | \boldsymbol{\varepsilon}^* \cdot \boldsymbol{\mu} | \xi_i v_i n'' \rangle}{E^{\xi' v' n'} - E^{\xi_i v_i n''} - \hbar\omega} [e^{i(E^{\xi' v' n'} - E^{\xi_i v_i n''})t/\hbar - \omega t} - 1] \Bigg\},
\end{aligned}
\tag{A.5}
$$

$$
\begin{aligned}
i\hbar \dot{F}^{\xi_i v_i n}(t) &= -\frac{1}{4} \sum_{n''} F^{\xi_i v_i n''} \sum_{\xi' v' n'} \Bigg\{ \frac{\langle \xi_i v_i n | \boldsymbol{\varepsilon} \cdot \boldsymbol{\mu} | \xi' v' n' \rangle \langle \xi' v' n' | \boldsymbol{\varepsilon}^* \cdot \boldsymbol{\mu} | \xi_i v_i n'' \rangle}{E^{\xi' v' n'} - E^{\xi_i v_i n''} - \hbar\omega} \\
&\quad + \frac{\langle \xi_i v_i n | \boldsymbol{\varepsilon}^* \cdot \boldsymbol{\mu} | \xi' v' n' \rangle \langle \xi' v' n' | \boldsymbol{\varepsilon} \cdot \boldsymbol{\mu} | \xi_i v_i n'' \rangle}{E^{\xi' v' n'} - E^{\xi_i v_i n''} + \hbar\omega} \Bigg\} e^{i(E^{\xi_i v_i n} - E^{\xi_i v_i n''})t/\hbar}.
\end{aligned}
\tag{A.6}
$$

Assuming that rotation–vibration coupling (Coriolis and centrifugal interactions) are not exhibited on the time-scale of relevance, $E^{\xi vn} = E^{\xi v} + E^n$, and using the fact that the rotational level spacing is much smaller than the vibronic, $E^{\xi' v'} - E^{\xi_i v_i} \gg E^{n'} - E^{n''}$, we can recast Eq. (A.6) in the form,

$$
i\hbar \dot{F}^{\xi_i v_i n}(t) = \sum_{n'} F^{v_i n'} \langle \xi_i v_i n | H_{\mathrm{ind}} | \xi_i v_i n' \rangle e^{i(E^{\xi_i v_i n} - E^{\xi_i v_i n'})t/\hbar},
\tag{A.7}
$$

where $H_{\text{ind}}^{\xi_i \mathbf{v}_i}$ is the induced dipole Hamiltonian of Eq. (5) in the text,

$$
\begin{aligned}
H_{\text{ind}}^{\xi_i \mathbf{v}_i} &= -\frac{1}{4} \sum_{\xi \mathbf{v}} \frac{2(E^{\xi' \mathbf{v}'} - E^{\xi_i \mathbf{v}_i})}{(E^{\xi' \mathbf{v}'} - E^{\xi_i \mathbf{v}_i})^2 - (\hbar\omega)^2} \\
&\quad \times \langle \xi_i \mathbf{v}_i | \boldsymbol{\varepsilon} \cdot \boldsymbol{\mu} | \xi' \mathbf{v}' \rangle \langle \xi' \mathbf{v}' | \boldsymbol{\varepsilon}^* \cdot \boldsymbol{\mu} | \xi_i \mathbf{v}_i \rangle \\
&= -\frac{1}{4} \sum_{\rho \rho'} \varepsilon_\rho \alpha_{\rho \rho'}^{\xi_i \mathbf{v}_i} \varepsilon_{\rho'}^*
\end{aligned}
\tag{A.8}
$$

and $\alpha^{\xi_i \mathbf{v}_i}$ is the AC polarizability tensor,

$$
\alpha_{\rho \rho'}^{\xi_i \mathbf{v}_i} = \sum_{\xi \mathbf{v}} \frac{2(E^{\xi \mathbf{v}} - E^{\xi_i \mathbf{v}_i})}{(E^{\xi \mathbf{v}} - E^{\xi_i \mathbf{v}_i})^2 - (\hbar\omega)^2} \langle \xi_i \mathbf{v}_i | \mu_\rho | \xi \mathbf{v} \rangle \langle \xi \mathbf{v} | \mu_{\rho'} | \xi_i \mathbf{v}_i \rangle.
\tag{A.9}
$$

9. References

[1] T.P. Rakitzis, A.J. van den Brom, M.H.M. Janssen, *Science* **303** (2004) 1852.
[2] V. Aquilanti, M. Bartolomei, F. Pirani, D. Cappelletti, F. Vecchiocattivi, Y. Shimizu, T. Kasai, *Phys. Chem. Chem. Phys.* **7** (2005) 291.
[3] C. Riehn, *Chem. Phys.* **283** (2002) 297.
[4] N. Mankoc-Borstnik, L. Fonda, B. Borstnik, *Phys. Rev. A* **35** (1987) 4132.
[5] T. Seideman, *J. Chem. Phys.* **113** (2000) 1677.
[6] T. Seideman, *Phys. Rev. A* **64** (2001) 042504.
[7] J.J. Larsen, I. Wendt-Larsen, H. Stapelfeldt, *Phys. Rev. Lett.* **83** (1999) 1123.
[8] G. Narayanan, T. Seideman (2005) in preparation.
[9] M.F. Gelin, C. Riehn, V.V. Matylitsky, B. Brutschy, *Chem. Phys.* **290** (2003) 307.
[10] J.C.H. Spence, K. Schmidt, J. Wu, G. Hembree, U. Weierstall, B. Doak, P. Fromme (2005) in preparation.
[11] R.A. Bartels, T.C. Weinacht, N. Wagner, M. Baertschy, C.H. Greene, M.M. Murnane, H.C. Kapteyn, *Phys. Rev. Lett.* **88** (2002) 013903.
[12] R. Velotta, N. Hay, M.B. Mason, M. Castillejo, J.P. Marangos, *Phys. Rev. Lett.* **87** (2001) 183901.
[13] M. Kaku, K. Masuda, K. Miyazaki, *Japan. J. Appl. Phys., Part 2* **43** (2004) L591.
[14] R. de Nalda, E. Heesel, M. Lein, N. Hay, R. Velotta, E. Spingate, M. Castillejo, J.P. Marangos, *Phys. Rev. A* **69** (2004) 031804.
[15] A.D. Bandrauk, H. Lu, *Internat. J. Quantum Chem.* **99** (2004) 431.
[16] A. Fleisher et al. (2005) in preparation.
[17] J. Itatani, J. Levesque, D. Zeidler, H. Niikura, H. Pepin, J.C. Kieffer, P.B. Corkum, D.M. Villeneuve, *Nature* **432** (2004) 867.
[18] S. Ramakrishna, T. Seideman, *Phys. Rev. Lett.* **95** (2005) 113001.
[19] K.F. Lee, D.M. Villeneuve, P.B. Corkum, E.A. Shapiro, *Phys. Rev. Lett.* **93** (2004) 233601.
[20] H. Stapelfeldt, T. Seideman, *Rev. Mod. Phys.* **75** (2003) 543.
[21] B. Friedrich, D. Herschbach, *J. Phys. Chem.* **99** (1995) 15686.
[22] H. Stapelfeldt, *Eur. Phys. J. D* **26** (2003) 15.
[23] H. Loesch, J. Moller, *Faraday Discussions* **113** (1999) 241.
[24] H.J. Loesch, J. Moller, *J. Phys. Chem.* **97** (1993) 2158.

[25] H.J. Loesch, A. Remscheid, *J. Chem. Phys.* **93** (1990) 4779.

[26] D. Herschbach, *Rev. Mod. Phys.* **71** (1999) S411.

[27] B. Friedrich, D.R. Herschbach, *Nature* **353** (1991) 412.

[28] R.E. Miller, et al., *Abstracts of Papers of the American Chemical Society* **222** (2001) 358.

[29] M. Wu, R.J. Bemish, R.E. Miller, *J. Chem. Phys.* **101** (1994) 9447.

[30] P.A. Block, E.J. Bohac, R.E. Miller, *Phys. Rev. Lett.* **68** (1992) 1303.

[31] R. González-Férez, P. Schmelcher, *Phys. Rev. A* **69** (2004) 023402.

[32] R. Kanya, Y. Ohshima, *Phys. Rev. A* **70** (2004) 013403.

[33] T. Seideman, *Annu. Rev. Phys. Chem.* **53** (2002) 41.

[34] P.M. Felker, A.H. Zewail, *J. Chem. Phys.* **86** (1987) 2460.

[35] P.M. Felker, *J. Phys. Chem.* **96** (1992) 7844.

[36] W.-M. Zhang, D.H. Feng, R. Gilmore, *Rev. Mod. Phys.* **62** (1990) 867.

[37] R.W. Robinett, *Phys. Rep.* **392** (2004) 1.

[38] O.V. Tikhonova, M.S. Molodenskii, *J. Exp. Theor. Phys.* **98** (2004) 1087.

[39] M.S. Child, *Mol. Phys.* **101** (2003) 637.

[40] M. Leibscher, I.S. Averbukh, P. Rozmej, R. Arvieu, *Phys. Rev. A* **69** (2004) 032102.

[41] C.A. Arango, W.W. Kennerly, G.S. Ezra, *Chem. Phys. Lett.* **392** (2004) 486.

[42] R. Baumfalk, N.H. Nahler, U. Buck, *J. Chem. Phys.* **114** (2001) 4755.

[43] H. Sakai, S. Minemoto, H. Nanjo, H. Tanji, T. Suzuki, *Phys. Rev. Lett.* **90** (2003) 083001.

[44] H. Sakai, S. Minemoto, H. Nanjo, H. Tanji, T. Suzuki, *Eur. Phys. J. D* **26** (2003) 33.

[45] S. Minemoto, H. Nanjo, H. Tanji, T. Suzuki, H. Sakai, *J. Chem. Phys.* **118** (2003) 4052.

[46] H. Ohmura, T. Nakanaga, *J. Chem. Phys.* **120** (2004) 5176.

[47] M. Renard, E. Hertz, B. Lavorel, O. Faucher, *Phys. Rev. A* **69** (2004) 043401.

[48] T. Seideman, *J. Chem. Phys.* **103** (1995) 7887.

[49] T. Seideman, *Phys. Rev. Lett.* **83** (1999) 4971.

[50] T. Seideman, *J. Chem. Phys.* **115** (2001) 5965.

[51] C.M. Dion, A. Keller, O. Atabek, A.D. Bandrauk, *Phys. Rev. A* **59** (1999) 1382.

[52] B. Friedrich, D. Herschbach, *Phys. Rev. Lett.* **74** (1995) 4623.

[53] B. Zon, B. Katsnel'son, *J. Exp. Theor. Phys.* **42** (1976) 595.

[54] F. Rosca-Pruna, M.J.J. Vrakking, *Phys. Rev. Lett.* **87** (2001) 153902.

[55] P.W. Dooley, I.V. Litvinyuk, K.F. Lee, D.M. Rayner, M. Spanner, D.M. Villeneuve, P.B. Corkum, *Phys. Rev. A* **68** (2003) 023406.

[56] V. Renard, M. Renard, S. Guèrin, Y.T. Pashayan, B. Lavorel, O. Faucher, H.R. Jauslin, *Phys. Rev. Lett.* **90** (2003) 153601.

[57] E. Péronne, M.D. Poulsen, C.Z. Bisgaard, H. Stapelfeldt, T. Seideman, *Phys. Rev. Lett.* **91** (2003) 043003.

[58] V. Renard, M. Renard, A. Rouzee, S. Guèrin, H.R. Jauslin, B. Lavorel, O. Faucher, *Phys. Rev. A* **70** (2004) 033420.

[59] M.D. Poulsen, E. Péronne, H. Stapelfeldt, C.Z. Bisgaard, S.S. Viftrup, E. Hamilton, T. Seideman, *J. Chem. Phys.* **121** (2004) 783.

[60] Z.-C. Yan, T. Seideman, *J. Chem. Phys.* **111** (1999) 4113.

[61] T. Seideman, *Phys. Rev. A* **56** (1997) R17.

[62] J.G. Underwood, M. Spanner, M.Y. Ivanov, J. Mottershead, B.J. Sussman, A. Stolow, *Phys. Rev. Lett.* **90** (2003) 223001.

[63] G. Herzberg, "Molecular Spectra and Molecular Structure", Prentice-Hall, Englewood Cliffs, NJ, 1939.

[64] A. Bohr, *Dan. Mat. Fys. Medd.* **26** (1952) 14.

[65] A.R. Edmonds, "Angular Momentum in Quantum Mechanics", Princeton Univ. Press, Princeton, 1957.

[66] R.N. Zare, "Angular Momentum", Wiley, New York, 1988.

[67] L.D. Landau, E.M. Lifshitz, "Mechanics", Pergamon Press, Oxford/New York, 1976.

[68] J.J. Larsen, K. Hald, N. Bjerre, H. Stapelfeldt, T. Seideman, *Phys. Rev. Lett.* **85** (2000) 2470.

[69] E.L. Hamilton, T. Seideman (2005) in preparation.

[70] B. Friedrich, D. Herschbach, *J. Phys. Chem.* **103** (1999) 10280.

[71] M. Machholm, N. Henriksen, *Phys. Rev. Lett.* **87** (2001) 193001.

[72] C.M. Dion, A.D. Bandrauk, O. Atabek, A. Keller, H. Umeda, Y. Fujimura, *Chem. Phys. Lett.* **302** (1999) 215.

[73] C. Dion, A. Keller, O. Atabek, *Eur. Phys. J. D* **14** (2001) 249.

[74] S. Guèrin, L.P. Yatsenko, H.R. Jauslin, O. Faucher, B. Lavorel, *Phys. Rev. Lett.* **88** (2002) 233601.

[75] A. Matos-Abiague, J. Berakdar, *Phys. Rev. A* **68** (2003) 063411.

[76] A. Matos-Abiague, J. Berakdar, *Chem. Phys. Lett.* **382** (2003) 475.

[77] O. Atabek, C.M. Dion, A.B.H. Yedder, *J. Phys. B: At. Mol. Opt. Phys.* **36** (2003) 4667.

[78] D. Sugny, A. Keller, O. Atabek, D. Daems, C.M. Dion, S. Guèrin, H.R. Jauslin, *Phys. Rev. A* **69** (2004) 033402.

[79] D. Sugny, A. Keller, O. Atabek, D. Daems, S. Guèrin, H.R. Jauslin, *Phys. Rev. A* **69** (2004) 043407.

[80] M. Spanner, E.A. Shapiro, M. Ivanov, *Phys. Rev. Lett.* **92** (2004) 093001.

[81] T. Kanai, H. Sakai, *J. Chem. Phys.* **115** (2001) 5492.

[82] W. Jarzeba, V.V. Matylitsky, C. Riehn, B. Brutschy, *Chem. Phys. Lett.* **368** (2003) 680.

[83] J. Ohkubo, T. Kato, H. Kono, Y. Fujimura, *J. Chem. Phys.* **120** (2004) 9123.

[84] T. Seideman, *J. Chem. Phys.* **111** (1999) 4397.

[85] P.S. Pershan, J.P. van der Ziel, L.D. Malmstrom, *Phys. Rev.* **143** (1966) 574.

[86] T. Seideman, *Chem. Phys. Lett.* **253** (1996) 279.

[87] E. Peronne, M.D. Poulsen, H. Stapelfeldt, C.Z. Bisgaard, E. Hamilton, T. Seideman, *Phys. Rev. A* **70** (2004) 063410.

[88] Y.-I. Suzuki, T. Seideman (2005) in preparation.

[89] G. Granucci, M. Persico, *J. Chem. Phys.* **120** (2004) 7438.

[90] E.A. Volkova, A.M. Popov, O.V. Tikhonova, *J. Exp. Theor. Phys.* **97** (2003) 702.

[91] S. Minemoto, H. Tanji, H. Sakai, *J. Chem. Phys.* **119** (2003) 7737.

[92] M. Leibscher, I.S. Averbukh, H. Rabitz, *Phys. Rev. A* **69** (2004) 013402.

[93] J. Ortigoso, *Phys. Rev. Lett.* **93** (2004) 073001.

[94] C.Z. Bisgaard, M.D. Poulsen, E. Péronne, S.S. Viftrup, H. Stapelfeldt, *Phys. Rev. Lett.* **92** (2004) 173004.

[95] K.F. Lee, I.V. Litvinyuk, P.W. Dooley, M. Spanner, D.M. Villeneuve, P.B. Corkum, *J. Phys. B: At. Mol. Opt. Phys.* **37** (2004) L43.

[96] A. Adelsward, S. Wallentowitz, *J. Opt. B: Quantum Semiclass. Opt.* **6** (2004) S147.

[97] L. Cai, J. Marango, B. Friedrich, *Phys. Rev. Lett.* **86** (2001) 775.

[98] P.M.A. Materny, N.E. Henriksen, V. Engel, *J. Chem. Phys.* **120** (2004) 5871.

[99] T. Shima, T. Nakayama, *Phys. Rev. A* **70** (2004) 013401.

[100] T. Kiljunen, M. Bargherr, M. Guhr, N. Schwentner, *Phys. Chem. Chem. Phys.* **6** (2004) 2185.

[101] T. Kiljunen, B. Schmidt, N. Schwentner, *Phys. Rev. Lett.* **94** (2005) 123003.

[102] R.A. Bartels, N.L. Wagner, M.D. Baertschy, J. Wyss, M.M. Murnane, H.C. Kapteyn, *Opt. Lett.* **28** (2003) 346.

[103] R.H. Havlin, G.H.J. Park, A. Pines, *J. Mag. Res.* **157** (2002) 163.

[104] K. Hoki, D. Kroner, J. Manz, *Chem. Phys.* **267** (2001) 59.

[105] T. Seideman, *J. Chem. Phys.* **106** (1997) 2881.

[106] T. Seideman, *J. Chem. Phys.* **107** (1997) 10420.

[107] H. Stapelfeldt, H. Sakai, E. Constant, P.B. Corkum, *Phys. Rev. Lett.* **79** (1997) 2787.

[108] H. Sakai, A. Tarasevitch, J. Danilov, H. Stapelfeldt, R.W. Yip, C. Ellert, E. Constant, P.B. Corkum, *Phys. Rev. A* **57** (1998) 2794.

[109] B.S. Zhao, et al., *Phys. Rev. Lett.* **85** (2000) 2705.
[110] H.S. Chung, B.S. Zhao, S.H. Lee, S. Hwang, K. Cho, S.-H. Shim, S.-M. Lim, W.K. Kang, D.S. Chung, *J. Chem. Phys.* **114** (2001) 8293.
[111] B.S. Zhao, et al., *J. Chem. Phys.* **119** (2003) 8905.
[112] P.F. Barker, M.N. Shneider, *Phys. Rev. A* **64** (2001) 033408.
[113] B. Friedrich, *Phys. Rev. A* **61** (2000) 025403.
[114] J. Karczmarek, J. Wright, P.B. Corkum, M. Ivanov, *Phys. Rev. Lett.* **82** (1999) 3420.
[115] D. Villeneuve, S.A. Aseyev, P. Dietrich, M. Spanner, M.Y. Ivanov, P.B. Corkum, *Phys. Rev. Lett.* **85** (2000) 542.
[116] R.J. Gordon, L.C. Zhu, W.A. Schroeder, T. Seideman, *J. Appl. Phys.* **94** (2003) 669.
[117] Y. Silberberg, *Nature* **430** (2004) 624.
[118] T. Brixner, G. Krampert, T. Pfeifer, R. Selle, G. Gerber, *Phys. Rev. Lett.* **92** (2004) 208301.
[119] T. Brixner, G. Gerber, *Opt. Lett.* **26** (2001) 557.
[120] T. Suzuki, S. Minemoto, H. Sakai, *Appl. Opt.* **43** (2004) 6047.
[121] C. Ellert, P. Corkum, *Phys. Rev. A* **59** (1999) R3170.
[122] E.B. Wilson Jr., J.C. Decius, P.C. Cross, "Molecular Vibrations", Dover, New York, 1980.
[123] G. Herzberg, "Molecular Spectra and Molecular Structure", D. Van Nostrand, Princeton, 1967.

RELATIVISTIC NONLINEAR OPTICS

DONALD UMSTADTER, SCOTT SEPKE AND SHOUYUAN CHEN

Physics and Astronomy Department, University of Nebraska, Lincoln, NE 68588-0111, USA

Abstract

The physics and applications of a new field of optics, involving the nonlinear interactions of ultra-intense light with matter, are discussed. They are characterized by many familiar nonlinear optical effects, such as self-focusing and harmonic generation, but based on a mechanism that became accessible experimentally only relatively recently, namely, the nonlinear motion of unbound electrons oscillating relativistically in strong electromagnetic fields. These phenomena form the basis of a new generation of compact and ultrashort-duration particle accelerators and X-ray light sources, with applications ranging from nuclear fusion to cancer therapy.

331
© 2005 Elsevier Inc. All rights reserved
ISSN 1049-250X
DOI 10.1016/S1049-250X(05)52007-X

1. Orientation and Background

1.1. INTRODUCTION TO RELATIVISTIC OPTICS AND HIGH FIELD SCIENCE

Immediately after the invention of the laser (1960), light could be focused to sufficient intensity to cause nonlinear optical effects in atomic media. In conventional nonlinear optics, where the electric field of a light wave E is much smaller than an atomic field, $E_{at} = 3 \times 10^9$ V/cm, various nonlinear phenomena, such as self-focusing, harmonic generation and Raman scattering arise due to the anharmonic motion of electrons in the combined fields of atom and laser. Approximate analytical solutions can be obtained by means of perturbation expansion methods, using E/E_{at} as the expansion parameter. At higher light fields, when E approaches E_{at}, this method breaks down and the medium becomes photo-ionized, creating a plasma, as illustrated in Fig. 1. Further increases in light intensity result in nonlinear optical effects of these free plasma electrons (see Fig. 2). The nonlinearity

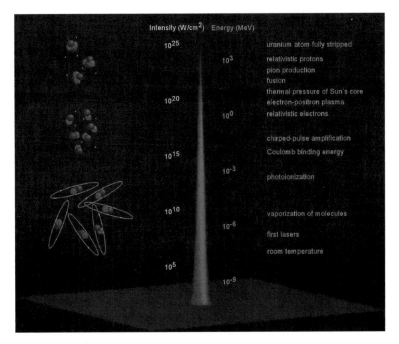

FIG. 1. The various regimes of laser-matter interactions. As the intensity of laser light increases, so does the energy of electrons accelerated in the light field resulting in the regime of conventional nonlinear optics with electrons bound to atoms replaced by the regime of relativistic nonlinear optics with free electrons in relativistic plasmas. At the highest intensities, even protons become relativistic, giving rise to what might be called the regime of nuclear optics, in which various nuclear processes, such as fusion, can take place. Reproduced from [72].

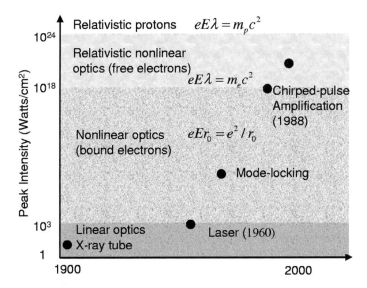

FIG. 2. History of light sources over the last century. Each advance in laser power enables a new regime of optics. Reproduced from [72].

arises in this case because the electrons oscillate at relativistic velocities in laser fields that exceed 10^{11} V/cm, resulting in relativistic mass changes and because the light's magnetic field becomes important. The work done by the electromagnetic field (E) on an electron $(eE\lambda)$ over the distance of a laser wavelength (λ) then approaches the electron rest mass energy $(m_e c^2)$, where e is the elementary charge of an electron, m_e is the electron rest mass and c is the speed of light. Effects analogous to those studied with conventional nonlinear optics—self-focusing, self-modulation, harmonic generation, and so on—are all found, but based on this entirely different physical mechanism. Thus, a new field of nonlinear optics, that of relativistic electrons, has been launched, as illustrated in Fig. 1.

One outcome of accessing this new optical regime is the generation of frequency-shifted light in a spectral region where there are no other compact sources. Another is the acceleration of other types of particles, such as positrons, ions and neutrons. These novel radiation sources have properties (femtosecond duration, micron source size, MeV energy) that make them suitable for numerous applications in imaging and spectroscopy in basic research, as well as medical diagnostics, cancer therapy, energy production and space propulsion. Rapid advancement is underway and new research tools, subfields and commercial products are on the horizon: e.g., compact and ultrashort pulse duration laser-based electron accelerators and X-ray sources.

FIG. 3. Photograph of the 33-fs duration 0.9-PW-peak power laser system at JAERI. Reproduced from [80].

Another physical regime will be encountered at even higher intensities ($I\lambda^2 \simeq 10^{24}$ W $\mu m^2/cm^2$), when even protons quiver relativistically: i.e., the work done on an proton over the distance of a laser wavelength approaches the rest mass energy. This might be called the nuclear regime of laser–plasma interactions, because of the fusion and fission reactions and the generation of pions, muons and neutrinos that should occur as nuclei collide in such energetic plasmas.

The recent dramatic increase in light intensity was made possible partly by the development, in the last decade, of compact lasers that have the ability to amplify shorter light pulses. For instance, solid-state lasers use chirped-pulse amplification [50] to generate femtosecond duration pulses. To accomplish this, an ultrashort, low-energy laser pulse is first stretched in time before it is amplified and then recompressed. Gas or dye lasers using solid-state switches have produced picosecond duration pulses. Present day advanced laser systems have multiterawatt peak powers and, when focused to micron spotsizes with adaptive optics, can produce electromagnetic intensities $I\lambda^2 \simeq 10^{21}$ W $\mu m^2/cm^2$. An example of a modern ultrahigh-power (0.9 PW) solid-state (Ti:Sapphire) laser system, located at the Advanced Photon Research Center, Kansai Research Establishment, Japan Atomic Energy Research Institute (JAERI), can be seen in the photograph shown in Fig. 3.

This paper addresses the fundamental concepts underlying what might be referred to more broadly as high-field science or relativistic optics. In many cases the phenomena of relativistic optics can be directly related to well-known, analogous effects in classical nonlinear optics. In that case, of course, these behaviors arise as a result of the nonlinear response of a medium generally described formally through the expansion of the polarization, **P**, as described above to yield

$$\mathbf{P} = \epsilon_0 \{\chi_1 \mathbf{E} + \chi_2 \mathbf{E}^2 + \cdots\},\tag{1}$$

where ϵ_0 is the free space permittivity, χ_n is the nth order susceptibility, and **E** is the response electric field. In relativistic optics, this response of a medium also often plays a significant role. Nonlinearity, however, arises as a result of the inherently nonlinear Lorentz force as well as from the time dependence of mass as a consequence of special relativity. For example, laser driven Raman and Brillouin scattering as well as light ion acceleration from thin films result directly from the interaction of a high intensity laser pulse with a dielectric material: in this case, a plasma. The presence of a dielectric is not necessary, however. After reviewing some fundamental concepts of electrodynamics, harmonic generation by a single oscillating electron is derived in detail. As the magnetic field force—the so-called Lorentz force—of the laser becomes important, the motion of an electron itself becomes nonlinear and significant harmonic content is added to the standard dipole oscillation exhibited at low field intensities. After touching on these topics, a few more recent developments in high-field science are discussed.

For more detailed descriptions of recent progress in experiments and theory, several review papers have been published on related topics: (1) relativistic nonlinear optics [52,69,72,38], (2) high-intensity laser development [50,81], (3) laser accelerators [19], (4) intense laser–plasma interactions [24,26,31,39,46,51,72,73, 78], (5) relativistic scattering [27,33], and (6) light ion acceleration [40].

1.2. GUIDING PRINCIPLES OF LASER PLASMA PHYSICS

All electrodynamics including, of course, intense laser–plasma interactions are governed by Maxwell's equations,

$$\nabla \cdot \mathbf{E} = 4\pi\rho,\tag{2}$$

$$\nabla \times \mathbf{E} = -\frac{1}{c}\frac{\partial \mathbf{B}}{\partial t},\tag{3}$$

$$\nabla \cdot \mathbf{B} = 0,\tag{4}$$

$$\nabla \times \mathbf{B} = \frac{4\pi}{c}\mathbf{J} + \frac{1}{c}\frac{\partial \mathbf{E}}{\partial t},\tag{5}$$

where **E** and **B** are the electric and magnetic field vectors and $\rho = \sum_i q_i n_i$ and $\mathbf{J} = \sum_i q_i n_i \mathbf{v}_i$ are the charge and current densities for species i having charge q_i, number density n_i and velocity $\mathbf{v}_i$. These can alternatively be presented in terms of the associated field potentials **A** and Φ by making use of the homogeneous

Maxwell equations through the relations

$$\mathbf{B} = \nabla \times \mathbf{A}, \tag{6}$$

$$\mathbf{E} = -\nabla\Phi - \frac{1}{c}\frac{\partial \mathbf{A}}{\partial t}. \tag{7}$$

Putting this result into the inhomogeneous Maxwell equations yields a set of two coupled equations

$$\nabla^2\Phi + \frac{1}{c}\frac{\partial}{\partial t}(\nabla \cdot \mathbf{A}) = -4\pi\rho, \tag{8}$$

$$\nabla^2\mathbf{A} - \frac{1}{c^2}\frac{\partial^2\mathbf{A}}{\partial t^2} - \nabla\left(\nabla \cdot \mathbf{A} + \frac{1}{c}\frac{\partial\Phi}{\partial t}\right) = -\frac{4\pi}{c}\mathbf{J} \tag{9}$$

that can be decoupled by specifying a gauge condition. For example, the Lorenz gauge is chosen to exploit the symmetry of the two potentials in vacuum which is readily apparent in the resulting wave equations for each under this choice

$$\nabla^2\mathbf{A} - \frac{1}{c^2}\frac{\partial^2\mathbf{A}}{\partial t^2} = 0, \tag{10}$$

$$\nabla^2\Phi - \frac{1}{c^2}\frac{\partial^2\Phi}{\partial t^2} = 0, \tag{11}$$

where the vacuum conditions $(\rho, |\mathbf{J}|) \to 0$ have been explicitly imposed [30]. These are then the equations that describe, for example, the fields of a focused, intense laser pulse.

Implicitly contained within Maxwell's equations are the fundamental principles: conservation of charge and energy. The first of these is simply stated for a given control volume with no sources as

$$\frac{\partial\rho}{\partial t} + \nabla \cdot \mathbf{J} = 0. \tag{12}$$

To understand the conservation of energy in a laser plasma system, one begins by summing the scalar product of $\mathbf{E}$ with Ampère's law and $\mathbf{B}$ with Faraday's law to yield after some rearranging

$$\frac{\partial}{\partial t}\left[\frac{|\mathbf{E}|^2 + |\mathbf{B}|^2}{8\pi}\right] + \left(\frac{c}{4\pi}\right)\nabla \cdot \mathbf{S} = -\mathbf{J} \cdot \mathbf{E}, \tag{13}$$

where the definition of the Poynting vector $\mathbf{S} = \mathbf{E} \times \mathbf{B}$ has been used. This expression is an exact statement of the conservation of energy within a control volume. The change in electromagnetic flux, $-\nabla \cdot \mathbf{S}$, which is equivalent to the rate at which electromagnetic energy flows out of the volume, is equal to the time rate of change of the electromagnetic field energy density pulse the rate at which the fields perform work, $\mathbf{J} \cdot \mathbf{E}$, on charged particles within the control volume.

In addition, the conservation of momentum is well described in many situations by the sum of the electric and Lorentz forces as

$$\frac{d\mathbf{p}}{dt} = m_0 \frac{\partial}{\partial t}(\gamma \mathbf{v}) = q\left[\mathbf{E} + \frac{1}{c}\mathbf{v} \times \mathbf{B}\right], \tag{14}$$

where $\mathbf{p}$ is the particle momentum, m_0 is the rest mass, $\gamma = (1 - v^2/c^2)^{-1/2}$ is the Lorentz factor, and q is the particle's charge. Considerable effort has been spent extending this equation of motion to a truly general result encompassing spatially and temporally nonuniform dielectric media [28]. The final Lorentz and gauge invariant force density relation has recently been derived by Hora and is given by

$$\mathbf{f} = \frac{1}{4\pi}\mathbf{E}\nabla \cdot \mathbf{E} + \frac{1}{c}\mathbf{J} \times \mathbf{H} + \frac{1}{4\pi}\left(1 + \frac{1}{\omega}\frac{\partial}{\partial t}\right)\nabla \cdot \mathbf{E}\mathbf{E}(\eta^2 - 1), \tag{15}$$

where ω is the angular frequency of the characteristic field oscillation and η is the index of refraction of the medium in question [57]. Note the similarity of this relation to the vacuum result of Eq. (14). The first two terms are again the electric field and Lorentz forces. The additional term represents the standard time averaged ponderomotive force of Lord Kelvin

$$\mathbf{f}_K = \left(\frac{\eta^2 - 1}{8\pi}\right)\left[\nabla \mathbf{E}^2 - 2\mathbf{E} \times (\nabla \times \mathbf{E})\right] \tag{16}$$

plus a small logarithmic term arising due to transient phenomena. For a more complete development of this topic see *Laser Plasma Physics* by Hora [28] as well as [57] and [29] and references therein.

2. Single-Particle Motion

The simplest laser–plasma system is the interaction of a single electron with an infinite laser pulse. Equation (14) governs this motion, and the fields are

$$\mathbf{E} = E_0\left[\cos\theta\hat{\mathbf{x}} + (1 - \delta)\sin\theta\hat{\mathbf{y}}\right], \tag{17}$$
$$\mathbf{B} = E_0\left[(\delta - 1)\sin\theta\hat{\mathbf{x}} + \cos\theta\hat{\mathbf{y}}\right], \tag{18}$$

where E_0 is the nominal field strength, $\theta = \omega_0 t - k_0 z$ is the phase, and the laser is linearly polarized for $\delta = 1$, right-hand circularly polarized for $\delta = 0$, and left-hand polarized for $\delta = 2$.

The electron orbit subject to a linearly polarized electromagnetic wave propagating in the $+z$ direction is governed by the Lorentz equation,

$$m\frac{d\mathbf{v}}{dt} = -e\left(\mathbf{E} + \frac{\mathbf{v}}{c} \times \mathbf{B}\right), \tag{19}$$

where **v** is the electron velocity, and E and B are the light's electric and magnetic fields.

A zeroth order solution to Eq. (19) is found by setting $v/c \ll 1$, which allows the term $(v/c) \times B$ to be neglected and γ is set to unity. Integrating Eq. (19) in this limit once over time (t) yields for the velocity.

$$\mathbf{v} = \text{Re}\left\{\frac{e\mathbf{E}}{im\omega}\right\} = a_0 c \begin{cases} \hat{e}_x \cos\psi, & \text{LP}, \\ (\hat{e}_x \cos\psi \mp \hat{e}_y \sin\psi), & \text{CP}, \end{cases} \tag{20}$$

where LP stands for linear and CP circular polarization.

An electron in low-intensity light oscillates with this velocity in a straight line along the polarization vector ($\hat{e}_x$) with an amplitude proportional to the normalized vector potential $a_0 \equiv eE_0/m_e\omega c$. Integrating again yields for the transverse displacement

$$\mathbf{x} = \text{Re}\left\{\frac{e\mathbf{E}}{m\omega^2}\right\} = \frac{a_0 c}{\omega} \begin{cases} \hat{e}_x \sin\psi, & \text{LP}, \\ (-\hat{e}_x \sin\psi \pm \hat{e}_y \cos\psi), & \text{CP}. \end{cases} \tag{21}$$

Thus, for $a_0 \sim 1$, the electron excursion during its oscillation is approximately λ.

A first order approximation for the electron motion can be found by substituting the zeroth order velocity Eq. (20) into the $v/c \times B$ term of (19). The latter then becomes proportional to

$$\mathbf{E} \times \mathbf{B} \propto \frac{a_0^2}{2}\left[1 + \cos(2\omega_0 t)\right]\hat{e}_z. \tag{22}$$

In the frame in which the electron is on the average at rest, the relativistic motion of an electron oscillates twice in the $\hat{e}_z$ or $\hat{k}$ direction for every once in the polarization direction ($\hat{e}_x$); i.e., a figure of eight motion. This originates from the fact that $v \times B \propto E \times B \propto E^2\hat{k}$, which is a product of two functions that vary sinusoidally at frequency ω and thus varies itself at frequency 2ω. In the lab frame, this transverse motion is superimposed upon a steady drift in the ($\hat{e}_z$) direction, originating from the DC term in Eq. (22). The next order approximation would include the mass shift $m = \gamma m_e$. As the field strength increases ($a_0^2 \gg 1$), the longitudinal motion ($\propto a_0^2$) begins to dominate the transverse motion ($\propto a_0$). This will be shown more formally in Section 2.1.

In the regime $a_0 \lesssim 1$, electrons radiate photons at harmonics of a modified laser frequency ω_0, with each harmonic order having its own unique angular distribution, as shown in Fig. 4. The radiation at the fundamental is the usual donut pattern, with a maximum in the direction perpendicular, and a minimum along, the polarization vector ($\hat{e}_x$). The second harmonic has two emission lobes with maxima pointing at angles between $\hat{e}_z$ and $\hat{e}_x$. An additional lobe is added for each additional harmonic order. This is referred to as nonlinear or relativistic Thomson scattering. The unique angular distributions of the second and third harmonics

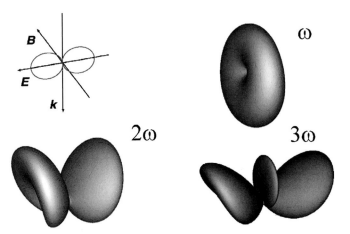

FIG. 4. Harmonics driven by relativistic Thomson scattering as the electrons in high-intensity laser fields (a_0^2) undergo figure-eight motion display unique angular distributions. Reproduced from [72].

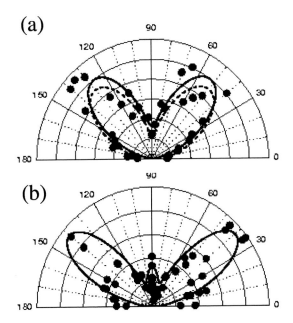

FIG. 5. Angular pattern of higher-order harmonic light. Shown are polar plots of the intensity of the second-harmonic light (top) and third harmonic (bottom) as a function of azimuthal angle. Filled circles, experimental data; solid and dashed lines, theoretical results. Reproduced from [9].

emitted from nonlinear relativistic Thomson scattering were observed experimentally [9] and are shown in Fig. 5. As will be shown in detail in Section 4, the situation is more complex for $a_0 \gtrsim 1$, and the scattered light is no longer simply harmonic.

2.1. CONSTANTS OF THE MOTION

Several constants of the motion can be found from an exact treatment for the motion. The derivation given in this section follows that of [46]. The starting point is the relativistically correct Lagrangian, which is written as

$$
L(\mathbf{r}, \dot{\mathbf{v}}, t) = -mc^2 \sqrt{1 - \frac{v^2}{c^2}} + \frac{q}{c} \mathbf{v} \cdot \mathbf{A} + q\phi,
\tag{23}
$$

where ϕ is the scalar potential associated with electrostatic fields. The Lorentz equation, Eq. (14), is derived from the Euler–Lagrange equation,

$$
\frac{d}{dt} \frac{\partial L}{\partial \mathbf{v}} - \frac{\partial L}{\partial \mathbf{r}} = 0.
\tag{24}
$$

For an infinite plane wave, in which the Lagrangian is independent of space in the transverse direction, Eq. (24) yields the conservation of transverse canonical momentum:

$$
\frac{\partial L}{\partial \mathbf{v}_\perp} = \mathbf{p}_\perp + \frac{q}{c} \mathbf{A}_\perp = \text{constant.}
\tag{25}
$$

The sum of the transverse momentum and field strength remain constant. The next constant follows from $dH/dt = -\partial L/\partial t$, yielding the relation

$$
\frac{dE}{dt} = -\frac{\partial L}{\partial t} = c\frac{\partial L}{\partial z} = c\frac{d}{dt}\frac{\partial L}{\partial v_z} = c\frac{dp_z}{dt},
\tag{26}
$$

where E is the time-dependent energy of the particle, which yields our second constant of the motion,

$$
E - cp_z = \text{constant.}
\tag{27}
$$

For a particle initially at rest, the kinetic energy E_{kin} becomes:

$$
E_{\text{kin}} = E - mc^2 = p_z c,
\tag{28}
$$

which yields

$$
E_{\text{kin}} = \frac{p_\perp^2}{2m} = p_z c = mc^2(\gamma - 1).
\tag{29}
$$

Thus, the electron scattering angle θ is related to the transverse and longitudinal momenta or the kinetic energy by:

$$\tan^2 \theta = \left(\frac{p_\perp}{p_z}\right)^2 = \frac{2m E_{\text{kin}}}{(E_{\text{kin}}/c)^2} = \frac{2}{\gamma - 1}. \tag{30}$$

In fact, this relation can be readily extended to the case of an electron with a nonzero initial kinetic energy, $E_{\text{kin}}(t = 0) = (\gamma_0 - 1)m_0c^2$ yielding

$$\tan^2 \theta = \frac{2(\gamma/\gamma_0 - 1)/(1 + \beta_0)}{[\gamma - \gamma_0(1 - \beta_0)]^2}, \tag{31}$$

where β_0 is the initial electron speed normalized to the speed of light. The angle of electrons produced by photo-ionization with intense lasers has been shown experimentally to obey the conservation of canonical momentum, as demonstrated in experiments that studied the angular distribution of relativistic electrons emitted from barrier-suppression ionization of atoms in intense laser fields, discussed in detail by Meyerhofer in [45, p. 30]. As the laser waist and pulse duration shrink and the intensity rises, this plane wave model eventually fails. This has demonstrated, for example, experimentally by Boreham and Hora [6], and Boreham and Luther-Davis [7] and numerically by Quesnel and Mora [54].

The orbit of an electron in the presence of a plane wave laser has been found using Hamilton–Jacobi formalism. For example, Sarachik and Schappert [60] first determined the trajectory for the special case of an electron initially at rest. This work was then extended to include an arbitrary initial velocity by Salamin and Faisal [58].

2.2. Role of Initial Phase

This previous work assumes that $\theta = 0$ at $t = 0$. The derivation given in this section follows that of [33]. If we wish to understand the role of the electron's initial phase, the relativistic Lorentz equation (Eq. (19)) may also be written, for an $\hat{x}$-polarized wave travelling in the $\hat{z}$ direction, as

$$\frac{d}{dt}(\gamma\boldsymbol{\beta}) = -(\hat{x} - \hat{z} \times \boldsymbol{\beta})a_0 \cos(t - z). \tag{32}$$

where we normalize time by $1/\omega_0$, velocity by c, and distance by c/ω_0. In Eq. (32), $a_0 = eE_0/m\omega_0c$ is the dimensionless parameter measuring the electric field strength, $\gamma = (1 - \beta_x^2 - \beta_y^2 - \beta_z^2)^{-1/2}$ is the relativistic mass factor and $\boldsymbol{\beta} = (\beta_x, \beta_y, \beta_z)$ is the electron velocity (in units of c). The electron orbit, subject to the following general initial conditions at time $t = 0$,

$$x = 0, \qquad y = 0, \qquad z = z_{\text{in}}, \tag{33}$$

$$\beta_x = \beta_{x0}, \qquad \beta_y = \beta_{y0}, \qquad \beta_x = \beta_{z0}, \tag{34}$$

$$B_{x0} = B_{y0} = 0, \tag{35}$$

has a closed form solution [27,33], when it is expressed parametrically: $t = t(\theta)$, $r = r(\theta)$, $\beta = \beta(\theta)$, where θ is the phase of the wave, defined by

$$\theta = t - z. \tag{36}$$

Note that $\beta_0 = (\beta_{x0}, \beta_{y0}, \beta_{z0})$ is the unperturbed velocity of the electron ($a_0 = 0$ limit) and that the initial phase that the electron sees is $\theta_{in} = -z_{in}$ according to Eqs. (33) and (36). This phase can be important in the ionization of the gas by an intense laser.

For the special case $\beta_{x0} = 0$, $\beta_{y0} = 0$, one finds $\beta_y = 0$, and the orbital equation yields the following closed form solution,

$$\gamma = \gamma_0 + \frac{a_0^2 (\sin\theta - \sin\theta_{in})^2}{2\gamma_0 (1 - \beta_{z0})}, \tag{37}$$

$$\gamma\beta_z = \gamma - \gamma_0 (1 - \beta_{z0}), \tag{38}$$

$$\gamma\beta_x = \frac{1}{\gamma_0 (1 - \beta_{z0})} a_0 (\sin\theta - \sin\theta_{in}), \tag{39}$$

$$x = \frac{a_0 [(\cos\theta_{in} - \cos\theta) - (\theta - \theta_{in})\sin\theta_{in}]}{\gamma_0 (1 - \beta_{z0})}, \tag{40}$$

$$t = \frac{(\theta - \theta_{in})}{1 - \beta_{z0}} \left[1 + \frac{a_0^2}{2} \left(\frac{1}{2} + \sin^2\theta_{in} \right)(1 + \beta_{z0}) \right]$$
$$+ \frac{a_0^2 (1 + \beta_{z0})}{2(1 - \beta_{z0})} \left[-\frac{\sin 2\theta}{4} + 2\cos\theta \sin\theta_{in} - \frac{3\sin 2\theta_{in}}{4} \right]. \tag{41}$$

In Eqs. (37)–(41), $\gamma_0 = (1 - \beta_{z0}^2)^{-1/2}$ and β_{z0} may either be negative (counter-propagating against the laser), or zero (Thomson scattering), or positive (co-propagating with the laser). Note that the velocity components β_x and β_z are given as explicit functions of θ according to Eqs. (38) and (39) upon using Eq. (37). They are periodic functions of θ having period 2π. The period, T, of this periodic ("figure-8") motion is thus equal to the increase in t as θ increases by 2π. Thus, we obtain from Eq. (41),

$$T = \frac{2\pi}{1 - \beta_{z0}} \left[1 + \frac{a_0^2}{2} (1 + \beta_{z0}) \left(\frac{1}{2} + \sin^2\theta_{in} \right) \right]. \tag{42}$$

The parametric solution for the z-coordinate of the electron orbit is given by $z = t - \theta$, in which t is given by Eq. (41). Over one orbital period, T, the electron undergoes a net displacement $r_0 = (x_0, 0, z_0)$, where x_0 is given by the increase in Eq. (40) as θ increases by 2π, and z_0 is simply $T - 2\pi$,

$$z_0 = T - 2\pi = \frac{2\pi}{1 - \beta_{z0}} \left[\beta_{z0} + \frac{a_0^2}{2} (1 + \beta_{z0}) \left(\frac{1}{2} + \sin^2\theta_{in} \right) \right], \tag{43}$$

$$x_0 = \frac{-2\pi a_0 \sin \theta_{in}}{\gamma_0(1 - \beta_{z0})}. \tag{44}$$

Note that the electron trajectory depends on a, β_{z0}, and θ_{in} in a rather complicated manner. Accordingly, the fundamental frequency ω_1 of the radiation spectrum depends on these three quantities. For backscattered radiation ($\hat{n} = -\hat{z}$), the dependence on the electron beam and on the laser becomes decoupled.

Following is a useful formula relating the change of t with respect to θ,

$$\frac{d\theta}{dt} = 1 - \beta_z = \frac{\gamma_0(1 - \beta_{z0})}{\gamma}, \tag{45}$$

which may be verified from Eqs. (36) and (38). Equation (45) is also valid for arbitrary values of β_{x0}, β_{y0}, β_{z0}, and θ_{in}, in which case γ_0 is the electron's relativistic mass factor in the absence of the laser.

2.3. DIRECT LASER ACCELERATION OF ELECTRONS IN VACUUM

In nonrelativistic, linear systems, the Lawson–Woodward theorem states that an electron can gain no net energy from a laser pulse. As the strength of the laser field increases, however, the Lorentz force introduces a nonlinearity to the problem, and energy can be transferred directly from the laser to the electron. The amount of energy is still limited by relativistic considerations, though. Since any charged mass necessarily travels at less than the speed of light, c, the laser light will overtake the particle. Thus, an electron cannot lock into a phase that constantly generates a positive acceleration. Instead, the electron alternates in a cycle of positive and negative acceleration yielding the figure-eight motion of an electron in a plane wave laser. When the laser is focused, the phase fronts become curved. Each point on these fronts travels at c along the propagation direction $\mathbf{k}$, but they do not cross a given plane normal to $\mathbf{k}$ at the same time. Therefore, an electron possessing the correct phase and velocity can observe a local phase velocity somewhat less than c by riding this curvature and can be accelerated to high energies by surfing these phase fronts without violating special relativity.

The gradients of focused laser fields can be used to accelerate and deflect electrons in vacuum for all angles of incidence. Before discussing a recent experiment demonstrated how an oblique collision between an electron beam and a high intensity laser pulse can be used as a tool to condition the beam both in terms of energy and longitudinal emittance [3]. Consider first the energetics of such a process. The full equation of motion illustrates an important and subtle issue. The ponderomotive force is associated with gradients in an electric field imposed on a dielectric material with refractive index η. In vacuum, $\eta = 1$, and the ponderomotive force is identically zero regardless of the fields present. Instead, forces in this case arise due to gradients in the electromagnetic field energy density. This can be seen more clearly from the tensor form of Eq. (15)

$$\mathbf{f} = \frac{1}{4\pi}\nabla \cdot \left[\mathbf{EE} + \mathbf{HH} - \frac{1}{2}\left(|\mathbf{E}|^2 + |\mathbf{H}|^2 \right)\underline{1} + \left(1 + \frac{1}{\omega}\frac{\partial}{\partial t} \right)\left(\eta^2 - 1 \right)\mathbf{EE} \right]$$
$$- \frac{1}{4\pi c}\frac{\partial \mathbf{S}}{\partial t}. \tag{46}$$

This affords a large amount of symmetry to these scattering processes.

In the adiabatic limit, the force density on an electron can then be identified directly with the gradient of the field energy: cf. Eq. (14). Proceeding from this premise and appealing only to Maxwell's equations, the standard vacuum equation of motion has been derived exactly

$$\mathbf{f} = \frac{1}{8\pi}\nabla\left(\mathbf{E}^2 + \mathbf{B}^2 \right) = \frac{1}{c}\mathbf{J} \times \mathbf{H} + \rho\mathbf{E}, \tag{47}$$

where $\mathbf{f}$ is the force density on a particle and $\mathbf{J}$ and ρ are the current and charge densities of the system, respectively [26]. Thus, Eq. (47) represents a good approximation of the force on a charged particle in vacuum or a low density collisionless plasma provided only that the scale lengths of variation of the laser are much larger than the scale length of the control volume $V_C^{1/3}$.

To demonstrate the scales over which acceleration occurs, first the TEM_{00} field model is expressed in a functionally convenient way by expanding in a power series about the focal parameter $\epsilon = 2/(k_0 w_0)$ [10,54]. Then to first order,

$$E_x = E_0 \frac{w_0}{w} \exp\left(-\frac{r^2}{w^2} \right) \sin\left(\phi_G^{(0)} \right), \tag{48}$$

$$E_z = E_0\epsilon \frac{x w_0}{w^2} \exp\left(-\frac{r^2}{w^2} \right) \cos\left(\phi_G^{(1)} \right), \tag{49}$$

$$B_y = E_0 \frac{w_0}{w} \exp\left(-\frac{r^2}{w^2} \right) \sin\left(\phi_G^{(0)} \right), \tag{50}$$

$$B_z = E_0\epsilon \frac{y w_0}{w^2} \exp\left(-\frac{r^2}{w^2} \right) \cos\left(\phi_G^{(1)} \right), \tag{51}$$

where $w = w_0\sqrt{1 + z^2/z_R^2}$ is the laser waist, $\phi_G^{(N)} = \omega_0 t - k_0 z + (N + 1) \times \arctan(z/z_R) - zr^2/z_R w^2 - \phi_0$, and $E_y = B_x \equiv 0$ [54]. This is, in fact, the first correction beyond the paraxial wave equation solution and represents a good approximation of many currently available lasers. Using these fields, including a cosine squared temporal envelope and following the model of Eq. (47), the time-averaged ponderomotive force density on an electron is given by

$$\langle f_\perp \rangle = \left(\frac{E_0^2}{2\pi w} \right)\left(\frac{w_0}{w} \right)^2 \left(\frac{x_\perp}{w} \right) e^{-2r^2/w^2}\left[1 + \epsilon^2\left(2\left[\frac{r}{w} \right]^2 - 1 \right) \right]\cos^4 \Phi,$$

$$\langle f_z \rangle = \left(\frac{E_0^2}{4\pi z_R} \right) \left(\frac{w_0}{w} \right)^4 \left(\frac{z}{z_R} \right) \Psi_0 \left(\left[1 - 2 \left(\frac{r}{w} \right)^2 \right] \right.$$

$$\left. + 4\epsilon^2 \left(\frac{r}{w} \right)^2 \left[1 - \left(\frac{r}{w} \right)^2 \right] \right)$$

$$- \left(\frac{E_0^2}{4c\Delta\tau} \right) \left(\frac{w_0}{w} \right)^2 e^{-2r^2/w^2} \left[1 + 2\epsilon^2 \left(\frac{r}{w} \right)^2 \right] \cos^3 \Phi \sin \Phi,$$

where $\Phi = (\pi/2\Delta\tau)[t - z/c]$. By adding the temporal envelope with no corrections, the assumption $\Delta\tau \gg \tau_0$ for laser period τ_0 has been made. The force perpendicular to the propagation direction, $\hat{\mathbf{z}}$, is denoted here as $f_\perp = \sqrt{f_x^2 + f_y^2}$. To find the force along either $\hat{\mathbf{x}}$ or $\hat{\mathbf{y}}$, simply replace the r in the prefactor with either x or y, respectively. Thus, the laser polarization appears nowhere in the force on a charge, and independence is demonstrated directly.

Each term can be understood physically. The transverse forces are driven by the radial focusing gradients which depend on the laser waist, w. As noted above, each of these exist in exactly the same form. The longitudinal force consists of two terms. The first is again caused by focusing now having the Rayleigh range, z_R, as the characteristic length. The second is due to the finite temporal envelope having scale length $c\Delta\tau$.

2.4. SELF-MODULATED LASER WAKEFIELD ELECTRON BEAM CHARACTERIZATION

Direct laser acceleration of electrons has recently been used to characterize a laser wakefield electron beam [3]. A hybrid Nd:glass and Ti:sapphire T^3 laser operating at a central wavelength of 1.053 μm is divided into a primary (accelerating) and a secondary (scattering) pulse using an 80/20 beam splitter. Each pulse retains the FWHM of 400 fs and is focused to a 10 μm waist. The primary pulse has a normalized vector potential of $a_0 = 1.7$ corresponding to an intensity of $I = 4 \times 10^{18}$ W/cm^2, and the scattering pulse has $a_0 = 0.85$ giving an intensity of $I = 1 \times 10^{18}$ W/cm^2.

The primary pulse is focused over a 400 psi supersonic helium gas jet and generates a plasma with a nominal background electron number density of 10^{19} cm^{-3}. This pulse also creates a self-modulated wakefield which, in turn, accelerates a Maxwellian electron beam with a temperature of 500 keV with an electron population of approximately 10^{11} particles. A newer gas jet has been measured as creating a temperature of up to 2000 keV.

To probe the temporal characteristics of the wakefield electron beam, the secondary laser pulse strikes the first beam at an angle of 135°. By using a variable optical delay, the electron beam intersects the laser pulse at different intensities along its 400 fs length. For example, when the delay is set to zero, the entire electron beam passes the path of the scattering laser pulse before the pulse arrives. Thus, no interaction occurs. For small delays, the electron beam begins to be affected by the front edge of the pulse. When the delay reaches 1.32 ps, the laser pulse crosses the electron beam path before the bunch arrives, and again, no interaction occurs. The effect of the laser on the electron bunch is to deflect those electrons with less than ~300 keV of energy along the laser wave vector. This energy limit was measured using a series of foil filters and has been verified through simulations. Thus, by spanning the 1.32 ps delay, the duration of the $\lesssim$300 keV portion of the wakefield electron beam is bounded at roughly 1 ps. This, in turn, bounds the duration of the MeV portion also at 1 ps since this is expected to be no longer than its lower energy counterpart.

Both analytical and numerical models have been developed to describe and explain the energy and intensity dependence of this deflection and to predict the laser parameters required to probe MeV range electrons and 100s keV protons. This deflection process was observed experimentally to be independent of the laser polarization.

2.5. A General Plane Wave Electron Scattering Model

A charged particle deflection model may be found easily for an arbitrary angle of incidence. For simplicity, consider first an arbitrary angle, ξ, in the $\hat{\mathbf{y}} - \hat{\mathbf{z}}$ plane. This implies that

$$\vec{\beta}_0 = \beta_0 \left[\hat{\mathbf{y}} \sin \xi + \hat{\mathbf{z}} \cos \xi \right], \tag{52}$$

where ξ is, of course, defined relative to the laser axis. The average drift velocity then becomes [58]

$$\vec{\beta}_D = \hat{\mathbf{y}} \left[\frac{(1 - \beta_0 \cos \xi) \beta_0 \sin \xi}{(1 - \beta_0 \cos \xi) + (a_0/(2\gamma_0))^2} \right]$$
$$+ \hat{\mathbf{z}} \left[\frac{(1 - \beta_0 \cos \xi) \beta_0 \cos \xi + (a_0/(2\gamma_0))^2}{(1 - \beta_0 \cos \xi) + (a_0/(2\gamma_0))^2} \right]. \tag{53}$$

The deflection angle is then defined as

$$\tan \left(\phi - \left[\xi - \frac{\pi}{2} \right] \right) = \frac{p_{Dz}}{p_{Dy}}, \tag{54}$$

where p_{Dj} is the drift momentum in the $j \in (y, z)$ direction. Putting Eqs. (53) and (54) together, the deflection angle predicted by the plane wave model is

$$\phi = \arctan\left[\cot\xi + \left(\frac{a_0^2}{4}\right)\left(\frac{1-\beta_0^2}{1-\beta_0\cos\xi}\right)\left(\frac{1}{\beta_0\sin\xi}\right)\right] + \left(\xi - \frac{\pi}{2}\right), \quad (55)$$

where all angles are measured in radians. This may also be written in terms of the particle energy, E,

$$\phi = \arctan\left[\cot\xi + \left(\frac{a_0^2}{4}\right)\left(\frac{(m_ec^2)^2}{E_1\sqrt{EE_2}\sin\xi + EE_2\sin\xi\cos\xi}\right)\right]$$
$$+ \left(\xi - \frac{\pi}{2}\right), \quad (56)$$

where $E_1 = E + m_0c^2$, $E_2 = E + 2m_0c^2$, and m_0c^2 is the particle rest mass energy.

In the Michigan deflection experiment, ξ was set to $135°$ to take advantage of the Doppler effect in generating EUV photons. When designing an experiment to only study charged particle deflection for beam conditioning, however, the full range of angles can be considered. Figure 6 shows the scattering angle ϕ as defined by Eq. (56) as a function of the initial electron energy for several angles of incidence and $a_0 = 10$ corresponding to a 1 μm laser intensity of 3.46×10^{21} W/cm^2. In the plane wave limit, electrons with an energy as high as 10 MeV can be deflected by a laser with intensity on the order of 10^{21} W/cm^2 which is now becoming available.

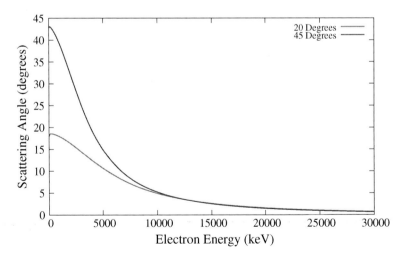

FIG. 6. The plane wave predicted scattering angle of an electron in a laser with $a_0 = 10$ for $\xi = 20°$ and $\xi = 45°$.

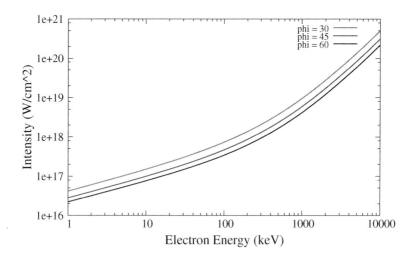

FIG. 7. Laser intensity required to scatter an electron 3° as a function of particle energy for $\xi = 30°, 45°$ and $60°$. Notice the stiffening of the electron as relativistic mass corrections occur.

Equation (56) can be inverted. That is, the intensity required to scatter a particle by ϕ radians is

$$I = (10^{-7}) \left(\frac{2\pi c}{e^2 \lambda_0^2} \right) \left[\tan \left(\phi - \xi + \frac{\pi}{2} \right) - \cot \xi \right]$$
$$\times \sin \xi \sqrt{E E_2} \left[E_1 + \cos \xi \sqrt{E E_2} \right] \tag{57}$$

in W/cm² where $e = 4.8032 \times 10^{-10}$ statCoulombs is the unit charge and λ_0 is the laser wavelength in centimeters. Figure 7 contains a plots of the intensity necessary to deflect an electron by 3°, for $\xi = 30°, 45°$ and $60°$, respectively. As determined above, 10 MeV electrons can be deflected by intensities on the order of 10^{21} W/cm², and 1 MeV electrons require only I on the order of 10^{19} W/cm², well within reach of currently available lasers. At these energies, electrons are, of course, relativistic. Their dressed mass demands a higher rate of increase in intensity with particle energy than in the classical case. This is shown clearly as the increase in the slope of Fig. 7 for all angles of incidence.

The ability to temporally probe MeV energy electron bunches is relevant to many applications. As shown in the Michigan experiment, even a pulse as short as 400 fs still follows this plane wave model. Thus, the duration of subpicosecond bunches of MeV electrons may be found by ponderomotively deflecting the beam as described above. Such time resolution is well beyond electronic methods and would allow direct measurement, for example, of the duration of the light burst produced by an all optical wiggler scheme. This time is the parameter that sets the temporal resolution of any imaging scheme using this source.

This model may be applied to any charged particle. Figure 8 shows the intensity necessary to deflect a proton by 3° as a function of energy and angle of incidence. To scatter protons with energies commensurate with the electrons in the Michigan experiment, a laser intensity on the order of 10^{22} W/cm^2 is needed.

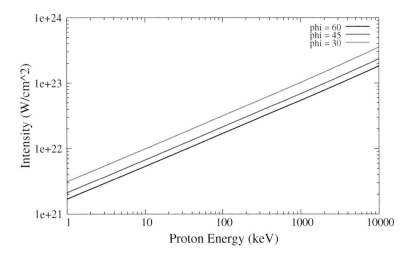

FIG. 8. Laser intensity required to scatter a proton 3° as a function of particle energy for $\xi = 30°$, 45° and 60°.

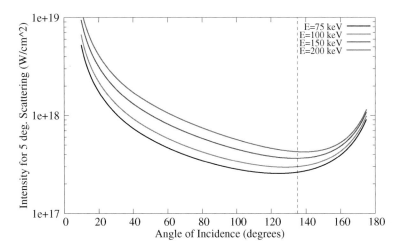

FIG. 9. The laser intensity required to scatter an electron 5° with given energy as a function of the angle of incidence. Fortuitously, the Michigan experiment operated at the optimum angle (135°) for the particle energies involved as denoted by the vertical dashed line.

The significant dependence on the angle of incidence exhibited in Figs. 6 to 8 implies an optimum angle exists. Figure 9 shows the intensity needed to deflect an electron by 5° as a function of ξ. The energies selected are those that were investigated in the Michigan experiment, which set $\xi = 135°$ as denoted by the vertical dashed line. Fortuitously, the optimum angle was used in this investigation. Observe that as the electron energy increases, the angle requiring the minimum intensity tends towards a counter propagating geometry corresponding to $\xi = 180°$. In this model, however, for ξ at exactly 180°, the deflection angle can be only 0 or 180 degrees since then the system collapses to one dimension.

3. Nonparaxial Solutions of the Maxwell Wave Equation

When investigating nonlinear laser particle interactions, accurate descriptions of all of the laser fields are needed. Many texts focus on the analytically convenient paraxial wave equation for describing the fields of a focused laser. Siegman, for example, spends considerable time developing two alternative formulations of the rectangular Hermite–Gaussian modes as well as the cylindrical Laguerre–Gaussian solutions [66]. These purely transverse modes, however, fail to capture the essential longitudinal components as first introduced in the laser literature by Lax et al. [34]. As field intensities rise and focusing approaches the diffraction limit ($w_0 \sim \lambda_0$), plane wave and even tried and true paraxial formulations fail to capture all of the relevant physics.

As early as 1979, Boreham et al. observed electron deflection angles in laser ionization experiments that deviated significantly from the theoretically predicted values based on a plane wave laser model [6,7]. To solve this discrepancy, following the model of Lax et al., Cicchitelli et al. developed an exact integral solution to the Maxwell wave equation that includes both transverse and longitudinal electric and magnetic fields [10]. Using this result, the correct deflection angles were then computed. This deviation in deflection angle can also be observed in the simulations by Quesnel and Mora [54] as well as those by Banerjee et al. [3].

The two primary laser models currently in use in the laser–plasma community are the integral solution of Cicchitelli et al. discussed above and a series representation of the TEM$_{00}$ mode. First, following the models of Davis [13] and Barton and Alexander [4] as well as Wang et al. [77], the Hermite–Gaussian (0, 0) mode is examined using a series expansion solution to the *full* Maxwell wave equation in vacuum for a laser polarized in the $\hat{\mathbf{x}}$ direction. The expansion is achieved by expanding the nonparaxial (∂_z^2) differential term about the small parameter $\epsilon = 2/k_0 w_0$, which measures the relative size of the spatial focusing. A single constant for each order arises in solving the wave equation, which must be found by imposing the boundary condition present in the actual laser system being modeled, which near the focus of a high intensity laser pulse is difficult, if not impossible, to know. Many articles in the past have omitted half of the solution yielding an asymmetric field description, which exhibits significant intensity astigmatism. The often more physical rotationally invariant intensity solution is here derived following the lead of Barton and Alexander [4]. Also, the infinite family of solutions is obtained here. Several earlier derivations omitted all but one particular choice; Wang et al., however, derived up to the fifth order fields including all possible solutions [76].

The aforementioned solution to the full Maxwell wave equation based on the spectral methodology employed by Cicchitelli et al. is also derived here in full. In this case, the boundary condition imposed declares that the transverse laser fields are purely Gaussian in the focal plane [10,54]. This solution is then compared to the series expansion TEM$_{00}$ mode and found to be an independent solution.

3.1. SERIES SOLUTION OF THE WAVE EQUATION FOR MONOCHROMATIC BEAMS

As a first approach to solving these wave equations for realistic laser pulses, a series expansion in the small parameter $\epsilon = 2/k_0 w_0$ is developed. The vector potential is written as

$$\mathbf{A} = \hat{\mathbf{x}} A_0 g(\eta) \Psi(\mathbf{r}) e^{i\eta}, \tag{58}$$

where A_0 is the magnitude of the lowest order term, $g(\eta)$ is the temporal envelope function that depends only on the phase $\eta = \omega_0 t - k_0 z$, $\Psi(\mathbf{r})$ is the spatial envelope function, and $e^{i\eta}$ is the plane wave propagator. Substituting Eq. (58) into Eq. (8) and performing some manipulation, one finds

$$\nabla^2 \Psi - 2ik_0 \frac{\partial \Psi}{\partial z}\left(1 - i\frac{g'}{g}\right) = 0, \tag{59}$$

where k_0 is the laser wave number and g' denotes the derivative of the envelope g with respect to the phase, η.

Most solutions then proceed by assuming that $g' \ll g$ thereby taking the pulse to be infinite in duration: that is, monochromatic. The series expansion can now be derived. This is accomplished by making the changes of variables

$$x \to \frac{x}{w_0}, \qquad y \to \frac{y}{w_0}, \qquad z \to \frac{z}{z_R}, \tag{60}$$

where w_0 is the laser waist at focus and $z_R = k_0 w_0^2/2$ is the Rayleigh range. Making this change in Eq. (59) yields

$$\nabla_\perp^2 \Psi - 4i\frac{\partial \Psi}{\partial z} + \epsilon^2 \frac{\partial^2 \Psi}{\partial z^2} = 0, \tag{61}$$

where $\nabla_\perp^2$ is the Laplacian operator acting only on the transverse coordinates x and y and $\epsilon \equiv \lambda_0/\pi w_0$ is the diffraction angle and is equal to 2 times the value of ϵ used by Cicchitelli et al. and by Quesnel and Mora [10,54]. Because diffraction limits laser focusing to roughly the laser wavelength, ϵ is necessarily smaller than π^{-1}. Thus, Ψ may, in general, be expanded in a series about the small parameter ϵ^2

$$\Psi = \Psi_0 + \epsilon^2 \Psi_2 + \epsilon^4 \Psi_4 + \cdots \tag{62}$$

which yields an explicit system of coupled, linear partial differential equations for all of the terms Ψ_N in the expansion:

$$\nabla_\perp^2 \Psi_0 - 4i\frac{\partial \Psi_0}{\partial z} = 0, \tag{63}$$

$$\nabla_\perp^2 \Psi_N - 4i\frac{\partial \Psi_N}{\partial z} + \frac{\partial^2 \Psi_{N-2}}{\partial z^2} = 0 \tag{64}$$

for $N = 2, 4, 6, \ldots$. The lowest order equation is homogeneous, and all subsequent equations are driven by the second derivative with respect to the axial coordinate z of the previous expansion term. Equation (63) is simply the so-called paraxial approximation in which

$$\left|\frac{\partial^2 \Psi}{\partial z^2}\right| \ll \left|2k_0\frac{\partial \Psi}{\partial z}\right|, \tag{65}$$

and the second z-derivative is omitted altogether. The higher order terms in this approach, however, do include this derivative term but only applied to the previous order term Ψ_{N-2} thereby taking $\partial_z^2 \Psi_N \ll \partial_z^2 \Psi_{N-2}$.

The scalar potential must still be determined before the electric and magnetic fields can be calculated. To do this, the scalar potential is assumed to have the same formal structure as the vector potential

$$\Phi \propto g(\eta)\Phi(\mathbf{r})e^{i\eta} \tag{66}$$

which—if $g'(\eta) \ll g(\eta)$—yields

$$\Phi = \frac{i}{k_0}\nabla \cdot \mathbf{A}. \tag{67}$$

To determine the electric and magnetic fields, recall the Lorenz gauge condition

$$\frac{1}{c}\frac{\partial \Phi}{\partial t} + \nabla \cdot \mathbf{A} = 0; \tag{68}$$

the relationship between the two potentials is known exactly to be

$$\Phi = \frac{i}{k_0}\nabla \cdot \mathbf{A}, \tag{69}$$

and the electric and magnetic fields can be formally defined using this relationship
and Eqs. (6) and (7)

$$\mathbf{E} = -ik_0\mathbf{A} - \left(\frac{i}{k_0}\right)\nabla[\nabla \cdot \mathbf{A}], \tag{70}$$

$$\mathbf{B} = \nabla \times \mathbf{A}. \tag{71}$$

This defines the first half of the general solution. The fully symmetric solution
is obtained by changing $\mathbf{A}$ to be polarized along the $\hat{\mathbf{y}}$ direction, inverting the
definitions of the electric and magnetic fields

$$\mathbf{B} = -ik_0\mathbf{A} - \left(\frac{i}{k_0}\right)\nabla[\nabla \cdot \mathbf{A}], \tag{72}$$

$$\mathbf{E} = \nabla \times \mathbf{A}, \tag{73}$$

and averaging these solutions, as was done in [4].

A lowest order solution as derived from Eq. (63) is the well known Hermite–
Gaussian TEM_{00} mode

$$\Psi_0 = b_1 f e^{-f\rho^2} + b_2 f^* e^{f^*\rho^2}, \tag{74}$$

where f^* denotes the complex conjugate of f, b_1 and b_2 are arbitrary constants,
$f = i(z + i)^{-1}$, $\rho^2 = x^2 + y^2$, x, y, and z retain the normalization of Eq. (60),
and $i = \sqrt{-1}$. Since the laser fields must remain bounded, $b_2 = 0$. Also, since
to lowest order the magnitude of the vector potential has been asserted to be con-
tained in the constant A_0, $|b_1|$ must be 1. Thus, let $b_1 = e^{i\phi_0}$ where ϕ_0 represents
some arbitrary phase added to the fields.

Using these values and appealing to the above derivations, the asymmetric $\bar{E}_x$
is found to order ϵ^2 to be

$$\bar{E}_x^{(2)}(x, y, z) = -iE_0 f e^{-f\rho^2}\left[\left(C_2 - \frac{1}{2}\right)f + \left(\left[\frac{1}{2} - C_2\right]\rho^2 + \bar{x}^2\right)f^2 \right.$$
$$\left. - \frac{1}{4}f^3\rho^4\right]e^{i(\eta+\phi_0)}, \tag{75}$$

where $E_0 = k_0 A_0$ and C_2 is a constant to be fixed by imposing a final boundary
condition. The realizable portion of the field is, of course, the real part of Eq. (75).

Following this technique, the higher order terms in Ψ, $\mathbf{E}$, and $\mathbf{B}$ follow directly with each order generating an infinite family of solutions each possessing
a single, indeterminate constant: C_j. The symmetric electric field components resulting from these potentials are summarized in Table I, and the magnetic fields
are identical except the role of x and y are reversed. The constants C_j must be
found by imposing a final boundary condition. Generally, this is done by asserting
that these field must resemble a spherical wave at large distances. This then sets
$C_2 = 2i$ and $C_4 = -6$.

Table I

The Hermite–Gaussian TEM$_{00}$ electric field mode to order ϵ^5. The parameters C_j are arbitrary constants to be determined by imposing an additional boundary condition. The function $A = A_0 \Psi_0 e^{i(\eta + \phi_0)}$, where A_0 is the lowest order vector potential amplitude, Ψ_0 is the lowest order mode as defined in the text, and ϕ_0 is an arbitrary phase constant also fixed by boundary condition. The physically realizable fields are the real part of the expressions shown here.

Symmetric Hermite–Gaussian TEM$_{00}$ mode electric field

$$E_x = -ik_0 A \left\{ \epsilon^0 [1] \right.$$

$$+ \epsilon^2 \left[\frac{1}{2}(2C_2 - 1)f + \frac{1}{2}([1 - 2C_2]\rho^2 + 2x^2)f^2 - \frac{1}{4}f^3\rho^4 \right]$$

$$+ \epsilon^4 \left[\frac{1}{4}(1 - 4C_2 + 4C_4)f^2 + \frac{1}{2}([4(C_2 - C_4) - 1]\rho^2 + [6C_2 - 3]x^2)f^3 \right.$$

$$+ \frac{1}{2}([C_4 - 3C_2 + 1]\rho^4 + [3 - 2C_2]x^2\rho^2)f^4$$

$$\left. \left. + \frac{1}{4}([C_2 - 1]\rho^6 - x^2\rho^4)f^5 + \frac{1}{32}f^6\rho^8 \right] \right\}$$

$$E_y = -ik_0 Axy \left\{ \epsilon^2 [f^2] \right.$$

$$\left. + \epsilon^4 \left[\frac{3}{2}(2C_2 - 1)f^3 + \frac{1}{2}(3 - 2C_2)f^4\rho^2 - \frac{1}{4}f^5\rho^4 \right] \right\}$$

$$E_z = k_0 Ax \left\{ \epsilon[f] \right.$$

$$+ \epsilon^3 \left[\frac{1}{2}(4C_2 - 3)f^2 + \frac{1}{2}(3 - 2C_2)f^3\rho^2 - \frac{1}{4}f^4\rho^4 \right]$$

$$+ \epsilon^5 \left[\frac{1}{4}(12C_4 - 18C_2 + 3)f^3 + \frac{1}{2}(15C_2 - 6C_4 - 6)f^4\rho^2 \right.$$

$$\left. \left. + \frac{1}{8}(4C_4 - 22C_2 + 18)f^5\rho^2 + \frac{1}{4}(C_2 - 2)f^6\rho^6 + \frac{1}{32}f^7\rho^8 \right] \right\}$$

3.2. A Spectral Method Solution of the Wave Equation

3.2.1. The Solution Assuming a Gaussian Laser at Focus

Another technique for solving such systems is the so-called spectral method. Cicchitelli et al. [10] and Quesnel and Mora [54] have derived the exact laser fields in this way by imposing a Gaussian field boundary condition in the focal plane: that is, at $z = 0$. Explicitly, the $\hat{\mathbf{x}}$-component of the electric field is formally specified in the focal plane as

$$\bar{E}_x(x, y, z = 0) = E_0 e^{-(x^2 + y^2)/w_0^2} \equiv E_0 e^{-\rho^2} \tag{76}$$

where E_0 is the field amplitude, and ρ retains the definition of the previous section. The Fourier transform of this is taken

$$
\begin{aligned}
\overline{E}_x(p, q) &= \frac{1}{\lambda_0^2} \iint_{\mathbb{R}^2} \overline{E}_x(x, y, z = 0) \exp\left(-ik_0[px + qy]\right) dx\, dy \\
&= \frac{E_0}{\pi \epsilon^2} e^{-(p^2+q^2)/\epsilon^2}
\end{aligned}
\tag{77}
$$

with $\epsilon = 2/k_0 w_0$ to be consistent with the series expansion method shown above. Finally, the field is found in all space by inverting the Fourier transform such that

$$
\begin{aligned}
\overline{E}_x(x, y, z) &= \iint_{\mathbb{R}^2} \overline{E}_x(p, q) \exp\left(ik_0\left[px + qy + z\sqrt{1 - p^2 - q^2}\right]\right) dp\, dq \\
&= \frac{2E_0}{\epsilon^2} \int_0^\infty e^{-b^2/\epsilon^2} e^{ik_0 z \sqrt{1-b^2}} J_0(k_0 r b) b\, db,
\end{aligned}
\tag{78}
$$

where $J_n(x)$ is the nth order Bessel function of the first kind, the inverse Fourier transform has been converted to polar coordinates such that

$$
b^2 = p^2 + q^2, \quad p = b \cos\phi, \quad q = b \sin\phi,
$$

and the angular integration has been performed. Repeating this calculation by imposing the boundary condition

$$
\widehat{B}_y(x, y, z = 0) = E_0 e^{-\rho^2}
\tag{79}
$$

and averaging the two results gives the exact symmetric laser fields

$$
E_x(x, y, z) = \frac{E_0}{\epsilon^2}\left[I_1 + \frac{x^2 - y^2}{k_0 r^3} I_2 + \frac{y^2}{r^2} I_3\right],
\tag{80}
$$

$$
E_y(x, y, z) = -\frac{E_0}{\epsilon^2} \frac{xy}{k_0 r^3}(k_0 r I_3 - 2I_2),
\tag{81}
$$

$$
E_z(x, y, z) = \frac{E_0}{\epsilon^2} \frac{x}{r} I_4,
\tag{82}
$$

$$
B_x(x, y, z) = -\frac{E_0}{\epsilon^2} \frac{xy}{k_0 r^3}(k_0 r I_3 - 2I_2),
\tag{83}
$$

$$
B_y(x, y, z) = \frac{E_0}{\epsilon^2}\left[I_1 + \frac{y^2 - x^2}{k_0 r^3} I_2 + \frac{x^2}{r^2} I_3\right],
\tag{84}
$$

$$
B_z(x, y, z) = \frac{E_0}{\epsilon^2} \frac{y}{r} I_4,
\tag{85}
$$

where the integral definitions

$$I_1 = \int_0^\infty e^{-b^2/\epsilon^2}\left(1 + \sqrt{1-b^2}\right)\sin(\phi_b)J_0(k_0 rb)b\,db, \tag{86}$$

$$I_2 = \int_0^\infty e^{-b^2/\epsilon^2}\frac{\sin(\phi_b)}{\sqrt{1-b^2}}J_1(k_0 rb)b^2\,db, \tag{87}$$

$$I_3 = \int_0^\infty e^{-b^2/\epsilon^2}\frac{\sin(\phi_b)}{\sqrt{1-b^2}}J_0(k_0 rb)b^3\,db, \tag{88}$$

$$I_4 = \int_0^\infty e^{-b^2/\epsilon^2}\left(1 + \frac{1}{\sqrt{1-b^2}}\right)\cos(\phi_b)J_1(k_0 rb)b^2\,db \tag{89}$$

have been made with $\phi_b = \omega_0 t - k_0 z\sqrt{1-b^2} + \phi_0$, ϕ_0 is an arbitrary phase constant analogous to that introduced above, and $r^2 = w_0^2\rho^2 = x^2 + y^2$ is the square of the unnormalized radial coordinate.

3.3. COMPARISON OF THE SERIES AND SPECTRAL SOLUTIONS

3.3.1. The Asymmetric Laser Field Solutions

Both solutions outlined above satisfy the full Maxwell wave equation to at least order ϵ^N, but since the expansion functions Ψ_N can, in principle, be found to arbitrary order, both solutions are, in fact, exact. These methods differ, however, in how each includes the nonparaxial $\partial_z^2(\cdot)$ operator. The spectral method includes this curvature term directly subject to the Gaussian boundary condition in the focal plane: cf. Eq. (76). Thus, to all orders in ϵ, the asymmetric field component $\overline{E}_x$ $(\widehat{B}_y)$ specified initially as purely Gaussian, is exactly Gaussian in the focal plane. This can be seen by evaluating the asymmetric $\overline{E}_x$ $(\widehat{B}_y)$ field component at $z = 0$ as shown in Eq. (78) using the relation

$$\frac{1}{n!}\int_0^\infty e^{-t}t^{n+k/2}J_k\left(2\sqrt{xt}\right)dt = e^{-x}x^{k/2}L_n^k(x), \tag{90}$$

which, indeed, yields

$$\overline{E}_x(x, y, z = 0) = E_0 e^{-\rho^2}. \tag{91}$$

The series expansion method does not afford this symmetry. As no formal boundary condition has been imposed, this method necessarily produces an infinite family of solutions for each order differentiated by the constants $C_2, C_4, \ldots$. Additionally, this method generates super-Gaussian corrections to the paraxial field of order ϵ^2 as shown in Eq. (75). Owing to these terms, regardless of the boundary condition imposed and, hence, the constants C_j chosen in the series method, these asymmetric field solutions are distinct sets of solutions of the wave equation.

3.3.2. The Symmetric Laser Field Solutions

Both techniques produce super-Gaussian terms to order ϵ^2 in the symmetric fields. To compare these solutions more concretely and determine whether they can be equated for any choice of the constants C_j, a series expansion of the spectral fields is required. To obtain this, note that the functions $(1 - \epsilon^2 t)^{\pm 1/2}$ can be expanded as

$$\sqrt{1 - \epsilon^2 t} = 1 - \frac{1}{2} \sum_{n=1}^{\infty} \frac{\Gamma_{n-1/2}}{n!} \epsilon^{2n} t^n, \tag{92}$$

$$\frac{1}{\sqrt{1 - \epsilon^2 t}} = \sum_{n=0}^{\infty} \frac{\Gamma_{n+1/2}}{n!} \epsilon^{2n} t^n \tag{93}$$

for $t \in [0, \epsilon^{-2})$ and $\Gamma_a \equiv \Gamma(a)/\sqrt{\pi}$. The integral I_1 from Eq. (86) can be rewritten as

$$I_1 = \frac{\epsilon^2}{2} \int_0^{\infty} e^{-t} \left(1 + \sqrt{1 - \epsilon^2 t}\right) J_0\left(2\sqrt{\rho^2 t}\right) dt \tag{94}$$

$$= \frac{\epsilon^2}{2} \left\{ \frac{1}{0!} \int_0^{\infty} e^{-t} J_0\left(2\sqrt{\rho^2 t}\right) dt + \frac{1}{0!} \int_0^{1/\epsilon^2} e^{-t} J_0\left(2\sqrt{\rho^2 t}\right) dt \right.$$

$$- \frac{1}{2} \sum_{n=1}^{\infty} \Gamma_{n-1/2} \epsilon^{2n} \frac{1}{n!} \int_0^{1/\epsilon^2} e^{-t} t^n J_0\left(2\sqrt{\rho^2 t}\right) dt$$

$$\left. + \int_{1/\epsilon^2}^{\infty} e^{-t} \sqrt{1 - \epsilon^2 t} J_0\left(2\sqrt{\rho^2 t}\right) dt \right\} \tag{95}$$

$$= \frac{\epsilon^2}{2} \left\{ 1 - \frac{1}{2} \sum_{n=1}^{\infty} \Gamma_{n-1/2} \epsilon^{2n} L_n(\rho^2) \right\} e^{-\rho^2}, \tag{96}$$

where the final equality holds in the limit of ϵ^2 small—as is naturally required for such an expansion—again following from Eq. (90). A similar process can be carried out for the remaining integrals to develop a series expansion for the $\hat{\mathbf{x}}$-component of the symmetric electric field, finally, yielding

$$E_x(x, y, z = 0) \approx E_0 e^{-\rho^2}\left[1 + \frac{1}{2}\sum_{n=1}^{\infty}\Gamma_{n-1/2}\left(n\frac{y^2}{r^2} - \frac{1}{2}\right)\epsilon^{2n}L_n^0(\rho^2)\right.$$

$$\left. + \frac{1}{4}\epsilon^2\left(\frac{x^2 - y^2}{r^2}\right)\sum_{n=0}^{\infty}\Gamma_{n+1/2}\epsilon^{2n}L_n^1(\rho^2)\right] \quad (97)$$

where $L_N^k(x)$ is the Nth order, kth associated Laguerre polynomial.

Observe that as in the asymmetric case, the purely Gaussian paraxial E_x field is obtained to zeroth order in ϵ from both solution techniques. In contrast to the asymmetric case, however, both methods generate higher order terms that produce super-Gaussian contributions. The series solution shown exhibits contributions from both asymmetric solutions. In the spectral method solution, however, such terms have arisen solely from the B_y boundary condition.

As an explicit comparison, the order ϵ^2 term of the symmetric, spectral solution E_x is

$$E_x^{(2)}(x, y, z = 0) = \frac{E_0}{4}(2x^2 - \rho^2)e^{-\rho^2}. \quad (98)$$

The series solution, however, gives the second order field as

$$E_x^{(2)}(x, y, z = 0) = \frac{E_0}{4}\left[(4C_2 - 2) + (2x^2 - (4C_2 - 3)\rho^2) - \rho^4\right]e^{-\rho^2}. \quad (99)$$

No value of the constant C_2 can be chosen to equate these two solutions. Thus, as intimated by the conclusion based on the asymmetric solution, the series and spectral methods do, indeed, generate distinct solutions of the Maxwell wave equation. The expression that is valid for a particular laser beam must be determined by the specific boundary condition present in the system: that is, by the exact form of E_x in the focal plane.

3.4. The Solution Assuming a Flattened Gaussian Laser at Focus

Several optical techniques can produce laser pulses that are distinctly non-Gaussian at the focus. Chirped pulse amplification (CPA), in fact, tends to amplify the wings of the pulse preferentially relative to the axis thereby creating a flat-top radial profile. In fact, any nonideal optical components or amplification lead to imperfect Gaussian laser profiles. To accurately model this, a more precise and flexible field description is required.

To allow flexibility in computing laser fields for such varied beam profiles and to generate a more complete set of solutions to compliment the specific forms illustrated above, consider a more general focal plane boundary condition

$$\bar{E}_x(x, y, z = 0) = E_0 \sum_{N=0}^{\infty} A_N \rho^{2N} e^{-\rho^2}, \tag{100}$$

where the A_N are arbitrary constants used to specify the exact form of the field desired. If the imposed field is of the form

$$\bar{E}_x(x, y, z = 0) = E_0 f(\rho^2) e^{-\rho^2}, \tag{101}$$

then the constants are simply the expansion coefficients for the MacLaurin series in ρ^2 of the function $f(\rho^2)$

$$A_N = \frac{f^{(N)}(\rho^2)}{\Gamma(N+1)} \bigg|_{\rho^2=0}, \tag{102}$$

where $f^{(N)}$ is the Nth derivative of $f(\rho^2)$ and $\Gamma(x)$ is the gamma (factorial) function. Of course, if $f(\rho^2) \equiv 1$, $A_0 = 1$ and all other coefficients are zero thereby reproducing the purely Gaussian result derived previously by Cicchitelli et al. [10].

The Fourier transform of Eq. (100) can be evaluated exactly as

$$\bar{E}_x(p, q) = \frac{1}{\pi^2 \epsilon^2} \iint_{\mathbb{R}^2} \bar{E}_x(x, y, z = 0) \exp(-ik_0 w_0 [p\bar{x} + q\bar{y}]) \, d\bar{x} \, d\bar{y}$$

$$= \frac{E_0}{\pi \epsilon^2} \exp\left(-\left[\frac{p^2 + q^2}{\epsilon^2}\right]\right) \sum_{N=0}^{\infty} A_N \Gamma(N+1) L_N\left(\frac{p^2 + q^2}{\epsilon^2}\right), \tag{103}$$

where $L_N(x)$ is the Nth order unassociated Laguerre polynomial [5].

The $\hat{\mathbf{x}}$-component of the electric field is found in all space by taking the inverse Fourier transform of this function to yield

$$\bar{E}_x(x, y, z) = \frac{2E_0}{\epsilon^2} \sum_{N=0}^{\infty} A_N \Gamma(N+1)$$

$$\times \int_0^{\infty} e^{-b^2/\epsilon^2} e^{ik_0 z \sqrt{1-b^2}} J_0(k_0 r b) L_N\left(\frac{b^2}{\epsilon^2}\right) b \, db. \tag{104}$$

The full, symmetric laser field solution for these flattened Gaussian fields is then formally identical to the definitions in Eqs. (80) to (85), but the integral definitions are now modified to

$$I_1 = \sum_{N=0}^{\infty} A_N N! \int_0^{\infty} e^{-b^2/\epsilon^2} \left(1 + \sqrt{1-b^2}\right) \sin(\phi_b) J_0(k_0 r b) L_N\left(\frac{b^2}{\epsilon^2}\right) b \, db,$$

$$I_2 = \sum_{N=0}^{\infty} A_N N! \int_0^{\infty} e^{-b^2/\epsilon^2} \frac{\sin(\phi_b)}{\sqrt{1-b^2}} J_1(k_0 r b) L_N\left(\frac{b^2}{\epsilon^2}\right) b^2 \, db,$$

$$I_3 = \sum_{N=0}^{\infty} A_N N! \int_0^{\infty} e^{-b^2/\epsilon^2} \frac{\sin(\phi_b)}{\sqrt{1-b^2}} J_0(k_0 r b) L_N\left(\frac{b^2}{\epsilon^2}\right) b^3 \, db,$$

$$I_4 = \sum_{N=0}^{\infty} A_N N! \int_0^{\infty} e^{-b^2/\epsilon^2} \left(1 + \frac{1}{\sqrt{1-b^2}}\right) \cos(\phi_b)$$

$$\times J_1(k_0 r b) L_N\left(\frac{b^2}{\epsilon^2}\right) b^2 \, db$$

for $\phi_b = \omega_0 t - k_0 z \sqrt{1-b^2} + \phi_0$ where again ϕ_0 is an arbitrary phase constant.

4. Radiation from Relativistic Electrons

The derivation given in this section follows that of [33], but other detailed discussions can also be found in Leemans et al. [38] and Meyerhofer [45].

An electron with displacement $r(t)$ and velocity $v(t)$ carries a current density $J(r, t) = ev(t)\delta[r - r(t)]$, whose Fourier transform $J_F(k, \omega)$ may be obtained easily. The total work done, W, in ergs, performed by the current $\mathbf{J}(\mathbf{r}, t)$ on the electric field $\mathbf{E}(\mathbf{r}, t)$ is given by

$$W \equiv \int d\omega \, d\Omega \frac{d^2 W}{d\omega \, d\Omega} = \int d\omega \, d\Omega \frac{\omega^2}{c^2} \left[|\mathbf{J}_F(\mathbf{k}, \omega)|^2 - |\mathbf{n} \cdot \mathbf{J}_F(\mathbf{k}, \omega)|^2\right], \tag{105}$$

where the $\mathbf{k}$-space differential volume $d^3\mathbf{k} = k^2 \, dk \, d\Omega$ is expressed in terms of the solid angle (Ω) in the direction of the unit vector $\mathbf{n} = c\mathbf{k}/\omega$. We obtain immediately from Eq. (105), in the far-field approximation,

$$\frac{d^2 W}{d\Omega \, d\omega} = \frac{e^2 \omega^2}{4\pi^2 c} |\mathbf{n} \times [\mathbf{n} \times \mathbf{F}(\omega)]|^2, \tag{106}$$

$$\mathbf{F}(\omega) = \int_{-\infty}^{\infty} dt \, \bar{\beta}(t) \, e^{i\omega[t - \mathbf{n} \cdot \mathbf{r}(t)/c]}. \tag{107}$$

Equation (106) gives the energy radiated by the electron in the direction of the unit vector, per unit solid angle, per unit frequency ω. Radiation damping is ignored throughout, because in Thomson scattering the energy of the scattered photon is much, much less than the electron rest mass in the average rest frame of the electron: i.e.,

$$\frac{\hbar \omega_L}{mc^2} \ll 1. \tag{108}$$

Therefore, the electron recoil is effectively zero. This inequality does not hold true in the case of Compton scattering.

As in Section 2.2, let us consider the simplest case where the electron orbit is strictly a periodic function of time with period T and a net displacement $\mathbf{r_0}$ per period. Thus, we have for all integers m (positive, negative or zero),

$$\bar{\beta}(t + mT) = \bar{\beta}(t), \qquad \mathbf{r}(t + mT) = m\mathbf{r_0} + \mathbf{r}(t). \tag{109}$$

Equation (107) may then be written as,

$$\mathbf{F}(\omega) = \sum_{m=-\infty}^{\infty} \int_{mT}^{(m+1)T} dt\,\bar{\beta}(t)\,e^{i\omega[t - \mathbf{n}\cdot\mathbf{r}(t)/c]} = \sum_{m=-\infty}^{\infty} \mathbf{f}(\omega)e^{im\omega[T - \mathbf{n}\cdot\mathbf{r_0}/c]}, \tag{110}$$

where

$$\mathbf{f}(\omega) = \int_0^T dt\,\bar{\beta}(t)\,e^{i\omega[t - \mathbf{n}\cdot\mathbf{r}(t)/c]}, \tag{111}$$

and we have used Eq. (109). Upon using $\sum_m e^{imx} = \sum_m 2\pi\delta(x - 2m\pi)$ in the last infinite sum in Eq. (110) and the property of the Dirac delta function, $\delta(ax) = (1/a)\delta(x)$, we obtain from Eqs. (110) and (111) the following expression for the spectrum,

$$\mathbf{F}(\omega) = \sum_{m=-\infty}^{\infty} \mathbf{F}_m \delta(\omega - m\omega_1), \tag{112}$$

$$\omega_1 = \frac{2\pi}{T - \mathbf{n}\cdot\mathbf{r_0}/c}, \tag{113}$$

$$\mathbf{F}_m = \frac{\omega_1}{2\pi}\int_0^T dt\,\bar{\beta}(t)e^{im\omega_1[t - \mathbf{n}\cdot\mathbf{r}(t)/c]}. \tag{114}$$

Note from Eq. (112) that the radiation spectrum is discrete, for strictly periodic motion of the electron. The base frequency of this spectrum, ω_1, depends on the periodicity (T) of the electron, on the electron's net displacement ($\mathbf{r_0}$) in one such period, and on the direction ($\hat{\mathbf{n}}$) in which the radiation is observed. Thus, the radiation spectrum is, in general, not at the harmonic frequency of the laser (nor at the harmonics of the electron orbital frequency, $2\pi/T$). It would thus be *wrong* to simply insert $\omega = n\omega_0$ in Eq. (111) and to replace the electron's orbital period T there by the laser's optical period $2\pi/\omega_0$ and consider the resultant value of that integral to give the spectral amplitude of the radiation at the nth harmonic of the laser frequency. Erroneous conclusions regarding generation of high laser harmonics have appeared in the literature based on such an intuitive (but incorrect) substitution.

The power, p_m (in erg/s), radiated at the harmonic frequency $\omega = m\omega_1$ per unit solid angle in the direction of the unit vector $\hat{\mathbf{n}}$ is then given by (cf. Eqs. (106) and (112)),

$$p_m = \frac{e^2 m^2 \omega_1^2}{4\pi^2 c} |\mathbf{n} \times \mathbf{F}_m|^2 , \tag{115}$$

where the dimensionless spectral amplitude $\mathbf{F}_m$ is given in Eq. (114). It is easy to show from Eqs. (109) and (113) that the integrand in Eq. (114) is a periodic function of t having period T. Integrals of this type are readily evaluated by the Romberg method. The radiation spectrum observed *exactly* in the forward direction of the laser ($\hat{n} = \hat{z}$) always has only one discrete frequency, $\omega = \omega_1 = \omega_o$, which is easily shown from Eq. (113) upon using the first equality of Eq. (43). This statement is true regardless of the velocity of the electron or the laser intensity, and may easily be deduced from Eq. (114) for this case. However, for an energetic electron beam that is almost co-propagating with the laser, such as that produced by the laser itself, high harmonics at the laser frequency may be observed in the direction just slightly off the laser direction.

In the backscattering direction of the laser ($\hat{\mathbf{n}} = -\mathbf{z}$), if we set $\theta_{in} = 0$, one obtains the following expressions for ω_1 and p_m, the backscatter power at $\omega = m\omega_1$ (cf. Eq. (115)),

$$\frac{\omega_1}{\omega_0} = \left(\frac{2}{2+a_0^2}\right)\left(\frac{1-\beta_{z0}}{1+\beta_{z0}}\right) = \left(\frac{2}{2+a_0^2}\right)\gamma_0^2(1-\beta_{z0})^2, \tag{116}$$

$$p_m = \frac{A}{\gamma_0^2(1-\beta_{z0})^2}\left(\frac{\omega_1}{\omega_0}\right)^4 s_m, \tag{117}$$

where $A = e^2\omega_0^2/4\pi^2 c = 0.69 \, [\lambda(1 \, \mu m)]^{-2}$ erg/s, $s_m = 0$ for $m = 0, \pm 2, \pm 4, \ldots$, and for $m = \pm 1, \pm 3, \pm 5, \ldots$,

$$s_m = (a_0\pi)^2 m^2 \big[J_{(m-1)/2}(m\kappa) - J_{(m+1)/2}(m\kappa) \big]^2, \tag{118}$$

$$\kappa = \frac{a_0^2}{2(a_0^2 + 2)}. \tag{119}$$

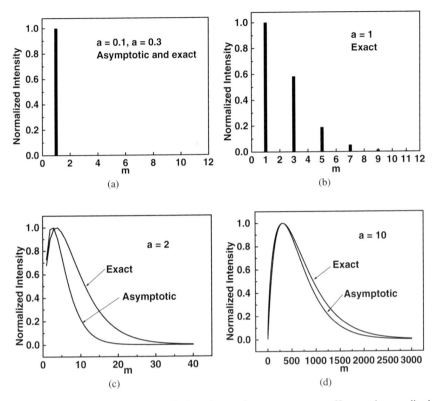

FIG. 10. Normalized spectral distribution of s_m, at frequency $\omega = m\omega_1$. Here, s_m is normalized with respect to the maximum value s_M, occurring at $m = M$. (Reproduced from [33].)

The $J_\nu(x)$ above are Bessel functions of the first kind of order ν. They also appear in the quantity "$[JJ]$" or "$F_m(K)$" in the FEL/synchrotron literature, where K is the undulator/wiggler parameter. Setting $K = a_0$, one finds $s_m = \pi^2(1 + K^2/2)^2 F_m(K)$. It is easy to show that $s_m = s_{-m}$ for all odd integers m. Note that the relative spectral shape of s_m depends only on a_0, and is independent of the electron beam energy. The discrete spectrum for small a_0 approaches a continuum for $a_0^2 \gg 1$.

The maximum value of s_m, occurring at $m = M$ with a value s_M, is shown in Fig. 10. Note that the frequency component $\omega = M\omega_1$ contains the highest backscattered power. In terms of the laser frequency, the frequency component $\omega = N\omega_0$ would contain the highest backscatter power, where $N = M\omega_1/\omega_0$. The total backscatter power, P_T (in ergs/s), per unit solid angle in the $\mathbf{n} = -\mathbf{z}$ direction is then given by $P_T = \Sigma p_m$ where the sum is taken over all odd values of m.

$$P_T \approx \frac{\gamma_0^6 (1 - \beta_{z0})^6}{(\lambda/1\ \mu m)^2} \times \begin{cases} 13.7 a_0^2, & a_0 < 0.3, \\ \\ 11.1/a_0, & a_0 \gg 1. \end{cases} \tag{120}$$

To summarize, due to the rapid acceleration of electrons in the direction of the light wave and decrease of the oscillation frequency in the deeply relativistic regime, the photon energy from relativistic nonlinear scattering scales only linearly with laser field strength, $\sim a_0$. However, the relativistic motion of the electron results in a reduction in the angle of the scattered light, such that the harmonics are generated in a low-divergence-angle forward-propagating beam. There is also a relativistic Doppler shift ($\sim \gamma^2$). For instance, this mechanism will be used in high-energy physics experiments to cleanly make constituent particles in the gamma-gamma ($\gamma\gamma$) collider, in which gamma rays with energy 200 GeV will be generated by Compton scattering 1 eV photons from 250-GeV energy conventionally accelerated electron beams ($\gamma = 5 \times 10^{11}$).

Using a much smaller, laser-based accelerator (discussed in Section 6.2), a 1 eV photon can be upshifted by this Doppler shift to an energy of 50 keV, corresponding to subatomic spatial resolution and of interest in medical diagnostics, by an electron beam with only $\gamma = 200$ (100 MeV). In this case, the maximum efficiency is obtained for laser fields a_0 of order unity. When compared with conventional light sources based on cm-wavelength magnetostatic wigglers, the electromagnetic wigglers of such an all-optical-laser-based EUV sources have ten-thousand times shorter wavelength (micron-scale). Thus, the total length of the wiggler region is correspondingly smaller (only mm in length). Another consequence of this is that the frequency upshift required to reach a given output wavelength is also ten-thousand times smaller. Also, given that the required electron energy scales as the square root of the upshift, the required electron energy can be one-hundred-times lower (10–100 MeV). It follows from this, and the fact that the field gradients of these accelerators can be ten-thousand-times higher (1 GeV/cm) than convention RF-based accelerators, that the size of the accelerating region can in principle be a million times smaller (only mm in length) [8,55,61]. Besides its small size, this EUV source can produce femtosecond duration pulses and be synchronized with a relatively low jitter with another femtosecond light pulse having a different wavelength (by virtue of the possibility of deriving the two pulses from the same laser pulse); this is advantageous for the study of ultrafast pump-and-probe photo-initiated processes. The exceptionally low transverse emittance of laser-accelerated electron beams may even make it possible to generate coherent XUV radiation by means of the self-amplified spontaneous emission (SASE) free-electron lasing (FEL) mechanism.

5. Collective Plasma Response

The derivation given in this section follows that of [19]. The above equations apply to electron motion in vacuum. In the case of electrons in plasmas, the collective plasma electron response needs to be considered. A theory for the 1D nonlinear interaction of intense laser fields with electrons has been developed [19].

The variables are again normalized such that:

$$\mathbf{a_0} = \frac{e\mathbf{A}}{mc^2}, \qquad \phi = \frac{e\Phi}{mc^2},$$

$$\mathbf{u} = \frac{\mathbf{p}}{mc}, \qquad \beta = \frac{\mathbf{v}}{c}, \quad \mathbf{u} = \gamma\beta,$$

$$\gamma = \left(1 + u^2\right)^{1/2} = \left(1 - \beta^2\right)^{-1/2},$$

where $\mathbf{A}$ is the vector potential.

The relativistic force equation (Lorentz equation) is given by

$$\frac{d}{dt}\mathbf{u} = \nabla\phi + \frac{\partial}{\partial t}\mathbf{a_0} - \bar{\beta} \times (\nabla \times \mathbf{a_0}).$$

The energy equation ($\mathbf{u}\cdot$ force equation) is given by

$$\frac{d}{dt}\gamma = \beta \cdot \left(\frac{\partial}{\partial z}\phi + \frac{\partial}{\partial t}\mathbf{a_0}\right).$$

The transverse force equation (1D) is

$$\frac{d}{dt}\mathbf{u}_\perp = \left(\frac{\partial}{\partial t} + \beta_z\frac{\partial}{\partial z}\right)\mathbf{a}_\perp,$$

where $\mathbf{a}_\perp$ represents the perpendicular components of the normalized vector potential. Since $\mathbf{a} = \mathbf{a}(z, t)$

$$\frac{d}{dt}\mathbf{a}_\perp = \left(\frac{\partial}{\partial t} + \beta_z\frac{\partial}{\partial z}\right)\mathbf{a}_\perp,$$

hence,

$$\frac{d}{dt}(\mathbf{u}_\perp - \mathbf{a}_\perp) = 0.$$

This is just conservation of transverse canonical momentum (cf. Section 2.1). Assuming $\mathbf{u}_\perp = 0$ prior to the laser interaction gives

$$\mathbf{u}_\perp = \mathbf{a}_\perp.$$

The electron's response to the normalized scaler and vector potentials of the form $\phi = \phi(z - ct)$ and $\mathbf{a_0} = \mathbf{a_0}(z - ct)$, which are a function of only $\xi = z - ct$, is completely described by the following constants of the motion:

$$\beta_\perp - \frac{a_\perp}{\gamma} = 0, \qquad \gamma(1 - \beta_z) - \phi = 1, \qquad n(1 - \beta_z) = n_0,$$

where $\gamma = (1 - \beta^2)^{-1/2}$ is the relativistic factor and n is the electron density. This allows the various electron quantities to be specified solely in terms of the potentials as

$$\beta_\perp = \frac{2(1 + \phi)a_\perp}{(1 + a_\perp^2) + (1 + \phi)^2}, \qquad \beta_z = \frac{(1 + a_\perp^2) - (1 + \phi)^2}{(1 + a_\perp^2) + (1 + \phi)^2},$$

$$\gamma = \frac{(1 + a_\perp^2) + (1 + \phi)^2}{2(1 + \phi)}, \qquad \frac{n}{n_0} = \frac{(1 + a_\perp^2) + (1 + \phi)^2}{2(1 + \phi)^2}.$$

In the single particle limit, $\phi \equiv 0$, the standard result is retrieved. For a long pulse interacting with a dense plasma, $\tau_L \gg \omega_p^{-1}$ (i.e., neglecting wakefield effects), where ω_p is the plasma frequency, it can be shown that $(1 + \phi) \simeq (1 + a_0^2)^{1/2}$ for circular polarization and $(1 + \phi) \simeq (1 + a_0^2/2)^{1/2}$ for linear polarization.

6. Propagation

To understand the propagation of high-intensity light in plasma, we need to understand how the dielectric properties of a plasma medium are affected by the relativistic electron mass change (see, e.g., Esarey in [20]). A wave equation is found by taking the curl of Faraday's law,

$$\nabla \times \mathbf{E} = -\frac{1}{c} \frac{\partial \mathbf{B}}{\partial t},$$

and using Ampere's law,

$$\nabla \times \mathbf{B} = \frac{4\pi \mathbf{J}}{c} + \frac{1}{c} \frac{\partial \mathbf{E}}{\partial t},$$

to obtain

$$\nabla \times \nabla \times \mathbf{E} = -\nabla^2 \mathbf{E} + \nabla(\nabla \cdot \mathbf{E}) = \frac{1}{c^2}\left[4\pi \frac{\partial \mathbf{J}}{\partial t} - \frac{\partial^2 \mathbf{E}}{\partial t^2}\right], \tag{121}$$

where $\mathbf{J} = \sum n_j q_j \mathbf{v}_j$ and the summation over the index j is done over all charged species of electrons and ions. Using Coulomb's law

$$\nabla \cdot \mathbf{E} = 4\pi \rho, \tag{122}$$

where $\rho = \sum n_j q_j$, and assuming a uniform plasma, the second term on the left vanishes. Substituting Eq. (20) for the velocity, we find

$$\left(-\frac{\partial^2}{\partial t^2} + c^2 \nabla^2 - \omega_p^2\right)\mathbf{E} = 0, \tag{123}$$

where

$$\omega_p = \left(\frac{4\pi n_j q_j^2}{m_j}\right)^{1/2}$$

is the plasma frequency. Assuming again plane waves and Fourier analyzing, yields the well-known dispersion relation for electromagnetic waves in plasma,

$$\omega^2 = \omega_p^2 + c^2 k^2. \tag{124}$$

The index of refraction can be written fully relativistically as

$$\eta^2 = \frac{c^2 k^2}{\omega^2} = \frac{c^2}{v_\phi^2} = 1 - \frac{\omega_p^2}{\omega^2}, \tag{125}$$

where v_ϕ is the phase velocity of the light wave. Assuming infinitely massive ions, the plasma frequency can be written as

$$\omega_p = \frac{\omega_{p0}}{\gamma^{1/2}} = \left(\frac{4\pi n_e e^2}{\gamma m_0}\right)^{1/2}, \tag{126}$$

where ω_{p0} is the plasma frequency in a quiescent plasma, e is the electron charge, m_0 is the electron rest mass, and n_e is the plasma electron density. A change in mass changes the plasma frequency, which in turn modifies the index of refraction and the velocity of the light wave. For an underdense plasma, $\omega_p^2 \ll \omega_0^2$, the plasma refractive index can be Taylor expanded, and written as

$$\eta = \sqrt{1 - \frac{\omega_p^2}{\omega_0^2}} = 1 - \frac{\omega_p^2}{2\omega_0^2} \simeq 1 - \frac{n_e}{2n_c \gamma}, \tag{127}$$

where n_c is the critical electron density and is equal to $\omega_0^2 m_e / 4\pi e^2$. When $a \leq 1$, the plasma refractive index can be further simplified by expanding about a_0, which results in

$$\eta = 1 - \left(1 - \frac{\langle a_0^2 \rangle}{2}\right)\frac{n_e}{2n_c} = n_1 + n_2 I, \tag{128}$$

where $n_1 = 1 - n_e/2n_c$ and $n_2 = (8.5 \times 10^{-10}\lambda\ [\mu m])^2 n_e/8n_c$. The light's phase velocity then depends on the laser intensity. This can be seen clearly if we expand

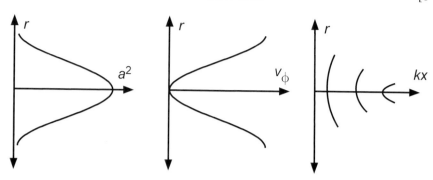

FIG. 11. Relativistic self-focusing occurs when an on-axis peak in light intensity (left) produces an on-axis dip in the phase velocity (middle), which acts like a positive lens to cause the phase fronts to curve inward (right). Reproduced from [72].

the phase velocity for small field strength ($a_0 \leq 1$),

$$v_\phi = \frac{c}{\eta} \sim c\left[1 + \frac{\omega_{p0}^2}{2\omega^2}\left(1 - \frac{\langle a_0^2 \rangle}{2}\right)\right], \tag{129}$$

where $\langle a_0^2 \rangle$ denotes the time averaged value of the normalized vector potential. An on-axis minimum of the phase velocity (i.e., $v_\phi(r) > v_\phi(0)$) can be created by a laser beam with an intensity profile peaked on axis, such as with a Gaussian beam, causing the wavefronts to curve inward and the laser beam to focus, as shown in Fig. 11. When this focusing effect just balances the defocusing caused by diffraction, the laser pulse can propagate over a longer distance than it could in vacuum, while maintaining a small cross section. This mechanism is referred to as relativistic self-guiding.

6.1. RELATIVISTIC SELF-FOCUSING

Following the derivation from [46], the threshold for relativistic self-focusing can be obtained using the fully relativistic formalism with the vector potential, as in our previous discussion of electron motion in vacuum. In this case, we can rewrite Eq. (123) as

$$\nabla^2 \mathbf{a_0} - \frac{1}{c^2}\frac{\partial^2 \mathbf{a_0}}{\partial t^2} = \frac{\omega_{p0}^2}{c^2}\frac{n_e \mathbf{a_0}}{n_0 \gamma}, \tag{130}$$

where $n_0 = n_i$ is the uniform background density. For a Gaussian beam focused in vacuum, where

$$\mathbf{a_0}(\mathbf{r}, t) = \text{Re}\left\{\mathbf{a_0}(\mathbf{r}, t) \exp\left[i(\mathbf{k} \cdot \mathbf{r} - \omega t)\right]\right\}, \tag{131}$$

and the amplitude $a_0(\mathbf{r}, t)$ varies much less in r and t than does the phase,

$$\left|\frac{\partial a_0}{\partial t}\right| \ll |\omega a_0|, \qquad \left|\frac{\partial a_0}{\partial z}\right| \ll |k a_0|, \tag{132}$$

then

$$a_0(r, z) = \frac{e^{-r^2/(r_0^2(1+z^2/L_r^2))}}{\sqrt{1 + z^2/L_R^2}} \exp\left\{-i \arctan\left(\frac{z}{L_R}\right) + i\left(\frac{r}{r_0}\right)^2 \frac{z/L_R}{1 + z^2/L_R^2}\right\} \tag{133}$$

satisfies the envelope equation (assuming second derivatives are small compared to first derivatives), where L_R is the Rayleigh range and r_0 is the focal radius. In plasma, Eq. (130) can then be written in this envelope (paraxial) approximation as

$$\left[\nabla_\perp^2 + 2ik \frac{\partial}{\partial z}\right] a_0 = -\frac{\omega_p^2}{c^2} \frac{|a_0|^2}{4} a_0. \tag{134}$$

The power at which the first and last terms of Eq. (134) are in balance is

$$P = \frac{\pi R^2 I_0}{2}. \tag{135}$$

An equation for the beam radius R can thus be found,

$$\frac{d^2 R(z)}{dz^2} = \frac{4}{k^2 r_0^3}\left[1 - \frac{1}{32} \frac{\omega_p^2}{c^2} |a_0|^2 r_0^2\right]. \tag{136}$$

Self-focusing occurs when the two terms in the brackets in Eq. (136) are equal, which gives for the critical power

$$P_{\text{crit}} \simeq 17.4 \cdot \left(\frac{\omega}{\omega_p}\right)^2 \text{GW}. \tag{137}$$

This corresponds to 1 TW for 1-μm light focused into a gas with electron density of $10^{19}/\text{cm}^3$. Note that this is a power threshold, not an intensity threshold, because the tighter the focusing, the greater the diffraction (the first term in Eq. (134)). Numerous recent experiments have confirmed this focusing mechanism when $P > P_{\text{crit}}$.

Any spatial variation of the laser intensity will act to push an electron to regions of lower intensity through the so-called ponderomotive force,

$$\mathbf{F} = \nabla \gamma = \nabla \sqrt{1 + a_0^2} = (2\gamma)^{-1} \nabla a_0^2, \tag{138}$$

$$\mathbf{F}_{\text{pond}} = -\frac{m_0 c^2}{8\pi} \frac{\nabla a_0^2}{\sqrt{1 + a_0^2/2}}\left[\left(\frac{2}{a_0^2} + 1\right) E(2\pi, \kappa) - \frac{2}{a_0^2} F(2\pi, \kappa)\right], \tag{139}$$

where

$$\kappa = \frac{a_0^2/2}{1 + a_0^2/2},$$

and F and E are the elliptic integrals of the first and second kind, respectively. In the low intensity limit, i.e., when $a_0^2 \ll 1$,

$$\mathbf{F}_{\text{pond}} = -\frac{m_0 c^2}{4} \nabla a_0^2. \tag{140}$$

That is, the laser ponderomotive force is roughly proportional to the gradient of laser intensity.

A Gaussian-shaped laser's ponderomotive force will tend to expel electrons radially from the region of the axis, so called "electron cavitation." If the ponderomotive force is high enough for long enough, the charge displacement due to expelled electrons (in either the lateral or longitudinal direction) will eventually cause the ions to move as well through the Coulomb electrostatic force, forming a density channel. Because $n_e(0) < n_e(r)$, and thus $v_\phi(0) < v_\phi(r)$, this enhances the previously-discussed relativistic self-guiding and can itself guide a laser pulse.

For plasmas created by photo-ionization of a gas by a Gaussian laser pulse, the density will be higher on the axis than off the axis. If we instead expand the phase velocity in terms of changes in density,

$$\frac{v_\phi}{c} \sim 1 + \frac{\omega_{p0}^2}{\omega^2} \frac{\Delta n_e}{n_e}. \tag{141}$$

The phase velocity will thus be higher on axis, which will tend to defocus the light and increase the self-guiding threshold. In order to avoid this, gases with low atomic number and thus fewer available electrons, such as H_2, are commonly used as targets.

6.2. Raman Scattering, Plasma Wave Excitation and Electron Acceleration

The local phase velocity, described in Eqs. (129) and (141), can also vary longitudinally if the intensity and/or electron density does. Local variation in the index of refraction can "accelerate" photons, i.e., shift their frequency, resulting in photon bunching, which in turn bunches the electron density through the ponderomotive force (F), and so on (e.g., see Mori in [47]). When the laser pulse duration is longer than an electron plasma period, $\tau \gg \tau_p = 2\pi/\omega_p$, this photon and electron bunching grows exponentially, leading to the stimulated Raman scattering instability. Energy and momentum must be conserved when the electromagnetic wave (ω_0, $\mathbf{k}_0$) decays into a plasma wave (ω_p, $\mathbf{k}_p$) and another light wave ($\omega_0 - \omega_p$, $\mathbf{k}_0 - \mathbf{k}_p$).

From an equivalent viewpoint, the process begins with a small density perturbation, Δn_e, which, when coupled with the quiver motion, Eq. (20), drives a current $J = \Delta n_e e v_e$. This current then becomes the source term for the wave equation (Eq. (121)), driving the scattered light wave. The ponderomotive force due to the beating of the incident and scattered light wave enhances the density perturbation, enhancing the plasma wave and the process begins anew. In three dimensions, a plasma wave can be driven when transverse self-focusing and stimulated Raman scattering occur together, a process called the self-modulated wakefield instability.

Two conditions must be satisfied for self-modulation to occur in the plasma. First, the laser pulse must be long compared to the plasma wave, $L \gg \lambda_p$. This allows the Raman instability time to grow, and it allows for feedback from the plasma to the laser pulse to occur. Second, the laser must be intense enough for relativistic self-focusing to occur, $P > P_c$, so that the laser can be locally modified by the plasma. Under these conditions, the laser can form a large plasma wave useful for accelerating electrons.

As the long laser pulse enters the plasma, it will begin to drive a small plasma wave due to either forward Raman scattering or the laser wakefield effect from the front of the laser pulse. This small plasma wave will have regions of higher and lower density with both longitudinal and radial dependence. That is, the plasma wave will be three-dimensional in nature with a modulation along the propagation direction of the laser and a decay in the radial direction to the ambient density (see Fig. 12). The importance of this lies with how it affects the index of refraction in the plasma. In the regions of the plasma wave where the plasma density is lower, the radial change in the index of refraction is negative, $\partial n(r)/\partial r < 0$. This means that this part of the plasma acts like a positive lens and focuses the laser. Whereas regions of the plasma wave where the density is higher, $\partial n(r)/\partial r > 0$, the opposite occurs and the laser defocuses. This has the effect of breaking up the laser pulse into a series of shorter pulses of length $\lambda_p/2$ which will be separated by the plasma period. The instability occurs because of how the plasma responds to this. Where the laser is more tightly focused, the ponderomotive force will be greater and will tend to expel more electrons. This decreases the density in these regions even further, resulting in more focusing of the laser. This feedback rapidly grows, hence the instability.

To study the Raman instability, it is convenient to use Eqs. (8) and (9) with the Coulomb gauge,

$$\nabla \cdot \mathbf{A} = 0. \tag{142}$$

The coupled equations of $\mathbf{A}$ and Φ can be written as

$$\nabla^2 \mathbf{A} - \frac{1}{c^2}\frac{\partial^2 \mathbf{A}}{\partial t^2} = -\frac{4\pi}{c}\mathbf{J}_t, \tag{143}$$

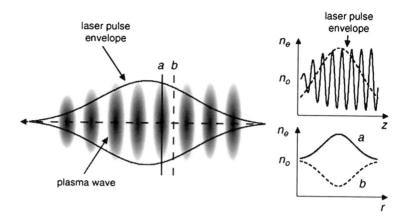

FIG. 12. The plasma wave generated by a SMLWFA is three-dimensional in nature. Note that the darker regions correspond to areas of higher plasma density. The graphs to the right represent lineouts of the plasma density longitudinally and radially at the indicated points. Reproduced from [75].

$$\nabla^2 \Phi = -4\pi\rho, \tag{144}$$

where $\mathbf{J}_t$ is the transverse component of the current density and is equal to

$$\mathbf{J}_t = \frac{1}{4\pi}\nabla \times \nabla \times \int \frac{\mathbf{J}}{|\mathbf{x}-\mathbf{x}'|}\, d^3\mathbf{x}'. \tag{145}$$

By assuming the electron density modulation is parallel to the laser propagation direction ($\mathbf{A} \cdot \nabla n_e = 0$) and the motion of the electron is nonrelativistic, we find

$$\mathbf{J}_t = n_e e \mathbf{v}_t, \tag{146}$$

where $\mathbf{v}_t$ is the quiver velocity of the electrons on the laser field and is equal to

$$\mathbf{v}_t = \frac{e\mathbf{A}}{m_e c}. \tag{147}$$

Equation (147) and (146) can be substituted into Eq. (145) to yield,

$$\left(\frac{\partial^2}{\partial t^2} - c^2\nabla^2\right)\mathbf{A} = -\frac{4\pi e^2 n_e}{m_e}\mathbf{A}. \tag{148}$$

To analyze the instability, we use a perturbation method and examine the scattering of a large amplitude light wave ($\mathbf{A}_L$) by a small amplitude density fluctuation ($\tilde{n}_e$) to see how the plasma wave grows,

$$\mathbf{A} = \mathbf{A}_L + \delta\mathbf{A}, \tag{149}$$

$$\mathbf{n}_e = n_0 + \delta n. \tag{150}$$

Substituting these equations into Eq. (148) and keeping the first order term, we then obtain

$$\left(\frac{\partial^2}{\partial t^2} - c^2\nabla^2 + \omega_p^2\right)\delta\mathbf{A} = -\frac{4\pi e^2 n_e}{m_e}\tilde{n}_e\mathbf{A}_L. \tag{151}$$

The dependence of the density fluctuation associated with the electron plasma wave can be derived from the fluid equations. Assuming ions are an immobile background, the continuity equation and force equation can be expressed as,

$$\frac{\partial n_e}{\partial t} + \nabla \cdot (n_e\mathbf{v}_e) = 0, \tag{152}$$

$$\frac{\partial \mathbf{v}_e}{\partial t} + \mathbf{v}_e \cdot \nabla\mathbf{v}_e = -\frac{e}{m}\left(\mathbf{E} + \frac{\mathbf{v}_e \times \mathbf{B}}{c} - \frac{\nabla p_e}{n_e m}\right). \tag{153}$$

Separating the velocity of the electrons into longitudinal and transverse components

$$\mathbf{v}_e = \mathbf{v}_L + \frac{e\mathbf{A}}{mc}, \tag{154}$$

substituting into the force equation, and using a vector identity, one finds

$$\frac{\partial \mathbf{v}_L}{\partial t} = \frac{e}{m}\nabla\phi - \frac{1}{2}\nabla\left(\mathbf{v}_L + \frac{e\mathbf{A}}{m_e c}\right) - \frac{\nabla p_e}{n_e m_e}. \tag{155}$$

Now we use the adiabatic equation of state (p_e/n_e^3 = constant) and linearize Eqs. (152) and (155). In particular, we use the following perturbation expansion parameters,

$$\mathbf{v}_L = \delta\mathbf{v}, \tag{156}$$

$$n_e = n_0 + \delta n, \tag{157}$$

$$\mathbf{A} = \mathbf{A}_L + \delta\mathbf{A}, \tag{158}$$

$$\phi = \delta\phi. \tag{159}$$

Keeping only the first order term gives

$$\frac{\partial \delta n}{\partial t} + n_0\nabla \cdot \delta\mathbf{v} = 0, \tag{160}$$

and

$$\frac{\partial \delta\mathbf{v}}{\partial t} = \frac{e}{m}\nabla\delta\phi - \frac{e^2}{m_e^2 c^2}\nabla(\mathbf{A}_L \cdot \delta\mathbf{A}) - \frac{3v_{th}^2}{n_0}\nabla\delta n, \tag{161}$$

where v_{th} is the thermal velocity. Taking a time derivative of Eq. (160), and then a divergence of Eq. (161), and using Poisson's equation

$$\nabla^2\tilde{\phi} = 4\pi e\tilde{n}_e, \tag{162}$$

a general equation describing the evolution of the plasma wave due to the laser pulse can be found:

$$\left(\frac{\partial^2}{\partial t^2} + \omega_p^2 - 3v_{th}^2 \nabla^2\right)\delta n = \frac{n_0 e^2}{m_e^2} c^2 \nabla^2 (\mathbf{A_L} \cdot \delta \mathbf{A}). \tag{163}$$

Equations (151) and (163) fully describe the feedback between the scattered light and electron density perturbation. For further analysis of the Raman instability, we need to derive the dispersion relation. Taking $\mathbf{A_L} = \mathbf{A}_0 \cos(\mathbf{k}_0 \cdot \mathbf{x} - \omega_0 t)$, and using Fourier analysis, we can recast Eqs. (151) and (163) as:

$$\left(\omega^2 - k^2 c^2 - \omega_p^2\right)\delta \mathbf{A}(\mathbf{k}, \omega)$$
$$= \frac{4\pi e^2}{2m_e} \mathbf{A}_0 \left[\delta n(\mathbf{k} - \mathbf{k}_0, \omega - \omega_0) + \delta n(\mathbf{k} + \mathbf{k}_0, \omega + \omega_0)\right] \tag{164}$$

and

$$\left(\omega^2 - \omega_{BG}^2\right)\delta(\mathbf{k}, \omega)$$
$$= \frac{k^2 e^2 n_0}{2m_e^2 c^2} \mathbf{A}_0 \cdot \left[\delta \mathbf{A}(\mathbf{k} - \mathbf{k}_0, \omega - \omega_0) + \delta \mathbf{A}(\mathbf{k} + \mathbf{k}_0, \omega + \omega_0)\right], \tag{165}$$

where $\omega_{BG} = (\omega_p + 3k_p^2 v_e^2)^{1/2}$ is the Bohm–Gross frequency. Taking $\omega \approx \omega_p$ and neglecting the term $\tilde{n}_e(k - 2k_0, \omega - 2\omega_0)$ and $\tilde{n}_e(k + 2k_0, \omega + 2\omega_0)$ as off resonant, we obtain the dispersion relationship:

$$\omega^2 - \omega_{BG}^2 = \frac{\omega_p^2 k^2 v_{os}^2}{4}\left[\frac{1}{D(\omega - \omega_0, \mathbf{k} - \mathbf{k_0})} + \frac{1}{D(\omega + \omega_0, \mathbf{k} + \mathbf{k_0})}\right], \tag{166}$$

where

$$D(\omega, \mathbf{k}) = \omega^2 - k^2 c^2 - \omega_p^2 \tag{167}$$

and v_{os} is the oscillatory velocity of the electron in the laser field.

The growth rate of the instability can be found from Eq. (166) by looking at how the noise will grow. The phase-matching condition will define which waves are resonant. For back or sidescatter, we can neglect the upshifted light wave as nonresonant, giving

$$\left(\omega^2 - \omega_{BG}^2\right)\left[(\omega - \omega_0)^2 - (\mathbf{k} - \mathbf{k}_0)^2 c^2 - \omega_{BG}^2\right] = \frac{\omega_p^2 k^2 v_{os}^2}{4}. \tag{168}$$

Taking $\omega = \omega_{BG} + \delta\omega$, where $\delta\omega \ll \omega_{BG}$ is the perturbation, the maximum growth occurs when

$$(\omega_{BG} - \omega_0)^2 - (\mathbf{k} - \mathbf{k}_0)^2 c^2 - \omega_p^2 = 0. \tag{169}$$

Letting $\delta\omega = i\gamma$, we find the growth rate of the noise is

$$\gamma_{bs} = \frac{kv_{os}}{4}\left[\frac{\omega_p^2}{\omega_{BG}(\omega_0 - \omega_{BG})}\right]. \tag{170}$$

The wavenumber k is determined by Eq. (169). For backscattering,

$$k = k_0 + \frac{\omega_0}{c}\left(1 - \frac{2\omega_p}{\omega_0}\right)^{1/2}. \tag{171}$$

The wave number starts from $k = 2k_0$ to $k = k_0$ depending on the plasma density. For forward scattering at very low intensity, $k \ll \omega_0/c$, the phase-matching condition is

$$D(\omega \pm \omega_0, \mathbf{k} \pm \mathbf{k_0}) \simeq 2(\omega_{BG} \pm \omega_0)\delta\omega, \tag{172}$$

where we have chosen $k = \omega_{BG}/c$ and let $\omega = \omega_{BG} + \delta\omega$, and $\delta\omega = i\gamma \ll \omega_{BG}$. Substituting into Eq. (166), we can find the maximum growth rate:

$$\gamma_{fs} \simeq \frac{\omega_p^2}{2\sqrt{2}\omega_0}\frac{v_{os}}{c}. \tag{173}$$

In the limit that $v_{os}/c \ll 1$, $v_{os}/c \simeq a_0$ and Eq. (173) can be written as

$$\gamma_{fs} \simeq \frac{\omega_p^2}{2\sqrt{2}\omega_0}a_0. \tag{174}$$

The bandwidth of the growth frequency is half of the growth rate, which is

$$\Delta\omega = 2\gamma \simeq \frac{\omega_p^2}{\sqrt{2}\omega_0}\frac{v_{os}}{c}. \tag{175}$$

Another interesting case is near-forward scattering where the anti-Stokes wave becomes nonresonant as the scattering angle increases [44]. This occurs when the diffracted frequency shift becomes comparable to the standard growth rate

$$\gamma_{fs} \approx \frac{c^2k_\perp^2}{\omega_0}, \tag{176}$$

where $k_\perp$ is the perpendicular component of the plasma wave. This occurs for small-angles (of order $1°$) relative to the laser field direction as can be seen from

$$\frac{k_\perp}{k_0} \approx \frac{\omega_p}{\omega_0}\sqrt{\frac{a_0}{2\sqrt{2}}}. \tag{177}$$

The growth rate in the near-forward case will be

$$\gamma_{nfs} = \frac{kv_{os}}{4}\sqrt{\omega_0\omega_p}. \tag{178}$$

This growth rate is comparable with back scattering growth rate and is much larger than the forward growth rate. Antenson and Mora [2] and Decker et al. [15] calculated the Raman forward scattering growth rate for various situations.

In the derivations above, the relativistic effect has been ignored. The growth rates with the relativistic effect have been derived by Mori et al. [48] and Decker et al. [15]. They found that the nonrelativistic growth rates can be transformed by recasting the plasma frequency and the normalized vector potential in terms of the relativistic factor γ,

$$\omega_p \to \frac{\omega_p}{\sqrt{\gamma}} \tag{179}$$

and

$$a_0 \to \frac{a_0}{\gamma}. \tag{180}$$

Substituting these equations into Eq. (174) yields:

$$\gamma_{fs} = \frac{\omega_p^2}{\sqrt{8}\omega_0} \frac{a_0}{\gamma^2}. \tag{181}$$

Another effect that needs to be taken into account for high intensity laser plasma interactions is the strongly coupled effect for Raman backscattering [12, 22,70]. Notice that we assume $\delta\omega = i\gamma \ll \omega_{BG}$ to derive the growth rate. To be self-consistent, the growth rate is valid when

$$\frac{v_{os}}{c} < 4\sqrt{\frac{\omega_p}{\omega_0}}. \tag{182}$$

This condition will be invalid for a relativistic laser intensity. When

$$\frac{v_{os}}{c} \gg 4\sqrt{\frac{\omega_p}{\omega_0}}, \tag{183}$$

we say the Raman instability is in the strongly coupled regime, where $\omega \gg \omega_{BG}$. The new solution for Eq. (168) is:

$$\omega^3 = -\frac{\omega_p^2 k^2 v_0^2}{8\omega_0}, \tag{184}$$

and

$$\omega = \frac{1 - i\sqrt{3}}{4}\left(\frac{\omega_p^2 k^2 v_0^2}{\omega_0}\right)^{1/3}. \tag{185}$$

The instability bandwidth for the strongly coupled regime is

$$\Delta\omega = \sqrt{3}\left(\frac{\omega_p^2 k^2 v_0^2}{8\omega_0}\right)^{1/3} \tag{186}$$

and this can be substantially larger than ω_p.

The phase velocity of the plasma wave in the case of forward scattering is equal to the group velocity of the beat wave, which for low-density plasma is close to the speed of light, as can be seen from the relation

$$v_\phi = \frac{\omega_p}{k_p} = \frac{\Delta\omega}{\Delta k} = v_g = c\eta \sim c,$$

where Eq. (125) and $\omega_p^2 \ll \omega^2$ were used to show that η is close to unity. Such relativistic plasma waves can also be driven by short pulses ($\tau \sim \tau_p$). In this case the process is referred to as laser-wakefield generation, referring to the analogy with the wake driven by the bow of a boat moving through water, but the mechanism is similar (except it has the advantage that the plasma wave is driven linearly instead of as an instability).

In either case, the resulting electrostatic plasma wave can continuously accelerate relativistic electrons with enormous acceleration gradients. The gradient can be estimated from Eq. (122) and the fact that because

$$\nabla \cdot \mathbf{E} \sim \frac{E\omega_p}{c} \propto E\sqrt{n_e},$$

then

$$E \sim \sqrt{n_e} \text{ eV/cm},$$

corresponding to 1 GeV/cm for $n_e = 10^{18}$ cm^3. Because this gradient is four orders-of-magnitude greater than achieved by conventional accelerators (based on fields driven by radio-frequency waves pumped into metal cavities), laser-driven plasma accelerators have received considerable recent attention.

In the self-modulated regime, they have been shown to accelerate electron charge (100 nC) comparable to that from conventional accelerators and to have superior transverse geometrical emittance (product of divergence angle and spotsize, similar to the $f/\#$ in light optics). However, their longitudinal emittance is currently much inferior, energy spreads of 100%. They have been shown to be useful for much of the same applications: radio-isotope production, radiation chemistry, as well as X-ray, proton and neutron generation. Once the longitudinal emittance can be reduced, they may be advantageous for, among other applications, injectors (especially of short-lived unstable particles) into larger conventional accelerators for high-energy physics research and light sources and, as discussed in Section 4, as stand-alone all-optically driven ultrashort-pulse duration X-ray sources. Dramatic reduction of the angular divergence of a laser accelerated electron beam was observed with increasing laser power above the relativistic self-focusing threshold [75], as shown in Fig. 13.

In the resonant regime, there have been some theoretical analysis and simulations done in the 80s and 90s [32,19]. In recent years, the Ti-sapphire laser using CPA technology can produce up to 50 TW of power with a pulse duration of

0.6 TW   1.1 TW   2.0 TW   2.9 TW

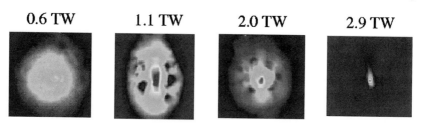

FIG. 13. Images of the spatial profiles of the electron beam measured by a ccd camera imaging a LANEX screen at a distance of 15 cm from the gas jet for various laser powers. The divergence angle of the beam decreases to a value of $\Delta\theta = 1°$ at a power of 2.9 TW, corresponding to a transverse geometrical emittance of just $\epsilon_\perp \lesssim 0.06\pi$ mm-mrad. Reproduced from [71].

30 fs. With the availability of high power and short duration laser pulses, there have been many breakthroughs in experiments in the resonant regime [77,41,43, 25,21]. Wang et al. first reported the experimental observation of an electron beam in this regime using an f/#3 parabola [77]. Their electron energy was continuous due to the filamentation induced by the hot spot in the laser beam. Malka et al. showed that the electron energy from LWFA can reach to 200 MeV with a long focal length parabolic mirror, and the energy spread was 100% [41]. In their experiment, the laser pulse duration was about 4 times longer than the plasma period, but, due to the group velocity difference in the front and back edge of the pulse (the front edge moves more slowly since more electrons were accumulated at the front), the laser pulse was self-compressed in the plasma and formed an optical shock. This resulted in an extremely sharp leading edge that was able to drive the relativistic plasma wave beyond its wave-breaking limit. In this case, there was no Raman scattering, but the laser spectrum became blue shifted and broadened. More recently, different research groups reported quasi-monoenergetic electron beams in this regime with 20% energy spread. The peak electron energy varies from 80 MeV to 120 MeV. Mangles et al. from Rutherford and Geddes et al. from Berkeley National Lab explained this as the match of the interaction length with the dephasing length. In this condition, the electrons did not experience the deceleration so the energy did not spread [43,25]. Faure et al. from LOA argued that the monoenergetic electrons accelerated by the electric field formed behind the laser pulse and this theory was supported by a PIC (particle-in-cell) simulation [21].

Most experiments on plasma wakefield acceleration are in the regime of $a_0^2 \sim 1$. The SIMLAC code has been used to study wakefield generation and laser propagation in the limit $a_0^2 \ll 1$ [68]. It draws from nonlinear optics models and treats propagation in the group velocity frame. In this idealized model (which assumes perfect Gaussian beams), the pulse and wake, are maintained over long enough propagation distance to accelerate an electron to GeV energy.

A three-dimensional envelope equation for the laser field was derived that includes nonparaxial effects, wakefields, and relativistic nonlinearities [68].

6.3. RELATIVISTIC PHASE MODULATION

Analogous to self-focusing, which is due to the spatial refractive index modulation, the relativistic phase modulation of a high-intensity laser pulse is due to the refractive index modulation in the time domain. The time dependence of a laser usually can be described as a Gaussian distribution

$$I = I_0 \exp\left(\frac{-T^2}{2T_0^2}\right), \tag{187}$$

where T_0 is the half-width (at $1/e$-intensity point), $T = t - z/v_g$ and v_g is the group velocity of the laser pulse in plasma.

As shown in Eq. (128), the time dependent laser intensity results in a time dependent refractive index. The frequency shift is given by

$$d\omega = -\frac{\omega}{c}\frac{\partial \eta}{\partial t}\,dx = -\frac{\omega}{c}n_2\frac{\partial I(t)}{\partial t}\,dx. \tag{188}$$

This equation shows that the frequency shift depends on the sign of $\partial I(t)/\partial t$. For the rising part of the laser pulse, $\partial I(t)/\partial t > 0$, and this causes the red-shifting of the laser pulse. Conversely, the falling edge of the laser pulse introduces a blue shift. A typical self-phase modulated spectrum is show in Fig. 14.

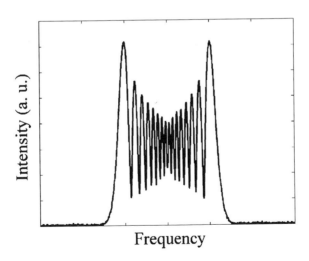

FIG. 14. Self-phase modulated spectrum of a Gaussian laser pulse.

A short pulse also has a finite frequency bandwidth and each component has different group velocity in the plasma, which causes the group velocity dispersion (GVD) effect that becomes severe when the laser pulse reaches tens of fs [14]. GVD effects also interplay with self-phase modulation and cause instabilities [67].

The frequency bandwidth of a transform limited Gaussian pulse can be estimated from Fourier analysis and is given by

$$\tau = \frac{2(\ln 2)\lambda^2}{\pi \, \Delta \lambda c},$$

(189)

where τ is the full width at half maximum (FWHM) of the laser pulse and c is the speed of light in vacuum. τ and T_0 are related by

$$\tau = 2(\ln 2)^{1/2} T_0 \approx 1.665 T_0.$$

(190)

For instance, the bandwidth of a 400 fs laser pulse at 1.053 µm is at least 4.1 nm. Due to the mismatch between the stretcher and compressor, the duration of a laser pulse is usually larger than that of the transform-limited pulse.

In comparison to self-focusing, fewer experimental results have been published on self-phase modulation (SPM) [79,82]. The difficulty of direct observation of SPM is due to the Gaussian spatial distribution of the laser pulse intensity, which gives different phase modulations at different positions of the focal spot. When collected by a lens, the accumulated spectrum is difficult to be explained. The inference of the SPM from the phase-modulation of the harmonics is not accurate since the phase-modulation of the harmonics is due to the relativistic cross-phase modulation (RXPM), which refers phase modulation of light pulse by another copropagating pulse of different wavelength.

To study the propagation of a short laser pulse in a plasma, we need to solve Eq. (123). Since there is no analytical solution for this equation, it needs to be solved numerically [67,18]. However, when the laser pulse is guided by the relativistic channel, the plasma channel can be treated as a fiber, thus the laser field can be described in a form [1]:

$$\widetilde{E}(\mathbf{r}, \omega - \omega_0) = F(x, y)\widetilde{A}(z, \omega - \omega_0) \exp(i\beta_0 z).$$

(191)

Using the Fourier transform, defined by,

$$\widetilde{E}(\mathbf{r}, \omega - \omega_0) = \int_{-\infty}^{\infty} E(\mathbf{r}, t) \exp\big[i(\omega - \omega_0)t\big] \, dt.$$

(192)

Eq. (123) can be expressed as,

$$\nabla^2 \widetilde{E} + \varepsilon(\omega)k_0^2 \widetilde{E} = 0,$$

(193)

where $\varepsilon(\omega) = 1 - \omega_p^2/\omega^2$, and ω_0 is the central frequency of laser pulse. Substituting Eq. (191) into Eq. (193), we can derive two equations for $F(x, y)$ and $A(z, \omega)$,

$$\frac{\partial^2 F}{\partial^2 x} + \frac{\partial^2 F}{\partial^2 y} + \left[\varepsilon(\omega)k_0^2 - \tilde{\beta}^2\right]F = 0 \tag{194}$$

and

$$2i\beta_0\frac{\tilde{A}}{z} + \left(\beta^2 - \beta_0^2\right)\tilde{A} = 0. \tag{195}$$

Equation (194) can be solved by using first-order perturbation theory [49]. The eigenvalue of $\tilde{\beta}$ is given by

$$\tilde{\beta}(\omega) = \beta(\omega) + \Delta\beta, \tag{196}$$

where

$$\Delta\beta = \frac{k_0 \int_{-\infty}^{\infty} \int_{-\infty}^{\infty} \Delta n |F(x, y)|^2 \, dx \, dy}{\int_{-\infty}^{\infty} \int_{-\infty}^{\infty} |F(x, y)|^2 \, dx \, dy} \tag{197}$$

and

$$\Delta n = n_2 |E|^2 + \frac{i\alpha}{2k_0}. \tag{198}$$

Then approximating $\tilde{\beta}^2 - \beta_0^2$ by $2\beta_0(\tilde{\beta} - \beta_0)$, we have

$$\frac{\partial \tilde{A}}{\partial z} = i\left[\beta(\omega) + \Delta\beta - \beta_0\right]\tilde{A}. \tag{199}$$

The inverse Fourier transform of Eq. (199) yields the propagation equation of $A(z, t)$. It is useful to expand $\beta(\omega)$ in a Taylor series about the carrier frequency ω_0,

$$\beta(\omega) = \beta_0 + (\omega - \omega_0)\beta_1 + \frac{1}{2}(\omega - \omega_0)^2\beta_2 + \frac{1}{6}(\omega - \omega_0)^3\beta_3 + \cdots, \tag{200}$$

where

$$\beta_n = \left[\frac{d^n \beta}{d\omega^n}\right]_{\omega=\omega_0}. \tag{201}$$

The cubic and higher-order terms in Eq. (200) can usually be neglected in laser–plasma interaction. Now using the inverse Fourier transform,

$$A(z, t) = \frac{1}{2\pi} \int_{-\infty}^{\infty} \tilde{A}(z, \omega - \omega_0) \exp\left[-i(\omega - \omega_0)t\right] d\omega. \tag{202}$$

In the Fourier transform, $(\omega - \omega_0)$ is replaced by the differential operator $i(\partial/\partial t)$ and the result is

$$\frac{\partial A}{\partial z} = -\beta_1 \frac{\partial A}{\partial t} - \frac{i}{2}\beta_2 \frac{\partial^2 A}{\partial t^2} + i\Delta\beta A. \tag{203}$$

$\Delta\beta$ can be evaluated by Eqs. (198) and (197), which represent the energy loss of EM wave in media and the nonlinear effects. Now Eq. (203) can be written as:

$$\frac{\partial A}{\partial z} + \beta_1 \frac{\partial A}{\partial t} + \frac{i}{2}\beta_2 \frac{\partial^2 A}{\partial t^2} + \frac{\alpha}{2}A = i\delta|A|^2 A, \tag{204}$$

where δ is the nonlinearity coefficient defined by

$$\delta = \frac{n_2 \omega_0}{c A_{\text{eff}}}. \tag{205}$$

In this equation, A is assumed to be normalized and $|A|^2$ corresponds to the laser power. The quantity $\delta|A|^2$ is then measured in units of m^{-1} if n_2 is expressed in units of m^2/W. The laser pulse will move with a group velocity of β_1 while β_2 represents the group velocity dispersion (GVD) and A_{eff} is known as the effective focus area and is given by

$$A_{\text{eff}} = \frac{[\int_{-\infty}^{\infty} \int_{-\infty}^{\infty} |F(x, y)|^2 \, dx \, dy]^2}{\int_{-\infty}^{\infty} \int_{-\infty}^{\infty} |F(x, y)|^4 \, dx \, dy}. \tag{206}$$

For a Gaussian distribution profile,

$$A_{\text{eff}} = \pi w^2, \tag{207}$$

where w can be evaluated from the focal spot of the laser pulse. Using Eq. (204), we can investigate phase-modulation of a guided laser pulse in a plasma.

In many cases of laser plasma interaction, Raman or high order harmonics will be generated and copropagate with the laser pulse. The Raman pulse or the harmonics experience the modulated refractive index induced by the laser pulse and are cross-phase modulated. The propagation of the laser and another pulse can be described by the coupled nonlinear Schrödinger equation as in nonlinear fiber optics [1]:

$$\frac{\partial A_p}{\partial z} + \frac{i}{2}\beta_{2p} \frac{\partial^2 A_p}{\partial T^2} = i\delta_p \left(|A_p|^2\right) A_p, \tag{208}$$

$$\frac{\partial A_s}{\partial z} - d\frac{\partial A_s}{\partial T} + \frac{i}{2}\beta_{2s} \frac{\partial^2 A_s}{\partial T^2} = i\delta_s \left(|A_p|^2\right) A_s + \frac{g_s}{2}|A_p|^2 A_s, \tag{209}$$

where $T = t - z/v_{\text{gp}}, d = (v_{\text{gs}} - v_{\text{gp}})/(v_{\text{gs}} v_{\text{gp}}), \delta = n_2 \omega_0/(c A_{\text{eff}}), \beta_2 = 1/c \times (2dn/d\omega + \omega d^2 n/d\omega^2)$.

In the equations, the subscripts p and s represent the laser and the Raman or high-order harmonic pulse, respectively. A is the electric field envelope amplitude of the optical pulse, A_{eff} is the effective focal area, which can be evaluated from the focal spot size of the laser, v_{g} is the group velocity of the optical field, T is the time measured in the moving frame of the pump pulse, g_{s} is the Raman gain coefficient, and δ is the nonlinear coefficient. d is the walk-off parameter by which we can define the walk-off distance $L_{\text{W}} = T_0/|d|$, which determines the importance the first derivative in Eq. (209). The dispersion distance $L_{\text{D}} = T_0^2/|\beta_2|$ is defined in order to determine the importance of the second derivative in Eqs. (208) and (209). The second derivative in the equation can be neglected if the dispersion distance is much larger than the interaction distance. In the laser underdense-plasma interaction, usually we have $L < L_{\text{W}} \ll L_{\text{D}}$, and the second derivative term in the coupled equation can be neglected. With these simplifications, the solution of the coupled equations can be written as:

$$A_{\text{p}}(L, T) = A_{\text{p}}(0, T) \exp(i\delta_{\text{p}}\phi_{\text{p}}), \tag{210}$$

$$A_{\text{s}}(L, T) = A_{\text{s}}(0, T + Ld) \exp\left(\left(\frac{g_{\text{s}}}{2} + i\delta_{\text{s}}\right)\phi_{\text{s}}\right), \tag{211}$$

$$\phi_{\text{p}} = P_0 \exp(-\tau^2)L, \tag{212}$$

$$\phi_{\text{s}} = P_0 \frac{\sqrt{\pi}}{2\tau}\left[\text{erf}(\tau + \varsigma) - \text{erf}(\tau)\right]L, \tag{213}$$

where $\tau = t/T_0$, $\varsigma = Ld/T_0$ and P_0 is the laser power.

If the effect of GVD can be neglected, the RXPM does not change the pulse shape, but adds only extra chirp to the pulse, and thus it can be exploited to compress the probe pulse [42]. There is no damage threshold in plasma as with other materials, so a plasma can easily support a high-intensity laser pulse, and the energy of the laser pulse is controllable. With proper spectral adjustment, it may be possible to obtain a high power, attosecond laser pulse after compression. Another advantage of the RXPM is that the pump will ionize the gas target first so the probe pulse will avoid spectral modulation during the ionization process [80].

RXPM in plasma also can be used as a diagnostic for the laser–plasma interaction. The precise intensity of the laser-beam focused into the plasma can be obtained from RXPM of the probe beam. Meanwhile, the plasma wave will also modulate the probe beam and generate a Raman sideband, from which the plasma density can be inferred [35]. The dependence of RXPM-induced chirp on the initial time delay of pump and probe pules can be used to determine the exact delay time between the two pulses.

If the laser intensity is sufficiently high, and the overlapping time of the two laser pulses is long, RXPM needs to be considered even for counterpropagation [74], especially for the backscattered Raman amplification [65]. It also can explain the broadened backscattered Raman spectrum observed by Rousseaux et al. in their high-power laser–plasma interaction experiment [56]. The backscattered Raman spectrum should be blue-shifted as the Raman signal rides on the falling edge of the laser pulse, due to the propagation geometry. Since the plasma is an anomalous GVD medium, the chirped Raman pulse will be self-compressed in the plasma [64]. This will affect the predicted gains from the Raman-seeding mechanism [53]. With amplification and compression occurring simultaneously, relativistic plasma is a promising medium for generating high-power, ultrashort laser pulses.

6.4. INTERACTIONS WITH SOLID-DENSITY TARGETS

The generation of electrons by high-intensity laser light interacting with solid targets can generate energetic X-rays, accelerate other types of particles and induce nuclear reactions. For instance, high-order harmonics have been generated by the oscillation of the critical density surface, in the so-called moving mirror model (as discussed by Gibbon in [26]). Bright X-rays, originating from Bremsstrahlung caused by electron collisions with high-Z atoms in solid-targets, have created isotopes by means of photofission [11,36]. Laser-accelerated electron energies and angular distributions have been inferred from analyzing (γ, n) and $(\gamma, 2n)$ reactions in composite Pb/Cu targets [37] and in Ta/Cu targets [59]. Positrons were created by colliding laser-accelerated electrons with a tungsten target [23].

When electrons are heated to high temperatures or accelerated to high energies, they can separate from plasma ions. Such charge displacement creates an electrostatic sheath, which eventually accelerates the ions. The ions are pulled by the charge of the electrons and pushed by the other ions' unshielded charges (similar to the "Coulomb explosion" that can occur during the ionization of atoms). When the charge displacement is driven by thermal expansion, as in long-pulse (low power) laser–plasma experiments, the maximum ion energies are limited to less than 100 keV. However, when the charge displacement is driven by direct laser heating, as in short-pulse high-power laser–plasma experiments, multi-MeV ion energies are possible [40]. This was first shown with gas jet targets, in which case the ions were accelerated radially into 2π, and then later with thin solid-density-films, in which case the ions were accelerated into collimated beams. In the latter case, hydrocarbons and water on the surface of the film can become ionized and provide a source of protons to be accelerated.

An intense laser can ponderomotively heat electrons. If the laser contrast is high, vacuum heating can occur in the following manner. When light encounters a sharp interface between vacuum and solid density, the electromagnetic field becomes evanescent in the region above the critical density. The instantaneous "$v \times B$" force can push electrons in the direction of the light's propagation vector; it also has a frequency twice that of the pump and a magnitude proportional to the square of the normalized vector potential, a_0^2. Thus, electrons can only complete half of their figure-eight orbits, on the vacuum side, gaining relativistic energies; they move through the overdense region without the electromagnetic field to pull them back. An electrostatic sheath can thus form, which will accelerate the ions left behind [62]. Another important heating mechanism is stochastic heating, which occurs when the light that is reflected from the critical surface beats with the incoming wave to create a standing wave. The motion of electrons in such a wave can become chaotic, resulting in a large increase in electron temperature (>100 keV) [63].

As the heated electrons propagate through a solid, they can instantaneously field-ionize the neutral atoms of the solid. This will both modify the solid's conductivity and provide a source of protons on the rear-side of the target. If the film is thin enough, the electrons can pass through, and create a sheath on the rear-side of the target. This latter mechanism has been dubbed the target normal sheath acceleration mechanism [62].

The results of these experiments indicate that a large number of protons (10^{13} p) can be accelerated, corresponding to source current densities (10^8 A/cm^2) that are nine orders-of-magnitude higher than produced by cyclotrons, but with comparable, or even lower, transverse emittances ($\epsilon_\perp \leq 1.0$ π mm-mrad). Proton energies up to 60 MeV have been observed in experiments at intensities exceeding 10^{20} W/cm^2 (using a petawatt power laser). The high end of the proton spectrum typically has a sharp cut-off, but is a continuum. In one experiment, protons were observed to be emitted in ring patterns, the radii of which depend on the proton energy, which was explained by self-generated magnetic fields.

The production of radionuclides have been used as an ion energy diagnostic. In another example of a nuclear reaction initiated by an intense laser, neutrons have been produced by the He fusion reaction in the focus of 200 mJ, 160 fs Ti:sapphire laser pulses on a deuterated polyethylene target. Optimizing the fast electron and ion generation by applying a well-defined prepulse led to an average rate of 140 neutrons per shot [16]. Neutrons have also been generated from cluster plasmas [17], which were produced by the cooled-nozzle, but with significantly lower laser intensities than required with planar solid targets.

7. Concluding Remark

The field of nonlinear optics with electrons bound to atoms has over the last forty years given rise to many scientific discoveries and technologies that are now commonplace. The relatively young field of relativistic nonlinear optics has already begun do the same. Some of the exciting phenomena and applications that have already been identified were discussed in this article. Others await discovery as ever higher laser intensities are reached, with no foreseeable limit.

8. Acknowledgements

The authors wish to thank the support of the National Science Foundation and the Chemical Sciences, Geosciences and Biosciences Division of the Office of Basic Energy Sciences, Office of Science, U.S. Department of Energy, as well as S. Banerjee, F. He, and Y.Y. Lau for their helpful comments.

9. References

[1] G.P. Agrawal, "Nonlinear Fiber Optics", Academic Press, New York, 1995, Chapter 8.

[2] T.M. Antonsen, P. Mora, Self-focusing and Raman scattering of laser pulses in tenuous plasmas, *Phys. Fluid B* **5** (1993) 1440.

[3] S. Banerjee, S. Sepke, R. Shah, A. Valenzuela, A. Maksimchuk, D. Umstadter, Optical deflection and temporal characterization of ultra-fast laser produced electron beams, *Phys. Rev. Lett.* **95** (2005) 035004.

[4] J.P. Barton, D.R. Alexander, Fifth order corrected electromagnetic field components for a fundamental Gaussian beam, *J. Appl. Phys.* **66** (1989) 2800.

[5] H. Bateman, "Higher Transcendental Functions", McGraw-Hill, New York, 1953.

[6] B.W. Boreham, H. Hora, Debye length discrimination of nonlinear laser forces acting on electrons in tenuous plasmas, *Phys. Rev. Lett.* **42** (1979) 776.

[7] B.W. Boreham, B. Luther-Davies, High-energy electron acceleration by ponderomotive forces in tenuous plasmas, *J. Appl. Phys.* **50** (1979) 2533.

[8] W. Brown, F. Hartemann, Three-dimensional time and frequency domain theory of femtosecond X-ray pulse generation through Thomson scattering, *Phys. Rev. STAB* **7** (2004) 060703.

[9] S.-Y. Chen, A. Maksimchuk, D. Umstadter, Experimental observation of relativistic nonlinear Thomson scattering, *Nature* **396** (1998) 653.

[10] L. Cicchitelli, H. Hora, R. Postle, Longitudinal field components for laser beams in vacuum, *Phys. Rev. A* **41** (1990) 3727.

[11] T.E. Cowan, A.W. Hunt, T.W. Phillips, S.C. Wilks, M.D. Perry, C. Brown, W. Fountain, S. Hatchett, J. Johnson, M.H. Key, T. Parnell, D.M. Pennington, R.A. Snavely, Y. Takahashi, Photonuclear fission from high energy electrons from ultraintense laser-solid interactions, *Phys. Rev. Lett.* **84** (2000) 903.

[12] C.B. Darrow, C. Coverdale, M.D. Perry, W.B. Mori, C. Clayton, K. Marsh, C. Joshi, Strongly coupled stimulated Raman backscatter from subpicosecond laser–plasma interactions, *Phys. Rev. Lett.* **69** (1992) 442.

[13] L.W. Davis, Theory of electromagnetic beams, *Phys. Rev. A* **19** (1979) 1177.

[14] C.D. Decker, W.B. Mori, Group velocity of large-amplitude electromagnetic waves in a plasma, *Phys. Rev. E* **51** (1995) 1364.

[15] C.D. Decker, W.B. Mori, K.-C. Tzeng, T. Katsouleas, The evolution of ultra-intense short-pulse lasers in underdense plasmas, *Phys. Plasmas* **72** (1996) 1482.

[16] L. Disdier, J.-P. Garconnet, G. Malka, J.-L. Miquel, Fast neutron emission from a high-energy ion beam produced by a high-intensity subpicosecond laser pulse, *Phys. Rev. Lett.* **82** (1999) 1454.

[17] T. Ditmire, Laser fusion on a tabletop, *Optics and Photonics News* **13** (2002) 28.

[18] E. Esarey, W.P. Leemans, Nonparaxial propagation of ultrashort laser pulses in plasma channels, *Phys. Rev. E* **59** (1998) 1082.

[19] E. Esarey, P. Sprangle, J. Krall, A. Ting, Overview of plasma-based accelerator concepts, *IEEE Trans. Plasma Sci.* **24** (1996) 252.

[20] E. Esarey, P. Sprangle, J. Krall, A. Ting, Self-focusing and guiding of short leser pulses in ionizing gases and plasmas, *IEEE J. Quantum Electronics* **33** (1997) 1879–1914.

[21] J. Faure, Y. Glinec, A. Pukhov, S. Kiselev, S. Gordienko, E. Lefebvre, J.-P. Rousseau, F. Burgy, V. Malka, A laser–plasma accelerator producing monoenergetic electron beams, *Nature* **431** (2004) 541.

[22] D.W. Forslund, J.M. Kindel, E.L. Lindman, Theory of stimulated scattering processes in laser-irradiated plasmas, *Phys. Fluids* **18** (1975) 1002.

[23] C. Gahn, G.D. Tsakiris, G. Pretzler, K.J. Witte, C. Delfin, C.-G. Wahlström, D. Habs, Generating positrons with femtosecond-laser pulses, *Appl. Phys. Lett.* **77** (2000) 2662.

[24] E.G. Gamiliy, *Laser and Part. Beams* **12** (1994) 185.

[25] C.G.R. Geddes, C. Toth, J.V. Tilborg, E. Esarey, C.B. Schroeder, D. Bruhwiler, C. Nieter, J. Cary, W.P. Leemans, High-quality electron beams from a laser wakefield accelerator using plasma-channel guiding, *Nature* **431** (2004) 538.

[26] P. Gibbon, E. Förster, Short-pulse laser–plasma interactions, *Plasma Phys. Controlled Fusion* **38** (1996) 769.

[27] F. Hartemann, "High-Field Electrodynamics", CRC Press, Boca Raton, FL, 2001.

[28] H. Hora, Theory of relativistic self-focusing of laser radiation in plasmas, *J. Opt. Soc. Amer.* **65** (1975) 882.

[29] H. Hora, The transient electrodynamic forces at laser–plasma interaction, *Phys. Fluids* **28** (1985) 3705.

[30] J.D. Jackson, "Classical Electrodynamics", Wiley, New York, 1975.

[31] C.J. Joshi, P.B. Corkum, Interactions of ultra-intense laser light with matter, *Phys. Today* **48** (1995) 36.

[32] T. Katsouleas, Physical mechanisms in the plasma wake-field accelerator, *Phys. Rev. A* **33** (1986) 2056.

[33] Y. Lau, F. He, D. Umstadter, Nonlinear Thomson scattering: a tutorial, *Phys. Plasmas* **10** (2003) 2155.

[34] M. Lax, W.H. Louisell, W.B. McKnight, From maxwell to paraxial wave optics, *Phys. Rev. A* **11** (1975) 1365.

[35] S.P. LeBlanc, M.C. Downer, R. Wagner, S.-Y. Chen, A. Maksimchuk, G. Mourou, D. Umstadter, Temporal characterization of a self-modulated laser wakefield, *Phys. Rev. Lett.* **77** (1996) 5381.

[36] K.W.D. Ledingham, I. Spencer, T. McCanney, R.P. Singhal, M.I.K. Santala, E. Clark, I. Watts, F.N. Beg, M. Zepf, K. Krushelnick, M. Tatarakis, A.E. Dangor, P.A. Norreys, R. Allott, D. Neely, R.J. Clark, A.C. Machacek, J.S. Wark, A.J. Cresswell, D.C.W. Sanderson, J. Magill, Photonuclear physics when a multiterawatt laser pulse interacts with solid targets, *Phys. Rev. Lett.* **84** (2000) 899.

[37] W.P. Leemans, D. Rodgers, P.E. Catravas, C.G.R. Geddes, G. Fubiani, E. Esarey, B.A. Shadwick, R. Donahue, A. Smith, Gamma-neutron activation experiments using laser wakefield accelerators, *Phys. Plasmas* **8** (2001) 2510.

[38] W.P. Leemans, R.W. Schoenlein, P. Volfbeyn, A.H. Chin, T.E. Glover, P. Balling, M. Zolotorev, K.-J. Kim, S. Chattopadhyay, C.V. Shank, Interaction of relativistic electrons with ultrashort laser pulses: generation of femtosecond X-rays and microprobing of electron beams, *IEEE J. Quant. Electron.* **33** (1997) 1925.

[39] B. Luther-Davies, E.G. Gamaly, Y. Wang, A.V. Rode, V.T. Tikhonchuk, Matter in ultrastrong laser fields, *Soviet J. Quant. Electron.* **22** (1992) 289.

[40] A. Maksimchuk, K. Flippo, H. Krause, G. Mourou, K. Nemoto, D. Shultz, D. Umstadter, R. Vane, V.Yu. Bychenkov, G.I. Dudnikova, V.F. Kovalev, K. Mima, V.N. Novikov, Y. Sentoku, S.V. Tolokonnikov, High-energy ion generation by short laser pulses, *Plasma Phys. Reports* **30** (2004) 473.

[41] V. Malka, S. Fritzler, E. Lefebvre, M.-M. Aleonard, F. Burgy, J.-P. Chambaret, J.-F. Chemin, K. Krushelnick, G. Malka, S.P.D. Mangles, Z. Najmudin, M. Pittman, J.-P. Rousseau, J.-N. Scheurer, B. Walton, A.E. Dangor, Electron acceleration by a wake field forced by an intense ultrashort laser pulse, *Science* **298** (2002) 1596.

[42] J.T. Manassah, Pulse compression of an induced-phase-modulated weak signal, *Opt. Lett.* **13** (1988) 755.

[43] S.P.D. Mangles, C.D. Murphy, Z. Najmudin, A.G.R. Thomas, J.L. Collier, A.E. Dangor, E.J. Divall, P.S. Foster, J.G. Gallacher, C.J. Hooker, D.A. Jaroszynski, A.J. Langley, W.B. Mori, P.A. Norreys, F.S. Tsung, R. Viskup, B.R. Walton, K. Krushelnick, Monoenergetic beams of relativistic electrons from intense laser–plasma interactions, *Nature* **431** (2004) 535.

[44] C.J. McKinstrie, R. Bingham, Stimulated Raman forward scattering and the relativistic modulational instability of light waves in rarefied plasma, *Phys. Fluid B* **4** (1992) 2626.

[45] D.D. Meyerhofer, High-intensity laser–electron scattering, *IEEE J. Quant. Electron.* **33** (1997) 1935.

[46] J. Meyer-ter-Vehn, A. Pukhov, Z.-M. Sheng, Relativistic laser-plasma interaction, in: D. Batani (Ed.), "Atoms, Solids and Plasmas in Super-Intense Laser Fields", Kluwer Academic/Plenum, New York, 2001, p. 167.

[47] W.B. Mori, The physics of the nonlinear optics of plasmas at relativistic intensities for short-pulse lasers, *IEEE J. Quant. Electron.* **33** (1997) 1942.

[48] W.B. Mori, C.D. Decker, D.E. Hinkel, T. Katsouleas, Raman forward scattering of short-pulse high-intensity lasers, *Phys. Rev. Lett.* **72** (1994) 1482.

[49] P.M. Morse, H. Feshbach, "Waves and Fields in Optoelectronics", Prentice-Hall, Englewood Cliffs, NJ, 1984, Chapter 10.

[50] G.A. Mourou, C.P.J. Barty, M.D. Perry, Ultrahigh-intensity lasers: physics of the extreme on a tabletop, *Phys. Today* (1998) 22.

[51] G. Mourou, D. Umstadter, Development and applications of compact high-intensity lasers, *Phys. Fluids B* **4** (1992) 2315.

[52] G. Mourou, D. Umstadter, Extreme light, *Sci. Amer.* (2002) 81.

[53] K. Nakajima, D. Fisher, T. Kawakubo, H. Nakanishi, A. Ogata, Y. Kato, Y. Kitagawa, R. Kodama, K. Mima, H. Shiraga, K. Suzuki, K. Yamakawa, T. Zhang, Y. Sakawa, T. Shoji, Y. Nishida, N. Yugami, M. Downer, T. Tajima, Observation of ultrahigh gradient electron acceleration by a self-modulated intense short laser pulse, *Phys. Rev. Lett.* **74** (1995) 4428.

[54] B. Quesnel, P. Mora, Theory and simulation of the interaction of ultraintense laser pulses with electrons in vacuum, *Phys. Rev. E* **58** (1998) 3719.

[55] A. Rousse, C. Rischel, J.-C. Gauthier, Colloquium: femtosecond X-ray crystallography, *Rev. Mod. Phys.* **73** (2001) 17.

[56] C. Rousseaux, G. Malka, J.L. Miquel, F. Amiranoff, S.D. Baton, Experimental validation of the linear theory of stimulated Raman scattering driven by a 500-fs laser pulse in a preformed underdense plasma, *Phys. Rev. Lett.* **74** (1995) 4655.

[57] T. Rowlands, The gauge and Lorentz invariance of the nonlinear ponderomotive 4-force, *Plasma Phys. Contr. Fusion* **32** (1990) 297.

[58] Y.I. Saliman, F.H.M. Faisal, Harmonic generation by superintense light scattering from relativistic electrons, *Phys. Rev. A* **54** (1996) 4383.

[59] M.I.K. Santala, Z. Najmudin, E.L. Clark, M. Tatarakis, K. Krushelnick, A.E. Dangor, V. Malka, J. Faure, R. Allott, R.J. Clarke, Observation of a hot high-current electron beam from a self-modulated laser wakefield accelerator, *Phys. Rev. Lett.* **86** (2001) 1227.

[60] E.S. Sarachik, G.T. Schappert, Classical theory of the scattering of intense laser radiation by free electrons, *Phys. Rev. D* **1** (1970) 2738.

[61] R.W. Schoenlein, W.P. Leemans, A.H. Chin, P. Volfbeyn, T.E. Glover, P. Balling, M. Zolotorev, K.-J. Kim, S. Chattopadhyay, C.V. Shank, Femtosecond X-ray pulses at 0.4 generated by 90

Thomson scattering: a tool for probing the structural dynamics of materials, *Science* **274** (1996) 236.

[62] Y. Sentoku, V.Yu. Bychenkov, K. Flippo, A. Maksimchuk, K. Mima, G. Mourou, Z.-M. Sheng, D. Umstadter, High-energy ion generation in interaction of short laser pulse with high-density plasma, *Appl. Phys. B* **74** (2002) 207.

[63] Z.-M. Sheng, K. Mima, Y. Sentoku, M.S. Jovanovic, T. Taguchi, J. Zhang, J. Meyer-ter-Vehn, Stochastic heating and acceleration of electrons in colliding laser fields in plasma, *Phys. Rev. Lett.* **88** (2002) 055004.

[64] O. Shorokhov, A. Pukhov, I. Kostyukov, Superradiant amplification of an ultrashort laser pulse in a plasma by a counterpropagating pump, *Phys. Rev. Lett.* **91** (2003) 265002.

[65] G. Shvets, et al., Superradiant amplification of an ultrashort laser pulse in a plasma by a counter-propagating pump, *Phys. Rev. Lett.* **81** (1998) 4879.

[66] A.E. Siegman, "Lasers", University Science Books, Mill Valley, CA, 1986.

[67] P. Sprangle, B. Hafizi, J.R. Peñano, Laser pulse modulation instabilities in plasma channels, *Phys. Rev. E* **61** (2000) 4381.

[68] P. Sprangle, J.R. Peñano, B. Hafizi, R.F. Hubbard, A. Ting, D.F. Gordon, A. Zigler, T.M. Antonsen, GeV acceleration in tapered plasma channels, *Phys. Plasmas* **9** (2002) 2364.

[69] T. Tajima, G. Mourou, Zettawatt-exawatt lasers and their applications in ultrastrong-field physics, *Phys. Rev. STAB* **5** (2002) 031301.

[70] J.J. Thomson, Stimulated Raman scatter in laser fusion target chambers, *Phys. Fluids* **21** (1978) 2082.

[71] D. Umstadter, Review of physics and applications of relativistic plasmas driver by ultra-intense lasers, *Phys. Plasmas* **8** (2001) 1774.

[72] D. Umstadter, Relativistic laser–plasma interactions, *J. Phys. D: Appl. Phys.* **36** (2003) R151–R165.

[73] D.P. Umstadter, Relativistic nonlinear optics, in: D. Robert, D.G.S. Guenther, L. Bayvel (Eds.), "Encyclopedia of Modern Optics", Elsevier, Oxford, 2004, p. 289.

[74] B.V. Vu, A. Szoke, O.L. Landen, Induced frequency shifts by counterpropagating subpicosecond optical pulses, *Opt. Lett.* **18** (1993) 723.

[75] R. Wagner, Laser–plasma electron accelerators and nonlinear relativistic optics. Ph.D. Thesis, University of Michigan, 1998.

[76] R. Wagner, S.-Y. Chen, A. Maksimchuk, D. Umstadter, Electron acceleration by a laser wakefield in a relativistically self-guided channel, *Phys. Rev. Lett.* **78** (1997) 3125.

[77] J.X. Wang, W. Sheid, M. Hoelss, Y.K. Ho, Fifth-order corrected field descriptions of the Hermite–Gaussian (0,0) and (0.1) mode laser beam, *Phys. Rev. E* **64** (2001) 066612.

[78] X. Wang, M. Krishnan, N. Saleh, H. Wang, D. Umstadter, Electron acceleration and the propagation of ultrashort high-intensity laser pulses in plasmas, *Phys. Rev. Lett.* **84** (2000) 5324.

[79] I. Watts, M. Zepf, E.L. Clark, M. Tatarakis, K. Krushelnick, A.E.D.R. Allott, R.J. Clarke, D. Neely, P.A. Norreys, Measurements of relativistic self-phase-modulation in plasma, *Phys. Rev. E* **66** (2002) 036409.

[80] W.M. Wood, C.W. Siders, M.C. Downer, Measurement of femtosecond ionization dynamics of atmospheric density gases by spectral blueshifting, *Phys. Rev. Lett.* **67** (1991) 3523.

[81] K. Yamakawa, Table-top lasers create ultrahigh peak powers, *Oyobuturi* **73** (2004) 186–193.

[82] H. Yang, J. Zhang, J. Zhang, L.Z. Zhao, Y.J. Li, H. Teng, Y.T. Li, Z.H. Wang, Z.L. Chen, Z.Y. Wei, J.X. Ma, W. Yu, Z.M. Sheng, Third-order harmonic generation by self-guided femtosecond pulses in air, *Phys. Rev. E* **67** (2003) 015401.

ADVANCES IN ATOMIC, MOLECULAR AND OPTICAL PHYSICS, VOL. 52

COUPLED-STATE TREATMENT OF CHARGE TRANSFER

THOMAS G. WINTER

Department of Physics, Pennsylvania State University, Wilkes-Barre Campus, Lehman,
PA 18627, USA

1. Introduction

Charge transfer in ion–atom collisions is a basic atomic collision process. Also called charge exchange, electron capture, and electron transfer, it is dealt with extensively in the literature both for single- and multi-electron targets, as reviewed in [1–9]. To focus on the fundamentals, we restrict ourselves to a bare nuclear projectile incident on a one-electron atom or ion. At projectile speeds on the order of the orbital speed of the target electron, charge transfer is typically significant, and competes with ionization and direct excitation as a mechanism for depleting the elastic channel. For proton–hydrogen collisions, "velocity" matching occurs for 25 keV projectiles; here, and at somewhat lower and higher energies, many intermediate states are potentially important during the collision, so coupled-state approaches are generally most appropriate. Such approaches have been extensively developed and applied in this so-called intermediate-energy range. They have also sometimes been carried out at higher and lower energies either to test perturbation theories or for want of other approaches. What follows is an outline of the theories in Section 2, a comparison of results in Section 3, and a short conclusion in Section 4. Atomic units are used unless noted otherwise.

ISSN 1049-250X
DOI 10.1016/S1049-250X(05)52008-1

2. Coupled-State Treatments

It is generally believed that at energies of at least about 100 eV/u the nuclear motion can be assumed to be classical [10,11]; there the nuclear deBroglie wavelength is significantly smaller than the size of the atomic target. At such energies an impact-parameter approach is appropriate, and the projectile follows a definite trajectory $\vec{R}(t)$, the position vector of the projectile nucleus relative to the assumed stationary target nucleus at a time t, taken to be zero at the point of closest approach. At still higher energies the trajectory can be assumed to be rectilinear; this additional assumption, which is not essential, will be discussed in Section 3.

2.1. IMPACT-PARAMETER APPROACHES

The time-dependent electronic wave function $\Psi(\vec{r}, t)$ is expanded in terms of orbitals $f_k(\vec{r}, t)$:

$$\Psi(\vec{r}, t) = \sum_k a_k(t) f_k(\vec{r}, t). \tag{1}$$

Here $\vec{r}$ is the electronic position vector, and both Ψ and the transition coefficients a_k depend implicitly on the impact parameter ρ, as well as on the initial projectile velocity $\vec{v}$. The various coupled-state approaches are distinguished by the choice of f_k. In all cases, substituting the expansion in the time-dependent Schrödinger equation

$$\left(H - i \frac{\partial}{\partial t} \right) \Psi(\vec{r}, t) = 0, \tag{2}$$

multiplying by a particular f_k^*, and integrating over the spatial electronic coordinates, we obtain coupled, first-order differential equations for the a_k's. In vector form,

$$i \underline{S}'(t) \frac{d\vec{a}}{dt} = \underline{G}'(t) \vec{a}(t), \tag{3}$$

where

$$\left[\vec{a}(t) \right]_k = a_k(t), \tag{4}$$

$$S'_{kk'}(t) = \left\langle f_k(\vec{r}, t) \middle| f_{k'}(\vec{r}, t) \right\rangle = S_{kk'}(t) P_{kk'}(t), \tag{5a}$$

$$G'_{kk'}(t) = \left\langle f_k(\vec{r}, t) \middle| H - i \left(\frac{\partial}{\partial t} \right)_{\vec{r}} \middle| f_{k'}(\vec{r}, t) \right\rangle = G_{kk'}(t) P_{kk'}(t). \tag{5b}$$

The coupled equations are solved subject to the initial condition

$$a_k(-\infty) = \delta_{1k} \tag{6}$$

assuming the target atom is initially in the state whose orbital is f_1. At a particular impact parameter ρ, the probability of a transition to a state f_k in the separated-atoms limit $t \to \infty$, such as a charge-transferring state, is[1]

$$P_k(\rho) = \left| a_k(\infty) \right|^2. \tag{7}$$

The integrated cross section is

$$Q_k = 2\pi \int\limits_0^\infty P_k(\rho)\rho \, d\rho. \tag{8}$$

The various impact-parameter approaches are distinguished by different choices of f_k, leading to different overlap and coupling matrix elements $S_{kk'}$ and $G_{kk'}$ and energy phases $P_{kk'}$.

2.1.1. Atomic-State and Atomic-Pseudostate Approaches

2.1.1.1. Two-center In the original two-center atomic-state approach of Bates [12], the orbitals f_k in Eq. (1) are two-center traveling atomic orbitals:

$$f_{k\alpha}(\vec{r}, t) = \psi_{k\alpha}(\vec{r}_\alpha) \exp\left[-i\left(\pm \frac{1}{2}\vec{v} \cdot \vec{r} + E_{k\alpha}t + \frac{1}{8}v^2 t \right) \right], \tag{9}$$

where the additional subscript α denotes the nuclear center, the target nucleus A or the projectile nucleus B. In atomic-state and atomic-pseudostate approaches, the projectile is generally assumed to move with constant velocity $\vec{v}$ with respect to a stationary target nucleus. The upper and lower signs in the translational factor correspond to $\alpha = A$ and B, respectively. The electronic coordinate $\vec{r}$ is in the laboratory frame and measured with respect to the midpoint of the internuclear line. However, under the assumption of constant $\vec{v}$, any origin along the internuclear line yields the same matrix elements and, hence, transition probabilities and cross sections; that is, observables are translationally invariant. In Bates' development, the atomic wave functions $\psi_{k\alpha}$ and corresponding energies $E_{k\alpha}$ in Eq. (9) are implicitly exact; however, in the expressions below, it is only necessary that

$$\langle \psi_{k\alpha} | \psi_{k'\alpha} \rangle = \delta_{kk'}, \tag{10a}$$

$$\langle \psi_{k\alpha} | H_\alpha | \psi_{k'\alpha} \rangle = E_{k\alpha}\delta_{kk'}, \tag{10b}$$

---

[1] The probability $P_j(\rho)$ is not to be confused with the energy phase $P_{jk}(t)$.

where H_α is the atomic Hamiltonian for center α. The overlap and coupling matrix elements in Eqs. (5a), (5b) are then of two forms: the "direct" matrix elements

$$S_{k\alpha k'\alpha} = \delta_{kk'}, \tag{11a}$$

$$G_{k\alpha k'\alpha} = \left\langle \psi_{k\alpha}(\vec{r}_\alpha) \left| -\frac{Z_\beta}{r_\beta} \right| \psi_{k'\alpha}(\vec{r}_\alpha) \right\rangle \tag{11b}$$

and the "charge-exchange" matrix elements

$$S_{k\alpha k'\beta}(t) = \left\langle \psi_{k\alpha}(\vec{r}_\alpha) \left| e^{\pm i\vec{v}\cdot\vec{r}} \right| \psi_{k'\beta}(\vec{r}_\beta) \right\rangle, \tag{12a}$$

$$G_{k\alpha k'\beta}(t) = \left\langle \psi_{k\alpha}(\vec{r}_\alpha) \left| e^{\pm i\vec{v}\cdot\vec{r}} \left(H_\beta - E_{k'\beta} - \frac{Z_\alpha}{r_\alpha} \right) \right| \psi_{k'\beta}(\vec{r}_\beta) \right\rangle, \tag{12b}$$

where $\alpha \neq \beta$, and the upper or lower sign in the translational factor $\exp(\pm i\vec{v}\cdot\vec{r})$ corresponds to $\alpha = A$ or B, respectively. For all α, β, the energy-phase matrix element is

$$P_{k\alpha k'\beta}(t) = e^{i(E_{k\alpha} - E_{k'\beta})t}. \tag{13}$$

There are several versions of the two-center approach of Bates, distinguished by the type of basis functions placed on each center.

Atomic-state. The first such approach carried out was a purely atomic-state approach, starting with the two-state calculation of McCarroll [13] for p–H electron transfer. This approach includes one or more exact, bound atomic states on each center, but neglects the continuum.

Atomic-plus-pseudostate. Most of the more recent calculations include also pseudostates which take some account of the continuum. The first such calculations, also applied to p–H transfer, were by Gallaher and Wilets [14] using a small Sturmian basis and Cheshire et al. [15] using a small bound-atomic-plus-pseudostate basis.

In typical recent, state-of-the-art calculations, the atomic-state functions $\psi_{k\alpha}(\vec{r}_\alpha)$ in Eq. (9) are expanded in a large, potentially complete set of functions $\varphi_{j\alpha}(\vec{r}_\alpha)$:

$$\psi_{k\alpha}(\vec{r}_\alpha) = \sum_j c_{kj\alpha}\varphi_{j\alpha}(\vec{r}_\alpha). \tag{14}$$

This expansion is determined by diagonalizing the overlap matrix and atomic Hamiltonians as in Eqs. (10a) and (10b). Several such systematic basis sets have been introduced: the Sturmian basis by Shakeshaft for the p–H system [16,17] and later Winter for asymmetric systems [18]; an atomic-state basis augmented with united-atoms orbitals (AO+ basis) by Fritsch and Lin [19]; other large pseudostate bases, particularly even-tempered bases with Slater-orbital exponents in a geometric sequence by Kuang and Lin [20]; Hermite bases by Reading et al. [21]; and

Gaussian bases by Toshima and Eichler [22]. The original approach of Reading et al. was a so-called one-and-a-half-center approach, in which a large basis was centered on the target, but only a single function on the projectile, the latter treated perturbatively. Recently, Reading et al. [23] designated as "one-and-a-half-center" a fully coupled two-center basis including bound states centered on both the projectile and target, but positive-energy states centered only on the target; however, most refer to this as a two-center basis.

It may be argued that with sufficiently large bases the particular choice is irrelevant, but with finite bases the basis type and size *is* important. Results with more recent, large bases will be presented in Section 3.

Continuum-distorted-wave. Following considerable earlier work by Brown and Crothers [24] and Brown [25], Ferreira da Silva and Serrão [26] recently presented a coupled-state, two-center, continuum-distorted-wave approach. Here the original traveling atomic orbital $f_{k\alpha}(\vec{r}, t)$ of Bates, given by Eq. (9), is multiplied by a distorted-wave factor

$$D_\alpha(\vec{r}, t) = \exp\left(\frac{\pi Z_\beta}{2v}\right) \Gamma\left(1 \mp \frac{iZ_\beta}{v}\right) {}_1F_1\left(\pm\frac{iZ_\beta}{v}, 1, \pm i\left[vr_\beta + \vec{v} \cdot \vec{r}_\beta\right]\right)$$
$$\times \exp\left(\pm\frac{iZ_A Z_B}{v} \ln\left[vR \mp \vec{v} \cdot \vec{R}\right]\right), \tag{15}$$

where $\alpha \neq \beta$ and the upper or lower sign corresponds to $\alpha = A$ or B, respectively. The resulting continuum distorted wave

$$f_{k\alpha}^{\text{cdw}}(\vec{r}, t) = f_{k\alpha}(\vec{r}, t) D_\alpha(\vec{r}, t)$$

is as in the original continuum-distorted-wave perturbative (CDW) approach of Cheshire [27]. That CDW approach allows for the Coulombic coupling between the electron and the more distant nucleus in *first-order* perturbation theory. Ferreira da Silva and Serrão fully coupled these CDW basis functions within a two-state approximation so as to extend the CDW theory to lower energies. The resulting matrix elements are considerably more complicated than those in a two-atomic-state approximation [13].

2.1.1.2. Three-center In the three-center approach introduced by Anderson et al. [28], a set of orbitals is also placed on the internuclear line at a center C:

$$f_{k\alpha}(\vec{r}, t) = \psi_{k\alpha}(\vec{r}_\alpha) \exp\left[-i\left(-q_\alpha \vec{v} \cdot \vec{r} + E_{k\alpha}t + \frac{1}{2}q_\alpha^2 v^2 t\right)\right], \tag{16}$$

where $\alpha = C$. The "nuclear charge" Z_C is chosen to be that of the united atom:

$$Z_C = Z_A + Z_B. \tag{17}$$

(Eq. (16) is a generalization of Eq. (9), with $q_\alpha = -1/2$ if $\alpha = A$ and $+1/2$ if $\alpha = B$. The origin of electronic coordinates is again placed at the midpoint of the internuclear line; however, the matrix elements, and, hence, observables continue to be independent of this choice.) Although the origin is arbitrary, the third center C is not. Anderson et al. chose C to be the center of charge. Winter and Lin [29] later chose the saddle point (equiforce point); in that case,

$$q_C = p - \frac{1}{2}, \tag{18}$$

where

$$p = \frac{\sqrt{Z_A}}{\sqrt{Z_A} + \sqrt{Z_B}}. \tag{19}$$

For a homonuclear system, such as p–H which they treated, the center of charge and saddle point coincide.

With the inclusion of a third center, there are several additional matrix elements, including some three-center matrix elements, for example, [30]:

$$G_{kCk'B} = -Z_A g^A_{kCk'B},$$

where

$$g^\gamma_{k\alpha k'\beta} = \left\langle \psi_{k\alpha}(\vec{r}_\alpha) \left| \exp[\mathrm{i}(-q_\alpha + q_\beta)\vec{v} \cdot \vec{r}] \left(\frac{1}{r_\gamma} \right) \right| \psi_{k'\beta}(\vec{r}_\beta) \right\rangle.$$

2.1.2. Molecular-State Approaches

The molecular-state, impact-parameter approaches were introduced in their present forms by Bates et al. [31]; Bates and McCarroll [32]; and, more recently, by others. The approach of Bates et al. is called the method of perturbed stationary states (pss); that of Bates and McCarroll is the plane-wave-factor, molecular-state method.

2.1.2.1. Perturbed-stationary-state In this, the first molecular-state approach, the orbital f_k in Eq. (1) is chosen to be

$$f_k(\vec{r}, t) = \psi_k(\vec{r}', R) \exp\left(-\mathrm{i} \int_{-\infty}^{t} E_k[R(t')] \, dt' \right). \tag{20}$$

Here ψ_k and E_k denote a *molecular* wave function and the corresponding energy, satisfying

$$H(\vec{r}', R)\psi_k(\vec{r}', R) = E_k(R)\psi_k(\vec{r}', R), \tag{21}$$

where

$$\vec{r}' = \vec{r}'[\vec{r}, \vec{R}(t)]$$

is the electronic position vector in the molecular frame relative to an *arbitrary* point on the internuclear line *assuming* rectilinear trajectories, and H is the electronic Hamiltonian of the molecule (without the internuclear term), the transition probabilities then depending on the particular origin chosen. As introduced by Piacentini and Salin [33], there is a major simplification in extracting charge-transfer probabilities if one places the origin at the target nucleus A, for then the target states are asymptotically defined, so that the direct excitation (plus elastic) probability is readily obtainable, and, by subtraction from unity, the charge-transfer probability as well, assuming ionization to be negligible at the lower energies where the molecular approach is customarily employed.

The matrix elements and energy phase in Eqs. (5a), (5b) are then

$$S_{kk'} = \delta_{kk'},$$ (22a)

$$G_{kk'}(t) = \left\langle \psi_k(\vec{r}', R) \left| -i\left(\frac{\partial}{\partial t}\right)_{\vec{r}} \right| \psi_{k'}(\vec{r}', R) \right\rangle,$$ (22b)

$$P_{kk'}(t) = \exp\left(i \int\limits_{-\infty}^{t} \left[E_k(R') - E_{k'}(R') \right] dt' \right).$$ (23)

If the origin is chosen to be at α, say $\alpha = A$ as in the approach of Piacentini and Salin, then the coupling matrix element in Eq. (22b) can be expressed [34] in terms of rotational and radial coupling matrix elements in the molecular frame as

$$G_{kk'}(t)_\alpha = -i\left\langle \psi_k(\vec{r}'_\alpha, R) \left| |\dot{\theta}|(-il_{y'_\alpha}) + \dot{R}\frac{\partial}{\partial R} \right| \psi_{k'}(\vec{r}'_\alpha, R) \right\rangle,$$ (24)

where the z and z' axes, separated by an angle θ, are along $\vec{v}$ and $\vec{R}$, respectively; the coincident y and y' axes are perpendicular to the collision plane; and $l_{y'}$ is the component of angular momentum along the y' axis.

2.1.2.2. Plane-wave-factor, molecular-state   The orbital f_k was chosen by Bates and McCarroll [32] to be

$$f_k(\vec{r}, t) = \psi_k(\vec{r}', R)$$
$$\times \exp\left[-i\left(\pm\frac{1}{2}\vec{v}\cdot\vec{r} + \int\limits_{-\infty}^{t} E_k[R(t')] dt' + \frac{1}{8}v^2 t \right) \right],$$ (25)

that is, the orbital in the pss approach (Eq. (20)) modified by the translational factor in Eq. (9) of Bates' atomic-state method. The exact molecular eigenfunctions and energies ψ_k and E_k are as in the pss approach (see Eq. (21)). For definiteness, the origin has again been chosen to be the midpoint of the internuclear line as in Bates' method; the results are again independent of this choice. The matrix elements here take two forms, direct and charge-exchange matrix elements analogous to those in the atomic-state approach described previously. The *direct* overlap and coupling matrix elements are exactly as in the pss approach; see Eqs. (22a) and (22b), where in Eq. (22b), $\vec{r} = \vec{r}_\alpha$, with $\alpha = A$ if both molecular states (labeled by k and k') correlate to states on A in the separated-atoms limit (or $\alpha = B$ if both correlate to states on B), leading to Eq. (24). The *charge-exchange* overlap and coupling matrix elements are [35]

$$S_{kk'}(t) = \left\langle \psi_k(\vec{r}', R) \middle| e^{\pm i\vec{v}\cdot\vec{r}} \middle| \psi_{k'}(\vec{r}', R) \right\rangle, \tag{26a}$$

$$G_{kk'}(t) = \left\langle \psi_k(\vec{r}', R) \middle| e^{\pm i\vec{v}\cdot\vec{r}} \left[-i\left(\frac{\partial}{\partial t}\right)_{\vec{r}_\beta} \right] \middle| \psi_{k'}(\vec{r}', R) \right\rangle, \tag{26b}$$

where the states labeled by k and k' correlate respectively to states on α and β in the separated atoms limit (where $\alpha \neq \beta$), the upper or lower sign in the translational factor $\exp(\pm i\vec{v}\cdot\vec{r})$ corresponds respectively to $\alpha = A$ or B, and

$$\left(\frac{\partial}{\partial t}\right)_{\vec{r}_\beta} = |\dot{\theta}|(-i l_{y'_\beta}) + \dot{R}\frac{\partial}{\partial R}, \tag{27}$$

much as in Eq. (24). Unlike in the perturbed-stationary-state approach, all inelastic matrix elements vanish in the separated-atoms limit, and the transition probabilities are easily defined and independent of the choice of origin.

2.1.2.3. Molecular-state with other a priori translational factors The other methods, developed more recently, employ different translational factors which reduce to plane-wave factors in the separated-atoms limit. For the most part, these methods may be distinguished according to whether the translational factors are chosen *a priori* and also are channel-independent, thus preserving the orthogonality of the basis functions; these were introduced by Schneiderman and Russek [36], Levy and Thorson [37], Thorson and Delos [38], and Taulbjerg et al. [39].

2.1.2.4. Molecular-state with optimized or variationally determined translational factors Otherwise they are either channel-dependent and optimized/variationally determined in some way, as introduced by Riley and Green [40] (recently extended and evaluated by McCaig and Crothers [41]), Rankin and Thorson [42], and Thorson et al. [43], or they are channel-independent and norm-optimized, as in Errea et al. [44].

The translational factor of McCaig and Crothers is expressible as

$$T_k(\vec{r}, \vec{R}) = \exp\left[\frac{1}{2} i s_k(\vec{R}) \vec{v}\cdot\vec{r}\right].$$

This translational factor is to be multiplied by the molecular wave function $\psi_k(\vec{r}', R)$ as well as a consistently chosen energy phase factor to obtain the traveling orbital f_k used in Eq. (1). To avoid confusion with our f_k, the switching function has been denoted by s_k. It approaches the appropriate Bates–McCarroll limit (± 1) as $R \rightarrow \infty$. In addition to being channel-dependent, this *variationally determined* function is noted to depend on the direction as well as magnitude of $\vec{R}$. The resulting matrix elements, which they evaluate, are substantially more complicated than those just summarized for the Bates–McCarroll approach.

The switching function of Thorson et al. is somewhat different, depending on R and the spheroidal angular variable $\eta = (r_A - r_B)/R$ but not on the direction of $\vec{R}$:

$$s_k = \tanh\left[b_k\left(\eta - \eta_k^0\right)\right],$$

where b_k and η_k^0 depend on R. It has been determined by *optimizing* both the bound and continuum couplings.

2.1.2.5. Hylleraas Lüdde and Dreizler [45] introduced a basis, which they termed a Hylleraas basis, which lacks translational factors but which is an expansion in spheroidal coordinates with the Slater exponent and number of basis functions (typically in the range 100 to 200) depending on the internuclear distance R.

2.1.2.6. Three-center molecular-state More recently, Errea et al. [46] developed and applied a three-center *molecular*-state method, in which a molecular-state basis with common translational factors is augmented by a set of Gaussian functions centered at the midpoint of the internuclear line. Test calculations were also carried out with Gaussians instead centered at the center of charge or the equiforce point as in previous three-center atomic-state calculations [30,29,47]. The ability of the three-center molecular-state basis to represent ionization was assessed partly by calculating the overlap of the molecular wave function with the classical ionization wave function, defined as the square root of the ionization density obtained from a classical calculation.

2.1.3. Two-Center, Momentum-Space Approach

Sidky and Lin [48] introduced a two-center expansion of the electronic wave function in momentum space:

$$\tilde{\Psi}(\vec{p}, t) = \sum_{lm} \tilde{\Upsilon}_{Alm}(p, t) Y_{lm}(\hat{p}) + e^{-i(\vec{p}\cdot\vec{R} - (1/2)v^2 t)} \sum_{lm} \tilde{\Upsilon}_{Blm}(q, t) Y_{lm}(\hat{q}),$$

where $\vec{q} = \vec{p} - \vec{v}$.[2] The radial momentum-space functions $\tilde{\Upsilon}_{\alpha lm}$ for centers $\alpha = A$ or B were in turn expressed in terms of B-spline basis functions. The

---

[2] The magnitudes of the electronic momenta p and q with respect to the target and projectile nuclei, respectively, are not to be confused with p and q_α in Eqs. (18) and (16) of the triple-center approach.

resulting expansion was then transformed to coordinate space to facilitate integration of the Schrödinger equation over time, in view of the potential-energy term. The equation was evaluated on a spatial grid. Thus the procedure is a hybrid of coupled-state and numerical approaches.

2.2. Quantum Approaches

In a fully quantum approach, both the nuclear and electronic motions are treated quantum mechanically, in contrast to the semiclassical (impact-parameter) approach, in which only the electronic motion is treated quantum mechanically: The full wave function is expanded as [49]

$$\Psi(\vec{r}, \vec{R}) = \sum_k F_k(\vec{R}) f_k(\vec{r}, \vec{R}),$$

which is a quantum generalization of Eq. (1). Ignoring terms of the order of $1/M_\alpha$ (where $\alpha = A$ or B), the time-independent Schrödinger equation for the nuclear-electronic wave function is

$$\left(-\frac{1}{2M}\nabla_R^2 + H + \frac{Z_A Z_B}{R}\right)\Psi(\vec{r}, \vec{R}) = E\Psi(\vec{r}, \vec{R}), \tag{28}$$

where

$$M = \frac{M_A M_B}{M_A + M_B}$$

is the reduced nuclear mass and H is the electronic molecular Hamiltonian. Multiplying by a particular electronic function $f_k^*(\vec{r}, \vec{R})$, and integrating over $\vec{r}$, one obtains a set of coupled *second-order* differential equations for the "nuclear wave functions" $F_k(\vec{R})$.

2.2.1. Translational Factor Approaches

In solving these equations subject to separated-atoms boundary conditions, the same problem of spurious long-range coupling arises as in the impact-parameter approach. Van Hemert et al. [50] addressed this problem in the context of He^{++}–H collisions by incorporating in $f_k(\vec{r}, \vec{R})$ a translational factor containing a channel-independent switching function: either the Vaaben–Taulbjerg switching function [51] or the Errea–Méndez–Riera switching function [52]. This translational factor multiplies the molecular electronic wave function $\psi_k(\vec{r}', R)$ (the same wave function as appears in the impact-parameter treatment). Van Hemert et al. approximated ψ_k for each of the four main states $2p\sigma, 3d\sigma, 2p\pi, 2s\sigma$ as a linear combination of atomic orbitals, yielding molecular energies noted to agree with the exact energies tabulated by Winter et al. [53] to within 10^{-4}.

2.2.2. Common-Reaction-Coordinate Method

A closely related quantum approach is the common-reaction-coordinate method, recently summarized by Le et al. [54]. In this method, the nuclear-plus-electronic wave function is expanded in terms of products of conventional molecular electronic wave functions ψ_k and functions of a common *reaction coordinate* $\vec{\xi}$, where Le et al. chose the particular form in Errea et al. [55]:

$$\vec{\xi} = \vec{R} + \frac{1}{M}\left[s(r, R)\vec{r} - \frac{1}{2}s^2(r, R)\vec{R}\right],$$

where s is a switching function which gives the appropriate boundary conditions and which was chosen specifically to be that of Harel and Jouin [56].

2.2.3. Hidden Crossing Approach

An alternate coupled-state quantum approach was recently devised by Krstić [57]. In this treatment, the molecular electronic Hamiltonian and wavefunction are extended to the complex R plane [58]. Two different real molecular energies of a given symmetry are actually branch energies of a single complex energy, the difference between them being expanded to first order in $R - R_c$, the branch point or hidden crossing in the complex plane, leading to a Lorentzian approximation for the nonadiabatic radial matrix element:

$$G_{kk'}(R) = -i\dot{R}\left\langle\psi_k\left|\frac{\partial}{\partial R}\right|\psi_{k'}\right\rangle \simeq -\frac{i}{2}\dot{R}\frac{\mathrm{Im}(R_c)}{[R - \mathrm{Re}(R_c)]^2 + [\mathrm{Im}(R_c)]^2}.$$

The rotational coupling matrix elements are approximated with a united-atom approximation. After transforming to a diabatic basis, Krstić solved the quantum Schrödinger equation (Eq. (28)) with up to 12 molecular basis functions.

2.2.4. Hyperspherical Approach

Recently, Liu et al. [59] introduced a hyperspherical coupled-state approach to low energy charge transfer. This approach begins with the usual molecular Jacobi vectors—the internuclear position vector $\vec{R}$ and the electron's position vector $\vec{r}'$ relative to the nuclear center of mass. Denoting the reduced electronic mass by μ, one defines a hyperradius

$$\mathcal{R} = \sqrt{R^2 + \frac{\mu}{M}r'^2} \simeq R$$

as well as hyperangles

$$\phi = \tan^{-1}\left(\sqrt{\frac{\mu}{M}}\frac{r'}{R}\right) \simeq \tan^{-1}\left(\frac{r'}{R\sqrt{M}}\right),$$
$$\theta' = \angle(\vec{r}', \vec{R}).$$

After scaling the full nuclear-electronic wave function by a factor $\frac{1}{2}\mathcal{R}^{3/2}\sin 2\phi$, one obtains a modified time-independent Schrödinger equation which is expanded in terms of "adiabatic basis functions" $\Phi_k(\phi, \theta', \mathcal{R})$ multiplied by D matrices relating the molecular and laboratory frames. These adiabatic basis functions, determined for each $\mathcal{R} \simeq R$, are analogous to the molecular wave functions $\psi_k(\vec{r}', R)$ in the coupled-molecular-state approaches outlined previously. Liu et al. expressed them in terms of B-spline basis functions. Their method is thus a hybrid coupled-state–numerical approach.

Finally, a very low energy coupled-state quantum theory was recently developed by Esry et al. [60] and applied to study the isotopic dependence of proton–hydrogen collisions.

3. Results

As indicated in the introduction, we restrict ourselves to one-electron collisional systems. Such systems are simpler, less ambiguous, and more exacting testing grounds of coupled-state theories of charge transfer. In terms of theoretical and experimental interest to date, there are three systems which should especially be addressed: the heteronuclear systems αH and pHe$^+$, and the homonuclear system pH, with the target atom (or hydrogenic ion) in each case usually assumed to be initially in the ground state. Each of these systems is amenable to treatment, in principle, with any of the coupled-state approaches outlined in the previous section, but each system is also distinctive: pHe$^+$ is nonresonant; pH is both resonant and has an extra degree of symmetry; and αH is accidentally resonant (in the separated-atoms limit)—i.e., resonant because of a match between the ratio of nuclear charges and the ratio of initial and final principal quantum numbers. Other one-electron systems which have also been treated can often be classified as one of these three types. Although the pH system was first treated about a decade before the other two systems, heteronuclear systems are in fact more numerous, and so we begin with these.

3.1. THE αH SYSTEM

Partly for its intrinsic interest and partly because of perceived relevance to nuclear fusion, there has been a great deal of work on the αH system which has continued to this day. We consider first the overall capture cross section, and secondly capture to the two principal states (2s and 2p) of He$^+$—i.e., those states which are asymptotically degenerate with the initial 1s state of H.

3.1.1. Charge Transfer to All States

3.1.1.1. Intermediate energies It is convenient to consider separately the low and intermediate projectile energy ranges, since the range of treated energies—and the range of cross sections—is great, and different coupled-state methods tend to be applicable in these energy ranges. We arbitrarily choose an α energy of 3 keV as the dividing point. A great deal of effort by many researchers has been focused on the low-intermediate-to-intermediate (henceforth called "intermediate" for short) energy range 3–200 keV, and particularly recently, a not inconsiderable effort on the low energy range 0.1–1 keV as well. (The corresponding center-of-mass energies are obtained by dividing by five.) It is the 3–200 keV α energy range which is our primary interest here, since, as will be seen, many states are necessarily coupled in this range. As suggested in the introduction, one might argue that coupled-state approaches are taken at lower energies for want of successful simpler calculations. The 3–200 keV range corresponds to α speeds 0.173–1.41, "velocity" matching ($v = 1$) occurring at 100 keV.

Shown in Table I are many sets of coupled-state results at energies spanning this intermediate energy range. The type of basis as well as number of functions

Table I

Coupled-state cross sections (in units of 10^{-17} cm^2) for electron transfer to all states of He$^+$ in α–H collisions at various intermediate α energies E.

Type of basis	Number of functions	Authors	E (keV)				
			3	8	20	70	200
hyperspherical	—	Liu et al. [59]	17.4	75.6	162		
perturbed stationary state	20–22	Winter et al. [34,35]	12.2	49.2	93		
molecular, common factor	10	Vaaben and Taulbjerg [51]	12.6	51.9	96		
molecular, variational factors	5	McCaig and Crothers [41]	15.1	60.9	110		
molecular, optimized factors	10	Kimura and Thorson [61]	14.3	61.5	112		
molecular, norm-optimized factors	10	Errea et al. [44]	14.2	61.9	120		
molecular, plane-wave factors	10	Hatton et al. [62,35]	15.0	63.5	123	121	
3-center atomic	24–34	Winter [47]		63.3	117	118	33.4
2-center Gaussian pseudo	200	Toshima [67]		63.7	118[a]	123[a]	36.2
Hylleraas	mixed	Lüdde and Dreizler [45]		64.7	109	98[a]	29.5
2-center Sturmian	19–24	Winter [18]			111	110	37.4
2-center atomic	14–20	Bransden et al. [63]			103[a]	122[a]	36[a]
2-center atomic	8	Winter [18][b]			101	99	22.1

[a]Graphically interpolated values.

[b]This author's previous results agree to 2% or one unit in the last digit with those of [64] with the same basis; they disagree with [65,66].

has been indicated. Although the numerical size of a basis is probably at best an indication of its completeness relative to other bases of the same type, this number has been included for want of a better measure. The first few bases may be better at energies lower than those in this table (as will be discussed when lower energies are considered in the next subsection), most of the last few may be better at the higher energies tabulated here, while the ones in the middle may be good over most of the energy range. The McCaig–Crothers basis [41] might have been superior to any of the bases over the intermediate energy range except that it is a smaller basis.

The molecular-state bases all include the "minimal basis" of three states [33] $2p\sigma$, $2p\pi$, and $3d\sigma$, as well as $2s\sigma$. The $2p\sigma$ state correlates to the $1s$ state of H in the separated-atoms limit, while the other three states correlate to $n = 2$ states of He^+; in this limit, they are all degenerate. The $2p\sigma$ state is strongly coupled rotationally to $2p\pi$ at small internuclear distances and radially to $3d\sigma$ at intermediate distances. Although the 20- or 22-state pss basis is larger than any of the other molecular-state bases, it is inferior to them due to a lack of any translational factors. With the exception of the pss and common-translation-factor results, there is close agreement (within 10%) among the molecular-state and three-center atomic-state results at the lower energies in this table (3–20 keV). Indeed, the plane-wave-factor, norm-optimized, and three-center results agree to at least 5%. The three-center calculation was not done at the lowest energy because of computing limitations; it would in principle be expected to be no less reliable there, since the inclusion of united-atoms functions on the third center incorporates a molecular character.

The two-center Sturmian and the larger of the two-center atomic-state calculations agree well (within 10%) at the calculated higher intermediate energies (20–200 keV). Surprisingly, this level of agreement holds even though only bound states (up to $n = 3$) were included in the atomic-state basis; it is expected that agreement would improve further if larger bases, such as are feasible today, were used. Results with the last basis—a relatively small 8-atomic-state basis—are included only to indicate the need for more states. The Hylleraas cross section agrees well with the other results except to some extent at the highest two energies, perhaps reflecting the omission of translational factors in that calculation.

Toshima [67] has reported large-basis calculations for electron transfer and ionization in α–H (and other bare-nuclei-H) collisions and has tabulated cross sections obtained with a two-center 200-Gaussian-pseudostate basis; the transfer cross sections are given in Table I for the intermediate energies being considered here. It is seen that the three-center cross section agrees within 1% with the Gaussian-pseudostate cross section at the two lower energies, 8 and 20 keV, while at 70 and 200 keV, the agreement is to within 4% and 8%, respectively. In the overlapping energy range (20–200 keV), the Sturmian and Gaussian cross sections agree to within 10%.

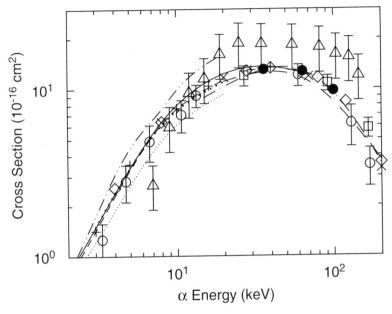

FIG. 1. Cross sections for electron transfer to all states of He^+ in α–H collisions. Coupled-state results: dash-double-dotted line, hyperspherical [59]; close-dotted line, pss [34]; dotted line, variational molecular [41]; dashed line, optimized molecular [61]; plus signs, norm-optimized molecular [44]; solid line, plane-wave-factor molecular [62,35]; crosses, three-center atomic [47]; diamonds, two-center Gaussian [67]; long-dashed line, two-center Sturmian [18]; dash-dotted line, two-center atomic [63]. Classical: solid circles, Illescas and Riera [68]. Experimental results: triangles, Bayfield and Khayrallah [69]; open circles, Shah and Gilbody [70], Nutt et al. [71]; squares, Olson et al. [72]. (Each curve in this and other figures has been generated with a cubic-spline fit.)

Most of these coupled-state cross sections are also graphed in Fig. 1 along with the experimental results of Bayfield and Khayrallah [69], Shah and Gilbody [70] and Nutt et al. [71], and Olson et al. [72]. In some cases, the data are for equivalent-velocity $^3He^{2+}$ projectiles. The error bars on the first two sets of measurements are total error limits. It is seen that there is generally excellent agreement between those coupled-state results just noted to be in accord and the experimental results except those of Bayfield and Khayrallah at the higher energies and at the lowest energy.

Also shown are the closely agreeing recent classical results of Illescas and Riera [68] for energies of at least 36 keV. Illescas and Riera emphasize the point made by Bransden and Janev [4] that classical results are surprisingly reliable at intermediate energies not because the classical, short wavelength limit is reached but because the special nature of the Coulomb interaction leads to symmetries and coincidences with the quantal description.

Not shown is the three-center *molecular-state*-plus-pseudostate cross section of Errea et al. [46], which is noted to agree closely up to 25 keV with experiment and with the two-center Gaussian-pseudostate cross section of Toshima, but is too high at higher energies, which is attributed by Errea et al. to the incompleteness of their own pseudostate basis on the third center.

Probability times impact parameter for the plane-wave-factor, the optimized-translational factor, and the norm-optimized-factor molecular bases and for the three-center atomic-state and Sturmian bases (not shown here) indicates that agreement persists even to this level. Further, differential cross sections probe the impact-parameter dependence not only of the magnitude but also the *phase* of the transition amplitude. These differential cross sections have been determined by Winter et al. [73] for the pss and plane-wave-factor, molecular-state approaches and by Winter [74] with a three-center atomic-state basis; there is very good agreement between the latter two approaches at 20 keV but some noticeable disagreement for larger angles at 70 keV. Unfortunately there are no experimental differential cross sections for this, the α–H process.

Before proceeding to the lower energies, the large value of the charge-transfer cross section, about 12×10^{-16} cm^2 at the broad 40 keV peak, should be noted. Indeed, the charge-transfer probability is on the order of 50% at impact parameters of a few Bohr that primarily determine the integrated cross section.

3.1.1.2. Lower energies Coupled-state cross sections are shown in Table II for α energies from 0.1 to 1 keV. Note the extreme falloff with decreasing energy for all except the straight-line pss cross section. At issue here more than the choice of basis states in a particular coupled-state approach is the nature of the approach itself—whether it be classical or quantum mechanical and, if classical, the particular trajectory assumed. There is virtually complete agreement between the top two, fully quantum approaches: the hyperspherical approach of Liu et al. [59] and the distorted-atomic-orbital approach of Fukuda and Ishihara [75]. Not shown are the coupled-state, hidden-crossing results of Krstić [57], which agree with those

Table II

Coupled-state cross sections (in units of 10^{-17} cm^2) for electron transfer to all states of He$^+$ in α–H collisions at various lower α energies E.

Type of approach	Authors	E (keV)		
		0.1	0.5	1
quantum, hyperspherical	Liu et al. [59]	7.1×10^{-8}	0.34	2.4
quantum, distorted atomic orbital	Fukuda and Ishihara [75]	7.1×10^{-8}	0.33	2.4
semiclassical, pss, $2p\sigma$ trajectory	Winter and Lane [34]	1.36×10^{-7}	0.356	2.22
semiclassical, pss, Coulombic trajectory	Winter and Lane [34]	$<10^{-8}$	0.0887	1.74
semiclassical, pss, rectilinear trajectory	Winter and Lane [34]	0.832	1.52	2.38
quantum, Vaaben–Taulbjerg factors	van Hemert et al. [50]	3.42×10^{-6}	0.114	1.32

of Liu et al. even in partial-wave (or impact-parameter) dependence. Also not shown is the partial-wave dependence of the common-reaction-coordinate results of Le et al. [54], which, at α energies between 0.25 and 3 keV, agree closely. Inexplicably, the third displayed set of quantum results—that of van Hemert et al. [50]—differs greatly at all displayed energies. Shown also are the semiclassical (that is, impact-parameter) three-state pss results of Winter and Lane [34] with three different trajectories: a $2p\sigma$ trajectory, a Coulombic trajectory, and a rectilinear trajectory. Presumably, where there is a difference, the $2p\sigma$ trajectory is to be preferred, since it is that of the elastic channel, which dominates at low energy. Above 1 keV, the results with the different trajectories [34] (not shown) agree closely: to within 4% at 3 keV and 1% at 8 keV. However, at low energies there are significant differences among cross sections using the three trajectories. The reason is that the curved trajectories at low energies prevent the nuclei from getting close enough for the $2p\sigma$ and $2p\pi$ states to couple rotationally; the low energy calculation with either curved trajectory is a two-state $2p\sigma$, $3d\sigma$ calculation, whereas with straight trajectories it is in effect a two-state $2p\sigma$, $2p\pi$ calculation. The superiority of the $2p\sigma$ trajectory is clearly born out by comparison with the quantum results of Liu et al. (or the nearly identical results of Fukuda and Ishihara): The $2p\sigma$-trajectory pss cross section agrees with the quantum result to within 5% at 0.5 keV and a factor of two at 0.1 keV; with the Coulombic trajectory the disagreement is at least a factor of four, and with the straight-line trajectory it is much worse. The good agreement at 0.5 keV using the $2p\sigma$ trajectory suggests it might be possible to extend the method to low energies in a simple, perturbative (two-state) *quantum* model.

3.1.2. Charge Transfer to the 2s and 2p States

Consider now charge transfer to individual excited states of He^+, which might be expected to be a more discriminating test of the coupled-state approaches. We begin with the larger of the two $n = 2$ cross sections, the $2p$ cross section, shown in Table III. Consider first the *molecular*-state and three-center atomic-state results. It is seen that with two exceptions, the agreement is extraordinarily good: to within *at least* 1.5% at all common energies. The exceptions are 8% disagreement (1) at 20 keV between the optimized-translational factor and the three other results and (2) at 70 keV between the three-center and single available molecular-state result. The two-center atomic-state cross section agrees within 10% with the above results. On the other hand, the Hylleraas cross section differs by up to 30%, with the largest difference being at the highest energy.

The 2s capture cross section in Table IV is several times smaller than the $2p$ cross section and so, at least in percentage terms, would be expected to be more sensitive to the choice of coupled-state approach. This is indeed the case: those molecular-state and three-center cross sections at 3–20 keV noted to agree very

Table III

Coupled-state cross sections (in units of 10^{-17} cm^2) for electron transfer to the $2p$ state of He$^+$ in α–H collisions at various intermediate α energies E.

Type of basis	Number of functions	Authors	E (keV)				
			3	8	20	70	200
molecular, optimized factors	10	Kimura and Thorson [61]		46.7	83.6		
molecular, norm-optimized factors	10	Errea et al. [44]	10.1	47.4	90.5		
molecular, plane-wave factors	10	Winter and Hatton [35]	10.18	46.9	90.2	90.8	
3-center atomic	24–34	Winter [47,78]		47.4	90.0	83.5	14.8
Hylleraas	mixed	Lüdde and Dreizler [45]		45.0	78.6	77[a]	19.4
2-center atomic	14–20	Bransden et al. [63]			81[a]	87[a]	15[a]

[a]Graphically interpolated values.

Table IV

Coupled-state cross sections (in units of 10^{-17} cm^2) for electron transfer to the $2s$ state of He$^+$ in α–H collisions at various intermediate α energies E.

Type of basis	Number of functions	Authors	E (keV)				
			3	8	20	70	200
molecular, variational factors	5	McCaig and Crothers [41]	5.1	12.9	20.5		
molecular, optimized factors	10	Kimura and Thorson [61]		12.1	21.8		
molecular, norm-optimized factors	10	Errea et al. [44]	3.6	12.4	22.0		
molecular, plane-wave factors	10	Hatton et al. [62]	3.97	12.7	23.7	11.5	
3-center atomic	24–34	Winter [47]		13.2	21.7	21.5	6.08
Hylleraas	mixed	Lüdde and Dreizler [45]		17.6	25.1	13[a]	2.6
2-center atomic	14–20	Bransden et al. [63]			22[a]	16[a]	4[a]

[a]Graphically interpolated values.

closely for $2p$ agree "only" to within 10% for $2s$, and there is almost a factor-of-two difference between the molecular-state and three-center results at 70 keV. The variational molecular cross section of McCaig and Crothers—reported only for $2s$—is about 30% above the other molecular-state results at 3 keV; the disagreement may be due to the smaller basis in that calculation. Differences between two- and three-center atomic-state results are up to 40%, substantially larger than for $2p$. Likewise, the difference from the Hylleraas cross section is larger—a factor of 2.3 at the highest energy.

Most of the $2s$ and $2p$ coupled-state cross sections are graphed in Fig. 2 along with the experimental results of Bayfield and Khayrallah [69] and Shah and

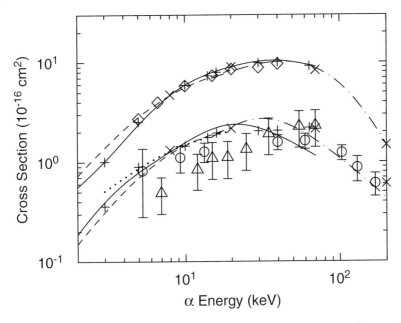

FIG. 2. Cross sections for electron transfer to the $2s$ state (lower curves and symbols) and $2p$ state (upper curves and symbols) of He$^+$ in α–H collisions. Coupled-state results: dashed line, optimized molecular [61]; dotted line ($2s$), variational molecular [41]; plus signs, norm-optimized molecular [44]; solid line, plane-wave-factor molecular ($2s$) [62], ($2p$) [35]; crosses, three-center atomic [47,78]; dash-dotted line, two-center atomic [63]. Experimental results: triangles ($2s$), Bayfield and Khayrallah [69]; circles ($2s$), Shah and Gilbody [70]; diamonds ($2p$), Ćirić et al. [76].

Gilbody [70] for $2s$ and Ćirić et al. [76] for $2p$. For $2p$, there is excellent agreement between theory and experiment. For $2s$, the coupled-state results lie above the preponderance of experimental data in the middle part of the displayed energy range. At the higher energies, the experimental results favor the two- and three-center atomic-state results over the Hylleraas result (not shown) and the molecular result [62], both of which would be expected to fail at high energies.

It should be added that Kuang and Lin [77] reported large-basis (209-atomic-state-plus-pseudostate, predominantly target-centered) cross sections for total capture and capture to the $2s$ and $2p$ states at α energies from 80 to 1200 keV; at the lower of these energies they noted agreement with the experimental results of Shah and Gilbody for total capture and capture to $2s$.

3.2. THE pHe$^+$ SYSTEM

The α–H electron-capture process just considered is dominated by capture to the $n = 2$ states of He$^+$; the ground state of He$^+$ is only weakly populated owing to

Table V
Coupled-state cross sections (in units of 10^{-18} cm^2) for electron transfer to all states of H in
p–He$^+$(1s) collisions at various center-of-mass energies E_{cm}.

Type of basis	Number of functions	Authors	E_{cm} (keV)				
			1.6	4	14	25	40
molecular, optimized factors	10	Kimura and Thorson [61]	0.00284	0.199			
molecular, plane-wave factors	10	Winter et al. [79]	0.00323	0.228	7.16		
molecular, norm-optimized factors	10	Errea et al. [44]		0.23	9.27	22[a]	28.0
3-center atomic	24 or 34	Winter [47]	0.00385[b]	0.220	9.03	22.4	26.7
2-center pseudo (AO+)	16	Fritsch and Lin [19]	0.0038[a]	0.19[a]	10[a]	19[a]	22[a]
Hylleraas	mixed	Lüdde and Dreizler [45]	0.04	0.7	5[a]	7[a]	21
2-center Sturmian	21 or 35	Winter [18,90]		0.211	9.64	23.6	27.2[c]
2-center pseudo A	23	Bransden et al. [63]		0.22[a]	8.0[a]	21[a]	27[a]
2-center atomic	8	Bransden and Noble [64]		0.23[a]	4.6[a]	11[a]	15[a]

[a] Graphically interpolated values.
[b] Transfer to the ground state.
[c] A 51-Sturmian value at 40 keV [91] agrees to 1.5%.

its large energy difference from the ground state of H. By detailed balancing, the cross section for electron transfer *from* the ground state of He$^+$ in p–He$^+$ collisions is the same, and hence also small, to the extent that capture to excited states of H can be neglected—a valid assumption at low energies. A considerable effort has been made both theoretically and experimentally to treat this quite different, nonresonant process. We will see that the smallness of the cross section makes this coupled-state cross section more challenging to determine than those for the dominant α–H processes, at least at higher intermediate energies.

Shown in Table V are coupled-state cross sections for electron transfer to all states of H in p–He$^+$(1s) collisions. Consider first the center-of-mass energy $E_{cm} = 4$ keV. Here, the plane-wave-factor, norm-optimized molecular, and three-center cross sections agree within 4%. This energy corresponds to an α energy of 20 keV, and close agreement among these three approaches has been noted there for the just considered α–H process. This agreement persists to the level of impact parameter (not shown here). The agreement in total cross section with the Sturmian result is about 10%. However, the optimized-translational-factor result only agrees to within 15%. At the lower energy $E_{cm} = 1.6$ keV, there is a significant difference of roughly 20% between the three-center or AO+ cross section and the two tabulated molecular-state results, which may be attributed [47] to the smaller, less converged molecular bases; however, additional calculations would be desirable to test this assertion.

The situation at the higher energy $E_{cm} = 14$ keV is very different from that at 4 keV: At 14 keV, the norm-optimized molecular, three-center, and Sturmian cross sections agree within 7%, but the plane-wave-factor molecular result is about 25% too low. This was attributed in 1982 by Winter [18] and Fritsch and Lin [19] to the neglect of the continuum in the *then existing* molecular-state results; inter-mediate states of ionization are important to bridge the large gap between the initial and final states even before ionization becomes important as an open chan-nel at higher energies. The agreement among the norm-optimized, three-center, two-center pseudostate A, and Sturmian results is within 5% at 40 keV. On the other hand, the relatively small AO+ basis is seen to be not as good in represent-ing this coupling to the continuum at 40 keV. The two-center purely atomic-state result is poor at all energies, particularly if one considers the impact-parameter dependence [18] (not shown here). Surprisingly, the Hylleraas cross section [45] is particularly bad at the *lower* energies. (The Hylleraas cross section is for cap-ture to the *ground* state, assumed, by detailed balancing, to be equivalent to the process $\alpha + H(1s) \rightarrow He^+(1s) + p$; an estimate of capture to all states may be obtained by multiplying by 1.2.)

Most of these coupled-state results are also displayed in Fig. 3 along with the experimental results of Peart et al. [82,83]; Rinn et al. [84]; and Watts et al. [85]. Total estimated error bars are given. It is seen that agreement between the experi-mental and *displayed* theoretical results is for the most part outstanding. Recalling the theoretical disparity at 14 keV, the experiments do not discriminate between the plane-wave-factor molecular and the other theoretical results there. The one-and-a-half-center cross section of Reading et al. [80] deviates from the other theoretical results and from experiment below 40 keV (corresponding to a projec-tile speed 71% that of the He^+ $1s$ electron); at higher energies, the agreement is excellent, as would be expected with a large, predominantly target-centered basis. Also shown is the numerical cross section of Tong et al. [81], which agrees with the experimental and with the preponderance of the coupled-state cross sections above 8 keV, but is somewhat below them at lower energies (not shown).

In closing this subsection, it should be stressed that p–$He^+(1s)$ capture is more than an order of magnitude less likely at its peak [at 40 keV cm] than α–H($1s$) capture is at *its* peak [at about 40 keV α energy]. There, the former (nonres-onant) process is strongly influenced by intermediate ionization states, unlike the latter (resonant) process. These ionization states *must* be accounted for there with whatever basis is chosen. Secondly, it should also be stated that the capture cross section *should* decrease rapidly with decreasing energy at low energies, as observed and as predicted by the above theories; there is no apparent physical reason for the curve flattening out, as recently predicted numerically by Melezhik et al. [86].

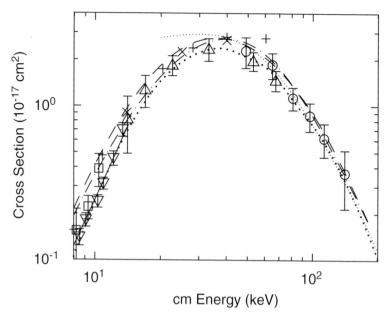

FIG. 3. Cross sections for electron transfer to all states of H in p–He$^+$(1s) collisions. Coupled-state results: dashed line, optimized molecular [61]; solid line, plane-wave-factor molecular [79]; plus signs, norm-optimized molecular [44]; crosses, three-center atomic [47]; dashed line, Sturmian [18,90,92]; dash-dotted line, two-center atomic-plus-pseudo [63]; close-dotted line, one-and-a-half-center [80]. Numerical: dotted line, Tong et al. [81]. Experimental results: inverted triangles, Peart et al. [82]; triangles, Peart et al. [83]; squares, Rinn et al. [84]; circles, Watts et al. [85].

3.3. THE pH SYSTEM

Charge transfer in p–H(1s) collisions, the prototype of resonant charge transfer in a symmetric system, has been studied even more extensively than charge transfer in the asymmetric α–H and p–He$^+$ collisions just considered. Coupled-state calculations have grown large since the two-atomic-state calculation of Mc-Carroll [13] and the two-molecular-state calculation of Ferguson [87] in 1961. Because of nuclear symmetry, the *gerade* and *ungerade* states decouple in *any* coupled-state calculation on this system. Thus, a two-state calculation in effect becomes two one-state calculations, the *gerade* and *ungerade* states merely being joined asymptotically to provide the correct boundary conditions. In the molecular picture, which is appropriate at low and perhaps intermediate energies, the initial and final states are combinations of the $1s\sigma_g$ and $2p\sigma_u$ molecular states. At smaller internuclear distances where molecular coupling may be strong, the $1s\sigma_g$ state is energetically removed from all the other molecular states. At very low energies, transitions between $1s\sigma_g$ and other (*gerade*) states (labeled below by k) are therefore unlikely because of the rapidly oscillating energy phase

$$P_{1s\sigma_g k}(t) = \exp\left(i \int_{-\infty}^{t} \left[E_{1s\sigma_g}(R') - E_k(R')\right] dt'\right)$$

$$= \exp\left(\frac{i}{v} \int_{-\infty}^{z} \left[E_{1s\sigma_g}(R') - E_k(R')\right] dz'\right)$$

as $v \to 0$. Thus transitions from $2p\sigma_u$ predominate, implying that capture to a particular state nlm and direct excitation to the same state nlm on the target are nearly equally likely at low energies [93]. Conflicting theoretical results [88] may be in error, as has recently been suggested [89].

3.3.1. Charge Transfer to All States

Shown in Table VI are coupled-state cross sections for electron transfer to all states in p–H($1s$) collisions. It is seen that cross sections with the three-center atomic and the relatively small two-center atomic-plus-pseudostate bases agree very closely (within 1%) from 3 to 15 keV, and even at the fairly low energy of 1.563 keV they agree to 8%. The agreement with the purely atomic-state cross section is within 6%. The large, predominantly resonant process is evidently well represented by a small number of atomic states. Oddly, the Hylleraas cross section only agrees with the three-center cross section to within 13% over this energy range. At 25 and 50 keV, the Sturmian, atomic-plus-pseudostate, and atomic-state cross sections agree to 5%. At 50 keV, the Hylleraas cross section differs by 40%; recall that the Hylleraas basis lacks translational factors, which may be important at the higher energies.

Shown also in Fig. 4 are the long-standing experimental results of McClure [101], with a stated estimated accuracy of 5%. The omitted error bars are thus

Table VI
Coupled-state cross sections (in units of 10^{-17} cm^2) for electron transfer to all states of H in
p–H($1s$) collisions at various intermediate proton energies E.

Type of basis	Number of functions	Authors	E (keV)						
			1.563	3	5.16	8	15	25	50
3-c atomic	28 or 36	Winter and Lin [30]	152	129	105	90.2	63.1		
Hylleraas	mixed	Lüdde and Dreizler [98]	133[a]	116[a]	94[a]	85.3	63[a]	37.1	5.1
2-c Sturmian	24	Shakeshaft [97]					64.1	36.0	8.6[a]
2-c atomic + pseudo	14	Cheshire et al. [15]	165[a]	130[a]	105[a]	90[a]	63.3	34.6	9.0[a]
2-c atomic	8	Cheshire et al. [15]	155[a]	125[a]	105[a]	90[a]	61.8	34.4	8.9[a]

[a]Graphically interpolated values.

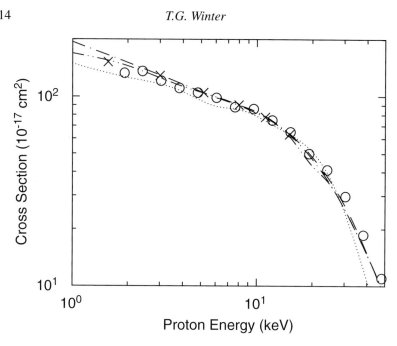

FIG. 4. Cross sections for electron transfer to all states of H in p–H(1s) collisions. Coupled-state results: crosses, three-center atomic [30]; dotted line, Hylleraas [98]; dashed line, Sturmian [97]; dash-dotted line, two-center atomic-plus-pseudo [15]; dash double-dotted line, two-center atomic [15]. Experimental results: circles, McClure [101].

comparable to the size of the circles on the graph. It is seen that except for the lowest energy experimental point and the Hylleraas cross section above 30 keV, there is excellent agreement between the theoretical results and experiment. Thus, as is well known, one should look to excited states for more exacting tests, particularly at lower intermediate energies.

3.3.2. Charge Transfer to the Ground State

It has just been noted that in the energy range 3 to 15 keV, the predominantly resonant process of *total* charge transfer is very well represented by a small number of atomic states. This is highlighted by considering capture to the *ground* state. For this entirely resonant process, the two-atomic-state cross section [13] agrees with the corresponding three-center, 8-atomic-state, and 14-atomic-plus-pseudostate cross sections to within 4% over this energy range. It is hard to improve on this agreement. Thus, it is perhaps not surprising that the continuum-distorted-wave version of the two-state approach, recently proposed as an improvement by Ferreira da Silva and Serrão [26], is actually a step backward in this lower intermediate energy range—the agreement being only 12% (although the improvement at energies above 100 keV is significant).

3.3.3. Charge Transfer to the 2s and 2p States

Consider first capture to the $2p$ state, as shown in Table VII. This is the more likely of the two $n = 2$ capture processes over the lower intermediate energy range. The difference of the two-center AO+ cross section from the three-center cross section—at most 1% at the two lowest energies—increases to 40% by 15 keV. Also shown at higher energies is the 40-atomic-plus-pseudostate cross section of Fritsch and Lin [96], which they regard as improving on the AO+ cross section at energies above 15 keV by taking fuller account of the continuum, and, indeed, the disagreement with the three-center cross section is reduced to 20% at 15 keV. The oscillating disagreement of the 14-atomic-plus-pseudostate cross section with the tabulated three-center cross section is about 1–20% over the range 1.6–15 keV, whereas for the atomic-state cross section the oscillating disagreement is about 5–50%—quite unsatisfactory. Below 10 keV, the closest agreement (within 18%) is to be found among the three-center, AO+, 40-atomic-plus-pseudostate, and Hylleraas curves. Representation of molecular states, and

Table VII

Coupled-state cross sections (in units of 10^{-18} cm^2) for electron transfer to the $2p$ state of H in p–H(1s) collisions at various intermediate proton energies E.

Type of basis	Number of functions	Authors	E (keV)						
			1.563	3	5.16	8	15	25	50
molecular, opt. factors	10	Kimura and Thorson [93]	23[a]	29[a]	37[a]	41[a]			
molecular, opt. factors	362	Zou et al. [94]	20[a]	28[a]	37[a]	44[a]			
3-c atomic	28–36	Winter and Lin [30]	26.3	29.3	25.1	34.5	24.4		
Hylleraas	mixed	Lüdde and Dreizler [98]	27[a]	28[a]	27[a]	31.5	40[a]	12.6	5.3
2-c augm. atomic (AO+)	22	Fritsch and Lin [95]	26.6	29.2	27.3	37.8	33.9		
2-c atomic + pseudo	40	Fritsch and Lin [96]				30[a]	34[a]	29[a]	18[a]	4.8[a]
2-c atomic + pseudo	96	Kuang and Lin [20]					31[a]	17[a]	3.8[a]
2-c Sturmian	24	Shakeshaft [97]					31.7	15.5	4.0[a]
2-c scaled Sturmian	68	Shakeshaft [17]					31.0	17.4	4.0
2-c atomic + pseudo	14	Cheshire et al. [15]	25[a]	29[a]	30[a]	33[a]	20.3	15.9	5.9[a]
2-c atomic	8	Cheshire et al. [15]	28[a]	39[a]	39[a]	31[a]	15.2	9.24	3.7[a]
2-c momentum	—	Sidky and Lin [99]						19.6	4.47

[a]Graphically interpolated values.

possibly some form of coupling to the continuum, is evidently needed to reproduce this smaller cross section at lower intermediate energies.

More recently, Toshima [100] has reported very large-basis calculations for
electron transfer, ionization, and excitation in p–H collisions, and has tabulated some of these cross sections at 1 and 4 keV with a symmetric two-
center 394-Gaussian-pseudostate basis. (His values for $2p$ with a predominantly
target-centered basis (not shown here) of the same size agree to better than
3% with his own symmetric-basis results.) The optimized-10-molecular, Hylleraas, and 14-atomic-plus-pseudostate cross sections—also tabulated at these two
energies—agree with Toshima's results to 24%, 4%, and 13%, respectively, for
capture to the $2p$ state. The three-center cross section has not been tabulated at
the same energies; however, graphic interpolation at 4 keV yields a value agreeing
with Toshima's result to 2%.

The optimized 10-molecular-state cross section of Kimura and Thorson [93]
lies below the AO+ and three-center curves by about 15% at our lowest tabulated
energy, crosses these curves at about 4 keV, and lies above them by $\gtrsim$10% at its
highest energy (8 keV). The recent optimized 362-molecular-state cross section
of Zou et al. [94] does nothing to resolve this discrepancy, since it crosses the
Kimura-Thorson cross section at about the same energy, lying more than 10%
lower at the lowest energy and almost another 10% higher at the highest energy.
Their basis consists of ten bound molecular states and 352 positive-energy states.

At higher intermediate energies 15–50 keV (i.e., including the velocity-
matching energy 25 keV), there is a spread of up to 44% among the scaled-
Sturmian, Sturmian, 40-atomic-plus-pseudostate, and 14-atomic-plus-pseudostate
$2p$ cross sections (an 18% spread without the last); at sufficiently high energy one
would expect the Sturmian and atomic-plus-pseudostate cross sections to merge
if fully converged. Also shown at energies $\geq$15 keV are the 96-atomic-plus-
pseudostate (even-tempered-basis) results of Kuang and Lin [20]. Their bases
include 76 atomic states and pseudostates centered on the projectile or target nucleus and 20 atomic states centered on the other nucleus; we have chosen their
cross section with the projectile-centered pseudostate basis. These results have
been obtained from a multi-decade graph; within that limitation of extracting their
cross section, there is agreement within 5% and 10% with the scaled-Sturmian and
Sturmian cross sections, respectively, in the 15–50 keV range.

At the higher energies, all these cross sections can perhaps be benchmarked
against the recent $\geq$25 keV two-center momentum-space calculations of Sidky
and Lin [99], inasmuch as the authors placed error bars of $\pm$5% on their results.
At 25 and 50 keV, the 96-atomic-plus-pseudostate, 40-atomic-plus-pseudostate,
scaled-Sturmian, Sturmian, 14-atomic-plus-pseudostate, and Hylleraas cross sections agree with the momentum-space cross section within 15%, 8%, 11%, 21%,
32%, and 36%, respectively.

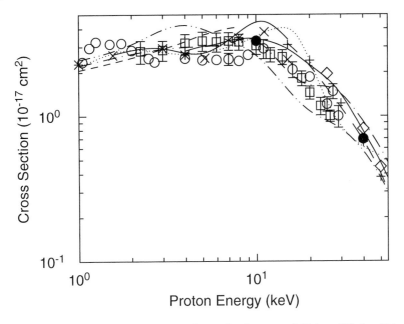

FIG. 5. Cross sections for electron transfer to the $2p$ state of H in p–H($1s$) collisions. Coupled-state results: dashed line, optimized molecular [93]; crosses, three-center atomic [30]; dotted line, Hylleraas [98]; solid line, two-center AO+ [95] (lower energies) and two-center 40-atomic-plus-pseudo [96] (higher energies); plus signs, two-center 96-atomic-plus-pseudo [20]; long-dashed line, Sturmian [97]; dash-dotted line, two-center 14-atomic-plus-pseudo [15]; dash-double-dotted line, two-center atomic [15]; asterisk (at 1 and 4 keV only), two-center Gaussian [100]; diamonds, two-center momentum [99]. Numerical: solid circles, Kolakowska et al. [103]. Experimental results: open circles, Kondow et al. [102]; squares, Morgan et al. [104].

Additional light may be shed on theory by comparison with the experimental results of Kondow et al. [102] and Morgan et al. [104] at the lower intermediate energies shown in Fig. 5. The error bars on the data of Kondow et al. (omitted here when no larger than the symbols) are one standard deviation; there is a systematic error, not shown, due to the polarization of the radiation, which, at energies below 15 keV, is believed to be within the indicated error bars. The error bars on the data of Morgan et al. omit a 30% absolute uncertainty. It is seen that the experimental cross section of Kondow et al.—but not of Morgan et al.— confirms the undulating shape of several of the curves at lower energies—namely, the AO+, three-center, atomic-plus-pseudostate, and Hylleraas curves—but not the amplitudes of the undulations. At the intermediate energy of 15 keV, both experimental cross sections favor the three-center and 14-atomic-plus-pseudostate results over the Sturmian, Hylleraas, and even-tempered-basis results.

Consider now the $2s$ capture cross section in Table VIII, which, at the lowest tabulated energy, is about one and a half orders of magnitude smaller than

Table VIII
Coupled-state cross sections (in units of 10^{-18} cm^2) for electron transfer to the 2s state of H in p–H(1s) collisions at various intermediate proton energies E.

Type of basis	Number of functions	Authors	E (keV)						
			1.563	3	5.16	8	15	25	50
molecular, opt. factors	10	Kimura and Thorson [93]	1.2[a]	3.4[a]	7.2[a]				
molecular, opt. factors	362	Zou et al. [94]	1.3[a]	3.2[a]	7.1[a]	13[a]			
3-c atomic	28–36	Winter and Lin [30]	0.833	2.60	6.97	10.7	36.5		
Hylleraas	mixed	Lüdde and Dreizler [98]	0.86[a]	2.4[a]	4.5[a]	12.0	23[a]	27.7	8.7
2-c augm. atomic (AO+)	22	Fritsch and Lin [95]	1.28	3.80	7.64	14.4	37.0		
2-c atomic + pseudo	40	Fritsch and Lin [96]				14[a]	34[a]	42[a]	14[a]
2-c atomic + pseudo	96	Kuang and Lin [20]					35[a]	39[a]	15[a]
2-c Sturmian	24	Shakeshaft [97]					34.1	42.2	15[a]
2-c scaled Sturmian	68	Shakeshaft [17]					34.1	39.8	13.9
2-c atomic + pseudo	14	Cheshire et al. [15]	2.6[a]	4.0[a]	4.9[a]	11[a]	30.9	37.3	17[a]
2-c atomic	8	Cheshire et al. [15]	5.3[a]	2.3[a]	4.9[a]	25[a]	31.2	36.6	17[a]
2-c momentum	—	Sidky and Lin [99]						39.9	13.8

[a]Graphically interpolated values.

the 2p cross section—the curves crossing at about 15 keV; hence, as for the αH system, we would expect the 2s cross section to be more sensitive there to the choice of coupled-state approach (and basis size) than the 2p cross section, and this is indeed the case: The AO+ cross section differs from the three-center cross section by 50% at the lowest energy, and the two-center 14-atomic-plus-pseudostate cross section differs from the three-center cross section there by a factor of 3; nor is agreement satisfactory at higher energies up to 15 keV. The oscillating atomic-state cross section differs even more dramatically and unacceptably from the three-center cross section than for 2p. The optimized-10-molecular, Hylleraas, and three-center (as well as AO+) cross sections differ somewhat in magnitude at these lower intermediate energies but have about the same shape, unlike the atomic-state and 14-atomic-plus-pseudostate curves. The optimized-10-molecular, Hylleraas, and 14-atomic-plus-pseudostate cross sections agree with Toshima's tabulated results [100] at 1 and 4 keV to 10%, 25%, and a factor of two, respectively. Graphic interpolation of the three-center cross section at 4 keV yields a value agreeing with Toshima's result to an estimated 2%.

Unlike for 2p, for 2s the recent large-basis optimized-molecular cross section [94] is above the earlier small-basis one [93] not only at the highest common energy (7 or 8 keV, again by 8%), but also at the lowest energy (1 keV, now by

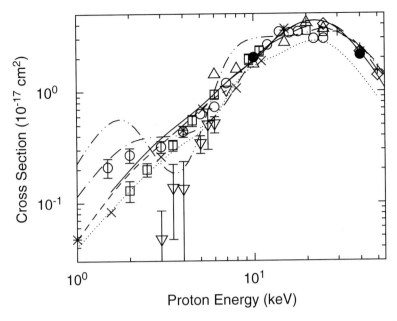

FIG. 6. Cross sections for electron transfer to the 2s state of H in p–H(1s) collisions. The theoretical results are labeled as in Fig. 5. Experimental results: open circles, Hill et al. [106]; squares, Morgan et al. [107]; triangles, Chong and Fite [108]; inverted triangles, Bayfield [109].

+13%); for the 2s state, the curves merge rather than cross in the middle of this energy range. At our lowest tabulated energy, it thus lies farther above the three-center cross section than does the earlier optimized-molecular cross section.

At higher intermediate energies 15–50 keV, there is at most a 20% spread among the scaled-Sturmian, Sturmian, 40-atomic-plus-pseudo-state, and 14-atomic-plus-pseudostate (and atomic-state) 2s cross sections (an 8% spread without the last)—less than half the spread for 2p noted in this energy range, probably because of the inversion of the cross sections and the likely smaller basis sensitivity of the larger cross section. (It may also be added that a 5-atomic-state cross section calculated by Winter and Lin [105] agrees closely (within 1%) at 25 keV with the 8-atomic-state value; the smaller calculation included only target-centered excited states, as well as both ground states.) Also shown at energies ≥15 keV is the 96-state, even-tempered-basis cross section of Kuang and Lin [20] and at energies ≥25 keV the two-center momentum-space cross section of Sidky and Lin [99], both of which agree to within 8% with either Sturmian cross section—the disagreement with the other earlier just-noted cross sections being somewhat greater. Differences with the Hylleraas cross section are much greater.

Also shown in Fig. 6 are the experimental results of Hill et al. [106], Morgan et al. [107], Chong and Fite [108], and Bayfield [109]. The error bars of Hill et al. and of Chong and Fite are one standard deviation (both sets of error bars being omitted where no larger than the size of the symbols); the error bars of Morgan et al. and Bayfield are also relative errors. An estimated absolute error of 25–30% is omitted from the displayed errors of Hill et al. and Bayfield, and an absolute error is also omitted from the displayed errors of Morgan et al. The three sets of data extending to low energies are seen to disagree there; the data are in accord at the higher energies, except for a Chong–Fite point at 6 keV. The data of Morgan et al. favors the three-center cross section at energies up to about 6 keV. At higher energies the experiments tend to confirm theory except the atomic-state and Hylleraas cross sections.

Also shown in Figs. 5 and 6 at 10 and 40 keV are the $2s$ and $2p$ cross sections obtained by solving the time-dependent Schrödinger equation numerically on a three-dimensional Cartesian lattice [103]. It is seen that there is generally good agreement with the other theoretical results, as well as with the experimental results.

4. Conclusion

Since 1961, charge transfer in intermediate-energy ion–atom collisions has been treated with many coupled-state calculations, differing both in type and size of basis. Although for brevity and simplicity we have restricted our attention to one-electron targets—and, indeed, the few most frequently treated ones—we have nonetheless revealed coupled-state results of varying quality. There has been a consensus of agreement among only some of them, at times supported by favorable comparison with experiment. (Consideration of the competing processes of ionization and excitation would have further highlighted these differences.) With noted exceptions, the larger, typically more recent calculations tend to be the more accurate.

The greatest interest appears to have been in the late 1970s and 1980s, and several benchmark calculations date from this period. Since these calculations were necessarily restricted by existing computing facilities, it may be useful to repeat some of them, and probe more thoroughly their numerical accuracy and basis convergence.

The question of basis convergence remains in many of the calculations. Particularly daunting is the lower-intermediate energy region and the extension to low energies, where some calculations have seemed to produce unphysical cross sections.

The present study has focussed on the dominant transitions. In the intermediate, nonperturbative energy range, such transitions must be treated to great accuracy—at least a few percent—if there is to be hope of determining the smaller, more highly excited-state cross sections to within 10–20%.

5. References

[1] D.R. Bates, in: D.R. Bates (Ed.), "Atomic and Molecular Processes", Academic Press, New York, 1962, pp. 549–621.

[2] N.F. Mott, H.S.W. Massey, "The Theory of Atomic Collisions", 3rd Edn., Clarendon, Oxford, 1965, Chapters XV, XIX.

[3] R.A. Mapleton, "Theory of Charge Exchange", Wiley-Interscience, New York, 1972.

[4] B.H. Bransden, R.K. Janev, *Adv. At. Mol. Phys.* **19** (1983) 1.

[5] W. Fritsch, C.D. Lin, *Phys. Rep.* **202** (1991) 1.

[6] B.H. Bransden, M.R.C. McDowell, "Charge Exchange and the Theory of Ion–Atom Collisions", Clarendon, Oxford, 1992.

[7] A.L. Ford, J.F. Reading, M. Gargaud, R. McCarroll, D.S.F. Crothers, F.B.M. Copeland, J.T. Glass, J.H. Macek, S.T. Manson, in: G.W.F. Drake (Ed.), "Atomic, Molecular, and Optical Physics Handbook", American Institute of Physics, New York, 1996, pp. 571–604.

[8] B.M. Smirnov, "Physics of Atoms and Ions", Springer, New York, 2003, Chapters 14 and 15.

[9] D. Belkić, "Principles of Quantum Scattering Theory", Institute of Physics, Bristol, 2004.

[10] M. Mittleman, *Phys. Rev.* **122** (1961) 499.

[11] L. Wilets, S.J. Wallace, *Phys. Rev.* **169** (1968) 84.

[12] D.R. Bates, *Proc. Roy. Soc. A* **247** (1958) 294.

[13] R. McCarroll, *Proc. Roy. Soc. A* **264** (1961) 547.

[14] D.F. Gallaher, L. Wilets, *Phys. Rev.* **169** (1968) 139.

[15] I.M. Cheshire, D.F. Gallaher, A.J. Taylor, *J. Phys. B* **3** (1970) 813.

[16] R. Shakeshaft, *J. Phys. B* **8** (1975) 1114.

[17] R. Shakeshaft, *Phys. Rev. A* **18** (1978) 1930.

[18] T.G. Winter, *Phys. Rev. A* **25** (1982) 697.

[19] W. Fritsch, C.D. Lin, *J. Phys. B* **15** (1982) 1255.

[20] J. Kuang, C.D. Lin, *J. Phys. B* **29** (1996) 1207.

[21] J.F. Reading, A.L. Ford, R.L. Becker, *J. Phys. B* **14** (1981) 1995.

[22] N. Toshima, J. Eichler, *Phys. Rev. Lett.* **66** (1991) 1050.

[23] J.F. Reading, J. Fu, M.J. Fitzpatrick, *Phys. Rev. A* **70** (2004) 032718.

[24] G.J.N. Brown, D.S.F. Crothers, *J. Phys. B* **27** (1994) 5309.

[25] G.J.N. Brown, *J. Phys. B* **32** (1999) 1009.

[26] M.F. Ferreira da Silva, J.M.P. Serrão, *J. Phys. B* **36** (2003) 2357.

[27] I.M. Cheshire, *Proc. Phys. Soc.* **84** (1964) 89.

[28] D.G.M. Anderson, M.J. Antal, M.B. McElroy, *J. Phys. B* **7** (1974) L118; **14** (1981) 1707(E); M.J. Antal, D.G.M. Anderson, M.B. McElroy, *J. Phys. B* **8** (1975) 1513.

[29] T.G. Winter, C.D. Lin, *Phys. Rev. A* **29** (1984) 3071; **30** (1984) 3323(E).

[30] T.G. Winter, C.D. Lin, *Phys. Rev. A* **29** (1984) 567.

[31] D.R. Bates, H.S.W. Massey, A.L. Stewart, *Proc. Roy. Soc. A* **216** (1953) 437.

[32] D.R. Bates, R. McCarroll, *Proc. Roy. Soc. A* **247** (1958) 175.

[33] R.D. Piacentini, A. Salin, *J. Phys. B* **7** (1974) 1666; **9** (1976) 563.

[34] T.G. Winter, N.F. Lane, *Phys. Rev. A* **17** (1978) 66.

[35] T.G. Winter, G.J. Hatton, *Phys. Rev. A* **21** (1980) 793.

[36] S.B. Schneiderman, A. Russek, *Phys. Rev.* **181** (1969) 311.

[37] H. Levy II, W.R. Thorson, *Phys. Rev.* **181** (1969) 252.

[38] W.R. Thorson, J.B. Delos, *Phys. Rev. A* **18** (1978) 117; **18** (1978) 135.

[39] K. Taulbjerg, J. Vaaben, B. Fastrup, *Phys. Rev. A* **12** (1975) 2325.

[40] M.E. Riley, T.A. Green, *Phys. Rev. A* **4** (1971) 619.

[41] C. McCaig, D.S.F. Crothers, *J. Phys. B* **33** (2000) 3555.

[42] J. Rankin, W.R. Thorson, *Phys. Rev. A* **18** (1978) 1990.

[43] W.R. Thorson, M. Kimura, J.H. Choi, S.K. Knudson, *Phys. Rev. A* **24** (1981) 1768.

[44] L.F. Errea, J.M. Gómez-Llorente, L. Méndez, A. Riera, *J. Phys. B* **20** (1987) 6089.

[45] H.J. Lüdde, R.M. Dreizler, *J. Phys. B* **15** (1982) 2713.

[46] L.F. Errea, C. Harel, C. Illescas, H. Jouin, L. Méndez, B. Pons, A. Riera, *J. Phys. B* **31** (1998) 3199.

[47] T.G. Winter, *Phys. Rev. A* **37** (1988) 4656.

[48] E.Y. Sidky, C.D. Lin, *J. Phys. B* **31** (1998) 2949.

[49] Mott and Massey, [2], pp. 428–430.

[50] M.C. van Hemert, E.F. van Dishoeck, J.A. van der Hart, F. Koike, *Phys. Rev. A* **31** (1985) 2227.

[51] J. Vaaben, K. Taulbjerg, in: K. Takayanzi, N. Oha (Eds.), "Abstracts of Contributed Papers, Eleventh International Conference on the Physics of Electronic and Atomic Collisions, Kyoto, Japan, 1979", The Society for Atomic Collision Research, Kyoto, 1979, p. 566, *J. Phys. B* **14** (1981) 1815.

[52] L.F. Errea, L. Méndez, A. Riera, *J. Phys. B* **15** (1982) 101.

[53] T.G. Winter, M.D. Duncan, N.F. Lane, *J. Phys. B* **10** (1977) 285.

[54] A.-T. Le, C.D. Lin, L.F. Errea, L. Méndez, A. Riera, B. Pons, *Phys. Rev. A* **69** (2004) 062703.

[55] L.F. Errea, C. Harel, H. Jouin, L. Méndez, B. Pons, A. Riera, *J. Phys. B* **31** (1998) 3527.

[56] C. Harel, H. Jouin, *Europhys. Lett* **11** (1990) 121.

[57] P.S. Krstić, *J. Phys. B* **37** (2004) L217.

[58] P.S. Krstić, C.O. Reinhold, J. Burgdörfer, *Phys. Rev. A* **63** (2001) 052702.

[59] C.-N. Liu, A.-T. Le, T. Morishita, B.D. Esry, C.D. Lin, *Phys. Rev. A* **67** (2003) 052705.

[60] B.D. Esry, H.R. Sadeghpour, E. Wells, I. Ben-Itzhak, *J. Phys. B* **33** (2000) 5329.

[61] M. Kimura, W.R. Thorson, *Phys. Rev. A* **24** (1981) 3019.

[62] G.J. Hatton, N.F. Lane, T.G. Winter, *J. Phys. B* **12** (1979) L571.

[63] B.H. Bransden, C.J. Noble, J. Chandler, *J. Phys. B* **16** (1983) 4191.

[64] B.H. Bransden, C.J. Noble, *J. Phys. B* **14** (1981) 1849.

[65] A. Msezane, D.F. Gallaher, *J. Phys. B* **6** (1973) 2334.

[66] D. Rapp, *J. Chem. Phys.* **61** (1974) 3777.

[67] N. Toshima, *Phys. Rev. A* **50** (1994) 3940.

[68] C. Illescas, A. Riera, *Phys. Rev. A* **60** (1999) 4546. The data points at 36 and 64 keV are tabular points. Their tabular data point at 100 keV, differing by more than a factor of two from their graphical point as well as the other results displayed in our Fig. 1, is evidently erroneous, and so the graphical point has instead been cited here.

[69] J.E. Bayfield, G.A. Khayrallah, *Phys. Rev. A* **12** (1975) 869.

[70] M.B. Shah, H.B. Gilbody, *J. Phys. B* **11** (1978) 121.

[71] W.L. Nutt, R.W. McCullough, K. Brady, M.B. Shah, H.B. Gilbody, *J. Phys. B* **11** (1978) 1457.

[72] R.E. Olson, A. Salop, R.A. Phaneuf, F.W. Meyer, *Phys. Rev. A* **16** (1977) 1867.

[73] T.G. Winter, G.J. Hatton, A.R. Day, N.F. Lane, *Phys. Rev. A* **36** (1987) 625.

[74] T.G. Winter, *Phys. Rev. A* **38** (1988) 1612.

[75] H. Fukuda, T. Ishihara, *Phys. Rev. A* **46** (1992) 5531.

[76] D. Ćirić, D. Dijkkamp, E. Vlieg, F.J. de Heer, *J. Phys. B* **18** (1985) 4745.

[77] J. Kuang, C.D. Lin, *J. Phys. B* **30** (1997) 101.

[78] T.G. Winter (1987), unpublished.

[79] T.G. Winter, G.J. Hatton, N.F. Lane, *Phys. Rev. A* **22** (1980) 930.

[80] J.F. Reading, A.L. Ford, R.L. Becker, *J. Phys. B* **15** (1982) 625.

[81] X.-M. Tong, D. Kato, T. Watanabe, S. Ohtani, *J. Phys. B* **33** (2000) 5585.

[82] B. Peart, R. Grey, K.T. Dolder, *J. Phys. B* **10** (1977) 2675.

[83] B. Peart, K. Rinn, K. Dolder, *J. Phys. B* **16** (1983) 1461.

[84] K. Rinn, F. Melchert, E. Salzborn, *J. Phys. B* **18** (1985) 3783.

[85] M.F. Watts, K.F. Dunn, H.B. Gilbody, *J. Phys. B* **19** (1986) L355.

[86] V.S. Melezhik, J.S. Cohen, C.-Y. Hu, *Phys. Rev. A* **69** (2004) 032709.

[87] A.F. Ferguson, *Proc. Roy. Soc. A* **264** (1961) 540.

[88] B.M. McLaughlin, T.G. Winter, J.F. McCann, *J. Phys. B* **30** (1997) 1043.

[89] T.G. Lee, A.-T. Le, C.D. Lin, *J. Phys. B* **36** (2003) 4081.

[90] T.G. Winter, *Phys. Rev. A* **35** (1987) 3799.

[91] C.D. Stodden, H.J. Monkhorst, K. Szalewicz, T.G. Winter, *Phys. Rev. A* **41** (1990) 1281.

[92] T.G. Winter, S.G. Alston, *Phys. Rev. A* **45** (1992) 1562.

[93] M. Kimura, W.R. Thorson, *Phys. Rev. A* **24** (1981) 1780.

[94] S. Zou, L. Pichl, M. Kimura, T. Kato, *Phys. Rev. A* **66** (2002) 042707.

[95] W. Fritsch, C.D. Lin, *Phys. Rev. A.* **26** (1982) 762. There are some errors in Table I of this paper which have been corrected in [30] and in the present review.

[96] W. Fritsch, C.D. Lin, *Phys. Rev. A.* **27** (1983) 3361.

[97] R. Shakeshaft, *Phys. Rev. A* **14** (1976) 1626.

[98] H.J. Lüdde, R.M. Dreizler, *J. Phys. B* **15** (1982) 2703.

[99] E.Y. Sidky, C.D. Lin, *Phys. Rev. A* **65** (2001) 012711.

[100] N. Toshima, *Phys. Rev. A* **59** (1999) 1981.

[101] G.W. McClure, *Phys. Rev.* **148** (1966) 47.

[102] T. Kondow, R.J. Girnius, Y.P. Chong, W.L. Fite, *Phys. Rev. A* **10** (1974) 1167.

[103] A. Kołakowska, M.S. Pindzola, F. Robicheaux, D.R. Schulz, J.C. Wells, *Phys. Rev. A* **58** (1998) 2872.

[104] T.J. Morgan, J. Geddes, H.B. Gilbody, *J. Phys. B* **6** (1973) 2118.

[105] T.G. Winter, C.C. Lin, *Phys. Rev. A* **10** (1974) 2141.

[106] J. Hill, J. Geddes, H.B. Gilbody, *J. Phys. B* **12** (1979) L341.

[107] T.G. Morgan, J. Stone, R. Mayo, *Phys. Rev. A* **22** (1980) 1460.

[108] Y.P. Chong, W.L. Fite, *Phys. Rev. A* **16** (1977) 933.

[109] J.E. Bayfield, *Phys. Rev.* **185** (1969) 105.

Index

CONTENTS OF VOLUMES IN THIS SERIAL